NOBEL LECTURES IN PHYSICS

1981–1990

NOBEL LECTURES

Including Presentation Speeches
And Laureates' Biographies

PHYSICS

CHEMISTRY

PHYSIOLOGY OR MEDICINE

LITERATURE

PEACE

ECONOMIC SCIENCES

NOBEL LECTURES

Including Presentation Speeches
And Laureates' Biographies

PHYSICS

1981–1990

Editor-In-Charge
Tore Frängsmyr

Editor
Gösta Ekspong

World Scientific
Singapore • New Jersey • London • Hong Kong

Published for the Nobel Foundation in 1993 by PHYSICS

World Scientific Publishing Co. Pte. Ltd.
P O Box 128, Farrer Road, Singapore 9128
USA office: Suite 1B, 1060 Main Street, River Edge, NJ 07661
UK office: 73 Lynton Mead, Totteridge, London N20 8DH

NOBEL LECTURES IN PHYSICS (1981–1990)

ISBN 981-02-0728-X
ISBN 981-02-0729-8 (pbk)

Printed in Singapore by Continental Press Pte Ltd

FOREWORD

Since 1901 the Nobel Foundation has published annually "Les Prix Nobel" with reports from the Nobel Award Ceremonies in Stockholm and Oslo as well as the biographies and Nobel lectures of the laureates. In order to make the lectures available to people with special interests in the different prize fields the Foundation gave Elsevier Publishing Company the right to publish in English the lectures for 1901–1970, which were published in 1964–1972 through the following volumes:

Physics 1901–1970	4 vols.
Chemistry 1901–1970	4 vols.
Physiology or Medicine 1901–1970	4 vols.
Literature 1901–1967	1 vol.
Peace 1901–1970	3 vols.

Elsevier decided later not to continue the Nobel project. It is therefore with great satisfaction that the Nobel Foundation has given World Scientific Publishing Company the right to bring the series up to date beginning with the Prize lectures in Economics in 2 volumes 1969–1990. Thereafter the lectures in all the other prize fields will follow.

The Nobel Foundation is very pleased that the intellectual and spiritual message to the world laid down in the laureates' lectures, thanks to the efforts of World Scientific, will reach new readers all over the world.

Lars Gyllensten   *Stig Ramel*
Chairman of the Board   Executive Director

Stockholm, June 1991

PREFACE

The present volume contains the lectures by the Nobel laureates in physics during the years 1981–1990. Included for each year is also a translation of the presentation speech by a member of the Nobel Committee for Physics within the Royal Academy of Sciences delivered during the ceremony on December 10 — the day of Alfred Nobel's death. The autobiographies of the laureates are reprinted with updated information according to the wishes of the respective author.

The number of laureates is 23, since during the decade, three prizewinners were honoured during each of the five years 1981, 1986, 1988, 1989, and 1990, two prizewinners in 1983, 1984, and 1987, whereas a single laureate was awarded the prize in 1981 and again in 1985.

Although there is no rule of rotation between the fields of physics, it is interesting in retrospect to note the following. Atomic physics was in focus in 1981 with the prize going to Nicolaas Bloembergen, Arthur Schawlow and Kai Siegbahn and again in 1989 to Norman Ramsey, Hans Dehmelt and Wolfgang Paul. In 1983 the prize to Subrahmanyan Chandrasekhar and William Fowler was awarded for researches in astrophysics, while achievements within the field of elementary particle physics was the motivation on three occasions, in 1984 to Carlo Rubbia and Simon van der Meer, in 1988 to Leon Lederman, Melvin Schwartz and Jack Steinberger, and in 1990 to Jerome Friedman, Henry Kendall and Richard Taylor. Discoveries or inventions within solid state physics appeared four times in the citations, namely in 1982 to Kenneth Wilson, in 1985 to Klaus von Klitzing, in 1986 to Ernst Ruska, Gerd Binnig and Heinrich Rohrer and in 1987 to Georg Bednorz and Alex Müller. The last two mentioned prizes went in consecutive years not only to the same field of physics but also to people working in the same research institute, something unique in the history of the Nobel prizes in physics.

The division of physics into subfields as done above is debatable, since the borderlines are not sharp. Take the case of lasers, which have importance and applications within large areas of science and technology, or take that of the electron microscope, which is of similar wide ranging importance. A new, important method or invention may be first used in conjunction with making a new discovery, but it may later turn out that the method or invention has much broader, significant applications within or outside the initial area of research.

Gösta Ekspong

CONTENTS

Physics 1981

NICOLAAS BLOEMBERGEN, ARTHUR L SCHAWLOW

for their contribution to the development of laser spectroscopy

and

KAI M SIEGBAHN

for his contribution to the development of high-resolution electron spectroscopy

processes. Furthermore, it has been possible in this way to generate laser light of shorter as well as longer wave lengths, which has extended the field of application for laser spectroscopy quite appreciably.

Professor Bloembergen, Professor Schawlow, Professor Siegbahn: you have all contributed significantly to the development of two spectroscopic methods, namely the laser spectroscopy and the electron spectroscopy. These methods have made it possible to investigate the interior of atoms, molecules and solids in greater detail than was previously possible. Therefore, your work has had a profound effect on our present knowledge of the constitution of matter.

On behalf of the Royal Swedish Academy of Sciences I wish to extend to you the heartiest congratulations and I now invite you to receive this year's Nobel prize in Physics from the hands of His Majesty the King!

Nicolaas Bloembergen

NICOLAAS BLOEMBERGEN

My parents, Auke Bloembergen and Sophia Maria Quint, had four sons and two daughters. I am the second child, born on March 11, 1920, in Dordrecht, the Netherlands. My father, a chemical engineer, was an executive in a chemical fertilizer company. My mother, who had an advanced degree to teach French, devoted all her energies to rearing a large family.

Before I entered grade school, the family moved to Bilthoven, a residential suburb of Utrecht. We were brought up in the protestant work ethic, characteristic of the Dutch provinces. Intellectual pursuits were definitely encouraged. The way of life, however, was much more frugal than the family income would have dictated.

At the age of twelve I entered the municipal gymnasium in Utrecht, founded as a Latin school in 1474. Nearly all teachers held Ph.D. degrees. The rigid curriculum emphasized the humanities: Latin, Greek, French, German, English, Dutch, history and mathematics. My preference for science became evident only in the last years of secondary school, where the basics of physics and chemistry were well taught. The choice of physics was probably based on the fact that I found it the most difficult and challenging subject, and I still do to this day. My maternal grandfather was a high school principal with a Ph.D. in mathematical physics. So there may be some hereditary factor as well. I am ever more intrigued by the correspondence between mathematics and physical facts. The adaptability of mathematics to the description of physical phenomena is uncanny.

My parents made a rule that my siblings should tear me away from books at certain hours. The periods of relaxation were devoted to sports: canoing, sailing, swimming, rowing and skating on the Dutch waterways, as well as the competitive team sport of field hockey. I now attempt to keep the body fit by playing tennis, by hiking and by skiing.

Professor L. S. Ornstein taught the undergraduate physics course when I entered the University of Utrecht in 1938. He permitted me and my partner in the undergraduate lab, J. C. Kluyver (now professor of physics in Amsterdam) to skip some lab routines and instead assist a graduate student, G. A. W. Rutgers, in a Ph.D. research project. We were thrilled to see our first publication, "On the straggling of Po-α-particles in solid matter", in print (*Physica* 7, 669, 1940).

After the German occupation of Holland in May 1940, the Hitler regime removed Ornstein from the university in 1941. I made the best possible use of the continental academic system, which relied heavily on independent studies. I took a beautiful course on statistical mechanics by L. Rosenfeld, did experimental work on noise in photoelectric detectors, and prepared the notes for a seminar on Brownian motion given by J. M. W. Milatz. Just before the Nazis closed the university completely in 1943, I managed to obtain the degree of Phil. Drs., equivalent to a M.Sc. degree. The remaining two dark years of the war I spent hiding indoors from the Nazis, eating tulip bulbs to fill the stomach and reading Kramers' book "Quantum Theorie des Elektrons und der Strahlung" by the light of a storm lamp. The lamp needed cleaning every twenty minutes, because the only fuel available was some left-over number two heating oil. My parents did an amazing job of securing the safety and survival of the family.

I had always harbored plans to do some research for a Ph.D. thesis outside the Netherlands, to broaden my perspective. After the devastation of Europe, the only suitable place in 1945 appeared to be the United States. Three applications netted an acceptance in the graduate school at Harvard University. My father financed the trip and the Dutch government obliged by issuing a valuta permit for the purchase of US$1,850. As my good fortune would have it, my arrival at Harvard occurred six weeks after Purcell, Torrey and Pound had detected nuclear magnetic resonance (NMR) in condensed matter. Since they were busy writing volumes for the M.I.T. Radiation Laboratory series on microwave techniques, I was accepted as a graduate assistant to develop the early NMR apparatus. My thorough Dutch educational background enabled me to quickly profit from lectures by J. Schwinger, J.H. Van Vleck, E.C. Kemble and others. The hitherto unexplored field of nuclear magnetic resonance in solids, liquids and gases yielded a rich harvest. The results are laid down in one of the most-cited physics papers, commonly referred to as BPP (N. Bloembergen, E. M. Purcell and R. V. Pound, Phys. Rev. *73*, 679, 1948). Essentially the same material appears in my Ph.D. thesis, "Nuclear Magnetic Relaxation", Leiden, 1948, republished by W. A. Benjamin, Inc., New York, in 1961. My thesis was submitted in Leiden because I had passed all required examinations in the Netherlands and because C. J. Gorter, who was a visiting professor at Havard during the summer of 1947, invited me to take a postdoctoral position at the Kamerlingh Onnes Laboratorium. My work in Leiden in 1947 and 1948 resulted in establishing the nuclear spin relaxation mechanism by conduction electrons in metals and by paramagnetic impurities in ionic crystals, the phenomenon of spin diffusion, and the large shifts induced by internal magnetic fields in paramagnetic crystals.

During a vacation trip of the Physics Club "Christiaan Huyghens" I met Deli (Huberta Deliana Brink) in the summer of 1948. She had spent the war years in a Japanese concentration camp in Indonesia, where she was born. She was about to start her pre-med studies. When I returned to Harvard in 1949 to join the Society of Fellows, she managed to get on a student hospitality exchange program and traveled after me to the United States on an immigrant ship. I proposed to her the day she arrived and we got married in Amsterdam in 1950. Ever since, she has been a source of light in my life. Her enduring encouragement has contributed immensely to the successes in my further career. After the difficult years as an immigrant wife, raising three children on the modest income of a struggling, albeit tenured, young faculty member, she has found the time and energy to develop her considerable talents as a pianist and artist. We became U.S. citizens in 1958.

Our children are now independent. The older daughter, Antonia, holds M.A. degrees in political science and demography, and works in the Boston area. Our son, Brink, has an M.B.A. degree and is an industrial planner in Oregon. Our younger daughter, Juliana, envisages a career in the financial world. She has interrupted her banking job to obtain an M.B.A. in Philadelphia.

In this family setting my career in teaching and research at Harvard unfolded: Junior Fellow, Society of Fellows 1949−1951; Associate Professor 1951−1957; Gordon McKay Professor of Applied Physics 1957−1980; Rumford Professor of Physics 1974−1980; Gerhard Gade University Professor 1980-present. While a Junior Fellow, I broadened my experimental background to include microwave spectroscopy and some nuclear physics at the Harvard cyclotron. I preferred the smaller scale experiments of spectroscopy, where an individual, or a few researchers at most, can master all aspects of the problem. When I returned to NMR in 1951, there were still many nuggets to be unearthed. My group studied nuclear quadrupole interactions in alloys and imperfect ionic crystals, discovered the anisotropy of the Knight shift in noncubic metals, the scalar and tensor indirect nuclear spin-spin coupling in metals and insulators, the existence of different temperatures of the Zeeman, exchange and dipolar energies in ferromagnetic relaxation, and a variety of cross relaxation phenomena. All this activity culminated in the proposal for a three-level solid state maser in 1956.

Although I was well aware of the applicability of the multilevel pumping scheme to other frequency ranges, I held the opinion − even after Schawlow and Townes published their proposal for an optical maser in 1958 − that it would be impossible for a small academic laboratory, without previous expertise in optics, to compete successfully in the realization of lasers. This may have been a self-fulfilling prophesy, but it is a matter of record that nearly all types of lasers were first reduced to practice in industrial laboratories, predominantly in the U.S.A.

I recognized in 1961 that my laboratory could exploit some of the new research opportunities made accessible by laser instrumentation. Our group started a program in a field that became known as "Nonlinear Optics". The early results are incorporated in a monograph of this title, published by W. A. Benjamin, New York, in 1965, and the program is still flourishing today. The principal support for all this work, over a period of more than thirty years, has been provided by the Joint Services Electronics Program of the U. S. Department of Defense, with a minimum amount of administrative red tape and with complete freedom to choose research topics and to publish.

My academic career at Harvard has resulted in stimulating interactions with many distinguished colleagues, and also with many talented graduate students. My coworkers have included about sixty Ph.D. candidates and a similar number of postdoctoral research fellows. The contact with the younger generations keeps the mind from aging too rapidly. The opportunities to participate in international summer schools and conferences have also enhanced my professional and social life. My contacts outside the academic towers, as a consultant to various industrial and governmental organizations, have given me an appreciation for the problems of socio-economic and political origin in the "real" world, in addition to those presented by the stubborn realities of matter and instruments in the laboratory.

Sabbatical leaves from Harvard have made it possible for us to travel farther and to live for longer periods of time in different geographical and cultural environments. Fortunately, my wife shares this taste for travel adventure. In 1957 I was a Guggenheim fellow and visiting lecturer at the École Normale Supérieure in Paris, in 1964—1965 visiting professor at the University of California in Berkeley, in 1973 Lorentz guest professor in Leiden and visiting scientist at the Philips Research Laboratories in the Netherlands. The fall of 1979 I spent as Raman Visiting Professor in Bangalore, India, and the first semester of 1980 as Von Humboldt Senior Scientist in the Institut für Quantum Optik, in Garching near Munich, as well as visiting professor at the Collège de France in Paris. I highly value my international professional and social contacts, including two exchange visits to the Soviet Union and one visit to the People's Republic of China, each of one-month duration. My wife and I look forward to continuing our diverse activities and to enjoying our home in Five Fields, Lexington, Massachusetts, where we have lived for 26 years.

Honors
Correspondent, Koninklijke Akademie van Wetenschappen, Amsterdam, 1956
Fellow, American Academy of Arts and Sciences, 1956
Member, National Academy of Sciences, Washington, D. C., 1959
Foreign Honorary Member, Indian Academy of Sciences, Bangalore, 1978
Associé Étranger, Académie des Sciences, Paris, 1980
Guggenheim Fellow, 1957
Oliver Buckley Prize, American Physical Society, 1958
Morris E. Liebman Award, Institute of Radio Engineers, 1959
Stuart Ballantine Medal, Franklin Institute, Philadelphia, 1961
National Medal of Science, President of the United States of America, 1974
Lorentz Medal, Koninklijke Akademie van Wetenschappen, Amsterdam, 1979
Frederic Ives Medal, Optical Society of America, 1979
Von Humboldt Senior Scientist, 1980

(*added in 1991*): In June 1990 I retired from the faculty of Harvard University and became Gerhard Gade University Professor Emeritus. During the past decade I was also a visiting professor or lecturer for extended periods at the California Institute of Technology, at Fermi Scuola Nationale Superiore in Pisa, Italy, and at the University of Munich, Germany.

In 1991 I serve as President of the American Physical Society. I became an honorary professor of Fudan University, Shanghai, People's Republic of China, and received honorary doctorates from Laval University, Quebec, the University of Connecticut and the University of Hartford. In 1983 I received the Medal of Honor from the Institute of Electrical and Electronic Engineers.

My research in nonlinear optics continued with special emphasis on interactions of picosecond and femtosecond laser pulses with condensed matter and of collision-induced optical coherences. My personal life and professional activities during the past decade have been a natural continuation of what I described in my autobiographical notes in 1981.

NONLINEAR OPTICS AND SPECTROSCOPY

Nobel lecture, 8 December, 1981

by

NICOLAAS BLOEMBERGEN

Harvard University, Division of Applied Sciences, Cambridge, Massachusetts 02138, USA

The development of masers and lasers has been reviewed in the 1964 Nobel lectures by Townes (1) and by Basov (2) and Prokhorov (3). They have sketched the evolution of the laser from their predecessors, the microwave beam and solid state masers. Lasers are sources of coherent light, characterized by a high degree of monochromaticity, high directionality and high intensity or brightness. To illustrate this last property, consider a small ruby laser with an active volume of one 1 cc. In the Q-switched mode it can emit about 10^{18} photons at 694 nm wavelength in about 10^{-8} sec. Because the beam is diffraction limited, it can readily be focused onto an area of 10^{-6}cm^2, about ten optical wavelengths in diameter. The resulting peak flux density is 10^{13} watts/cm^2. Whereas 0.1 Joule is a small amount of energy, equal to that consumed by a 100 watt light bulb, or to the heat produced by a human body, each one-thousandth of a second, the power flux density of 10 terawatts/cm^2 is awesome. It can be grasped by noting that the total power produced by all electric generating stations on earth is about one terawatt. (The affix "tera" is derived from the Greek $\tau\epsilon\varrho\alpha\sigma$ = monstrosity, not from the Latin "terra"!) Indeed, from Poynting's vector it follows that the light amplitude at the focal spot would reach 10^8 volts/cm, comparable to the electric field internal to the atoms and molecules responsible for the binding of valence electrons. These are literally pulled out of their orbits in multiphoton tunneling processes, and any material will be converted to a highly ionized dense plasma at these flux densities. It is clear that the familiar notion of a linear optical response with a constant index of refraction, i.e., an induced polarization proportional to the amplitude of the light field, should be dropped already at much less extreme intensities. There is a nonlinearity in the constitutive relationship which may be expanded in terms of a power series in the electric field components.

$$P_i = \chi_{ij}^{(1)} E_j + \chi_{ijk}^{(2)} E_j E_k + \chi_{ijkl}^{(3)} E_j E_k E_l + \ldots \qquad (1)$$

Such nonlinearities have been familiar at lower frequencies for over a century. For example, power and audio engineers knew about the nonlinear relationship between magnetic field and induction, $B = \mu(H)H$, in transformers and solenoids containing iron. Waveform distortion results (4). Such nonlinear phenomena at optical frequencies are quite striking and can readily be calculated by combining the nonlinear constitutive relation (1) with Maxwell's equations. In the first decade of this century Lorentz (5) calculated $\chi^{(1)}$ with the electron modeled as a harmonic oscillator. If he had admitted some anharmonicity, he could have developed the field of nonlinear optics seventy years ago. It was, however, not experimentally accessible at that time, and Lorentz lacked the stimulation from stimulated emission of radiation.

Nonlinear effects are essential for the operation of lasers. With dye lasers it is possible to cover the range of wavelengths from 350–950 nm continuously, including the entire visible spectrum. A variety of nonlinear processes, including harmonic generation, parametric down conversion and the stimulated Raman effects extend the range for coherent sources throughout the infrared and into the vacuum ultraviolet. Thus the field of nonlinear laser spectroscopy could be developed rapidly during the past two decades, aided considerably by previous investigations of related phenomena at radiofrequencies. It is, therefore, appropriate to start this review by recalling some nonlinear phenomena first discovered in the field of magnetic resonance.

NONLINEAR PRECURSORS IN MAGNETIC RESONANCE

As a graduate student of Professor E. M. Purcell at Harvard University, I studied relaxation phenomena of nuclear magnetic resonance in solids, liquids and gases. A radiofrequency field at resonance tends to equalize the population of two spin levels, while a relaxation mechanism tries to maintain a population difference, corresponding to the Boltzmann distribution at the temperature of the other degrees of freedom in the sample. The reduction in population difference is called saturation. It is a nonlinear phenomenon, as the magnitude of the susceptibility tends to decrease with increasing field amplitude. In 1946 we found that "a hole could be eaten", or a saturation dip could be produced, in an inhomogeneously broadened line profile (6). Figure 1a shows the proton spin resonance in water, broadened by field inhomogeneities of the available magnet. Figures 1b and 1c show saturation of a particular packet in the distribution, which is subsequently probed by sweeping through the resonance with a weaker signal after various time intervals. The disappearance rate of the hole is determined by the spin lattice relaxation time. This was also the first indication of the extremely sharp features of NMR lines in liquids, due to motional narrowing, on which the widespread use of NMR spectroscopy is founded.

If two pairs of levels have one level in common, saturation of one resonance may influence the susceptibility at another resonance. This was also observed early in NMR in spin systems with quadrupole splitting and quadrupolar relaxation (7). The detection of Hertzian resonances by optical methods described by Kastler (8) is another manifestation of this phenomenon. A change in the population of sublevels with different values of the spatial quantum number m_J induced by a radiofrequency field produces a change in the polarization of the emitted light. The Overhauser effect (9) describes the change in population of nuclear spin levels in metals (10) due to an application of a

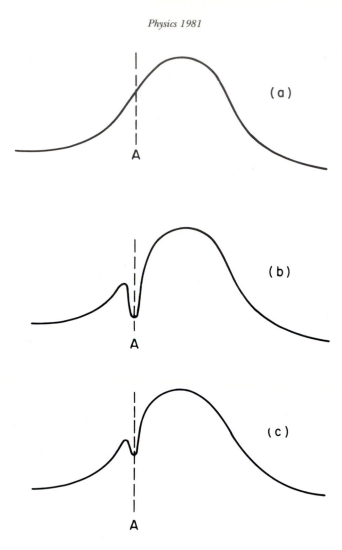

Fig. 1. (after reference 6)
a) Inhomogeneous broadened profile of NMR in water.
b) Saturation dip in inhomogeneous profile, observed in 1946.
c) As in b), but with longer delay between pump signal and probing scan.

microwave field at the electron spin resonance. Both optical and microwave pumping methods have been used to obtain nuclear spin polarized targets (11).

It is possible to maintain a steady state inverted population, in which a level with higher energy is more populated than another level with lower energy (12). This pair of levels may be said to have a negative temperature. The principle of the method, displayed in Fig. 2, is based on frequency selective pumping between a pair of nonadjacent energy levels, with the simultaneous action of a suitable relaxation mechanism. The pump tends to establish a high temperature for a pair of levels separated by a higher frequency, while at the same time relaxation maintains a low temperature between a pair with a smaller frequency separation. Stimulated emission will dominate over absorp-

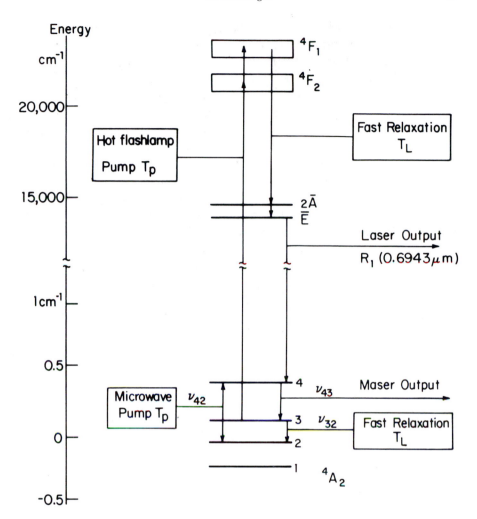

Fig. 2. Energy level diagram of Cr^{3+} in ruby. Note change in vertical scale between microwave maser and optical laser action.

tion at the third pair of a three-level system. Basov and Prokhorov (13) had proposed a frequency selective excitation mechanism for molecular beam masers without explicit discussion of relaxation.

The spin levels of paramagnetic ions in crystals are useful to obtain maser action at microwave frequencies. The stimulated emission may be considered as the output of a thermodynamic heat engine (14), operating between a hot pump temperature and a low relaxation bath temperature. These two temperatures occur in the same volume element in space, while in a conventional heat engine there is, of course, a spatial separation between the hot and cold parts. The question of thermal insulation between the paramagnetic spin transitions is based on frequency differences and differentials in relaxation rates. This question was addressed in a study of cross-relaxation phenomena (15), which determine the heat transfer between different parts of the spin Hamiltonian. It turns out that concentrated paramagnetic salts cannot be used in masers, because no large thermal differentials can be maintained in the magnetic energy level system. As a historical curiosity I may add that the biggest hurdle for me in working out the pumping sheme was the question of how to obtain a nonvanishing matrix element between nonadjacent spin levels. This, of course, is resolved by using states which are a superposition of several magnetic quantum numbers m_s. This can be obtained by applying the external magnetic field at an arbitrary angle with respect to the axis of the crystal field potential. The multilevel paramagnetic solid state maser is useful as an extremely low noise microwave amplifier. Such a maser, based on the energy levels of the Cr^{3+} ion in ruby, was used, for example, by Penzias and Wilson in their detection of the cosmic background radiation (16).

The same principle has subsequently been used to obtain a medium with gain in most lasers. It was incorporated in the basic proposal for an optical maser by Schawlow and Townes (17). It is noteworthy that the first operating laser by Maiman (18) also used the Cr^{3+} ions in ruby as the active substance. Of course, a different set of energy levels is involved in the two cases, and the change in frequency scale in the top and bottom part of Fig. 2 should be noted. The amplitude of the laser output is limited by a nonlinear characteristic, as for any feed-back oscillator system. It is the onset of saturation by the laser radiation itself which tends to equalize the populations in the upper and lower lasing levels.

NONLINEAR OPTICS

With the development of various types of lasers, the stage was set for a rapid evolution of the study of nonlinear optical phenomena. The demonstration by Franken and coworkers of second harmonic generation of light by a ruby laser pulse in a quartz crystal marks the origin of nonlinear optics as a new separate subfield of scientific endeavor (19). The straightforward experimental arrangement of this demonstration is shown in Fig. 3.

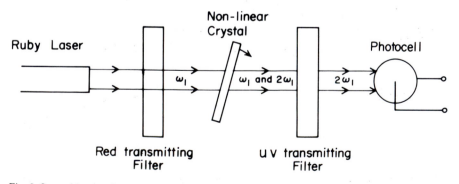

Fig. 3. Second harmonic generation of light.

The lowest order nonlinear susceptibility $\chi^{(2)}$ in equation (1) has only nonvanishing tensor elements in media which lack inversion symmetry. The polarization quadratic in the field amplitude leads to the optical phenomena of second harmonic generation, sum and difference frequency mixing, as well as to rectification of light. These properties of a device with a quadratic response were, of course, well known in radio engineering. The photoelectric emission current is a quadratic function of the light field amplitudes, and it is modulated at a difference frequency when two light beams with a small frequency difference are incident on it (20).

In general, the terms in $\chi^{(2)}$ provide a coupling between sets of three electromagnetic waves. Each wave has its own frequency ω_i, wave vector k_i, state of polarization $\hat{e}_i$, as well as a complex amplitude $E_i = A_i \exp(i\phi_i)$. In the same manner the term in $\chi^{(3)}$ causes a coupling between four electromagnetic waves. A general formulation of three- and four-wave light mixing was developed by our group at Harvard (21). The quantum mechanical calculation of the complex nonlinear susceptibilities, based on the evolution of the density matrix, was also quickly applied to optical problems (22). Generalizations of the Kramers-Heisenberg dispersion formula result. The nonlinear susceptibilities are functions of several frequencies and have more than one resonant denominator. They are tensors of higher order, and each element has a real and an imaginary part in the presence of damping. They describe a large variety of nonlinear optical effects. At the same time Akhmanov and Khokhlov (23) also extended the formulation of parametric nonlinearities from the radiofrequency to the optical domain.

Returning to the generation of optical second harmonics in transparent piezo-electric crystals, the problem of momentum matching in the conversion

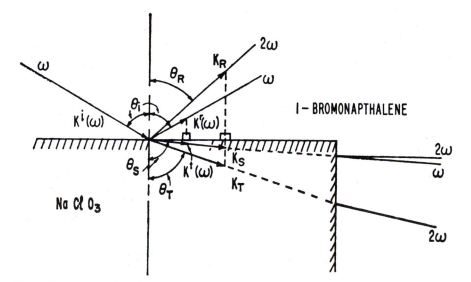

Fig. 4. Wave vectors of fundamental and second harmonic light waves at the boundary of a cubic piezoelectric crystal immersed in an optically denser fluid.

of two fundamental quanta to one quantum at the second harmonic frequency presents itself. Due to color dispersion, one usually has $k_2 - 2k_1 = \Delta k \neq 0$. The mismatch in phase velocities between the second harmonic polarization and the freely propagating wave at 2ω leads to the existence of two waves at 2ω in the nonlinear crystal, a forced one with wave vector $k_s = 2k_1$ and another with wave vector $k_T = k_2$, for a freely propagating wave at 2ω. In addition, there is a reflected second harmonic wave with wave vector k_R. Figure 4 depicts the geometry for the case that the nonlinear crystal is embedded in a liquid with a higher linear index of refraction. Conservation of the components of momentum parallel to the surface determines the geometry (24). The amplitudes of the free waves, which are solutions of the homogeneous wave equations, are determined by the condition that the tangential components of the second harmonic electric and magnetic field at the boundary are continuous. Thus a very simple procedure, based on conservation of the component of momentum parallel to the boundary, yields the generalizations of the familiar optical laws of reflection and refraction to the nonlinear case (24). Table 1 illustrates the enormous compression in the time scale of the development of linear and nonlinear geometrical optics. This compression is made possible, of course, by the establishment of a general formulation of electromagnetic phenomena by Maxwell in the second half of the nineteenth century. Lorentz showed in his Ph.D. thesis (25) how the laws of linear reflection, recorded by Hero of Alexandria (first century A.D.), Snell's laws (1621) and Fresnel's laws (1823) for the intensities and polarizations all followed from Maxwell's equations.

It is also suggested by the geometry of Fig. 4 that, on increasing the angle of incidence θ_i, nonlinear analogues for total reflection and evanescent surface waves should occur. Indeed, all such predictions have been verified (26), and in

Table 1. Historical dates of linear and nonlinear optical laws.

	Linear	*Nonlinear*
Law of Reflection	1st century (Hero of Alexandria)	1962 (Bloembergen and Pershan)
Law of Refraction	1621 (Snell)	1962 (Bloembergen and Pershan)
Intensity of Reflected and Refracted Light	1823 (Fresnel)	1962 (Bloembergen and Pershan)
Conical Refraction		
Theory	1833 (Hamilton)	1969 (Bloembergen) and Shih)
Experiment	1833 (Lloyd)	1977 (Schell and Bloembergen)

particular the nonlinear coupling between surface excitations is of active current interest (27). In 1833 Hamilton, who was to formulate Hamiltonian mechanics three years later, predicted the phenomenon of conical refraction based on Fresnel's equations of light propagation in biaxial optical crystals. The experimental confirmation in the same year by Lloyd was considered a triumph of the Fresnel equations for the elastic nature of optical propagation! The time lag between the prediction of nonlinear conical refraction and its experimental confirmation was much longer (28), as shown in Table I. In the twentieth century the description of electromagnetic propagation is not in doubt, and most researchers were too busy with more important applications of laser beams than the rather academic problem of nonlinear conical refraction.

The parametric coupling of light waves in a nonabsorbing medium may be considered as the scattering of photons between eigenmodes or waves of the electric field by the material nonlinearity. Heisenberg (29) and others had discussed an intrinsic nonlinearity of the vacuum.

The virtual intermediate states in that process are the electron-hole pair creation states which lie about a million times higher in energy than the excited states of electrons bound in a material medium. Since the energy mismatch of the intermediate states enters as the cube in the expression of $\chi^{(3)}$, the vacuum nonlinearity has not been detected. It would be difficult to exclude the nonlinear action of one atom or molecule in the focal volume of extremely intense laser beams used in attempts to detect the nonlinearity of vacuum.

In parametric, nondissipative processes the energy and momentum between incident and emerging photons must be conserved, $\Sigma_i \hbar \omega_i = 0$ and $\Sigma_i \hbar k_i = 0$, where the frequencies and wave vectors of the incident photons are taken to be negative. As noted above, color dispersion generally gives rise to a momentum mismatch $\Delta k = k_2 - 2k_1$. This limits the active volume of emission to a layer of thickness $|\Delta k|^{-1}$. It is possible, however, to compensate the color dispersion by optical birefringence in anisotropic crystals. This was demonstrated independently by Giordmaine (30) and by Terhune (31). For $\Delta k = 0$, the polarization in all unit cells in the crystal contributes in phase to the second harmonic field, and if the crystal is long enough and the light intensity high enough, the fundamental power may be quantitatively converted to second harmonic power (21). Phase coherence is essential. For random phases the final state would be one of equipartition with equal power in the fundamental and the second harmonic mode. More than eighty percent of the fundamental power at 1.06 μm wavelenght in a large pulsed Nd-glass laser system has recently been converted to third harmonic power (32) at 0.35 nm. In the first step two-thirds of the fundamental power is converted to second harmonic power. Then equal numbers of fundamental and second harmonic photons are combined to third harmonic photons in another crystal. This conversion may be important for the inertial confinement of fusion targets, as the laser-plasma coupling is improved at higher frequencies. The Manley-Rowe relations, which describe the balance in the photon fluxes of the beams participating in a parametric process, are here put to practical use. A few simple conservation laws thus determine many fundamental features of nonlinear optics.

NONLINEAR SPECTROSCOPY

Processes which are described by the imaginary part of the nonlinear susceptibility, $\chi^{(3)''}$, include saturation and cross saturation, twophoton absorption
and stimulated Raman effect. The corresponding real part $\chi^{(3)'}$ describes the
intensity dependent index of refraction. It plays a role in self-focusing and
defocusing of light, and in creating dynamic optical Stark shifts.

Saturation dip spectroscopy is used extensively to eliminate the effects of
Doppler broadening in high resolution spectroscopy and in frequency stabilization of lasers. Consider the case of two traveling waves incident on a gas sample
with the same frequency ω, but with opposite wave vectors, $\underline{k} = -\underline{k}'$. The wave
with $\underline{k}$ produces a saturation dip in the Doppler profile for the velocity packet of
molecules satisfying the relation $\omega = \omega_{ba} - \underline{k} \cdot \underline{v}$, where ω_{ba} is the atomic
resonance frequency. The beam in the opposite direction probes the packet
satisfying $\omega = \omega_{ba} - \underline{k}' \cdot \underline{v}' = \omega_{ba} + \underline{k} \cdot \underline{v}'$. The two packets coincide only for
$\omega = \omega_{ba}$. If ω is scanned across the Doppler profile, the probe beam will
register a saturation dip exactly at the center. The correspondence with the
NMR situation described earlier is clear. At optical frequencies the effect was
first demonstrated as a dip in the output of a helium neon laser (33, 34), and is
known as the Lamb dip (35). It is experimentally advantageous to observe the
effect in an external absorption cell with a strong pump beam in one direction
and a weak probe beam in the opposite direction. While the Doppler width of

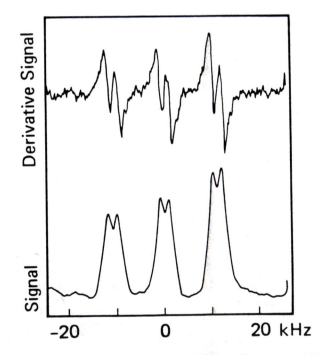

Fig. 5. High resolution (< 1 kHz) saturation spectroscopy of a $^{12}CH_4$ spectral line near 3.39 μm
wavelength. Each of the three hyperfine components is split into a doublet from the optical recoil
effect. The upper curve is the experimental derivative trace (after reference 36).

the vibrational rotational transition of methane near 3.39 μm wavelength is about 300 MHz, spectral features of about 1 kHz have been resolved by Hall and Borde (36). Figure 5 shows the features of the saturation dip, as the frequency of the probe beam was modulated. Saturation spectroscopy reveals not only a hyperfine structure of the molecular transition due to spin-rotational interaction, but also the infrared photon recoil effect which doubles each individual component. With a resolution approaching one part in 10^{11}, many

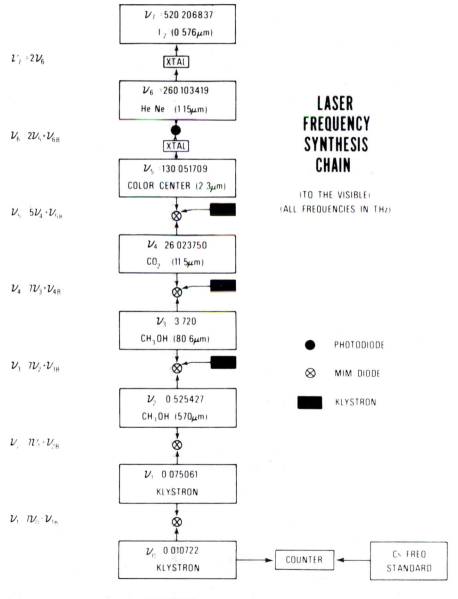

Fig. 6. Laser frequency synthesis chain (after reference 38).

other effects, such as curvature of the optical phase fronts and dwell time of the molecules in the beam, must be considered.

The frequencies of lasers throughout the infrared, each stabilized on the saturation dip of an appropriate molecular resonance line, have been compared with each other by utilizing the nonlinear characteristics of tungsten whisker-nickel oxide-nickel point-contact rectifiers (37, 38). The difference frequency between one laser and a harmonic of another laser is compared with a microwave frequency, which in turn is calibrated against the international frequency standard. Thus it has been possible to extend absolute frequency calibrations to the visible part of the spectrum (38), as shown by the chain in Fig. 6. Since the wavelength of the laser is independently compared with the krypton source length standard, it is possible to determine the velocity of light (39) with a precision set by the length standard definition, $c = 299\ 792\ 458.98 \pm 0.2$ m/s. It is proposed to define the velocity of light by international agreement, with length measurements then being tied directly to the frequency standard.

The application of saturation spectroscopy to a determination of the Rydberg constant and many other spectroscopic advances are discussed by Schawlow (40). Further details may be found in several comprehensive books on the subject (41–43). Optical saturation spectroscopy has also been carried out in solids, for example for Nd^{3+} ions in a crystal of LaF_3. Here the analogy with NMR techniques is more striking (44).

Two-photon absorption spectroscopy at optical frequencies, predicted by Goeppert-Mayer (45), was first demonstrated by Kaiser and Garrett (46) for Eu^{2+} ions in CaF_2. When the two photons have different wave vectors, an excitation with energy $2\hbar\omega$ and wave vector $\underline{k}$ $(\omega) + \underline{k}'$ (ω) may be probed. Fröhlich (47) applied wave vector-dependent spectroscopy by varying the angle between $\underline{k}$ and $\underline{k}'$ to the longitudinal and transverse excition branches in CuCl.

It was suggested by Chebotayev (48) that Doppler-free two-photon absorption features may be obtained in a gas. Consider again two counter-propagating beams. Tune the frequency ω so that $2\omega = \omega_{ba}$ corresponds to the separation of two levels with the same parity. For processes in which one photon is taken out of the beam with wave vector $\underline{k}$ and the other photon is taken out of the beam with wave vector $\underline{k}' = -\underline{k}$, all atoms regardless of their velocity are resonant. The apparent frequencies $\omega + \underline{k}\cdot\underline{v}$ and $\omega - \underline{k}\cdot\underline{v}$ of the photons in the two beams in the rest frame of an atom always add up to ω_{ba}. The two-photon absorption signal thus exhibits a very sharp Doppler-free feature, which was demonstrated experimentally in three independent laboratories (49–51). Thus high energy levels, including Rydberg states, of the same parity as the ground state may be studied in high resolution (52). The reader is again referred to the literature of further details (41, 43).

There is, of course, a close correspondence between two-photon absorption and Raman processes. A medium with a normal population difference between two levels |a > and |b >, which permit a Raman active transition, will exhibit a gain at the Stokes frequency, $\omega_s = \omega_L - \omega_{ba}$, in the presence of a strong pump

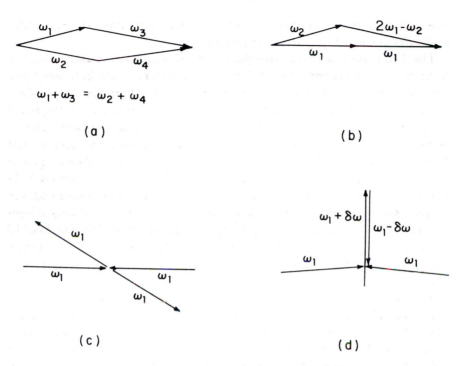

Fig. 7. Some typical wave vector geometries of four-wave light mixing.

beam at ω_L. Owyoung (53), for example, has resolved the fine structure in the Q-branch of a vibrational-rotational band of the methane molecule by the technique of stimulated Raman scattering. It is also possible to compare directly the Raman gain and a two-photon absorption loss with these nonlinear techniques.

FOUR-WAVE MIXING SPECTROSCOPY

The nonlinearity $\chi^{(3)}$ describes a coupling between four light waves, and some typical wave vector geometries which satisfy both energy and momentum conservation of the electromagnetic fields are shown in Fig. 7. The generation of a new beam at the frequency $2\omega_1 - \omega_2$, due to one incident beam at ω_1 and another at ω_2, corresponding to the geometry in Fig. 7b, was first demonstrated by Maker and Terhune (54, 55). They detected coherent antistokes raman scattering in organic liquids, where the nonlinear coupling constant $\chi^{(3)}$ exhibits a Raman-type resonance at the intermediate frequency $\omega_1 - \omega_2$, as shown schematically in Fig. 8b. Enchancement can also occur by a resonance at the intermediate frequency $2\omega_1$. It is thus possible, using light beams at visible wavelengths in a transparent crystal, to obtain information about resonance and dispersive properties of material excitations in the infrared (56, 57) and the utltraviolet (58). An example of this type of nonlinear spectroscopy is shown in Fig. 9. The two-dimensional dispersion of $\chi^{(3)}$ $(-2\omega_1 + \omega_2, \omega_1, \omega_1, -\omega_2)$ in

CuCl is measured as $2\omega_1$ is varied in the vicinity of the sharp Z_3 exciton resonance, while at the same time $\omega_1 - \omega_2$ is varied in the vicinity of the infrared polariton resonance. The interference of two complex resonances with each other, as well as the interference of their real parts with the nonresonant background contribution to $\chi^{(3)}$, leads to a direct comparison of these nonlinearities.

Wave vector-dependent four-wave mixing spectroscopy by variation of the angle between the incident beams was first performed by De Martini (59). The case of enhancement of the CARS process by one-photon absorptive resonances was investigated by several groups (60–62). Figure 8d shows an example of this situation. The CARS technique is used to monitor the composition and temperature profile in flames. In this and other situations with a large incandescent or fluorescent background, the coherent technique provides additional discrimination (63).

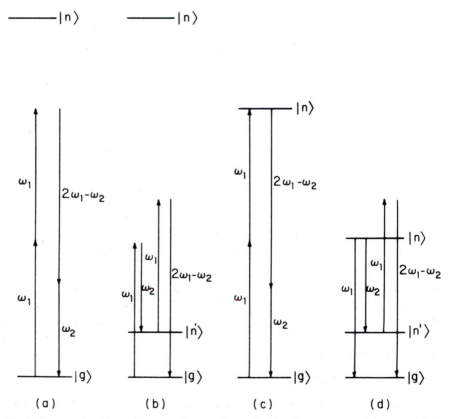

Fig. 8. The creation of a new beam at $2\omega_1-\omega_2$ by two incident beams at ω_1 and ω_2, respectively, according to the geometry of Fig. 7b.
a) Nonresonant mixing
b) Intermediate Raman resonance (coherent antistokes Raman scattering, or CARS)
c) Intermediate two-photon absorption resonance
d) One-photon resonantly enhanced CARS

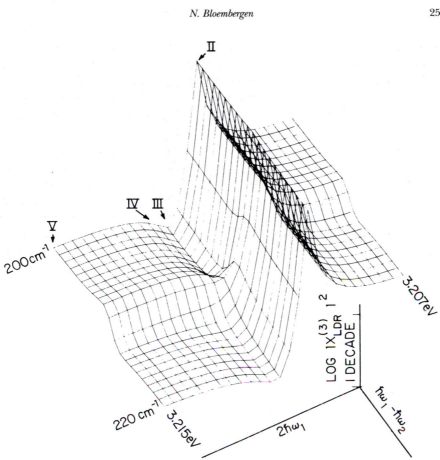

Fig. 9. Two-dimensional frequency dispersion of the nonlinear susceptibility χ^3 in cuprous chloride (after reference 58)

An important recent application of four-wave mixing is phase conjugation (64). Time-reversed phase fronts are obtained by the frequency-degenerate scattering geometry depicted in Fig. 7c. A strong standing wave pump field provides two beams at ω with equal and opposite wave vectors, $\underset{\sim}{k}_1 = -\underset{\sim}{k}_3$. The nonlinear medium may be liquid CS_2, Na vapor, InSb, an absorbing fluid, a molecular gas, or any other medium (65). A signal beam at the same frequency ω has a wave vector $\underset{\sim}{k}_2$, which makes an arbitrary small angle with $\underset{\sim}{k}_1$. In the four-wave scattering process a new beam with wave vector $\underset{\sim}{k}_4 = -\underset{\sim}{k}_2$ is created by the nonlinear polarization

$$P_4(\omega) = \chi^{(3)}(-\omega, \omega, \omega, -\omega)E_1 E_2^* E_3 e^{-i k_2 \cdot r}.$$

Note that not only the wave vector but also the phase is reversed, because $E_2^* = |E_2| \exp(-i\phi_2)$. This implies that the backward wave is the time reverse of the signal wave. If the phase front of the latter has undergone distortions in propagation through a medium, these will all be compensated as the backward wave returns through the same medium. The amplitude of the backward wave may show gain, because the parametric process involved takes one photon each out of the two pump beams and adds one each to the signal and its phase-conjugate beam. The process may also be viewed as real-time instant hologra-phy (66). The signal wave forms an intensity interference pattern with each of the pump beams. The physical cause for the grating may be a variation in temperature, in carrier density, in bound space charges, in molecular orienta-tion, depending on the material medium. The other pump reads out this hologram and is scattered as the phase-conjugate wave.

Another variation of nearly degenerate frequency four-wave light mixing has resulted in the recent demonstration of collision-induced coherence (67). Two beams at frequency ω_1 were incident in a vertical plane on a cell containing Na vapor and helium buffer gas. A third beam at a variable frequency ω_2 is incident in the horizontal plane. The generation of a beam in a new direction in the horizontal plane is observed at frequency $2\omega_1 - \omega_2$. The intensity of this new beam displays resonances for $\omega_1 = \omega_2$ and $\omega_1 - \omega_2 = 17 \text{ cm}^{-1}$, corresponding to the fine structure splitting of the 3P doublet of the Na atom. These reson-ances, however, occur only in the presence of collisions. Their intensity varies linearly or quadratically with the partial pressure of helium (68). The paradox that a phase-destroying collisional process can give rise to the generation of a coherent light beam is resolved as follows. In four-wave mixing many different scattering diagrams contribute to the final result (60, 61). These different coherent pathways happen to interfere destructively in the wave mixing case under consideration. Collisions of the Na atoms destroy this destructive inter-ference.

HIGHER ORDER NONLINEARITIES

Higher order terms in the perturbation expansion of equation (1) are responsi-ble for the generation of higher harmonics and multiphoton excitation pro-cesses. Akhmanov (69) has studied the generation of fourth harmonics in a crystal of lithium formate and the fifth harmonic in calcite. Reintjes et al. (70) have generated coherent radiation in the vacuum ultraviolet at 53.2 nm and 38.02 nm, as the fifth and seventh harmonic of a laser pulse at 266 nm which was focused into helium gas. The intensity at 266.1 nm was itself derived by two consecutive frequency doublings from a Nd^{3+} glass laser at 1.06 µm. Radiation at this infrared wavelength can induce photoelectric emission from tungsten. The energy of four photons is necessary to overcome the work function. This photoelectric current is proportional to the fourth power of the laser intensity (71).

Studies of multiphoton ionization of atoms and molecules has been pioneered by Prokhorov and coworkers (72). There is clear evidence for ionization of xenon by eleven photons at 1.06 µm. The ion current increases as the eleventh power of the intensity (73). The required laser intensities are so high that extreme care must be taken to avoid avalanche ionization started by electrons created from more readily ionizable impurities.

Atoms and molecules may, of course, also be ionized stepwise. A real excited bound state may be reached, whence further excitation beyond the ionization limit proceeds. The spectroscopy of auto-ionizing states has also been furthered by multiphoton laser excitation (74).

The intermediate resonances in the stepwise ionization process are species selective. The ionization of single atoms may be detected with a Geiger-Müller counter. Resonance ionization spectroscopy (75) uses this device in combination with one or more tunable dye lasers. The presence of a single atom amidst 10^{20} atoms of other species may be detected. Thus rare stable or unstable daughter atoms may be identified in coincidence with the decay of the parent atom. Ultralow level counting may also aid in measuring inverse β-decay products induced by the solar neutrino flux (75).

Many polyatomic molecules with absorption features near the infrared emission lines from pulsed CO_2 lasers can be dissociated without collisions in a true unimolecular reaction (76,77). In many cases more than thirty infrared quanta at $\lambda = 9.6$ or 10.6 μm wavelength are needed to reach the dissociation limit. Nevertheless the rate determining the step appears to be a succession of one-photon absorption (and emission) processes (78). The dissociation yield depends on the total energy fluence in the pulse and is largely independent of pulse duration (or peak intensity). This may be understood in terms of the large density of states in polyatomic molecules with a high degree of vibrational excitation. The energy absorbed by one mode is rapidly shared (equipartitioned) with the other degrees of freedom. Intramolecular relaxation times in highly excited polyatomic molecules are often quite short, on the order of one picosecond (10^{-12} sec). Infrared photochemistry of molecules in highly excited states has been stimulated by the availability of high power lasers. Both multiphoton dissociation and ionization processes can be applied to laser isotope separation (77).

OPTICAL TRANSIENTS

The perturbation expansion in equation (1) converges only if the Rabi frequency, $\hbar^{-1}|ex|_{ba}|E|$, proportional to the magnitude of the electric dipole matrix element and the field amplitude, is small compared to the detuning from resonance $\omega - \omega_{ba}$, or small compared to the homogeneous width or damping constant Γ_{ba} of the resonance. When this condition is not satisfied, very interesting nonlinear optical phenomena occur. They again have their precoursors in magnetic resonance and include, among others, free induction decay (79), optical nutation (79), optical echoes (80, 81) and split field resonances (82). The one-to-one correspondence of the evolution of any two-level system with the motion of a spin 1/2 system in magnetic resonance offers a convenient basis for description and also has heuristic value (83, 84).

Self-induced transparency (85) describes the propagation of a solitary optical wave or "soliton" which develops when an intense light pulse enters a material medium at a sharp absorbing resonance. The front part of the pulse excites the resonant transition; then the excited resonant state feeds back the energy to the trailing part of the pulse. The net result is that each two-level member of the ensemble executes a complete revolution around the effective field in the rotating frame of reference (83). In this 2π pulse no electromagnetic energy is dissipated in the medium, but the propagation velocity of the energy is slowed down. The fraction of energy stored in the medium does not contribute to the propagation.

The spontaneous emission process in the presence of a large coherent driving field (86, 87), the cooperative radiation phenomena associated with the super-radiant state (88), and the statistical properties (89, 90) of electromagnetic fields with phase correlations have increased our understanding of the concept of the photon.

Short optical pulses have been used extensively for time-resolved studies of transient phenomena and the measurement of short relaxation times. Very powerful pulses of about 10 picosecond (10^{-11} sec) duration are readily obtained by the technique of mode locking. Generally, the medium is excited by the first short pulse and probed by a second pulse with a variable time delay. The first pulse, for example, may excite a molecular vibration by stimulated Raman scattering. This coherent vibration will interact with the second pulse to give an antistokes component. A picosecond pulse traversing a cell of water generates a nearly continuous white spectrum due to phase modulation. This white picosecond pulse may be used to probe variations in absorption due to the first pulse. These techniques have been developed in depth by Kaiser (91) and others (92). More recently, the creation of light pulses as short as 4×10^{-14} sec has been achieved.

It is also possible, with a picosecond pulse, to melt a thin surface layer of a metal, alloy or semiconductor. After the light pulse is gone, this layer ($10-20$ nm thick) resolidifies rapidly by thermal conduction to the cool interior. Cooling rates of 10^{13}°C/sec are attainable. Thus it is possible to freeze in amorphous phases or other normally unstable configurations (93). New regimes of solid state kinetics are thus opened up for investigation.

CONCLUSION

Nonlinear optics has developed into a significant subfield of physics. It was opened up by the advent of lasers with high peak powers. The availability of tunable dye lasers has made detailed nonlinear spectroscopic studies possible throughout the visible region of the spectrum, from 0.35 to 0.9 nm. Conversely, nonlinear techniques have extended the range of tunable coherent radiation. Harmonic generation, parametric down conversion, and stimulated Raman scattering in different orders have all extended the range from the vacuum ultraviolet (94) to the far infrared (95). The soft X-ray region still presents a challenge.

Nonlinear optical processes are essential in many applications. Modulators and demodulators are used in optical communications systems. Saturable absorption and gain play an essential role in obtaining ultrashort pulses. The domain of time-resolved measurement may be extended to the femtosecond domain. This opens up new possibilities in materials science and chemical kinetics. A detailed understanding of nonlinear processes is essential in pushing the frontiers of time and length metrology, with applications to geological and cosmological questions.

The field of nonlinear spectroscopy has matured rapidly but still has much potential for further exploration and exploitation. The applications in chemistry, biology, medicine, materials technology, and especially in the field of communications and information processing are numerous. Alfred Nobel would have enjoyed this interaction of physics and technology.

I wish to express my indebtedness to my coworkers and graduate students, past and present, as well as to many colleagues, scattered in institutions around the globe, whose work in nonlinear optics and spectroscopy, cited or uncited, is also honored by this award.

REFERENCES

1. Townes, C. H., in *Nobel Lectures in Physics* (Elsevier, Amsterdam, 1972), Vol. 4, p. 58.
2. Basov, N. G., *ibid*, p. 89.
3. Prokhorov, A. M., *ibid.*, p. 110.
4. Salinger, H., Archiv. f. Elektrotechnik *12*, 268 (1923).
5. Lorentz, H. A., *Theory of Electrons* (Teubner, Leipzig, 1909).
6. Bloembergen, N., Purcell, E. M. and Pound, R. V., Phys. Rev. *73*, 679 (1948).
7. Pound, R. V., Phys. Rev. *79*, 685 (1950).
8. Kastler, A., in *Nobel Lectures in Physics* (Elsevier, Amsterdam, 1972), Vol. 4, p. 186.
9. Overhauser, A. W., Phys. Rev. *91*, 476 (1953).
10. Carver, T. R. and Slichter, C. P., Phys. Rev *102*, 975 (1956).
11. See, for example, Abragam,
    A., *Principles of Nuclear Magnetism* (Oxford University Press, London, 1960).
    Goldman, M., *Spin Temperature and Nuclear Magnetic Resonance in Solids* (Oxford University Press, London, 1970).
12. Bloembergen, N., Phys. Rev. *104*, 324 (1956).
13. Basov, N. G. and Prokhorov, A. M., Zh. Eksp. Teor. Fiz. *28*, 249 (1955).
14. Scovil, H. E. D. and Schulz-du Bois, E. O., Phys. Rev. Lett. *2*, (1955).
15. Bloembergen, N., Shapiro, S., Pershan, P. S. and Artman, J. O., Phys. Rev. *114*, 445 (1959).
16. Wilson, R. W., "Nobel Lecture 1978," in Rev. Mod. Phys. *51*, 767 (1979).
17. Schawlow, A. L., and Townes, C. H., Phys. Rev. *112*, 1940 (1958).
18. Maiman, T. H., Nature *187*, 493 (1960).
19. Franken, P., Hill, A. E., Peters, C. W. and Weinreich, G., Phys. Rev. Lett. *7*, 118 (1961).
20. Forrester, A. T., Gudmundsen, R. A. and Johnson, P. O., Phys. Rev. *99*, 1961 (1955).
21. Armstrong, J. A., Bloembergen, N., Ducuing, J. and Pershan, P. S., Phys. Rev. *128*, 606 (1962).
22. Bloembergen, N. and Shen, Y. R., Phys. Rev. *133*, A 37 (1963).
23. Akhmanov, C. A. and Khokhlov, R. V., *Problems in Nonlinear Optics* (Academy of Sciences, USSR, Moscow, 1964).
24. Bloembergen, N. and Pershan, P. S., Phys. Rev. *128*, 606 (1962).
25. Lorentz, H. A., *Collected Papers* (Martinus Nyhoff, The Hague, 1935), Vol. 1.
26. Bloembergen, N., Simon, H. J. and Lee, C. H., Phys. Rev. *181*, 1261 (1969).
27. DeMartini, F., Colocci, M., Kohn, S. E. and Shen, Y. R., Phys. Rev. Lett. *38*, 1223 (1977).
28. Schell, A. J. and Bloembergen, N., Phys. Rev. A *18*, 2592 (1978).
29. Heisenberg, W. and Euler, H., Z., Phys. *98*, 714 (1936).
30. Giordmaine, J. A., Phys. Rev. Lett. *8*, 19 (1962).
31. Maker, P. D., Terhune, R. W., Nisenhoff, M. and Savage, C. M., Phys. Rev. Lett. *8*, 21 (1962).

32. Seka, W., Jacobs, S. D., Rizzo, J. E., Boni, R. and Craxton, R. S., Opt. Commun. *34*, 469 (1980).
33. Szoke, A. and Javan, A., Phys. Rev. Lett. *10*, 521 (1963).
34. McFarlane, R. A., Bennett, W. R. Jr. and Lamb, W. E. Jr., Appl. Phys. Lett. *2*, 189 (1963).
35. Lamb, W. E., Phys. Rev. *134*, 1429 (1964).
36. Hall, J. L., Bordé, C. J. and Uehara, K., Phys. Rev. Lett. *37*, 1339 (1976).
37. Hocker, L. O., Javan, A. and Ramachandra Rao, D., Appl. Phys. Lett. *10*, 147 (1967).
38. Jenning, D. A., Petersen, F. R. and Evenson, K. M., in *Laser Spectroscopy IV*, edited by H. Walther and K. W. Rothe (Springer, Heidelberg, 1979), p. 39.
39. Baird, K. M., Smith, D. S. and Whitford, B. C., Opt. Commun. *31*, 367 (1979); Rowley, W. R. C., Shotton, K. C. and Woods, P. T., *ibid. 34*, 429 (1980).
40. Schlawlow, A. L., *Les Prix Nobel 1981* (Almqvist and Wiksell International, Stockholm, 1982); also Rev. Mod. Phys., to be published (1982).
41. *High-Resolution Laser Spectroscopy*, edited by K. Shimoda, Topics in Applied Physics *13* (Springer-Verlag, Berlin, Heidelberg, 1976).
42. Letokhov, V. S. and Chebotayev, V. P., *Nonlinear Laser Spectroscopy*, Springer Series in Optical Sciences *4* (Springer-Verlag, Berlin, 1977).
43. Levenson, M. D., *Introduction to Nonlinear Laser Spectroscopy* (Academic Press, New York, 1982). The author is indebted to Dr. M. D. Levenson for having received a preprint of his manuscript.
44. McFarlane, R. M. and Shelby, R. M., Opt. Lett. *6*, 96 (1981).
45. Goeppert-Meyer, M., Ann. Phys. *9*, 273 (1931).
46. Kaiser, W. and Garrett, G. C. B., Phys. Rev. Lett. *7*, 229 (1961).
47. Fröhlich, D., Staginnus, B. and Schönherr, E., Phys. Rev. Lett. *19*, 1032 (1967).
48. Vasilinko, L. S., Chebotayev, V. P. and Shishaev, A. V. JETP Lett. *12*, 113 (1970).
49. Cagnac, B., Grynberg, G. and Biraben, F., Phys. Rev. Lett. *32*, 643 (1974).
50. Levenson, M. D. and Bloembergen, N., Phys. Rev. Lett. *32*, 645 (1974).
51. Hänsch, T. W., Harvey, K. C., Meisel, G. and Schawlow, A. L., Opt. Commun. *11*, 50 (1974).
52. Stoicheff, B. P. and Weinberger, E., in *Laser Spectroscopy IV*, edited by H. Walther and K. W. Rothe (Springer, Heidelberg, 1979), p. 264.
53. Owyoung, A., Patterson, C. W. and McDowell, R. S., Chem. Phys. Lett, *59*, 156 (1978). For a historical review of the early work in the stimulated Raman effect, see: Bloembergen, N., Am. J. of Phys. *35*, 989 (1967).
54. Terhune, R. W., Solid State Design *4*, 38 (1963).
55. Maker, P. D. and Terhune, R. W., Phys. Rev. *137*, A801 (1965).
56. Levenson, M. D. and Bloembergen, N., Phys. Rev. B *10*, 4447 (1974).
57. Lotem, H., Lynch, R. T. Jr., and Bloembergen, N., Phys. Rev. A *14*, 1748 (1976).
58. Kramer, S. D. and Bloembergen, N., Phys. Rev. B *14*, 4654 (1976).
59. Coffinet, J. P. and DeMartini, F., Phys. Rev. Lett. *22*, 60 (1969).
60. Bloembergen, N., Lotem, H. and Lynch, R. T. Jr., Ind. J. of Pure & Appl. Phys. *16*, 151 (1978).
61. Attal, B., Schnepp, O. O., Taran, J.-P. E., Opt. Commun. *24*, 77 (1978). Druet, S. A. J., Taran, J.-P. E. and Bordé, Ch. J., J. Phys. (Paris) *40*, 819 1979).
62. Carreira, L. A., Goss, J. P. and Malloy, T. B., Jr., J. Chem. Phys. *69*, 855 (1978).
63. Eckbreth, A. C., Appl. Phys. Lett. *32*, 421 (1978).
64. Hellwarth, R. W., J. Opt. Soc. Am. *67*, 1 (1977).
65. See, for example, a number of papers by various authors published in Opt. Lett. *5*, 51, 102, 169, 182 and 252 (1980), and other references quoted therein.
66. Gabor, D., *Les Prix Nobel 1971* (P. A. Norstedt & Söner, Stockholm, 1972), p. 169.
67. Prior, Y., Bogdan, A. R., Dagenais, M. and Bloembergen, N., Phys. Rev. Lett. *46*, 111 (1981).
68. Bloembergen, N., Bogdan, A. R. and Downer, M., in *Laser Spectroscopy V*, edited by A. R. W. McKellan, T. Oka and B. Stoicheff, (Springer-Verlag, Heidelberg, 1981), p. 157.
69. Akhmanov, S. A., in *Nonliner Spectroscopy*, edited by N. Bloembergen, Fermi, Course 64 (North-Holland Publishing Co., Amsterdam, 1977), p. 239.
70. Reintjes, J., Eckardt, R. C., She, C. Y., Karangelen, N. E., Elton, R. C. and Andrews, R. A., Phys. Rev. Lett. *37*, 1540 (1976); Appl. Phys. Lett. *30*, 480 (1977).

71. Bechtel, J. H., Smith, W. L. and Bloembergen, N., Opt. Commun. *13*, 56 (1975).
72. Bunkin, F. V. and Prokhorov, A. M., Zh. Eksp. Teor. Fiz. *46*, 1090 (1964); Bunkin, F. V., Karapetyan, R. V. and Prokhorov, A. M., *ibid. 47*, 216 (1964).
73. LeCompte, C., Mainfray, G., Manus, C. and Sanchez, F., Phys. Rev. Lett. *32*, 265 (1972).
74. Armstrong, J. A. and Wynne, J. J., in *Nonliner Spectroscopy,* edited by N. Bloembergen, E. Fermi, Course 64 (North-Holland Publishing Co., Amsterdam, 1977), p. 152.
75. Hurst, G. S., Payne, M. G., Kramer, S. D. and Young, J. P., Rev. Mod. Phys. *51*, 767 (1979).
76. Isenor, N. R. and Richardson, M. C., Appl. Phys. Lett. *18*, 225 (1971).
77. Ambartsumian, R. V., Letokhov, V. S., Ryabov, E. A. and Chekalin, N. V., JETP Lett. *20*, 273 (1974).
78. Bloembergen, N., Opt. Commun. *15*, 416 (1975).
79. Brewer, R. G., in *Nonlinear Spectroscopy,* edited by N. Bloembergen, E. Fermi, Course 64 (North-Holland Publishing Co., Amsterdam, 1977), p. 87.
80. Kurnit, N. A., Abella, I. D. and Hartman, S. R., Phys. Rev. Lett. *13*, 567 (1964); Phys. Rev. *141*, 391 (1966).
81. Mossberg, T. W., Whittaker, F., Kachru, R. and Hartmann, S. R., Phys. Rev. A *22*, 1962 (1980).
82. Salour, M. M. and Cohen-Tannoudji, C., Phys. Rev. Lett. *38*, 757 (1977).
83. Rabi, I. I., Ramsey, N. F. and Schwinger, J. S., Rev. Mod. Phys. *26*, 167 (1954).
84. Feynman, R. P., Vernon, F. L. and Hellwarth, R. W., J. Appl. Phys. *28*, 49 (1957).
85. McCall, S. L. and Hahn, E. L., Phys. Rev. Lett. *18*, 908 (1967); Phys. Rev. *183*, 457, (1969).
86. Mollow, B. R., Phys. Rev. *188*, 1969 (1969).
87. Cohen-Tannoudji, C. and Reynaud, S., J. Phys. B *10*, 345 (1977).
88. Gibbs, H. M., Vrehen, Q. H. F. and Hikspoors, H. M. J., Phys. Rev. Lett. *39*, 547 (1977).
89. Glauber, R., Phys. Rev. *131*, 2766 (1963).
90. See, for example, *Coherence and Quantum Optics IV,* edited by L. Mandel, and E. Wolf, (Plenum Press, New York, 1978).
91. von der Linde, D., Laubereau, A. and Kaiser, W. Phys. Rev. Lett. *26*, 954 (1971).
92. See, for example, *Ultrashort Light Pulses,* edited by Shapiro, S. L., Topics in Applied Physics *18* (Springer-Verlag, Heidelberg, 1977).
93. Liu, J. M., Yen, R., Kurz, H. and Bloembergen, N., Appl. Phys. Lett. *39*, 755 (1981).
94. Harris, S. E., Appl. Phys. Lett. *31*, 398 (1977); Harris, S. F. *et al.,* in *Laser Spectroscopy V,* edited by A. R. W. McKellan, T. Oka, and B. Stoicheff, (Springer-Verlag, Heidelberg, 1981), p. 437.
95. Byer, R. L. and Herbst, R. L., in *Nonlinear Infrared Generation,* edited by Y. R. Shen, Topics in Applied Physics *16* (Springer-Verlag, Heidelberg, 1977), p. 81; V. T. Nguen, and T. J. Bridges, *ibid.,* p. 140; J. J. Wynne, and P. P. Sorokin, *ibid.,* p. 160.

Arthur L. Schawlow

ARTHUR L. SCHAWLOW

I was born in Mount Vernon, New York, U.S.A. on May 5, 1921. My father had come from Europe a decade earlier. He left his home in Riga to study electrical engineering at Darmstadt, but arrived too late for the beginning of the term. Therefore, he went on to visit his brother in New York, and never returned either to Europe or to electrical engineering. My mother was a Canadian and, at her urging, the family moved to Toronto in 1924. I attended public schools there, Winchester elementary school, the Normal Model School attached to the teacher's college, and Vaughan Road Collegiate Institute (high school).

As a boy, I was always interested in scientific things, electrical, mechanical or astronomical, and read nearly everything that the library could provide on these subjects. I intended to try to go to the University of Toronto to study radio engineering, and my parents encouraged me. Unfortunately my high school years, 1932 to 1937, were in the deepest part of the great economic depression. My father's salary as one of the many agents for a large insurance company could not cover the cost of a college education for my sister, Rosemary, and me. Indeed, at that time few high school graduates continued their education. Only three or four out of our high school class of sixty or so students were able to go to a unversity.

There were, at that time, no scholarships in engineering, but we were both fortunate enough to win scholarships in the faculty of Arts of the University of Toronto. My sister's was for English literature, and mine was for mathematics and physics. Physics seemed pretty close to radio engineering, and so that was what I pursued. It now seems to me to have been a most fortunate chance, for I do not have the patience with design details that an engineer must have. Physics has given me a chance to concentrate on concepts and methods, and I have enjoyed it greatly.

With jobs as scarce as they were in those years, we had to have some occupation in mind to justify college studies. A scientific career was something that few of us even dreamed possible, and nearly all of the entering class expected to teach high school mathematics or physics. However, before we graduated in 1941 Canada was at war, and all of us were involved in some way. I taught classes to armed service personnel at the University of Toronto until 1944, and then worked on microwave antenna development at a radar factory.

In 1945, graduate studies could resume, and I returned to the University. It was by then badly depleted in staff and equipment by the effects of the depression and the war, but it did have a long tradition in optical spectroscopy. There were two highly creative physics professors working on spectroscopy,

Malcolm F. Crawford and Harry L. Welsh. I took courses from both of them, and did my thesis research with Crawford. It was a very rewarding experience, for he gave the students good problems and the freedom to learn by making our own mistakes. Moreover, he was always willing to discuss physics, and even to speculate about where future advances might be found.

A Carbide and Carbon Chemicals postdoctoral fellowship took me to Columbia University to work with Charles H. Townes. What a marvelous place Columbia was then, under I. I. Rabi's leadership! There were no less than eight future Nobel laureates in the physics department during my two years there. Working with Charles Townes was particularly stimulating. Not only was he the leader in research on microwave spectroscopy, but he was extraordinarily effective in getting the best from his students and colleagues. He would listen carefully to the confused beginnings of an idea, and join in developing whatever was worthwhile in it, without ever dominating the discussions. Best of all, he introduced me to his youngest sister, Aurelia, who became my wife in 1951.

From 1951 to 1961, I was a physicist at Bell Telephone Laboratories. There my research was mostly on superconductivity, with some studies of nuclear quadrupole resonance. On weekends I worked with Charles Townes on our book *Microwave Spectroscopy*, which had been started while I was at Columbia and was published in 1955. In 1957 and 1958, while mainly still continuing experiments on superconductivity, I worked with Charles Townes to see what would be needed to extend the principles of the maser to much shorter wavelengths, to make an optical maser or, as it is now known, a laser. Thereupon, I began work on optical properties and spectra of solids which might be relevant to laser materials, and then on lasers.

Since 1961, I have been a professor of physics at Stanford University and was chairman of the department of physics from 1966 to 1970. In 1978 I was appointed J. G. Jackson and C. J. Wood Professor of Physics. At Stanford, it has been a pleasure to do physics with an outstanding group of graduate students, occasional postdoctoral research associates and visitors. Most especially the interaction with Professor Theodor W. Hänsch has been continually delightful and stimulating. Our technicians, Frans Alkemade and Kenneth Sherwin have been invaluable in constructing apparatus and keeping it in operation. My secretary for the past nineteen years, Mrs. Fred-a Jurian, provides whatever order that can be found amidst the chaos of my office. Much of the time, my thoughts are stimulated there by the sounds of traditional jazz from my large record collection.

My wife is a musician, a mezzo soprano and choral conductor. We have a son, Arthur Keith, and two daughters, Helen Aurelia and Edith Ellen. Helen has studied French literature at Stanford, the Sorbonne, and at the University of California in Berkeley, and is now on the staff of Stanford University. Edith graduated from Stanford this year with a major in psychology.

Awards

Stuart Ballantine Medal (1962); Thomas Young Medal and Prize (1963); Morris N. Liebmann Memorial Prize (1964); California Scientist of the Year (1973); Frederick Ives Medal (1976); Marconi International Fellowship (1977).

Honorary doctorates from University of Ghent, Belgium (1968), University of Toronto, Canada (1970), University of Bradford, England (1970). Honorary professor, East China Normal University, Shanghai (1979).

Member, U.S. National Academy of Sciences.
Fellow, American Academy of Arts and Sciences.
President, Optical Society of America (1975)
President, American Physical Society (1981)

(*added in 1991*): I retired from teaching and became Professor Emeritus in 1991. My wife died in an automobile accident in May, 1991. My daughter Helen is now Assistant Professor of French at the University of Wisconsin. From Helen and her sister Edith, I now have four grandchildren.

Awards:

Arthur Schawlow Medal, Laser Institute of America (1982)

U.S. National Medal of Science (1991)

Honorary doctorates from University of Alabama, U.S.A. (1984), Trinity College, Dublin, Ireland (1986), University of Lund, Sweden (1988)

SPECTROSCOPY IN A NEW LIGHT

Nobel lecture, 8 December, 1981

by

ARTHUR L. SCHAWLOW

Department of Physics, Stanford University, Stanford, California 94305, USA

INTRODUCTION

Scientific spectroscopy really began in Uppsala, Sweden, where Anders Ångström in 1853 showed that some of the lines in the spectrum of an electric spark come from the metal electrodes and others from the gas between them.[1] Even earlier, Joseph Fraunhofer had charted the dark lines in the spectrum of the sun, and had measured their wavelengths.[2] But it was Ångström who first identified some of these lines as corresponding to bright lines emitted by particular substances in the spark. Most importantly, he showed the red line of hydrogen, now known as H\α. In subsequent years, Ångström found several more visible lines from hydrogen, and measured their wavelengths accurately. When W. Huggins[3] and H. W. Vogel[4] succeeded in photographing the spectra of stars in 1880, they found that these visible lines were part of a longer series extending into the ultraviolet. J. J. Balmer[5] in 1885 was able to reproduce the wavelengths of these lines by a formula, which we might write as

$$\frac{1}{\lambda} = \nu = \frac{R}{n^2} - \frac{R}{2^2}.$$

Balmer obtained the values of the constants from Ångström's measurements.

Five years later Johannes Rydberg[6], without knowing of Balmer's work, developed a more general formula for the atomic spectra of alkali metals such as sodium.

$$\frac{1}{\lambda} = \nu = \frac{R}{(n-d)^2} - \text{Constant},$$

Rydberg's formula includes Balmer's as a special case where $d = 0$. The constant R is now universally known as the Rydberg constant. We know now that it measures the strength of the binding between electrons and nuclei in atoms.

It is well known that the Balmer equation for the hydrogen spectrum helped Niels Bohr to introduce a quantum theory of atoms. In the 1920s, atomic and molecular spectroscopy was the principal experimental tool leading to the discoveries of the laws of quantum mechanics, from which comes most of our understanding of modern physics and chemistry.

In the 1940s, when I was a graduate student at the University of Toronto, nuclear physics seemed to be the most active branch of the subject but we had no accelerator. Therefore, I worked with two other students, Frederick M. Kelly and William M. Gray, under the direction of Professor M. F. Crawford to use high-resolution optical spectroscopy to measure nuclear properties from their effects on the spectra of atoms. The shifts and splittings of spectral lines from the interactions between electrons and nuclei were so small that they are known as hyperfine structures. To resolve them, we needed to build high resolution spectroscopic equipment. We also had to reduce the widths of the spectral lines from our light source, because broad lines cause overlapping that could completely hide much of the detail that we were seeking. When the gas density is so low that collisions could be neglected, the principal source of the line widths is the Doppler-broadening from the thermal motions of the atoms. Atoms moving toward the observer emit light that is shifted upward in frequency, while atoms moving away emit light of lower frequency than atoms at rest. Since there is a distribution of velocities, the line is broadened, with a Doppler width given by

$$\Delta v = \frac{2v}{c} \sqrt{2k \ N_o \ \ell n \ 2} \ \sqrt{\frac{T}{M}} \ ,$$

where v is the line frequency, k is Boltzmann's constant, N_o is Avogadro's number, T is the absolute temperature, and M is the molecular weight.

This Doppler width, as a fraction of the line frequency is of the order of $\bar{v}/c$, where $\bar{v}$ is the atomic velocity and c is the velocity of light, or typically about 10^{-5}. We were able to reduce it by a factor of ten or so, by using a roughly collimated beam of atoms, excited by an electron beam, as had been done earlier by Meissner and Luft[7] and by Minkowski and Bruck[8], and by observing the emitted light from a direction perpendicular to the atomic beam. The hyperfine structures we sought could be resolved, but four hours exposure time on our photographic plates was required. It seemed that there really ought to be an easier method that would give still sharper spectral lines, and indeed a large part of our work in laser spectroscopy has been devoted to finding such methods.

LASER SPECTROSCOPY

By the time that Charles Townes and I were working to see if it was possible to make a laser[9], in 1957 and 1958, both of us had experience in microwave spectroscopy. Thus, it was a familiar idea to us that spectra could be observed without a spectrograph, by tuning a narrowband source across the spectrum. At some wavelengths the light would be absorbed, at others it would be transmitted. This was one of the few applications we could then foresee for lasers. But each of the early lasers gave its own characteristic wavelength, depending on the material used. They could not be tuned very far, just across the width of the laser line. More tuning could be obtained sometimes by applying an external magnetic field, or by changing the temperature of a solid laser material. For instance, a ruby laser emits at 6943 Angstrom units at room temperatures and 6934 Angstrom units at liquid nitrogen temperature of 77K. William Tiffany, Warren Moos and I used a temperature-tuned ruby laser to map out a small portion of the absorption spectrum of gaseous bromine and see how changing the laser wavelength affected the chemical reactivity of the bromine.[10] We studied bromine because it had a rich spectrum with many absorption lines within the range of the available laser.

Others studied the spectra of the atoms used in a gas laser, particularly neon, and interesting phenomena were discovered. To understand them, we must note that the output wavelength of a laser is determined only roughly (within the Doppler line width) by the amplifying medium, and more precisely by the tuning of the laser resonator. By changing the spacing between the end mirrors, the laser can be tuned over the frequency range for which amplification is enough to overcome the losses. One might expect that in the center of the atomic line, where the gain is largest, the laser output will be greatest. But Willis E. Lamb, Jr. predicted, from a detailed theoretical analysis of laser principles, that there would be a dip in power output at the center of the line.[11] It would occur because light beams travel in both directions inside the laser resonator. At the center of the line both beams stimulate the same excited atoms, those with zero velocity, as indicated in Figure 1. At any other tuning, the light waves interact with those atoms whose velocity provides just the

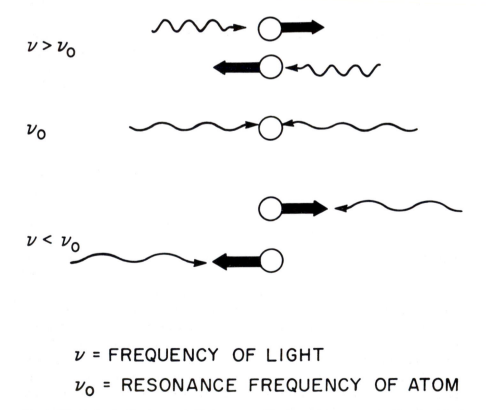

$$\nu = \text{FREQUENCY OF LIGHT}$$
$$\nu_0 = \text{RESONANCE FREQUENCY OF ATOM}$$

Fig. 1. Moving molecules interact with an approaching lower frequency wave Doppler-shifted upward in frequency, or a following higher-frequency wave shifted downward.

Doppler shift needed to bring the light into resonance with them. Thus there are two groups of atoms, with the same speed but opposite directions, which can be stimulated to provide the laser output. This "Lamb dip" at the center of the laser line was very soon observed by Bennett, Macfarlane and Lamb.[12] It was used for spectroscopy by Szoke and Javan[13], who also showed that the narrow resonance at the dip, free from Doppler-broadening, is sensitive to broadening by collision unless the gas pressure is quite low.

Paul Lee and M. L. Skolnick[14] also showed that if an absorbing gas is present inside the laser resonator, an "inverse Lamb dip" can occur, in which the laser output shows a peak at the center of the absorption line where the absorption of the molecules is saturated by the beams from both directions. The narrow, Doppler-free, optical resonances revealed by the Lamb dip and its inverse have been used for stabilizing the wavelength of lasers.

Thus by the middle of the 1960s it could be seen that, for spectroscopy, laser light possesses several advantages in addition to monochromaticity. Its intensity makes it possible to at least partially saturate absorption or stimulated emission transitions, and so to burn a narrow absorption or emission hole in a Doppler-broadened line. The directionality permits us to observe the combined effects of oppositely directed beams. Thus we could recognize the absorption from just those atoms or molecules which have zero velocity component along a chosen direction, and observe spectral details without Doppler-broadening. But at that time, we could only do so inside the resonator of some laser, and we could only work at those wavelengths where there happened to be laser lines. Later in the decade, Theodor W. Hänsch and Peter Toschek prepared the way for the subsequent advances by using the beam from a second laser on a cascade transition to probe the distribution of molecules as it was affected by saturation inside a laser.[15]

SATURATION SPECTROSCOPY

Laser spectroscopy became much more widely useful when, in 1970, Theodor W. Hänsch,[16, 17] and Christian Bordé[18] independently introduced a method which uses these properties of laser light to give Doppler-free spectra of gases external to the laser. As shown in Figure 2, the light from the laser is divided by a partial mirror into two beams which pass through the sample in nearly opposite directions. The stronger "pump" beam is chopped at an audio frequency. When it is on, it is strong enough to partially saturate the absorption of the molecules in the region through which it passes. The probe beam then is less attenuated by its passage through the gas, and a stronger signal reaches the detector. When the chopper obstructs the pump beam, the gas absorption returns, and less of the probe's light reaches the detector. Thus the probe beam is modulated as the pump is alternately turned on and off by the chopper. However, this modulation occurs only when the two beams interact with the same molecules, and that happens only when the laser is tuned to interact with molecules at rest, or at least with zero velocity component along the direction of the beams. Any molecule moving along the beams sees one wave as shifted up

SATURATION SPECTROMETER

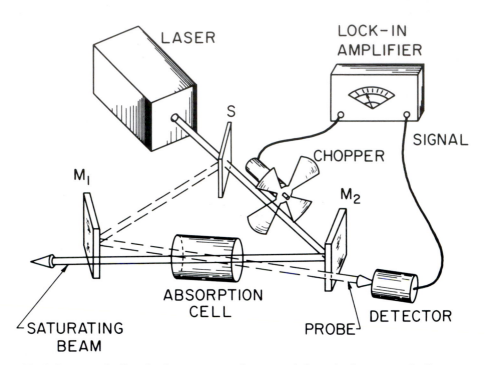

Fig. 2. Apparatus for Doppler-free spectroscopy by saturated absorption in an external cell.

in frequency and the other shifted down, and so a moving molecule cannot be simultaneously in resonance with both beams.

Hänsch and Marc Levenson applied this method first using a single-mode krypton ion laser which could be tuned over a range of about a twentieth of a reciprocal centimeter, that is, 1500 Megahertz, around the wavelengths of a few visible lines. This tuning range, although still quite limited, was enough to explore the details of several of the many lines in the dense visible absorption spectrum of the iodine molecule, I_2. For example, Figure 3 shows the hyperfine structure of a single line in that spectrum, produced by the interaction between the molecular axial field gradient and the quadrupole moments of the two iodine nuclei. Although other workers subsequently attained considerably better resolution by more carefully stabilizing the laser wavelength, the power of the method was already a spectacular advance over what could be done before. Thus, if Figure 3 is projected on a screen two meters wide, on the same scale the visible portion of the spectrum would have a width of more than 500 kilometers! Moreover, the individual lines in the pattern, although still limited by pressure broadening and laser frequency jitter, had a width of about 6 MHz, or about one part in 10^8. The hyperfine structures revealed, which had up to then

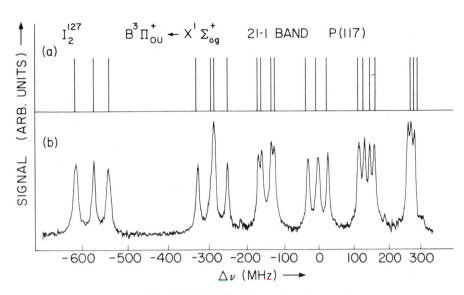

Fig. 3. Hyperfine structure of the P (117) 21-1 B ← X transition of molecular iodine at 568.2 nm (a) theoretical (b) experimental.

always been obscured by the Doppler-broadening of 600 MHz even with the best spectrographs, can be interpreted as in microwave spectroscopy to provide information about the distribution of electrons in the molecule.[19]

Narrow as these lines were, they were still broadened by intermolecular collisions at the operating pressure of about one torr. It was easy to reduce the vapor density by cooling the iodine cell, but then the absorption of the probe beam would be negligible whether the pump beam was on or not. However, C. Freed and Ali Javan[20] had shown, in some infrared spectroscopic studies, that when absorption is saturated, any fluorescence that follows from the absorption also shows a saturation. That is, the fluorescence intensity is not linearly proportional to the laser power, but levels off when the laser intensity is enough to deplete the number of molecules contributing to it. In our case the Javan-Freed method was not immediately applicable because, if the two oppositely directed beams were to work together to saturate one of the hyperfine-structure lines, it would cause only a very small change in the unresolved fluorescence in all of the components. Sorem and I, therefore, introduced a method of inter-modulated fluorescence.[21] We chopped both of the counterpropagating beams at different audio frequencies, by using two rings with different numbers of holes on the chopping wheel. Our fluorescence detector was tuned to respond to modulation at the sum of the two chopping frequencies, which arose when the stationary molecules were simultaneously excited by the two laser beams. Thus we obtained a good signal, free from Doppler-broadening, even at pressures as low as one millitorr or a thousand times less than we had been able to use with the saturated absorption method.

This is still far from the sensitivity that could ultimately be attained. When continuous wave, broadly tunable lasers became available, William M. Fairbank, Jr. and Theodor W. Hänsch tuned the laser to the orange-yellow wavelength of the sodium resonance lines. With a sodium cell designed to avoid stray light from the walls, we were able to measure the intensity of the light scattered from the sodium atoms, down to as few as a hundred atoms per cubic centimeter.[22] At that density, attained when the cell was cooled to $-30°$ C, there was on the average only one or two atoms at a time in the beam. With this method, we were able to measure the vapor density of sodium metal with a million times greater sensitivity than could previously be obtained. It was evident that laser methods could be much more sensitive than other techniques, such as radioactive methods, for detecting small amounts of suitable substances. A single atom can scatter very many light quanta without being destroyed, and so it should be possible to observe and study a single atom or molecule of a substance. Indeed, in favorable cases this sensitivity can already be achieved, by a method that uses resonant laser excitation followed by ionization.[23] The principal difficulty in making such methods broadly applicable is the lack of suitable lasers at some of the wavelengths needed, especially in the ultraviolet regions.

BROADLY TUNABLE LASERS

During the 1960s, there was a rapid growth in discoveries of new laser materials and ways to excite them. Solids, liquids, and gases were made to produce laser action under optical excitation, as well as electrical discharges in gases and semiconductors. But each of them operated at its own characteristic wavelength, determined by the properties of the material, and there was no way to obtain an arbitrary wavelength even in the same spectral region. We did not at first expect to be able to produce laser operation over a continuous band of wavelengths, because we knew that the available optical amplification was inversely proportional to the width of the lasing line or band. Nevertheless, Peter Sorokin and J. R. Lankard[24] and, independently, Fritz Schäfer[25] were able to use intense flashlamps to excite laser action in organic dyes, whose emission bands could be as wide as a hundred Angstrom units or more.

A further advance came when it was realized that the high light intensity to pump broadband laser materials could be best obtained from another laser. I must admit that at first I wondered why anyone would want to compound the inefficiencies by using one laser to pump another. But when you need concentrated light for pumping, a suitable laser is a good way to get it. Thus J. A. Myer, C. L. Johnson, E. Kierstead, R. D. Sharma, and I. Itzkan[26] used a pulsed ultraviolet nitrogen laser to pump a tunable dye laser. As shown in Figure 4, the dye laser can consist of just a cell containing a dye in solution, an output mirror and a diffraction grating. The grating replaces the second mirror of the ordinary laser structure, and acts as a good mirror for one wavelength that changes as the grating is rotated. By now, dyes are available to give laser action at all wavelengths in the visible, extending into the near ultraviolet and

Fig. 4. Photograph of a simple pulsed dye laser, pumped by an ultraviolet beam from a nitrogen laser, and tuned by a diffraction grating.

infrared. When pumped by a nitrogen laser, the dye laser typically gives pulses of a few nanoseconds duration. The amplification is very high, so that the end mirror and the diffraction grating or other tuning element do not need to have very high reflectivity.

But such a simple dye laser of this kind typically gives an output too widely spread in wavelength to be useful for high resolution spectroscopy. Hänsch was able to obtain narrow line output by adding a telescope between the dye cell and grating. Then the light at the grating was spread over more of the rulings, and was better collimated, so that the sharpness of its tuning was improved. To get output with sub-Doppler narrowness limited only by the length of the light pulse, or about 300 Megahertz, he placed a tilted etalon in front of the grating (Figure 5).[27] Even more monochromatic output, with a corresponding increase in pulse length could be obtained by filtering the output through a passive resonator.

Continuous-wave dye laser operation was obtained in 1970 by Peterson, Tuccio, and Snavely,[28] who used an argon ion laser to excite it. The pumping laser and dye laser beams were collinear, as had been used for c.w. ruby lasers by Milton Birnbaum,[29] rather than in the transverse arrangement used in most earlier lasers. Refining the output of continuous-wave lasers presented difficulties because the available amplification was small, so that any tuning

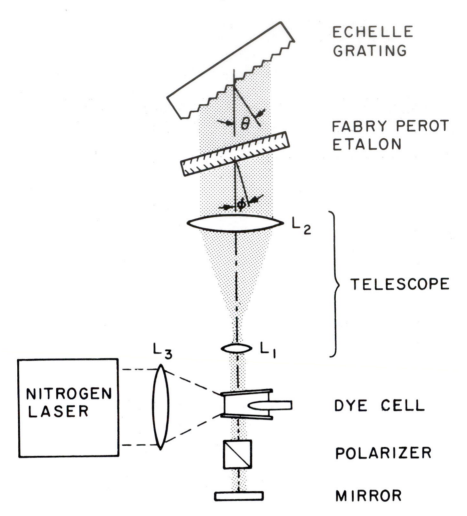

Fig. 5. Diagram of an improved dye laser, with a telescope and an etalon between the dye cell and the diffraction grating.

elements had to present low losses. But by now, extremely stable, narrowband lasers have been made, with line widths much less than one Megahertz.

Once a powerful, narrowband but broadly tunable, laser was available, it became possible to adapt the laser to the problem rather than the reverse. The methods of saturation spectroscopy could be applied to examine in detail the spectral lines of atoms simple enough to be of theoretical interest. With the pulsed dye laser, Hänsch and Issa S. Shahin first obtained Doppler-free spectra of the sodium atom's D lines at 5890 and 5896 Angstrom units, with the ground state hyperfine structure clearly resolved.[30] Then they applied it to study the fine structure of the red line H_α of atomic hydrogen.[31] For this purpose, they constructed an electric glow discharge tube with end windows through which the two beams could pass. One beam was, as before, the saturating beam, while the other, weaker beam was the probe to detect the absorption from atoms with no velocity component along the beam direction.

HYDROGEN TERMS

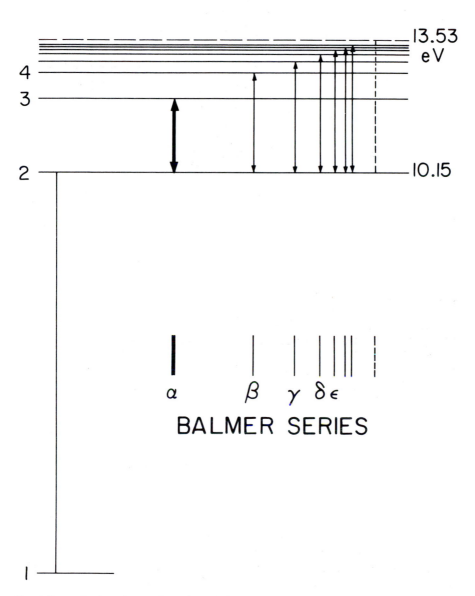

Fig. 6. Energy levels and transitions of atomic hydrogen.

Figure 6 recalls the energy levels and spectral lines of the hydrogen atom according to the quantum theory of Niels Bohr. The transitions from higher levels to the level with principal quantum number, n, equal to 2 give rise to the Balmer series spectrum, drawn at the top of Figure 7. Below it is shown the fine structure of the red line, on a scale expanded by a factor of 40000, as it would be

SPECTRUM OF HYDROGEN

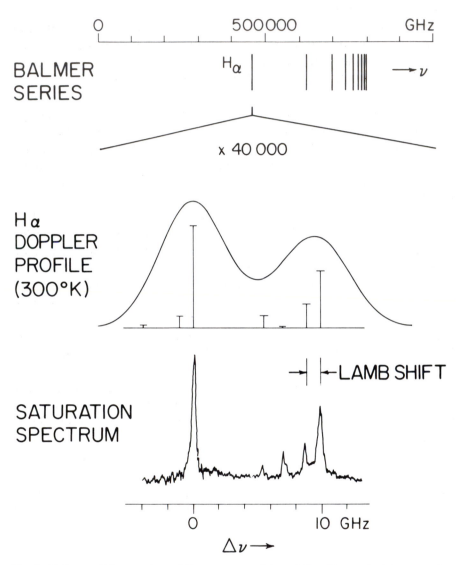

Fig. 7. Hydrogen Balmer series and fine structure of the red line Hα, resolved by saturation spectroscopy.

revealed by a perfect conventional spectrograph. The line is known, from theory and from radiofrequency measurements, to have the several fine structure components indicated, but they would be nearly obscured by the large Doppler width. At the bottom of the Figure is shown the fine structure of this line revealed by laser saturation spectroscopy. The improvement is dramatic, and most of the details of the fine structure can be clearly seen. In particular, the Lamb shift between the $2s_{1/2}$ and the $2p_{1/2}$ levels is clearly revolved, which had not been possible previously in hydrogen, although Gerhard Herzberg had resolved the Lamb shift in the corresponding line of ionized helium, where the shift is four times greater.[32]

Microwave measurements had already given an accurate account of all of these details, and the optical resolution had little hope of improving on them. What could be done much better than before was to make an accurate determination of the wavelength of one of the components, and thereby obtain an improved value of the Rydberg constant. Of course, if the positions and relative intensities of the components are known, it is possible to compute the line shape and compare it precisely with the shape and position determined by optical spectroscopy. But the relative intensities are determined by the detailed processes of excitation and deexcitation in the gas discharge, and their uncertainty was the principal source of error in earlier measurements of the line wavelength.

Hänsch and his associates Issa S. Shahin, Munir Nayfeh, Siu Au Lee, Stephen M. Curry, Carl Wieman, John E. M. Goldsmith, and Erhard W. Weber, have refined the measurement of the wavelength of the line and thereby of the Rydberg constant, through a series of careful and innovative researches.[33] They have improved the precision by a factor of about eighty over previous work, so that the Rydberg is now one of the most accurately known of the fundamental constants. The value obtained, $R_\infty = 109737.3148 \pm .0010$ cm^{-1} is in good agreement with that obtained in a recent experiment using laser excitation of an atomic beam, by S. R. Amin, C. D. Caldwell, and W. Lichten.[34]

In the course of these investigations, Wieman and Hänsch[35] found a new method to increase the sensitivity of the saturation method for avoiding Doppler-broadening. As shown in Figure 8, they used a polarized pump beam. It preferentially excites molecules with some particular orientation, leaving the remainder with a complementary orientation. The probe beam is sent through two crossed polarizers, one before and one after the sample region, so that no light reaches the detector except at the wavelengths where the light is depolarized by the oriented molecules. The saturation signal then appears as an

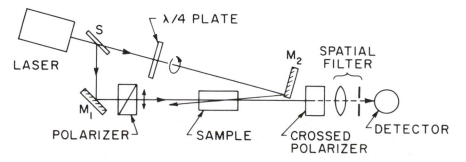

Fig. 8. Apparatus for Doppler-free polarization spectroscopy.

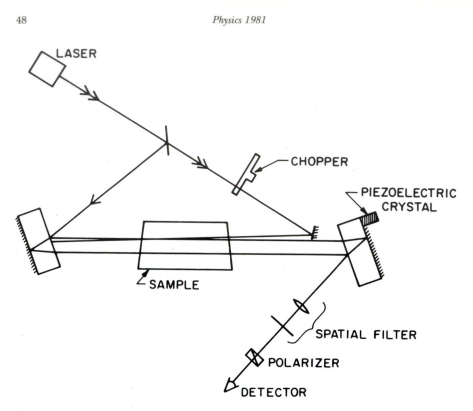

Fig. 9. Apparatus for Doppler-free saturated interference spectroscopy.

increased transmission with nearly no interfering background. Thus the noise caused by fluctuations in intensity of the probe laser is nearly eliminated, and the spectra can be observed at lower density or with lower light intensity. This method is now known as polarization spectroscopy.

Another way of balancing out the background was introduced by Frank V. Kowalski and W. T. Hill[36] and, independently, by R. Schieder.[37] They used a configuration like a Jamin interferometer, in which the probe beam is split into two parts which travel parallel paths through the sample cell, as shown in Figure 9. The beams are recombined in such a way that they cancel each other. Then when a saturating beam reduces the absorption along one of the paths, the interferometer becomes unbalanced and a Doppler-free signal is seen. In a way, polarization spectroscopy can be thought of as a special case of saturated-interference spectroscopy. The plane polarized probe wave is equivalent to two waves circularly polarized in opposite senses. They combine to produce a plane wave of the original polarization, which is stopped by the second polarizer, unless one of the two circularly polarized components experiences a different absorption or a different effective path length than the other.

SIMPLIFYING SPECTRA BY LASER LABELING

Spectra of molecules are very much more complicated than those of atoms. Even a diatomic molecule such as Na_2 has dozens of vibrational and hundreds of rotational levels for every electronic level. We have, therefore, sought systematic ways to use lasers to simplify molecular spectra so as to identify their various states. Even before lasers, something like this could be done by using a monochromatic light source, such as a filtered mercury lamp, to excite just one level, and observing the fluorescence from it to lower levels. With monochromatic, tunable lasers this can be used, for instance, to explore the vibrational and rotational structure of the ground electronic state of molecules. The upper state in this case may be said to be "labeled," since it is identified by having molecules excited to it, while neighboring states have none.

But if anything at all is known about a molecule, it is likely to be the constants of the ground electronic state, which can also be studied by microwave, infrared and Raman spectroscopy. Mark E. Kaminsky, R. Thomas Hawkins, and Frank V. Kowalski therefore inverted the process, by using a laser to pump molecules out of a chosen lower level.[38] All of the absorption lines originating on this chosen level were, then, weakened. If the pumping laser was chopped, the absorption lines from the labeled level were modulated at the chopping frequency. Thus when a high-resolution optical spectrometer was scanned across the spectrum, the lines from the labeled level could be recognized by their modulation, even if perturbations displaced them far from their expected position.

Almost as soon as Hänsch and Wieman introduced the method of polarization spectroscopy, it was apparent to us that it could be adapted for searching for and identifying the levels of molecules or complex atoms.[39] Apparatus for the polarization labeling method is shown in Figure 10. A polarized beam from a repetitively pulsed dye laser is used to pump molecules of a particular orientation from a chosen lower level, and leave the lower level with the complementary orientation. A broadband probe from a second laser is directed through two crossed polarizers, before and after the sample, and then into a

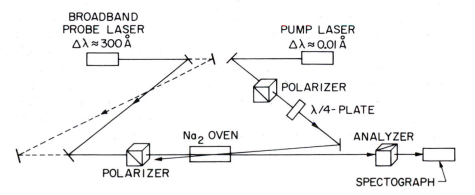

Fig. 10. Apparatus for simplifying spectra by polarization labeling.

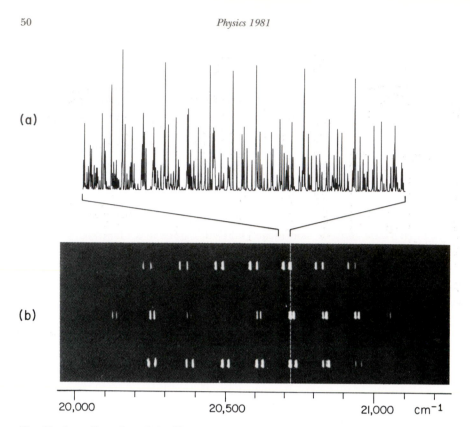

Fig. 11. A small section of the Na$_2$ spectrum revealed by conventional spectroscopy and by polarization labeling.

photographic spectrograph. Figure 11 shows the spectra observed by Richard Teets and Richard Feinberg,[39] as several neighbouring lines of N$_2$ are pumped. It is seen that from each labeled level, there are just two rotational lines for each vibrational level (J' = J" + 1 and J' = J" − 1). A small portion of the spectrum as obtained by simple absorption spectroscopy is shown for comparison. As the tuning of the pump laser is changed slightly, different groups of lines appear. For each of them, the upper state vibrational quantum number can be inferred simply by recognizing that the lines of lowest frequency end on the v' = 0 level of the upper electronic state.

As the molecules raised to the excited electronic level by the polarized pump laser are also oriented, the probe can record transitions from that to still higher levels. Nils W. Carlson, Antoinette J. Taylor, and Kevin M. Jones[40, 41, 42] have identified 24 excited singlet electronic states in Na$_2$ by this method, whereas all previous work on this molecule had only produced information about 6 excited states. The new levels include Σ, Π, and $\triangle$ states from the electron configurations 3sns and 3snd, as indicated in Figure 12. For larger values of n, these are molecular Rydberg states, with one electron far outside the core of two Na$^+$ions bound mostly by the single 3s electron. In the $\triangle$ states the outer electron contributes something to the bonding, so that the depth of

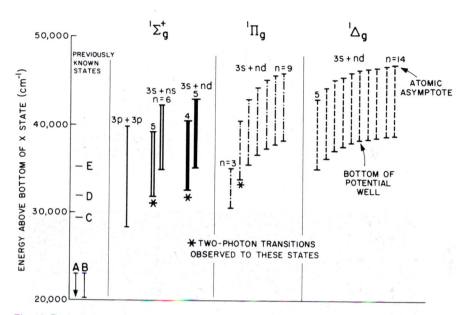

Fig. 12. Excited electronic states of Na₂, as revealed by two-step polarization labeling.

the potential well increases as n is decreased which brings the outer electron closer to the core, as is seen in Figure 13. In the Π states, the outer electron is antibonding, and so it decreases the molecular bonding when it is close to the core. Corresponding behaviors are observed for the other molecular constants, vibrational energy and bond length. Thus they can all be extrapolated to obtain good values for the constants of the ground state of the Na_2^+ ion.[36] This method and the several related techniques of optical-optical double resonance are making increasing contributions to the analysis of complex atomic and molecular spectra.

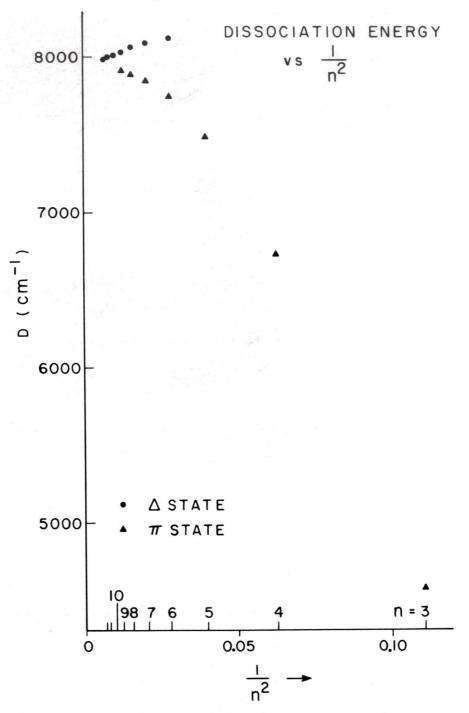

Fig. 13. Dissociation energy for Na_2 molecular Rydberg states as a function of $1/n^2$.

TWO-PHOTON DOPPLER-FREE SPECTROSCOPY

In 1970, L. S. Vasilenko, V. P. Chebotayev, and A. V. Shishaev proposed a method for obtaining two-photon spectral lines without Doppler-broadening.[43] As shown schematically in Figure 14, a molecule moving along the

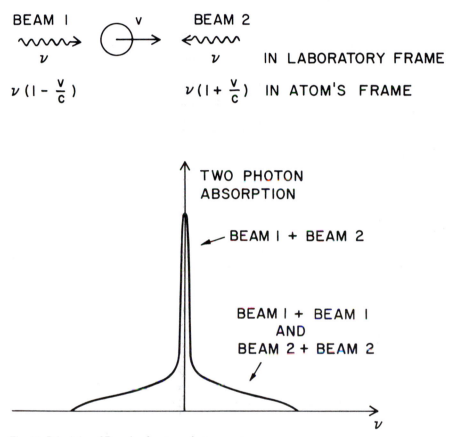

Fig. 14. Principles of Doppler-free two-photon spectroscopy.

direction of one of two oppositely-directed light beams, from the same laser, sees one of them shifted up in frequency and the other shifted down by just the same amount. Thus the sum of the photon frequencies is, to first order, unaffected by the Doppler shifts. All molecules contribute equally to the Doppler-free two-photon line. The predicted effect was observed in atomic sodium vapor by B. Cagnac, G. Grynberg, and F. Biraben,[44] by M. D. Levenson and N. Bloembergen,[45] and by Hänsch, K. C. Harvey, and G. Meisel.[46] They observed transitions from the 3s ground state to 4d or 5s levels (Figure 15). The lines were not only sharp, but remarkably prominent and easy to detect because of the presence of very strong allowed transitions, the well known sodium D lines, less than 100 Angstroms from the wavelength needed

TWO – PHOTON SPECTROSCOPY OF
Na 4d²D STATE

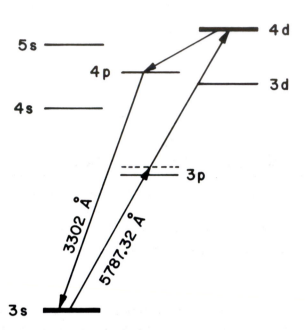

Fig. 15. Energy levels and some two-photon transitions in sodium atoms.

for the two-photon transitions. The allowed transition makes the atom more polarizable at the light frequency. Thus it enhances the two-photon absorption coefficient by a factor proportional to the inverse square of the offset, or frequency difference between the light frequency and the frequency of the allowed transition. Subsequently, R. T. Hawkins, W. T. Hill, F. V. Kowalski, and Sune Svanberg[47] were able to use two lasers of different frequencies in the beams, and so to take advantage of different enhancing lines to reach a number of other levels in the sodium atom, and to measure the Stark shifts caused by an applied electric field. They used a roughly collimated atomic beam illuminated transversely by the lasers, to provide further reduction of the Doppler-broadening.

It was rather surprising, in Kenneth Harvey's early work, that some other two-photon lines were seen in the neighborhood of the expected atomic lines.[48] Since they did not exhibit the well known hyperfine structure of the sodium ground state, they could only come from molecular sodium, Na_2 (Figure 16). But that was remarkable, because the number of molecules at that temperature was very small in comparison with the number of atoms, and there would be still fewer in any individual level. Yet the molecular lines were as strong as the atomic lines. We realized that the explanation must be a more or less accidental close coincidence with some allowed, and therefore enhancing, molecular line.

DOPPLER FREE TWO PHOTON SPECTROSCOPY IN Na VAPOR

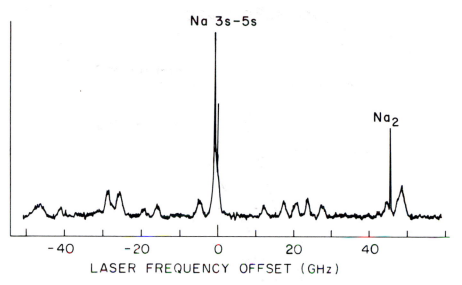

Fig. 16. Atomic and molecular two-photon lines in Na_2.

J. P. Woerdman[49] also observed some of these lines, and was able to identify the rotational quantum number through the nearby enhancing line of the A ←X band of Na_2. Recently Gerard P. Morgan, Hui-Rong Xia, and Guang-Yao Yan[50, 51] have found and identified a large number of these strong two-photon lines in Na_2. The offsets from neighboring enhancing lines have been measured by simultaneous one-photon and two-photon Doppler-free spectroscopy. They are indeed small, ranging from 0.1 cm^{-1} to as little as 38 Mega-hertz or about .001 cm^{-1}. Thus we see how it is possible to have the probability of two-photon absorption, and thus two steps of excitation, nearly as strong as that of a single step.

On the other hand, two-photon Doppler-free lines may also be observed if there is no enhancing state anywhere near, if there is enough laser intensity and sufficiently sensitive detection. Thus Hänsch, Siu Au Lee, Richard Wallenstein, and Carl Wieman[52, 53] have observed the 1s to 2s two-photon transition in atomic hydrogen, excited by the second harmonic (2430 Angstroms) of a visible dye laser which simultaneously scans the blue H_β Balmer line. They have, thus, made an accurate comparison between the 1s to 2s interval and four times the 2s to 4s interval in hydrogen. According to the Bohr theory, the ratio of these level spacings should be exactly 4 to 1. The deviation observed is a measure of the Lamb shift in the ground 1s state, which is otherwise not measurable.

The 1s to 2s transition is particularly intriguing, because the lower state is stable, and the upper state is metastable so that it has a lifetime of 1/7 second. Thus the lifetime width need to be no more than one Hertz, or a part in 10^{15}. Since we can usually locate the center of a line to one percent of the linewidth, it

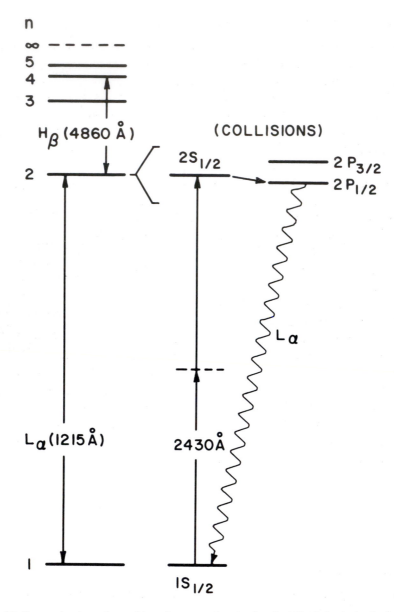

Fig. 17. Energy levels and transitions for measuring the Lamb shift of the 1s level of atomic hydrogen.

should be possible eventually to measure this line to one part in 10^{17} or so. But nobody measures anything to a part in 10^{17}! Before we can hope to achieve that, such things as second order Doppler effect, transit time broadening, radiation recoil and power broadening will have to be eliminated. The challenge is great, and should occupy experimental physicists for some years.

OTHER METHODS

There is not enough room to discuss all of the laser spectroscopic methods that have interested our colleagues. Serge Haroche and Jeffrey A. Paisner have used short, broadband laser pulses to produce quantum beats in fluorescence, by exciting a coherent superposition of several hyperfine levels.[54] James E. Lawler extended the methods of optogalvanic detection of laser absorption[55] to detect Doppler-free intermodulation and two-photon lines.[56] In turn, this method has been extended by Donald R. Lyons and Guang-Yao Yan,[57] to use electrodeless radiofrequency detection of Doppler-free resonances.

Even less is it possible to begin to describe the many exciting discoveries and developments from other laboratories. Some indication of them can be obtained from the proceedings of the five biennial conferences on Laser Spectroscopy.[58] The field has had an almost explosive growth, and laser spectroscopy in some form or other extends from the submillimeter wavelengths in the far infrared to the vacuum ultraviolet and soft x-ray regions.

Thus in the powerful, directional, coherent and highly monochromatic new light of lasers, we are learning to do entirely new kinds of spectroscopy. We can resolve fine details hitherto obscured by thermal broadening, can observe and study very small numbers of atoms, and can simplify complex spectra. We can take the measure of simple atoms with a precision that is providing a real challenge to the best theoretical calculations. Our experimental capabilities have been extended so rapidly in the past few years, that there has not been time to bring them fully to bear on the interesting, fundamental problems for which they seem so well suited. But the spectroscopy with the new light is illuminating many things we could not even hope to explore previously, and we are bound to encounter further intriguing surprises.

REFERENCES

1. Ångström, A. J., Kungl. Svenska Vetensk. Akad. Handl., p. 327 (1852); Poggendorff's Annalen 94, 141 (1855); Phil. Mag. 9, 327 (1855).
2. Fraunhofer, J., Gilbert's Annalen 56, 264 (1817).
3. Huggins, W., Phil. Trans. Roy. Soc. London 171, 669 (1880).
4. Vogel, H. W., Monatsbericht der Königl. Academie der Wissenschaften zu Berlin, July 10, 1879, reported by Huggins, op. cit.
5. Balmer, J. J., Poggendorff's Annalen, N. F. 25, 80 (1885).
6. Rydberg, J. R., Kungl. Svenska Vetensk. Akad. Handl. 32, 80 (1890); Phil. Mag. 29, 331 (1890).
7. Meissner K. W., and Luft, K. F., Ann. der Physik, 28, 667 (1937).
8. Minkowski, R., and Bruck, G., Zeits. f. Phys. 95, 284 (1935).
9. Schawlow, A. L., and Townes, C. H., Phys. Rev. 112, 1940 (1958).
10. Tiffany, W. B., Moos, H. W., and Schawlow A. L., Science, 157, 40 (1967).
11. Lamb, W. E. talk at the Third International Conference on Quantum Electronics, Paris. February 11–15, 1963; Phys. Rev. 134A, 1429 (1964).

12. McFarlane, R. A., Bennett, W. R. Jr., and Lamb, W. E., Jr, Appl. Phys. Letters, *2*, 189 (1963).
13. Szoke, A., and Javan, A., Phys. Rev. Letters *10*, 521 (1963).
14. Lee, P. H., and Skolnick, M. L., Appl. Phys. Letters *10*, 303 (1967).
15. Hänsch, T. W. and Toschek, P., I.E.E.E. J. Quant. Electr. *4*, 467 (1968).
16. Smith, P. W., and Hänsch, T. W., Phys. Rev. Letters *26*, 740 (1971).
17. Hänsch, T. W., Levenson, M. D., and Schawlow, A. L., Phys. Rev. Letters *27*, 707 (1971).
18. Bordé, C., C. R. Acad. Sci. Paris *271*, 371 (1970).
19. Levenson, M. D., and Schawlow, A. L., Phys. Rev. *A6*, 10 (1972).
20. Freed, C., and Javan, A., Appl. Phys. Letters *17*, 53 (1970).
21. Sorem, M. S., and Schawlow, A. L., Opt. Comm. *5*, 148 (1972).
22. Fairbank, W. M., Hänsch, T. W., Jr., Schawlow, A. L., J. Opt. Soc. Am. *65*, 199 (1975).
23. Hurst, G. S., Payne, M. G., Kramer, S. D., and Young, J. P., Rev Mod. Phys. *51*, 767 (1979).
24. Sorokin, P. P., and Lankard, J. R., IBM J. Res. Develop. *10*, 306 (1966).
25. Schafer, F. P., Schmidt, W., and Volze, J., Appl. Phys. Letters *9*, 306 (1966).
26. Myer, J. A., Johnson, C. L., Kierstead, E., Sharma, R. D., and Itzkan, I., Appl. Phys. Letters *16*, 3 (1970).
27. Hänsch, T. W., Appl. Opt. *11*, 895 (1972).
28. Peterson, O. G., Tuccio, S. A., and Snavely, B. B., Appl. Phys. Letters *17*, 245 (1970).
29. Birnbaum, M., Wendizowski, P. H., and Fincher, C. L., Appl. Phy. Letters *16*, 436 (1970).
30. Hänsch, T. W., Shahin, I. S., and Schawlow, A. L., Phys. Rev. Letters *27*, 707 (1971).
31. Hänsch, T. W., Shahin, I. S., and Schawlow, A. L., Nature *235*, 63 (1972).
32. Herzberg, G., Physik, *146*, 269 (1956).
33. Goldsmith, J. E. M., Weber, E. W. and Hänsch, T. W., Phys. Rev. Letters *41*, 1525 (1978).
34. Amin, S. R., Caldwell, C. D., and Lichten, W., Phys. Rev. Letters *47*, 1234 (1981).
35. Wiemann C., and Hänsch, T. W., Phys. Rev. Letters *36*, 1170 (1976).
36. Kowalski, F. V., Hill, W. T., and Schawlow, A. L., Optics Letters, *2*, 112 (1978).
37. Schieder, R., Opt. Comm. *26*, 113 (1978).
38. Kaminsky, M. E., Hawkins, R. T., Kowalski, F. V., and Schawlow, A. L., Phys. Rev. Letters *36*, 671 (1976).
39. Teets, R. E., Feinberg, R., Hänsch, T. W., and Schawlow, A. L., Phys. Rev. Letters *37*, 683 (1976).
40. Carlson, N. W., Kowalski, F. V., Teets, R. E., and Schawlow, A. L., Opt. Comm. *18*, 1983 (1979).
41. Carlson, N. W., Taylor, A. J., Jones, K. M., and Schawlow A. L., Phys. Rev. *A24*, 822 (1981).
42. Taylor, A. J., Jones, K. M., and Schawlow, A. L., Opt. Comm. *39*, 47 (1981).
43. Vasilenko, L. S., Chebotayev, V. P., and Shishaev, A. V., JETP Letters *12*, 113 (1970).
44. Cagnac, B., Grynberg, G., and Biraben, F., J. Phys. (Paris) *34*, 845 (1973).
45. Levenson, M. D., and Bloembergen, N., Phys. Rev. Letters *32*, 645 (1974).
46. Hänsch, T. W., Harvey, K. C., Miesel, G. and Schawlow, A. L., Opt. Comm. *11*, 50 (1974).
47. Hawkins, R. T., Hill, W. T., Kowalski, F. V., Schawlow, A. L., and Svanberg, S., Phys. Rev. *A15*, 967 (1977).
48. Harvey, K. C., thesis, Stanford University, M. L. Report No. 2442 (1975).
49. Woerdman, J. P., Chem. Phys. Letters *43*, 279 (1976).
50. Xia, H.-R., Yan, G.-Y., and Schawlow, A. L., Opt. Comm. *39*, 153 (1981).
51. Morgan, G. P., Xia, H.-R., and Schawlow A. L., J. Opt. Soc. Am., *72*, 315 (1982).
52. Lee, S. A., Wallenstein, R., and Hänsch, T. W., Phys. Rev. Letters *35*, 1262 (1975).
53. Wieman, C., and Hänsch, T. W., Phys. Rev. *A22*, 1 (1980).
54. Haroche, S., Paisner, J. A., and Schawlow, A. L., Phys. Rev. Letters *30*, 948 (1973).
55. Green, R. B., Keller, R. A., Luther, G. G., Schenck, P. K., and Travis, J. C., Appl. Phys. Letters *29*, 727 (1956).
56. Lawler, J. E., Ferguson, A. I., Goldsmith, J. E. M., Jackson, D. J., and Schawlow, A. L., Phys. Rev. Letters *42*, 1046 (1979).
57. Lyons, D. R., Schawlow, A. L., and Yan, G.-Y., Opt. Comm. *38*, (1981).
58. The most recent of these is *Laser Spectroscopy V*, McKellar, A. R. W., Oka, T., and Stoicheff, B. P., editors, Springer-Verlag, Berlin, Heidelberg and New York (1981).

Kai Siegbahn

KAI SIEGBAHN

Born April 20, 1918, in Lund, Sweden. Parents: Manne Siegbahn and Karin Högbom. Married May 23, 1944, to Anna Brita Rhedin. Three children: Per (1945), Hans (1947) and Nils (1953). Attended the Uppsala Gymnasium; Studied physics, mathematics and chemistry at the University of Uppsala from 1936 until 1942. Graduated in Stockholm 1944. Docent in physics that year. Research associate at the Nobel Institute for Physics 1942–1951. Professor of physics at the Royal Institute of Technology in Stockholm from 1951 to 1954. Professor and head of the Physics Department at the University of Uppsala since 1954. Member of the Royal Swedish Academy of Sciences, Royal Swedish Academy of Engineering Sciences, Royal Society of Science, Royal Academy of Arts and Science of Uppsala, Royal Physiographical Society of Lund, Societas Scientiarum Fennica, Norwegian Academy of Science, Royal Norwegian Society of Sciences and Letters, Honorary Member of the American Academy of Arts and Sciences, Membre du Comité International des Poids et Mesures, Paris, President of the International Union of Pure and Applied Physics (IUPAP).

Awards
The Lindblom Prize 1945, Björkén Prize 1955, Celsius Medal 1962, Sixten Heyman Award, University of Gothenburg 1971, Harrison Howe Award, Rochester 1973, Maurice F. Hasler Award, Cleveland 1975, Charles Frederick Chandler Medal, Columbia University, New York 1976, Björkén Prize 1977, Torbern Bergman Medal 1979, Pittsburgh Award of Spectroscopy 1982. Doctor of Science, honoris causa: University of Durham 1972, University of Basel 1980, University of Liège 1980, Upsala College, New Jersey, 1982.

Research in physics covering atomic and molecular physics, nuclear physics, plasma physics and electron optics. Main research activity in the field of electron spectroscopy, ESCA. Books: Beta- and Gamma-Ray Spectroscopy, 1955; Alpha-, Beta- and Gamma-Ray Spectroscopy, 1965; ESCA – Atomic, Molecular and Solid State Structure Studied by Means of Electron Spectroscopy, 1967; ESCA Applied to Free Molecules, 1969.

Surveys on ESCA:
Electron Spectroscopy for Chemical Analysis, Phil. Trans. Roy. Soc. London A, 33–57, 1970.
Electron Spectroscopy, Encyclopedia of Science and Technology, McGraw-Hill, 1971.

Perspectives and Problems in Electron Spectroscopy, Proc. Asilomar Conference 1971, Ed. D. A. Shirley, North Holland, 1972.

Electron Spectroscopy—A New Way of Looking into Matter, Endeavor 32, 1973.

Electron Spectroscopy for Chemical Analysis, Proc. of Conf. on Atomic Physics 3, Boulder, 1972, Ed. S. J. Smith and G. K. Walters, Plenum, 1973.

Electron Spectroscopy for Chemical Analysis (together with C. J. Allan), MTP Int. Rev. of Science, Vol. 12, Analytical Chemistry, Part 1, Butterworths, 1973.

Electron Spectroscopy — An Outlook, Proc. Namur Conference 1974, Elsevier 1974.

Electron Spectroscopy and Molecular Structure, Pure and Appl. Chem. 48, Pergamon, 1976.

Electron Spectroscopy for Solids, Surfaces, Liquids and Free Molecules, in Molecular Spectroscopy, Ch. 15, Heyden 1977.

ELECTRON SPECTROSCOPY FOR ATOMS, MOLECULES AND CONDENSED MATTER

Nobel lecture, 8 December, 1981

by

KAI SIEGBAHN

Institute of Physics, University of Uppsala, Sweden

In my thesis [1], which was presented in 1944, I described some work which I had done to study β decay and internal conversion in radioactive decay by means of two different principles. One of these was based on the semi-circular focusing of electrons in a homogeneous magnetic field, while the other used a big magnetic lens. The first principle could give good resolution but low intensity, and the other just the reverse. I was then looking for a possibility of combining the two good properties into one instrument. The idea was to shape the previously homogeneous magnetic field in such a way that focusing should occur in two directions, instead of only one as in the semi-circular case. It was known that in betatrons the electrons performed oscillatory motions both in the radial and in the axial directions. By putting the angles of period equal for the two oscillations Nils Svartholm and I [2, 3] found a simple condition for the magnetic field form required to give a real electron optical image i.e. we established the two-directional or double focusing principle. It turned out that the field should decrease radially as $\frac{1}{\sqrt{R}}$ and that double focusing should occur after $\pi \cdot \sqrt{2} \sim 255°$. A simple mushroom magnet was designed, the circular pole tips of which were machined and measured to fit the focusing condition. ThB was deposited on a wire net and put into position in the pole gap. A photographic plate was located at the appropriate angle and the magnet current set to focus the strong F line of ThB on the plate. Already the first experiment gave a most satisfactory result. Both the horizontal and the vertical meshes of the wire net were sharply imaged on the plate. A more detailed theory for the new focusing principle was worked out and a large instrument with R = 50 cm was planned and constructed [4]. Due to the radially decreasing field form an additional factor of two was gained in the dispersion, compared to the homogeneous field form. Since all electrons, for reasonably small solid angles, were returning to the symmetry plane of the field at the point of focus no loss in intensity was experienced by increasing the radius of curvature of the instrument. Very large dispersion instruments with good intensity and much improved resolving power could therefore be designed to record β spectra and internal conversion spectra from radioactive sources. The magnetic double focusing was convenient for the fairly high energy electrons (50 keV−2 MeV) normally occurring in radioactive decay and the field form could easily be achieved by means of shaping the poles of an iron magnet. In my laboratory and in many other nuclear physics laboratories double focusing spectrometers frequently became used for high resolution work [5]. This type of focusing was also used subsequently by R. Hofstadter [6] in his well-known work on high energy electron scattering from nuclei and nucleons.

During the late forties, the fifties and the early sixties I was much involved in nuclear spectroscopy. This was a particularly interesting and rewarding time in nuclear physics since the nuclear shell model, complemented with the collective properties, was then developed, which to a large extent was founded on experimental material brought together from nuclear decay studies. Nuclear disintegration schemes were thoroughly investigated, and the spins and parities of the various levels were determined, as well as the intensities and multipole characters of the transitions. During this period the discovery of the non-conservation of parity added to the general interest of the field. Also the form of the interaction in β decay, appearing originally in Fermi's theory, was exten-sively investigated. A large part of my own and my students' research was therefore concerned with nuclear spectroscopy of radioactive decay [7–23]. In 1955 I edited a volume [24] on "Beta and Gamma Ray Spectroscopy". In 1965 I concluded my own career as a nuclear spectroscopist by publishing [25] "Alpha, Beta and Gamma Ray Spectroscopy". In this extensive survey of nuclear spectroscopy I had been able to collect the prodigious number of 77 coauthors, all being prominent authorities and in many cases the pioneers in the various fields. Although my own scientific activity at that time had almost entirely become directed towards the new field which is the subject of the present article I have still kept my old interest in nuclear physics much alive as the editor for the journal Nuclear Instruments and Methods in Physics Re-search (NIM) ever since its start in 1957.

Let me now return to the situation around 1950. At that time my coworkers and I had already for some time been exploring the high resolution field by means of our large dispersion double focusing instrument and other methods, such as the high transmission magnetic lens spectrometer and coincidence techniques. Often I found, however, that my experimental work had to stop and wait for radioactive samples, the reason being a capricious cyclotron. It then came to my mind that I should try to simulate the radioactive radiation by a substitute which I could master better than the cyclotron. I had found that a very convenient way to accurately investigate gamma radiation from radioac-tive sources was to cover them with a γ-ray electron converter, i.e. a thin lead foil which produced photoelectrons to be recorded in the spectrometer. I now got the idea that I should instead use an X-ray tube to expel photoelectrons from ordinary materials, in order to measure their binding energies to the highest possible accuracy. In my nuclear physics work such binding energies had to be added to the energy values of the internal conversion lines from the radioactive sources in order to get the energies of the nuclear transitions. I studied what had been done along these directions before [26, 27], and I finally got a vague feeling that I could possibly make an interesting and perhaps big step forward in this field if I applied the experience I had from nuclear spectroscopy using the above mentioned external photoeffect and my high resolution instruments. The previous investigations had confirmed that the atomic electrons were grouped in shells, and by measuring on the photographic plates the high energy sides of the extended veils from the various electron distributions approximate values of the binding energies could be deduced. On the other hand, since the observed electron distributions had no line structure and consequently did not correspond to atomic properties, the attained precision and the actual information was far inferior to what could be obtained by X-ray emission and absorption spectroscopy. I realized that electron spectroscopy for atoms and solids could never become competitive with X-ray emission or absorption spectroscopy unless I was able to achieve such a high resolution that really well-defined electron lines were obtained with linewidths equal to or close to the inherent atomic levels themselves.

I thought of these problems considerably and started to make plans for a new equipment which should fulfil the highest demands on resolution at the low electron energies I had to be concerned with, ten to a hundred times smaller than in radioactive work. I recall I sat down for some days early in 1950 to try to make a thorough calculation about the expected intensities. I designed [28, 29, 30] an ironfree double focusing spectrometer with R = 30 cm, in which I should be able to measure the current with a precision of better than 1 part in 10^4. The spectrometer was surrounded by a big, three-component Helmholtz coil system to eliminate, to better than 1 part in 10^3, the earth's magnetic field over the entire region of the spectrometer. If I had an X-ray tube with a $K\alpha$ radiation in the region of 5 keV, this would enable me to measure expelled photoelectrons with a precision of a fraction of an electron volt. This I thought was about sufficient in atomic physics. I also hoped to observe phenomena of chemical interest provided I could realize the resolution I aimed at, but at that time my ideas in this latter respect were of course very vague, centering around atomic level shifts in alloys, etc. When I calculated the expected intensities of the photoelectron lines, I started from the very beginning, i.e. with a certain number of mA·s in the X-ray tube, I then calculated by means of existing knowledge the number of $K\alpha$ X-ray photons, next I put in all solid angles both in the X-ray tube and the electron spectrometer and made some assumptions about the effective photoelectron cross sections to expel electrons from the outermost layers in a solid surface. Those electrons could not be expected to suffer much energy loss and were the interesting ones upon which I should base my spectroscopy. In retrospect, this last stage in my considerations was of course of some interest, in view of the later development of electron spectroscopy into a *surface* spectroscopy. I guessed that what is now called the "escape depth" of the electrons should be less than a light wavelength and more than a few atomic layers and so I used 100 Å in my calculations. This was not too bad a guess in consideration of later studies, indicating a lower figure for metals and 100 Å for organic multilayers. I finally arrived at an estimated counting rate on a photoline in my apparatus of several thousands of electrons per minute as recorded in the Geiger-Müller (G-M) counter placed at the focal plane in the double focusing spectrometer. Afterwards I was satisfied to find that this calculation turned out to correspond fairly well to reality. This step, however, in fact took several years to make.

The equipment which I had to build and test was at that time very complicated. The resolution ultimately achieved turned out to be high enough to enable recording even of the inherent widths of internal conversion lines [24]. This was done in 1954 and in 1956 I published [31] together with my collaborator, Kay Edvarson, an account of this phase of the work under the title "β-ray Spectroscopy in the Precision Range of $1:10^5$". In the next phase I had, however, to overcome many difficulties in handling the low energy electrons excited by X-rays and to record them by the G-M counter. This had an extremely thin window through which gas diffused continuously and so was compensated for by an automatic gas inlet arrangement. I did not realize, to start with, the precautions I had to take when dealing with surfaces of solids in order to record resolved line structures.

After some further testing of the equipment, concerning the influence of the finite nuclear size on the conversion lines in some nuclei [32, 33], my two new coworkers, Carl Nordling and Evelyn Sokolowski, and I finally made the transition to atomic physics and were able to record our first photoelectron spectrum [34, 35] with extremely sharp lines and with the expected intensities. These electron lines definitely had all the qualities which I had set as my first goal. They were symmetric, well defined and had linewidths which could be deduced from the linewidth of the X-ray line used and the width of the atomic level of the element under study, plus of course a small additional broadening due to the resolution of the instrument. Fig. 1 shows an early recording of MgO. The exact position of the peak of the electron lines could be measured

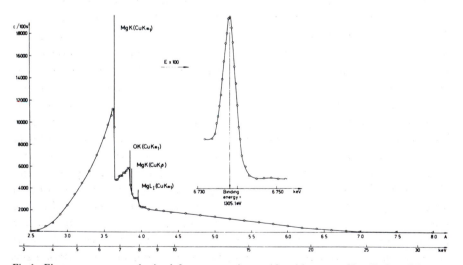

Fig.1. Electron spectrum obtained from magnesium oxide with copper X-radiation. Edges are found at energies corresponding to atomic levels of magnesium and oxygen. A very sharp electron line can be resolved from each edge. Such an electron line is shown in the insert figure with the energy scale expanded by a factor of one hundred to bring out the finite width of the line.

with considerable accuracy. Electron spectroscopy for atoms could be developed further with confidence.

Fig. 2 illustrates the steps which we took from the earlier recording of the photoelectrons expelled from a gold foil by Robinson [36] in 1925 to the introduction of the electron line spectroscopy in 1957. The dotted line inserted in Robinson's spectrum should correspond to the place where our spin doublet $N_{VI}N_{VII}$ to the right in the figure is situated. The distance between the two well resolved lines in our spectrum would correspond to about 0.1 mm in the scale of Robinson's spectrum. Below this spectrum the black portion has been enlarged (the gray scale) to show the corresponding part in our spectrum. This spectrum was taken at a later stage of our development. Within this enlarged spectrum a further enlargement of the $N_{VI}N_{VII}$ doublet is inserted. The spin-orbit doublet has now a distance between the lines which corresponds to a magnification of 600 times the scale in Robinson's photographic recording.

A comparison between the middle and the lower spectrum in the figure further demonstrates the extreme surface sensitivity of electron spectroscopy. The difference between the two spectra is caused by a slight touch of a finger. At the beginning this sensitivity caused us much trouble, but later on, when electron spectroscopy was applied as a surface spectroscopy, it turned out to be one of its most important assets.

In 1957 we published some papers [34, 35, 37] describing our first results, which really did indicate great potential for the future. We also obtained our first evidences of chemical shifts [37−42] for a metal and its oxides and for Auger electron lines. I thought, however, that we should first improve our techniques and explore purely atomic problems until we had achieved a greater knowledge to enable us to progress to molecular problems. We therefore systematically measured atomic binding energies for a great number of elements with much improved accuracy compared to previous methods, in particular X-ray absorption spectroscopy [40, 41, 43−61]. We were surprised to find how inaccurate previously accepted electron binding energies for various shells and elements could be. We made so-called "modified" Moseley diagrams. We were bothered by uncertainties due to the chemical state and therefore tried to use only metals or at least similar compounds of the elements in our systematic studies. We also devoted much effort to the investigation of the Auger electron lines [62−69] which appeared in our spectra with the same improved resolution as our photoelectron lines. As one of the results of such studies we were able to observe for a group of elements around $Z = 40$ all the nine lines expected in the intermediate coupling theory as compared to the observed six lines in pure j-j coupling [63]. In general, in the spectroscopy we developed photoelectron and Auger electron lines were found side by side. Later on, therefore, we avoided any notation for this spectroscopy which could give the false impression that only one of the two types of electron lines were present. Auger electron lines can *in addition* to the X-ray mode of excitation also be produced by electrons. Much of the above basic work on atomic energy levels is described in theses by E. Sokolowski [70], C. Nordling [71], P. Bergvall [72], O. Hörnfeldt [73], S. Hagström [74] and A. Fahlman [75].

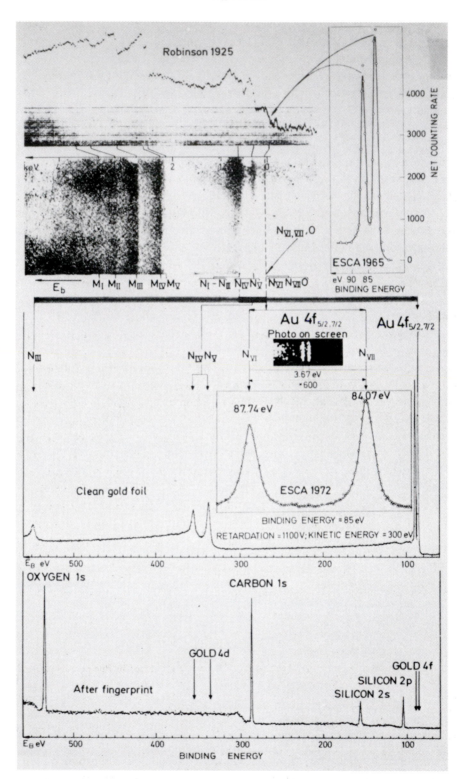

After some years' work in electron spectroscopy on problems in atomic physics the next step came to the fore, namely to make systematic studies of the chemical binding. This step was taken together with my coworkers Stig Hagström and Carl Nordling when $Na_2S_2O_3$ was found to give two well resolved K photoelectron lines from the sulphur [76]. This showed that two differently bonded sulphur atoms could be separated in the molecule, which according to classical chemistry were in the -2 and the $+6$ valence states, respectively. This was a more clear-cut case than the copper-copperoxide case we had studied before, since the reference level for the two sulphur atoms could be traced to the same molecule. The systematic investigation of chemical binding by means of electron spectroscopy is described in theses by S.-E. Karlsson [77], R. Nordberg [78], K. Hamrin [79], J. Hedman [80], G. Johansson [81], U. Gelius [82] and B. Lindberg [83].

Fig. 3 shows the chemically shifted C1s spectrum of ethyl trifluoroacetate [84, 85]. Fig. 4 [86] shows how the chemical shift effect can be used to identify groups linked together in branched chains in polymers [87−90]. The intensities of the lines are correlated to the different branchings in the two viton polymers.

In the interpretation of the electron spectra the first step is to consider the electron structure as 'frozen' under the photoelectron emission process. In this approximation the measured electron binding energies can be identified with the Hartree-Fock energy eigenvalues of the orbitals. One then disregards the fact that the remaining electronic structure, after electron emission, is relaxing to a new hole state. This relaxation energy is by no means negligible and an accurate calculation of the relevant binding energies has to include both the ground state and the hole state energies as the difference between them. The inclusion of relativistic effects in this treatment is essential for inner core ionization and heavier elements (I. Lindgren [91]). More recently, methods have been devised to describe the photoelectron emission by means of a transition operator which properly accounts for the relaxation process [92−94]. Various conceptual models complement the computational procedures on an ab initio level.

For chemical shifts in free molecules, it is usually sufficiently accurate to consider only the ground state properties [95−106]. This is due to the circum-

Fig.2. Electron spectra of gold.
Upper left: Spectrum recorded by Robinson in 1925 /36/. (Reproduced by due permission from Taylor & Francis Ltd, London.)
Upper right: ESCA spectrum recorded in Uppsala before 1965 by non-monochromatized $MgK\alpha$ excitation. The N_{VI}, N_{VII} levels are seen as two completely resolved lines in this spectrum whereas the $N_{VI}N_{VII}$ and O levels appear together as a hump in the photometric recording by Robinson and are only barely visible on the photographic plate.
Middle: ESCA spectrum recorded in Uppsala 1972 by monochromatized $AlK\alpha$ excitation. The magnification of this spectrum is 600 times that of Robinson.
Lower Part: ESCA spectrum of a gold foil with a fingerprint on the surface. The electron lines are entirely due to the fingerprint whereas the gold lines are missing.

Physics 1981

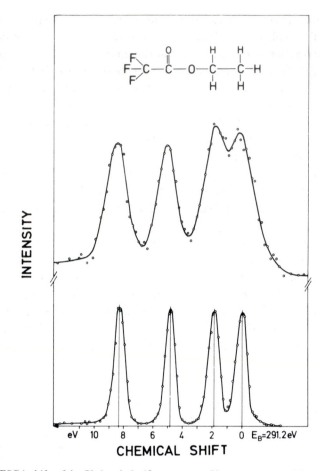

Fig.3. The ESCA shifts of the Cls in ethyl trifluoroacetate. Upper spectrum without and lower with X-ray monochromatization /84, 85).

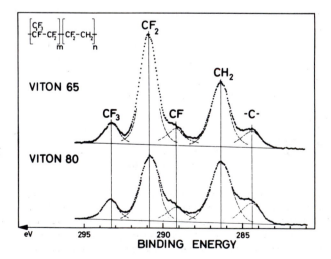

Fig.4. ESCA spectra of Viton 65 and Viton 80 polymers.

stance that relaxation energies for a series of similar electronic systems vary only marginally. This can be described by the division of the relaxation energy into two contributions, one connected to the atomic contraction at ionization, the other to the 'flow' of charge from the rest of the molecule [107, 108]. The atomic part, which is very nearly constant for one specific element, is the dominating contribution to the relaxation energy. The 'flow' part varies generally marginally for free molecules of similar structure, leading to constant relaxation energies. There are cases, however, where the 'flow' part can significantly change from one situation to another. One example is when a molecule is adsorbed on a metal surface. In such a situation the flow of conduction electrons from the metal substrate will contribute to the relaxation of the core hole. This can increase the relaxation energy by several eV [109, 110]. Other cases are pure metals and alloys where the conduction electrons are responsible for the screening of the hole. [111−116]. These are treated in theses work by N. Mårtensson [117] and R. Nyholm [118].

In view of the interesting applications which the chemical shift effect offered for chemistry and the fact that at that time we had found that electron spectroscopy was applicable for the analysis of all elements in the Periodic System, we coined the acronym ESCA, Electron Spectroscopy for Chemical Analysis. If one is particularly interested in conduction bands for metals or alloys (Fig. 5

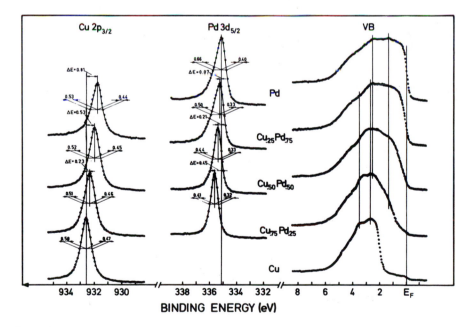

Fig. 5. Core and valence electron spectra (excited by monochromatized AlKα-radiation) of some Cu_xPd_{1-x} alloys, including the pure constituents. The binding energies undergo positive chemical shifts with increasing Cu-content. The asymmetries of the lines are due to creation of soft electron-hole pairs at the Fermi edge upon core ionization. The magnitude of the asymmetry is thus related to the (local) density of states at the Fermi level. The Pd lines are seen to become more symmetric as the Cu content increases (Pd local density of states decreases).

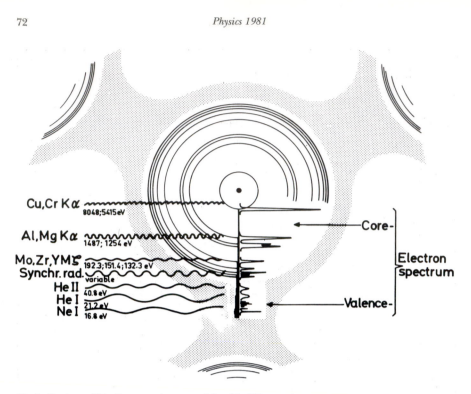

Fig.6. Regions of binding energies accessible with different photon sources.
Solid circles: localized, atom-like orbitals.
Shaded area: more or less delocalized, molecular orbitals.

[119]) or valence electron structures of solid material in general, or free
molecules, more detailed notations can be preferred. One useful distinction is
between core and valence electron spectra (Fig. 6). Obviously, a further ground
of classification is due to the different origin of the photoelectron and Auger
electron lines, which both always occur in ESCA, as mentioned before. The
corresponding chemical shift effect for the Auger electron lines we established
soon afterwards [69] in the case of $Na_2S_2O_3$. Further studies [111, 112, 120–
133] have shown that the two shift effects are complementary. The combination of the two shifts provides insight into the mechanism of relaxation in the
photoionization process. Auger electron spectra are given in Fig. 7 for a clean
Mg metal, and when it is partly and finally fully oxidized to MgO [86].

Apart from the ordinary core electron lines and the Auger electron lines from
the various shells, all characteristic of each element, the electron spectra also
contain additional features. Satellites situated close to (~10 eV) the main core
lines at the low energy sides are often observed with intensities around 10 % of
the latter. Fig. 8 [114] shows the electron spectrum of gaseous Hg. Inserted are
the satellites to the $N_{VI}N_{VII}$ lines. Strong satellites were first observed [68] in
our spectra in the KLL Auger electron spectrum of potassium in some compounds. Satellites have been found to occur frequently for core lines and
occasionally they can even have intensities comparable to the main lines, e.g.
paranitroaniline [134–136], transition metal compounds [137–146] and var-

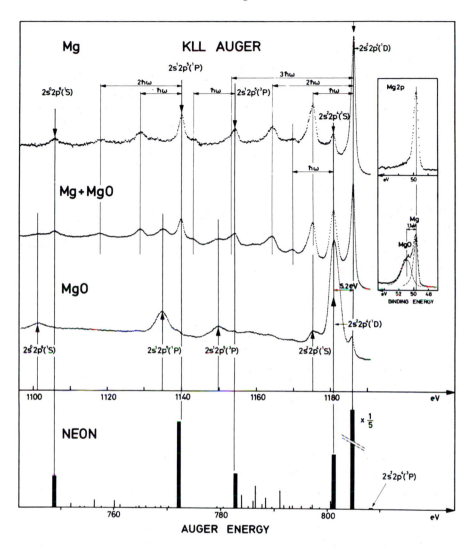

Fig. 7. MgKLL Auger electron spectra at different stages of oxidation as obtained in ESCA. Upper spectrum is from a clean metal surface, lower spectrum from the oxidized metal (with only a trace of metal) and the middle spectrum is from an intermediate oxidation. Volume plasmon lines are observed. For comparison, the positions of the NeKLL Auger electron lines are given below, as recorded in the ESCA instrument by means of electron beam excitation.

ious adsorbed molecules on surfaces [148, 149]. Since these electrons can be visualized as being emitted from excited states, the satellites were given the name "shake-up" lines.

Molecules like O_2 or NO contain unpaired electrons and are therefore paramagnetic. Large classes of solid materials have similar properties. In such cases core electron spectra show typical features called spin-, multiplet-, or exchange splitting. We first observed this phenomenon [150] in oxygen when

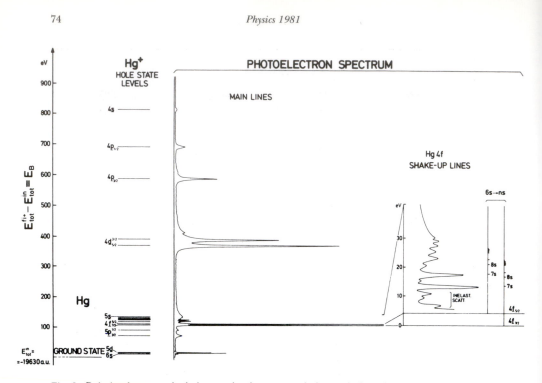

Fig. 8. Relation between the hole-state level system and observed photoelectron spectrum for the mercury atom. The figure illustrates that the main lines are connected with states of the ion where (in a one-electron picture) an atomic orbital has been removed from the neutral ground state. The energy region close to the 4f-lines has been expanded (far right) to show that the additional satellite lines observed (shake-up lines) are due to excitations (6s→ns) above the 4f hole ground states. (Note that the intensities of the 4f-lines have been truncated to fall into the scale of the figure.) (From refs. 114 and 260.)

air was introduced into the gas cell in our ESCA instrument (Fig. 9). The ls line of O_2 is split in the intensity ratio of 2:1. This spin splitting is due to the exchange interaction between the remaining ls electron and the two unpaired electrons in the π_g2p orbital, which are responible for the paramagnetism of this gas. The resulting spin can be either 1/2 or 3/2. The corresponding electrostatic exchange energies can be calculated and correspond well with the measured splitting of 1.11 eV [151]. Apart from oxygen and nitrogen, argon and CO_2 can also be seen in air in spite of the low abundances of these gases. A statistical treatment of the data even exhibits the presence of neon (0.001%).

Other particular features in the spectra occur in the valence electron region, i.e. at binding energies extending from zero binding energy to say 50 eV. Our first study of this entire region concerned ionic crystals like the alkali halides [152].

In a later study [153] (1970) of a single crystal of NaCl we discovered the phenomenon of *ESCA diffraction*. We investigated the angular distribution of emitted Auger electrons from the NaKLL ($^1D_2$) transition and the photoelec-

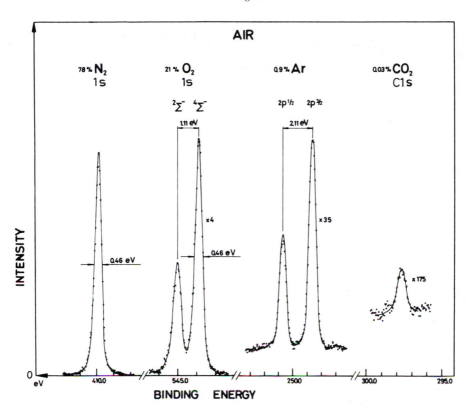

Fig. 9. Electron spectrum of air. The O1s is split into two components due to 'spin' or 'multiplet' splitting. Excitation was performed by means of monochromatized AlKα ($\Delta h\nu = 0.2$ eV) radiation.

trons from the Na1s, Na2s, Cl2p$_{3/2}$ and the Cl3p levels, the latter being the outermost valence orbital of the crystal. For excitation both AlKα and MgKα were used. The crystal could from outside be set at different angular positions relative to the emission direction of the electrons, which in turn was defined by the slit system of the ESCA instrument. For comparison, the angular distributions from polycrystalline samples were also recorded. In all cases typical diffraction patterns were found. In the control experiments on the polycrystalline samples there were no such patterns. Fig.10 shows two of the diffraction patterns recorded. Subsequent measurements [154−158] on other single crystals have shown agreement with the above investigation.

ESCA diffraction has more recently been applied to surface studies giving interesting information on the geometry of adsorbed molecules [159−161] on single crystals. This field is under development and should have a promising future in surface science.

In X-ray diffraction there is an incoming photon wave and an outgoing diffracted photon wave. In electron diffraction there is an incoming electron wave and an outgoing diffracted electron wave. In ESCA diffraction there is an

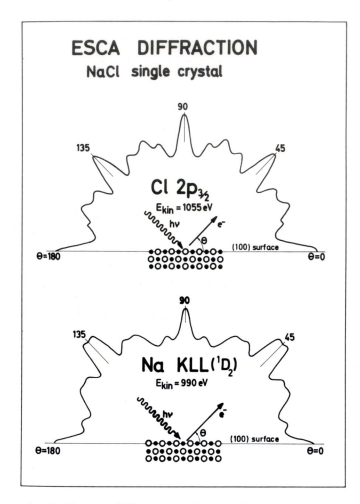

Fig. 10. Angular distributions of Cl2p$_{3/2}$ photoelectrons (MgKα) and NaKLL (^{1}D$_2$) Auger electrons from a NaCl single crystal.

incoming photon wave and an outgoing diffracted electron wave with different energies. These are three distinctly different physical phenomena which obviously require both different experimental equipment to observe and different theoretical treatments to evaluate. With more suitably built instruments for this purpose and with the addition of stronger X-ray sources and synchrotron radiation [161−164] the development can proceed further.

In order to study gases and vapours from liquids, we first introduced a freezing technique [165] to condense the gases onto the specimen plate. In this way we obtained the valence or molecular orbital spectrum of solidified benzene [166]. Soon afterwards we found that we could study the gaseous phase just as well by introducing differential pumping in the instrument. Acetone was our first study with this technique for gases, i.e. for free molecules, revealing two well separated C1s core lines, one for the ketocarbon and one for the methyl carbon in the intensity ratio of 1:2 [167].

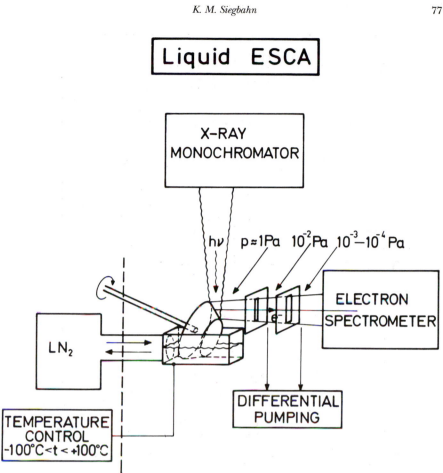

Fig. 11. Schematic diagram of the liquid-sample arrangement.

Since solids, surfaces, gases and vapours from liquids were all found to be suitable samples in electron spectroscopy the question arose whether also liquids could be studied. This turned out to be quite possible and several satisfactory methods have been developed in our laboratory [168−173]. The early methods and applications are described in theses by H. Siegbahn [174], L. Asplund [175] and P. Kelfve [176]. Recently a new, more convenient arrangement (H. Siegbahn) has been developed which is shown in Fig. 11 [173]. A small trundle is rotating in the sample cell in which the liquid is introduced. A slit transmits the exciting radiation, e.g. X-radiation and the expelled electrons from the continuously wetted trundle can leave the house through a slit where differential pumping reduces the gas pressure. Cooling of the sample has been introduced which has enabled a vast increase of the number of liquids that can be studied. Fig. 12 shows part of a recent [173] spectrum of ethanol as a solvent in which iodine and sodium iodide are dissolved. One observes here a well resolved spin-orbit doublet of iodine $3d_{3/2}$

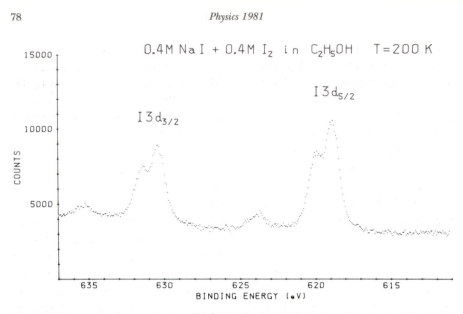

Fig. 12. I3d spectrum from a solution of NaI (0.4M)+I$_2$ (0.4M) in ethanol obtained at T = 200K. The doublet for each spin-orbit component is due to ionization of the central atom (lower peak) and outer atoms (higher peak) of the I$_3^-$ ion. The extra peaks at the high-binding-energy sides of each spin-orbit component are interpreted as shake-up structures.

and 3d$_{5/2}$. Each of these electron lines is chemically split in the ratio of ~1:2. The interpretation is that I$_3^-$ has been formed in the solution. The centrally located iodine has the highest binding energy. The correct intensity ratio of 1:2 is obtained when the shake-up satellites are ascribed to the two externally situated iodine atoms, a conclusion which is in agreement with what we have found for similar configurations in other electron spectra. Liquid ethanol itself is shown in Fig. 13. Here one observes the oxygen 1s core line, the chemically split carbon 1s line and the valence electron spectrum. The field of liquids is presently in a state of rapid development.

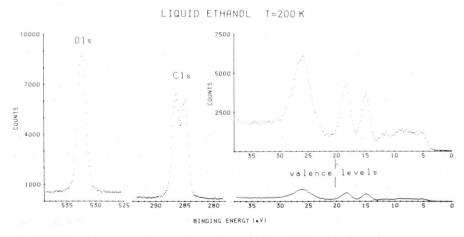

Fig. 13. ESCA spectrum of liquid ethanol obtained at T = 200K.

In the valence region for free molecules it was possible to achieve much improved resolution if UV light, especially the He resonance radiation at 21 eV was used for excitation. Development work in this field was performed by D. W. Turner [177−182] and W. C. Price [183, 184] and their coworkers in England. The conduction bands of metals could be studied by a corresponding technique using ultra-high vacuum (UHV) which was done by W. E. Spicer and coworkers [185−188] in the USA.

In my laboratory a large electrostatic sector focusing instrument was designed in the early part of the sixties for exciting electron spectra in the gaseous phase by VUV radiation and also by electrons. High resolution valence electron spectra were thus obtained and furthermore Auger and autoionization spectra of rare gases and organic molecules could be investigated at a resolution which enabled vibrational structures to appear also in the latter type of spectra. Studies of angular distributions were initiated by using polarized radiation. This was produced by VUV-polarizers which we developed in my laboratory. Much of the above work is described in theses by T. Bergmark [189], L. Karlsson [190], R. Jadrny [191] and L. Mattsson [192]. The Auger electron spectroscopy was further explored in more recent publications [128, 132, 193−196].

The source of excitation was for a time confined to either the soft X-ray region or the UV region with a gap between them from ∼50 eV (HeII) to 1250 eV (MgKα). Fig.14 shows the valence spectrum of SF_6 excited by HeI, HeII and AlKα [197, 198]. Some intermediate X-ray lines were later on added [199−205], such as YMζ at 132 eV but the main step of development was the introduction during the seventies of the variable synchrotron radiation [e.g. 206−210] which partly bridged the gap. The previous strong distinction between X and UV excited electron spectra is therefore not so easy to maintain any more unless one is emphasizing the particular technique at hand for exciting the spectra. This is naturally not a trivial point for most researchers, however, and excellent work can be done with one or the other technique alone or in combination.

In 1967 we had gone through most of the basic features of the spectroscopy, designed several new spectrometers (electrostatic double focusing ones included), developed new radiation sources in the soft X-ray and UV region, made theoretical investigations of the process of electronic relaxation at ionization and applied the spectroscopy to a variety of different research fields. We then decided to present the new spectroscopy in a more consistent and complete way than we had done before. At the end of that year our book "ESCA−Atomic, Molecular and Solid State Structure Studied by Means of Electron Spectroscopy" appeared [211]. Two years later we published a second book [212], this time on "ESCA, Applied to Free Molecules". At that time several instrumental firms started to develop commercial instruments. I took part in one of these developments at Hewlett-Packard in Palo Alto during a leave of abscence from my laboratory in 1968. I spent that year at the Lawrence Berkeley Laboratory with which we had had a long cooperation both in nuclear spectroscopy and then in ESCA. The Hewlett-Packard instrument [213] was designed to include

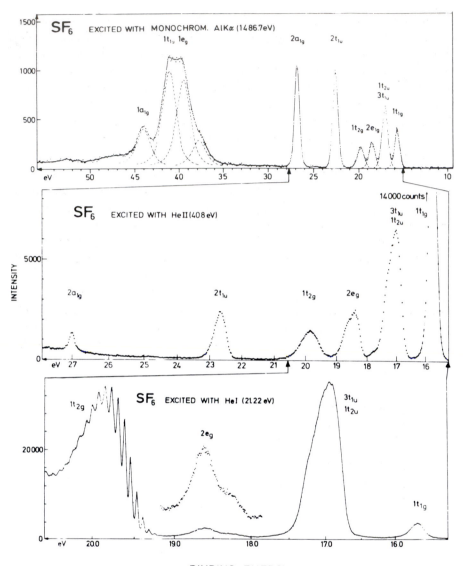

BINDING ENERGY

Fig. 14. Valence electron spectra of the SF₆ molecule excited with different photon energies (AlKα-, HeII and HeI-radiations). This figure illustrates the complementary nature of the various excitation sources. The AlKα-excited spectrum enables a recording of the full valence region (including the innermost orbitals), which is not possible with the lower photon energies. The higher resolution in the spectra excited with the He resonance radiations allows the study of finer details of each of the outer electron bands. Note also the strong variations in the relative intensities of the bands as a function of photon energy. This can be used as an aid in assigning the spectrum / 197,198/.

a monochromator for the AlKα radiation consisting of three spherically bent quartz crystals and a retarding electrostatic lens system to match the dispersion of these crystals to the electron spectrometer.

The spherically bent quartz crystal monochromator, having the property of being double focusing, was invented in my laboratory in 1958 [214] in quite another connection, namely in low angle scattering of X-rays against latex and other particles of biological interest [215]. The combination of double focusing both in the X-ray monochromator and in the electron analyzer has turned out to be essential for the further progress in ESCA. Other important technical developments have been the introduction of swiftly rotating water cooled anodes (U. Gelius), multidetector systems by means of electron channel plates and computerization of the instruments (E. Basilier).

In 1972 my coworker Ulrik Gelius and I made a new instrument [85, 216] with all these components included, in particular designed for the studies of gases. With the improved resolution of this instrument new structures could be resolved. One of principal interest was the discovery of the vibrational fine structure of core lines [82, 217]. Fig. 15 shows the line profile of the C1s in CH_4. It turns out that this line can be separated into three components caused by the symmetric vibration when photoionization occurs in the 1s level of the central carbon atom. When the photoelectron leaves the methane molecule the latter shrinks about 0.05 A. The minimum of the new potential curve for the ion will consequently be displaced by the corresponding amount and Franck-Condon transitions which take place will then give rise to the observed vibra-

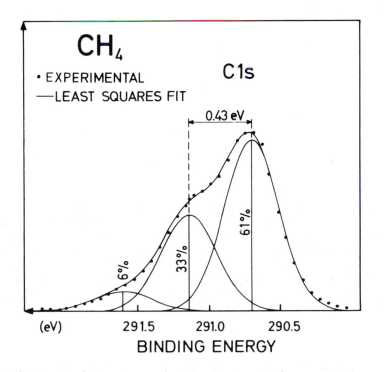

Fig. 15. Vibrational structure of the core electron line C1s in CH_4. The line structure can be quantitatively explained as a consequence of the shrinkage of the equilibrium distances upon core electron emission /217/.

tional fine structure of the electron line and with the intensities given by the Franck-Condon factors. This finding can be correlated with the simultaneously made discovery in our laboratory of vibrational fine structures in soft X-ray emission lines [218—233]. This development is further described in theses work by L. O. Werme [234], J. Nordgren [235] and H. Ågren [236]. Combined, these results show that vibrations occur in these molecules during X-ray emission both in the initial and the final states.

The above high resolution instrument designed together with U. Gelius was planned to be a prototype instrument for a new generation of advanced instruments which have now been constructed in a recently built laboratory for electron spectroscopy in Uppsala [237]. These have just been finished and are the sixth generation in the sequence from my laboratory since 1954. Two of the new instruments are designed for molecular studies and the third for surface

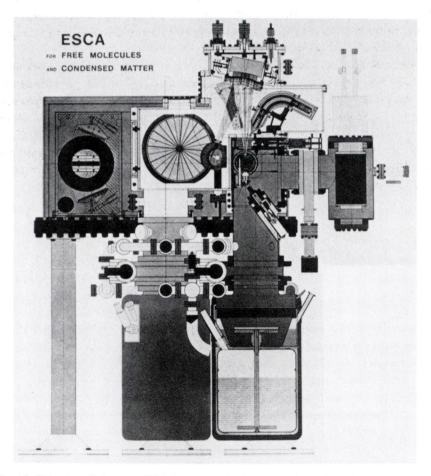

Fig. 16. Side view of the new ESCA instrument for free molecules and condensed matter. The instrument is UHV compatible and includes four different excitation modes (Monochromatized AlKα; Monochromatized and polarized UV; electron impact; monochromatized electron impact).

studies. The spherical electrostatic analyzer (R=36 cm) is provided with an electrostatic lens system due to B. Wannberg [238]. The modes of excitation included in the instruments are: monochromatic AlKα radiation ($\triangle h\upsilon=0.2$ eV) at 1486.6 eV; a UV light source with a grating providing selected UV lines between 10 eV to ~50 eV; an electron monochromator of variable energy with an energy homogeneity of $\leqslant 10$ meV and an additional electron gun for Auger electron excitation, also variable in energy (Fig. 16). A polarizer for the UV light at different wavelengths can alternatively be used in angular distribution studies.

The new instrument has been put into operation. As an illustration of its improved qualities Fig. 17 shows a new investigation of the previously recorded spectrum of methane (compare with Fig. 15) under increased resolution [261]. The vibrational structure is now well resolved. A convolution of the recorded spectrum using the window curve of the spectrometer and a computer program yields a remaining spectrum consisting of three narrow lines, the widths (~ 100 meV) of which can be measured with good accuracy. For calibration purposes the Ar 2 $p_{1/2, \, 3/2}$ lines are simultaneously recorded. This investigation approaches an accuracy of a few meV in the determination of binding energies around 300 eV, i.e. close to $1:10^5$.

The brief account I have given above concerns the work which was done in my laboratory in the development of electron spectroscopy. During the seventies several reviews and books on electron spectroscopy have been written and for a complete account the reader has to go to such sources [239−258]. From my own laboratory two new books have just been completed authored by Hans Siegbahn and Leif Karlsson [259, 260], which cover the developments mainly after 1970 and present current experimental and theoretical aspects on electron spectroscopy.

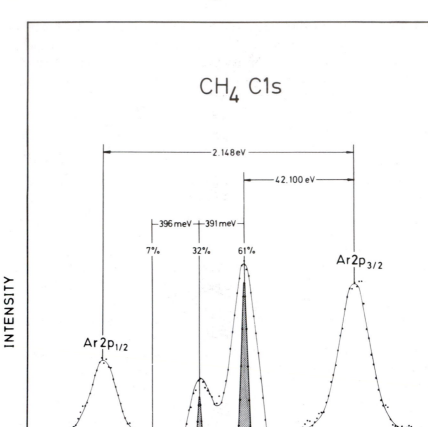

Fig. 17. New study of the methane C1s core vibrational structure (compare Fig. 15) by means of the instrument acc. to Fig. 16. The structure is resolved and deconvoluted into three narrow lines which yield the binding energies and widths of the components to a high degree of accuracy. Argon is used as a calibration gas, mixed with methane.

REFERENCE LIST

1. Siegbahn K., Studies in β-Spectroscopy (Thesis), Arkiv f. Mat. Astr., Fys. *30A*, 1 (1944).
2. Siegbahn K., and Svartholm N., Nature *157*, 872 (1946).
3. Svartholm N., and Siegbahn K., Arkiv f. Mat. Astr., Fys. *33A*, 21 (1946).
4. Hedgran A., Siegbahn K., and Svartholm N., Proc. Phys. Soc. *A63*, 960 (1950).
5. Siegbahn K., in Alpha-, Beta- and Gamma-Ray Spectroscopy, ref. 25, p. 79.
6. Hofstadter R., Les Prix Nobel, 1961.
7. Lindström G., Nuclear Resonance Absorption Applied to Precise Measurements of Nuclear Magnetic Moments and the Establishment of an Absolute Energy Scale in β-Spectroscopy (Thesis), Ark. Fysik *4*, 1 (1951).
8. Hedgran A., Precision Measurements of Nuclear Gamma-Radiation by Techniques of Beta-Spectroscopy (Thesis), Almqvist & Wiksell, Uppsala 1952.
9. Bergström I., The Isomers of Krypton and Xenon; An Investigation with Eletromagnetically Separated Radioactive Isotopes (Thesis), Ark. Fysik *5*, 191 (1952).
10. Thulin S., Studies in Nuclear Spectroscopy with Electromagnetically Separated Gaseous Isotopes (Thesis), Almqvist & Wiksell, Uppsala 1954.
11. Gerholm T.R., Coincidence Spectroscopy; Studies of Nuclear Magnetic Dipole Transition Probabilities and of Positron Annihilation Lifetimes. Design and Construction of an Electron-Electron Coincidence Spectrometer and Its Application to the Investigation of Some Complex Nuclear Disintegrations (Thesis), Almqvist & Wiksell, Uppsala 1956.
12. Lindqvist T., Angular Correlation Measurements; Studies of Relative Transition Probabilities of Gamma Radiation and of Internal Compton Effect. Experiments on Nuclear g-Factors (Thesis), Almqvist & Wiksell, Uppsala 1957.
13. Bäckström G., Levels and Transitions of Atomic Nuclei: Experiments and Experimental Methods (Thesis), Almqvist & Wiksell, Uppsala 1960.
14. Pettersson B.-G., Experimental Studies of Electron-Gamma Directional Correlations (Thesis), Almqvist & Wiksell, Uppsala 1961.
15. Karlsson E., Studies of Excitation Modes of Atomic Nuclei by Means of Magnetic Moment Determinations and General Decay Investigations (Thesis). Acta Universitatis Upsaliensis 11, 1962.
16. Thun J.E., Studies of Nuclear Properties of Decay Modes of Radioactive Nuclei; Directional Correlation and Coincidence Investigations (Thesis), Acta Universitatis Upsaliensis 15, 1962.
17. Lindskog J., Studies of Nuclear Properties by Coincidence Spectrometer Methods; Decay Schemes and Transition Probabilities (Thesis), Acta Universitatis Upsaliensis 22, 1963.
18. Matthias E., Experimental and Theoretical Investigations of Perturbed Angular Correlations (Thesis), Almqvist & Wiksell, Uppsala 1963.
19. Schneider W., Applications of Digital Computer Techniques to Theoretical Investigations and the Analysis of Experimental Data in Nuclear Spectroscopy (Thesis), Almqvist & Wiksell, Uppsala 1963.
20. Bergman O., Experimental Studies of Nuclear States and Transitions (Thesis), Acta Universitatis Upsaliensis 41, 1964.
21. Sundström T, Electromagnetic Transitions in Atomic Nuclei; Experimental Studies of Transition Probabilities and Decay Schemes (Thesis), Acta Universitatis Upsaliensis 47, 1964.
22. Kleinheinz P., Studies of Internal Conversion Transitions in the Decay of Excited Nuclear States (Thesis), Almqvist & Wiksell, Uppsala 1965.
23. Bäcklin A., Nuclear Levels Excited in Neutron Capture Reactions and Radioactive Decays (Thesis), Acta Universitatis Upsaliensis 96, 1967.
24. Siegbahn K. (Editor), Beta- and Gamma Ray Spectroscopy, North-Holland Publ. Co., Amsterdam, 1955.
25. Siegbahn K. (Editor), Alpha-, Beta and Gamma Ray Spectroscopy, North-Holland Publ. Co., Amsterdam, 1965.
26. Jenkin J.G., Leckey R.C.G. and Liesegang J., J. Electron Spectrosc. *12*, 1 (1977).

27. Carlson T.A. (Editor), X-ray Photoelectron Spectroscopy, Benchmark Papers in Phys. Chem. and Chem. Phys. Vol. 2, Dowden, Hutchinson and Ross (1978).
28. Siegbahn K., Conf. of the Swedish Nat. Committee for Physics, Stockholm, 1952, Ark. Fys. *7*, 86 (1954).
29. Siegbahn K., Conf. of the Swedish Nat. Committee for Physics, Stockholm, 1953, Ark. Fys. *8*, 19 (1954).
30. Siegbahn K., Introductory Talk at the Int. Conf. on Beta and Gamma Radioactivity, Amsterdam, 1952; Physica *18*, 1043 (1952).
31. Siegbahn K. and Edvarson K., Nucl. Phys. *1*, 137 (1956).
32. Sokolowski E., Edvarson K. and Siegbahn K., Nucl. Phys. *1*, 160 (1956).
33. Nordling C., Siegbahn K., Sokolowski E. and Wapstra A.H., Nucl. Phys. *1*, 326 (1956).
34. Nordling C., Sokolowski E. and Siegbahn K., Phys. Rev. *105*, 1676 (1957).
35. Sokolowski E., Nordling C. and Siegbahn K., Ark. Fysik. *12*, 301 (1957).
36. Robinson H., Phil. Mag. *50*, 241 (1925).
37. Siegbahn K., Nordling C. and Sokolowski E., Proc. Rehovoth Conf. Nucl. Structure, North-Holland Publ. Co., Amsterdam, 1957, p. 291.
38. Sokolowski E., Nordling C. and Siegbahn K., Phys. Rev. *110*, 776 (1958).
39. Nordling C., Sokolowski E. and Siegbahn K., Arkiv Fysik *13*, 282 (1958).
40. Sokolowski E., Nordling C. and Siegbahn K., Arkiv Fysik *13*, 288 (1958).
41. Nordling C., Sokolowski E. and Siegbahn K., Arkiv Fysik *13*, 483 (1958).
42. Sokolowski E. and Nordling C., Arkiv Fysik *14*, 557 (1959).
43. Sokolowski E., Arkiv Fysik *15*, 1 (1959).
44. Nordling C., Arkiv Fysik *15*, 241 (1959).
45. Nordling C., Arkiv Fysik *15*, 397 (1959).
46. Nordling C. and Hagström S., Arkiv Fysik *15*, 431 (1959).
47. Bergvall P. and Hagström S., Arkiv Fysik *16*, 485 (1960).
48. Nordling C. and Hagström S., Arkiv Fysik *16*, 515 (1960).
49. Bergvall P. and Hagström S., Arkiv Fysik *17*, 61 (1960).
50. Bergvall P., Hörnfeldt O. and Nordling C., Arkiv Fysik *17*, 113 (1960).
51. Nordling C., Hagström S. and Siegbahn K., Arkiv Fysik *22*, 428 (1962).
52. Fahlman A., Hörnfeldt O. and Nordling C., Arkiv Fysik *23*, 75 (1962).
53. Hagström S., Hörnfeldt O., Nordling C. and Siegbahn K. Arkiv Fysik *23*, 145 (1962).
54. Hagström S., Nordling C. and Siegbahn K., "Alpha-, Beta- and Gamma-Ray Spectroscopy", Appendix 2. Ed.Siegbahn K., North-Holland Publ. Co., Amsterdam 1965.
55. Hagström S. and Karlsson S.-E., Arkiv Fysik *26*, 252 (1964).
56. Hagström S. and Karlsson S.-E., Arkiv Fysik *26*, 451 (1964).
57. Fahlman A. and Hagström S., Arkiv Fysik *27*, 69 (1964).
58. Andersson I. and Hagström S., Arkiv Fysik *27*, 161 (1964).
59. Nordling C. and Hagström S., Z. Phys. *178*, 418 (1964).
60. Fahlman A., Hamrin K., Nordberg R., Nordling C. and Siegbahn K., Phys. Rev. Lett. *14*, 127 (1965).
61. Fahlman A., Hamrin K., Nordberg R., Nordling C., Siegbahn K. and Holm L.W., Phys. Lett. *19*, 643 (1966).
62. Hörnfeldt O. and Fahlman A., Arkiv Fysik *22*, 412 (1962).
63. Hörnfeldt O., Fahlman A. and Nordling C., Arkiv Fysik *23*, 155 (1962).
64. Hörnfeldt O., Arkiv Fysik *23*, 235 (1962).
65. Hagström S., Z. Phys. *178*, 82 (1964).
66. Fahlman A. and Nordling C. Arkiv Fysik *26*, 248 (1964).
67. Fahlman A., Nordberg R., Nordling C. and Siegbahn K., Z. Phys. *192*, 476 (1966).
68. Fahlman A., Hamrin K., Axelson G., Nordling C. and Siegbahn K., Z. Phys. *192*, 484 (1966).
69. Fahlman A., Hamrin K., Nordberg R., Nordling C. and Siegbahn K., Phys. Lett. *20*, 159 (1966).
70. Sokolowski E., Investigations of Inner Electron Shells by Spectroscopic Studies of Photo- and Auger Electrons (Thesis), Almqvist & Wiksell, Uppsala 1959.

71. Nordling C., Experimental Studies of Electron Binding Energies and Auger Spectra (Thesis), Almqvist & Wiksell, Uppsala 1959.

72. Bergvall P., Investigations of Atomic and Nuclear Energy Levels by Crystal Diffraction and Photo Electron Spectroscopy (Thesis), Almqvist & Wiksell, Uppsala 1960.

73. Hörnfeldt O., Studies of Atomic Level Energies and Auger Spectra (Thesis), Almqvist & Wiksell, Uppsala 1962.

74. Hagström S., Studies of Some Atomic Properties by Electron Spectroscopy (Thesis), Almqvist & Wiksell, Uppsala 1964.

75. Fahlman A., Electron Spectroscopy of Atoms and Molecules (Thesis), Acta Universitatis Upsaliensis 73, 1966.

76. Hagström S., Nordling C. and Siegbahn K., Phys. Lett. *9*, 235 (1964).

77. Karlsson S.-E., Precision Electron Spectroscopy for Studies of Nuclear and Atomic Energy Levels (Thesis), Acta Universitatis Upsaliensis 99, 1967.

78. Nordberg R., ESCA Studies of Atoms and Molecules (Thesis), Acta Universitatis Upsaliensis 118, 1968.

79. Hamrin K., ESCA Applied to Solids and Gases (Thesis), Acta Universitatis Upsaliensis 151, 1970.

80. Hedman J., ESCA Studies of Electron Levels and Bands in Atoms, Molecules and Solids (Thesis), Acta Universitatis Upsaliensis 196, 1972.

81. Johansson G., Chemical Bonding and Electronic Structure of Molecules and Solids Studied by ESCA (Thesis), Acta Universitatis Upsaliensis 205, 1972.

82. Gelius U., Molecular Spectroscopy by Means of ESCA; Experimental and Theoretical Studies (Thesis), Acta Universitatis Upsaliensis 242, 1973.

83. Lindberg B., Some Aspects of the Sulphur-Oxygen Bond in Organic Sulphur Groups (Thesis), Acta Universitatis Upsaliensis 173, 1970.

84. ESCA I, ref. 211, p. 21.

85. Gelius U., Basilier E., Svensson S., Bergmark T. and Siegbahn K., J. Electron Spectrosc. *2*, 405 (1974).

86. Siegbahn K., ibid. *5*, 3 (1974).

87. Clark D.T., Dilks A., Shuttleworth S. and Thomas H.R., ibid., *14*, 247 (1978).

88. Clark D.T. and Dilks A., J. Polymer Sci. *16*, 911 (1978).

89. Clark D.T., Phys. Scr. *16*, 307 (1977).

90. Clark D.T., in Handbook of X-ray and UV Photoelectron Spectroscopy (Ed. D. Briggs) Heyden 1977, p. 211.

91. ESCA I, ref. 211, sect III-9, p. 63.

92. Goscinski O., Howat G. and Åberg T., J. Phys. *B8*, 11 (1975).

93. Pickup B. and Goscinski O., Mol. Phys. *26*, 1013 (1973).

94. Goscinski O., Hehenberger M., Roos B. and Siegbahn P., Chem. Phys. Lett. *33*, 427, (1975).

95. ESCA II, ref. 212, Sect. 5.4.

96. Gelius U., Hedén P.-F., Hedman J., Lindberg B.J., Manne R., Nordberg R., Nordling C. and Siegbahn K., Phys. Scr. *2*, 70 (1970).

97. Lindberg B.J., Hamrin K., Johansson G., Gelius U., Fahlman A., Nordling C. and Siegbahn K., Phys. Scr. *1*, 286 (1970).

98. Allison D.A., Johansson G., Allan C.J., Gelius U., Siegbahn H., Allison J. and Siegbahn K., J. Electron Spectrosc. *1*, 269 (1972/73).

99. Shirley D.A., Adv. Chem. Phys. *23*, 85 (1973).

100. Clark D.T., in Electron Emission Spectroscopy. (Eds. Dekeyser et al.) D. Reidel Publ. Co. (1973).

101. Davis D.W., Banna M.S. and Shirley D.A., J. Chem. Phys. *60*, 237 (1974).

102. Wyatt J.F., Hillier I.H., Saunders V.R., Connor J.A. and Barber M., J. Chem. Phys. *54*, 5311 (1971).

103. Jolly W.L. and Perry W.B., Inorg. Chem. *13*, 2686 (1974).

104. Jolly W.L. and Perry W.B., J. Amer. Chem. Soc. *95*, 5442 (1974).

105. Carver, J. C., Gray R. C. and Hercules, D. M., J. Amer. Chem. Soc. *96*, 6851 (1974).

106. Perry, W. B. and Jolly, W. L., Inorg. Chem. *13*, 1211 (1974).

107. Gelius, U., Johansson, G. Siegbahn, H. Allan, C. J., Allison, D. A., Allison J. and Siegbahn, K., J. Electron Spectrosc. *1*, 285 (1972/73).
108. Siegbahn, H., Medeiros R. and Goscinski, O. ibid. *8*, 149 (1976).
109. Bagus P. S. and Hermann, K., Solid State Comm. *20*, 5 (1976).
110. Ellis, D. E., Baerends, E. J., Adachi H. and Averill, F. W., Surf. Sci. *64*, 649 (1977).
111. Ley, L. Kowalczyk, S. P., McFeely, F. R., Pollak R. A., and Shirley, D. A. Phys. Rev. *B8*, 2392 (1973).
112. Kowalczyk, S. P., Pollak, R. A., McFeely, F. R., Ley L. and Shirley, D. A., ibid. *B8*, 2387 (1973).
113. Shirley, D. A., Martin, R. L., Kowalczyk, S. P., McFeely F. R. and Ley, L., Phys. Rev. *B15*, 544 (1977).
114. Svensson, S., Mårtensson, N., Basilier, E., Malmquist, P. Å., Gelius U. and Siegbahn, K., J. Electron Spectrosc. *9*, 51 (1976).
115. Jen, J. S. and Thomas, T. D., Phys. Rev. *B13*, 5284 (1976).
116. Johansson B. and Mårtensson, N., ibid. B21, 4427 (1980).
117. Mårtensson, N., Atomic, Molecular and Solid State Effects in Photoelectron Spectroscopy (Thesis), Acta Universitatis Upsaliensis 579, Uppsala 1980.
118. Nyholm. R., Electronic Structure and Photoionization Studied by Electron Spectroscopy (Thesis), Acta Universitatis Upsaliensis 581, Uppsala 1980.
119. Mårtensson, N., Nyholm, R., Calén. H., Hedman, J. and Johansson. B. Uppsala University, Institute of Physics Report, UUIP-1008 (July 1980).
120. Kowalczyk, S. P., Pollak. R. A., McFeely, F. R. and Shirley, D. A., Phys. Rev. *B8*, 3583 (1973).
121. Kowalczyk, S. P., Ley, L., McFeely, F. R., Pollak, R. A. and Shirley, D. A., Phys. Rev. *B9*, 381 (1974).
122. Wagner, C. D. and Biloen, P. Surf. Sci. *35*, 82 (1973).
123. Matthew, J. A. D., Surf. Sci. *40*, 451 (1973).
124. Utriainen, J., Linkoaho, M. and Åberg, T., in Proc. Int. Symp. X-ray Sp. and El. Str. of Mat. *1*, 382 (1973).
125. Asplund, L., Kelfve, P., Siegbahn, H., Goscinski. O., Fellner-Feldegg, H., Hamrin, K., Blomster. B. and Siegbahn, K., Chem. Phys. Lett. *40*, 353 (1976).
126. Keski-Rahkonen, O. and Krause, M. O., J. Electron Spectrosc. *9*, 371 (1976).
127. Siegbahn, H. and Goscinski, O., Phys. Scr. *13*, 225 (1976).
128. Asplund, L., Kelfve, P., Blomster, B., Siegbahn, H., Siegbahn, K., Lozes, R. L. and Wahlgren U. I., Phys. Scr. *16*, 273 (1977).
129. Kim, K. S., Gaarenstroom, S. W. and Winograd, N., Chem. Phys. Lett. *41*, 503 (1976).
130. Wagner, C. D., Far. Disc. Chem. Soc. *60*, 291 (1975).
131. Wagner, C. D., in Handbook of X-ray and Ultraviolet Photoelectron Spectroscopy (Ed. D. Briggs) Chapt. 7 (Heyden, 1977).
132. Kelfve, P., Blomster, B., Siegbahn, H., Siegbahn, K., Sanhueza, E. and Goscinski O., Phys. Scr. *21*, 75 (1980).
133. Thomas, T. D., J. Electron Spectrosc, 20, 117 (1980).
134. ESCA I, ref. 211, p. 118.
135. Pignataro, S. and Distefano, G., J. Electron Spectrosc, *2*, 171 (1973), Z. Naturforsch. *30a*, 815 (1975).
136. Ågren, H., Roos, B., Bagus, P. S., Gelius, U., Malmquist, P. Å., Svensson, S., Maripuu, R. and Siegbahn, K., Uppsala University, Institute of Physics Report, UUIP-1057 (1982).
137. Yin, L., Adler, I., Tsang, T., Mantienzo, L. J. and Grim, S. O., Chem. Phys. Lett. *24*, 81 (1974).
138. Pignataro, S., Foffani, A. and Distefano, G., Chem. Phys. Lett. *20*, 350 (1973).
139. Johansson, L. Y., Larsson, R., Blomquist, J., Cederström, C., Grapengiesser, S., Helgeson, U., Moberg, L. C. and Sundbom, M., Chem. Phys. Lett. 24, 508 (1974).
140. Barber, M., Connor, J. A. and Hillier, I. H., Chem. Phys. Lett. *9*, 570 (1971).
141. Hüfner, S. and Wertheim, G. K., Phys. Rev. *B7*, 5086 (1973).
142. Rosencwaig, A., Wertheim, G. K. and Guggenheim, H. J., Phys. Rev. Lett. *27*, 479 (1971).

143. Rosencwaig, A. and Wertheim, G. K., J. Electron Spectrosc. *1*, 493 (1973).

144. Frost, D. C., McDowell, C. A. and Woolsey, I. S., Chem. Phys Lett. *17*, 320 (1972).

145. Kim, K. S., Chem. Phys. Lett. *26*, 234 (1974).

146. Carlson, T. A., Carver, J. C., Saethre, L. J., Santibanez, F. G. and Vernon, G. A., J. Electron Spectrosc, *5*, 247 (1974).

147. Fuggle, J. C., Umbach, E., Menzel, D., Wandelt, K. and Brundle, C. R., Solid State Comm. *27*, 65 (1978).

148. Lang, N. D. and Williams, A. R., Phys. Rev. *B16*, 2408 (1977).

149. Schönhammer, K. and Gunnarsson, O., Solid State Comm. *23*, 691 (1978) and *26*, 399 (1978) and Z., Phys. *B30*, 297 (1978).

150. Hedman, J., Hedén, P. F., Nordling, C. and Siegbahn, K., Phys. Lett. *A29*, 178 (1969).

151. ESCA II, ref. 212, section 5.1, p. 56.

152. ESCA I, ref. 211, section IV, p. 72.

153. Siegbahn, K., Gelius, U., Siegbahn, H. and Olsson, E., Phys. Lett. *32A*, 221 (1970) and Phys. Scr. *1*, 272 (1970).

154. Fadley, C. S. and Bergström S. Å. L., Phys. Lett. *35A*, 375 (1971).

155. Baird, R. J., Fadley, C. S. and Wagner, L. F., Phys. Rev. *B15*, 666 (1977).

156. Zehner, D. M., Noonan, J. R. and Jenkins, L. H., Phys. Lett. *62A*, 267 (1977).

157. Briggs, D., Marbrow, R. A. and Lambert, R. M., Solid State Comm. *25*, 40 (1978).

158. Erickson, N. E., Phys. Scr. *16*, 462 (1977).

159. Kono, S., Goldberg, S. M., Hall, N. F. T. and Fadley, C. S., Phys. Rev. Lett. *41*, 1831 (1978).

160. Petersson, L. -G., Kono. S., Hall, N. F. T., Fadley C. S. and Pendry, J. B., Phys. Rev. Lett *42*, 1545 (1979).

161. Woodruff, D. P., Norman, D., Holland, B. W., Smith. N. W., Farrell, H. H. and Traum, M. M., Phys. Rev. Lett. *41*, 1130 (1978).

162. Kevan. S. D., Rosenblatt, D. H., Denley, D. R., Lee, B.-C. and Shirley. D. A., Phys. Rev. *B20*, 4133 (1979); Phys. Rev. Lett. *41*, 1565 (1978).

163. McGovern, I. T., Eberhardt. W. and Plummer, E. W., Sol. St. Comm. *32*, 963 (1979).

164. Guillot, C., Jugnet, Y., Lasailly, Y., Lecante, J., Spanjaard, D. and Duc, T. M. (private communication, 1980).

165. Bergmark, T., Magnusson, N. and Siegbahn, K., Ark. Fys. *37*, 355 (1968).

166. ESCA I, ref. 211, p. 20.

167. ESCA I, ref. 211, p. 21.

168. Siegbahn, H. and Siegbahn, K., J. Electron Spectrosc. *2*, 319 (1973).

169. Siegbahn, H., Asplund, L., Kelfve, P., Hamrin, K., Karlsson, L. and Siegbahn, K., J. Electron Spectrosc. *5*, 1059 (1974).

170. Siegbahn, H., Asplund, L., Kelfve, P. and Siegbahn, K., J. Electron Spectrosc, *7*, 411 (1975).

171. Fellner-Feldegg, H., Siegbahn, H., Asplund, L., Kelfve, P. and Siegbahn, K., J. Electron Spectrosc. *7*, 421 (1975).

172. Lindberg, B. Asplund, L., Fellner-Feldegg, H., Kelfve, P., Siegbahn, H. and Siegbahn K., Chem. Phys. Lett. *39*, 8 (1976).

173. Siegbahn, H., Svensson, S. and Lundholm. M., J. Electron Spectrosc. *24*, 205 (1981). The first considerations of continuously renewable liquid samples in vacuum for ESCA and experiments on liquid beams were reported at the Asilomar conference 1971; K. Siegbahn: Perspectives and Problems in Electron Spectroscopy, (Editor: D. Shirley), North Holland, 1972. Further accounts for these developments are found in ref. /168/−/172/. Different suggested alternatives for handling the renewable samples (liquid beams, wetted rotating wires and rotating disks) were presented at the Namur Conference 1974; K. Siegbahn: Electron Spectroscopy—An Outlook (secs 3 and 4, figs 5−10). Editors: R. Caudano and J. Verbist, Elsevier, 1974. References to recent other works on liquids are found in H. Siegbahn et. al. ref. /173/ (R. Ballard et. al. see ref. ibid. 6 and 7; P. Delahay et. al. ref. ibid. 8−10; K. Burger et. al. ref. ibid. 11 and 12; S. Avanzino and W. Jolly ref. ibid. 13).

174. Siegbahn, H., ESCA Studies of Electronic Structure and Photoionization in Gases, Liquids and Solids (Thesis), Acta Universitatis Upsaliensis 277, Uppsala 1974.

175. Asplund, L., Electron Spectroscopy of Liquids and Gases (Thesis), Acta Universitatis Upsaliensis 409, Uppsala 1977.

176. Kelfve, P., Electronic Structure of Free Atoms, Molecules and Liquids Studied by Electron Spectroscopy (Thesis), Acta Universitatis Upsaliensis 483, Uppsala 1978.

177. Turner, D. W. and Al-Joboury, M. I., J. Chem. Phys. *37*, 3007 (1962).

178. Turner, D. W. and Al-Joboury, M. I., Chem. Soc., London, 5141 (1963).

179. Al-Joboury, M. I., May, D. P. and Turner, D. W., J. Chem. Soc., 616, 4434, 6350 (1965).

180. Radwan, T. N. and Turner D. W., J. Chem. Soc., 85 (1966).

181. Al-Joboury, M. I. and Turner, D. W., J. Chem. Soc., 373 (1967).

182. Turner, D. W., Baker, C., Baker, A. D. and Brundle, C. R., Molecular Photoelectron Spectroscopy, Wiley-Interscence, London (1970).

183. Price, W. C., Endeavour *26*, 78 (1967).

184. Lempka, H. J., Passmore, T. R. and Price, W. C., Proc. Roy. Soc. *A304*, 53 (1968).

185. Berglund, C. N. and Spicer, W. E., Phys. Rev. *136*, A1030; A1044 (1964).

186. Spicer, W. E., J. Appl. Phys. *37*, 947 (1966).

187. Blodgett, A. J., Jr., and Spicer, W. E., Phys. Rev. *146*, 390 (1966).

188. Blodgett, A. J., Jr. and Spicer, W. E. Phys. Rev. 158, 514 (1967).

189. Bergmark, T., Electronic Structure of Atoms and Molecules Studied by Means of Electron Spectroscopy (Thesis), Acta Universitatis Upsaliensis 274, 1974.

190. Karlsson, L., High Resolution Studies of the Valence Electron Structure of Atoms and Molecules by Means of Electron Spectroscopy (Thesis), Acta Universitatis Upsaliensis 429, 1977.

191. Jadrny, R., Electronic Structure Studied by Electron Spectroscopy and Angular Distributions (Thesis), Acta Universitatis Upsaliensis 463, 1978.

192. Mattsson, L., High Resolution Valence Electron Spectroscopy. Development of a VUV-Polarizer for Angular Distributions (Thesis), Acta Universitatis Upsaliensis 562, 1980.

193. Asplund, L., Kelfve, P., Blomster, B., Siegbahn, H. and Siegbahn, K., Phys. Scr. *16*, 268 (1977).

194. Siegbahn, H., Asplund, L. and Kelfve, P., Chem. Phys. Lett. *35*, 330 (1975).

195. Siegbahn, H., in Excited States in Quantum Chemistry (Eds. Nicolaides C. A. and Beck, D. R.), Plenum (1978) p. 273.

196. Darko, T., Siegbahn, H. and Kelfve, P., Chem. Phys. Lett. *81*, 475 (1981).

197. Karlsson, L., Mattsson, L., Jadrny, R., Bergmark, T. and Siegbahn, K., Phys. Scr. *14*, 230 (1976).

198. Gelius, U., J. Electron Spectrosc. *5*, 985 (1974).

199. Krause, M. O., Chem. Phys. Lett. *10*, 65 (1971).

200. Wuilleumier, F. and Krause, M. O., Phys. Rev. A10, 242 (1974).

201. Banna, M. S. and Shirley, D. A., Chem. Phys. Lett. *33*, 441 (1975), J. Electron Spectrosc. *8*, 23 (1976); ibid. *8*, 255 (1976).

202. Nilsson, R., Nyholm, R., Berndtsson, A., Hedman, J. and Nordling, C., J. Electron Spectrosc. *9*, 337 (1976).

203. Berndtsson, A., Nyholm, R. Nilsson, R. Hedman J. and Nordling, C. ibid. *13*, 131 (1978).

204. Cavell, R. G. and Allison, D. A., Chem. Phys. Lett. *36*, 514 (1975).

205. Allison, D. A. and Cavell, R. G., J. Chem. Phys. *68*, 593 (1978).

206. Kunz, C. (Ed.), Synchrotron Radiation, Techniques and Applications, Topics in Current Physics *10*, Springer-Verlag (1979).

207. Synchrotron Radiation Instrumentation and New Developments, Nuclear Instr. Meth. *152*, (Conf. Vol., Eds: Wuilleumier, P. and Y. Farge) 1978.

208. Synchrotron Radiation Instrumentation, Nuclear Instr. Meth. *172* (Conf. Vol., Eds: Ederer, D. L. and West, J. B.) 1980.

209. Synchrotron Radiation Facilities, ibid. *177* (Conf. Vol., Ed. Howells, M. R.) 1980.

210. Synchrotron Radiation Research, Eds: Winick H. and Doniach, S., Plenum (1980).

211. Siegbahn, K., Nordling, C., Fahlman, A., Nordberg, R., Hamrin, K., Hedman, J., Johansson, G., Bergmark, T., Karlsson, S.-E., Lindgren., I. and Lindberg, B., ESCA − Atomic, Molecu-

lar and Solid State Structure Studied by Means of Electron Spectroscopy, Nova Acta Regiae Soc. Sci. Upsaliensis, Ser. IV, Vol. 20 (1967).

212. Siegbahn, K., Nordling, C., Johansson, G., Hedman, J., Hedén, P. F., Hamrin, K., Gelius, U., Bergmark, T., Werme, L. O., Manne R. and Baer Y., ESCA Applied to Free Molecules, North-Holland Publ. Co., Amsterdam-London, 1969.

213. Siegbahn, K., Hammond, D., Fellner-Feldegg, H. and Barnett, E. F., Science *176*, 245 (1972).

214. Wassberg. G. and Siegbahn, K., Ark. Fys. *14*, 1 (1958).

215. Hagström, S. and Siegbahn, K., J. Ultrastructure Research *3*, 401 (1960).

216. Gelius, U. and Siegbahn, K., Far. Disc. Chem. Soc. *54*, 257 (1972).

217. Gelius, U., Svensson, S., Siegbahn H., Basilier, E., Faxälv, Å and Siegbahn, K., Chem. Phys. Lett. *28*, 1 (1974).

218. Siegbahn, K. in Electron Spectroscopy (Ed. D.A. Shirley), Proc. Intern. Conf. Asilomar, California, USA, 1971. North-Holland, Amsterdam, 1972.

219. Siegbahn, K., Werme, L.O., Grennberg, B., Lindeberg, S. and Nordling, C. Uppsala University Institute of Physics Report, UUIP-749 (Oct. 1971).

220. Siegbahn, K., Werme, L.O., Grennberg, B., Nordgren, J. and Nordling, C, Phys. Lett *41A* (2), 111 (1972).

221. Werme, L.O., Grennberg, B., Nordgren, J., Nordling, C. and Siegbahn, K., Phys. Lett *41A* (2), 113 (1972).

222. Werme, L.O., Grennberg, B., Nordgren, J., Nordling, C. and Siegbahn, K., Phys. Rev. Lett. *30* (12), 523 (1973).

223. Werme, L.O., Grennberg, B., Nordgren, J., Nordling, C. and Siegbahn, K., Nature *242* (5348), 453 (1973).

224. Werme, L.O., Grennberg, B., Nordgren, J., Nordling, C. and Siegbahn, K., J. Electron Spectrosc. *2*, 435 (1973).

225. Werme, L.O., Nordgren, J., Nordling, C. and Siegbahn, K., C.R. Acad Sci. Paris, *B-119*, 279 (1974).

226. Nordgren, J., Werme, L.O., Ågren, H., Nordling, C. and Siegbahn, K., J. Phys. *B8* (1), L18 (1975).

227. Werme, L.O., Nordgren, J., Ågren, H., Nordling, C. and Siegbahn, K., Z. Physik *272*, 131 (1975).

228. Nordgren, J., Ågren, H., Werme, L.O., Nordling, C. and Siegbahn, K., J. Phys. *B9* (2), 295 (1976).

229. Nordgren, J., Ågren, H., Selander, L., Nordling, C. and Siegbahn, K., paper presented at the Nordic Symposium on Atomic and Molecular Transition Probabilities in Lund, March 28–29, 1977.

230. Nordgren, J., Ågren, H., Nordling, C. and Siegbahn, K., Ann. Acad. Reg. Scientiarum Upsaliensis *21*, 23 (1978).

231. Ågren, H., Nordgren, J., Selander, L., Nordling, C. and Siegbahn, K., Phys. Scr. *18*, 499 (1978).

232. Nordgren, J., Ågren, H., Pettersson, L., Selander, L., Griep, S., Nordling, C. and Siegbahn, K., Phys. Scr. *20*, 623 (1979).

233. Ågren, H., Selander, L., Nordgren, J., Nordling, C., Siegbahn, K. and Müller, J., Chem. Phys. *37*, 161)1979).

234. Werme, L.O., Electron and X-ray Spectroscopic Studies of Free Molecules (Thesis), Acta Universitatis Upsaliensis 241, 1973.

235. Nordgren, J., X-ray Emission Spectra of Free Molecules (Thesis) Acta Universitatis Upsaliensis 418, 1977.

236. Ågren, H., Decay and Relaxation of Core Hole States in Molecules (Thesis), Uppsala 1979.

237. Siegbahn, K. in Electron Spectroscopy, Eds. Hedman, J. and Siegbahn, K., Phys. Scr. *16*, 167, (1977).

238. Wannberg, B. and Sköllermo, A., J. Electron Spectrosc. *10*, 45 (1977).

239. Phil. Trans. Roy. Soc. (London) *268*, 1-175 (1970).

240. Shirley, D.A. (Ed.) Electron Spectroscopy, North-Holland, Amsterdam (1972).

241. Faraday Disc. Chem. Soc. *54* (1972).

242. Dekeyser, W., et al. (Eds.) Electron Emission Spectroscopy, Reidel (1973).

243. Caudano, R. and Verbist, J. (Eds.) Electron Spectroscopy, Elsevier Amsterdam (1974).

244. Electron Spectroscopy of Solids and Surfaces, Far. Disc. Chem. Soc. *60* (1975).

245. Proc. Int. Symp. Electron Spectroscopy, Uppsala (May 1977), (Eds. Hedman, J. and Siegbahn, K.) Phys. Scr. *16* (5—6) (1977).

246. Proc. Australian Conf. on Electron Spectroscopy, J. Electron Spectrosc. *15* (1979).

247. Baker, A.D. and Betteridge, D., Photoelectron Spectroscopy, Pergamon Press (1972).

248. Eland, J.H.D., Photoelectron Spectroscopy, Butterworths, London (1974).

249. Carlson, T.A., Photoelectron and Auger Spectroscopy, Plenum Press, New York (1975).

250. Brundle, C.R. and Baker, A.D. (Eds.) Electron Spectroscopy, Academic Press Vol.1 (1977), Vol. 2 (1978), Vol. 3 (1979), Vol. 4 (1981).

251. Briggs, D. (Ed.), Handbook of X-ray and Ultraviolet Photoelectron Spectroscopy, Heyden (1977).

252. Rabalais, J.W., Principles of Ultraviolet Photoelectron Spectroscopy, Wiley-Interscience (1977).

253. Ghosh, P.K., A Whiff of Photoelectron Spectroscopy, Swan Printing Press, New Delhi (1978).

254. Photoemission and the Electronic Properties of Surfaces (Eds. Feuerbacher, B., Fitton, B. and Willis, R.F.) Wiley-Interscience (1978).

255. Photoemission in Solids (Eds. Cardona, M. and Ley, L.), Topics in Applied Physics Vol. 26, 27, Springer-Verlag (1978, 1979).

256. Roberts, M.W. and McKee, C.S., Chemistry of the Metal-Gas Interface, Clarendon Press (1978).

257. Ballard, R.E., Photoelectron Spectroscopy and Molecular Orbital Theory, Adam Hilger (1978).

258. Berkowitz, J., Photoabsorption, Photoionization and Photoelectron Spectroscopy, Academic Press (1979).

259. Siegbahn, H. and Karlsson, L., Photoelectron Spectroscopy, Handbuch d. Physik, Vol.31 (Ed. Mehlhorn, W.) Springer-Verlag (1982).

260. Siegbahn, H. and Karlsson, L., Electron Spectroscopy for Atoms, Molecules and Condensed Matter, North-Holland, Publ. Co. (to be published).

261. Gelius, U., Helenelund, K., Hedman, S., Asplund, L., Finnström, B. and Siegbahn, K. (to be published).

Physics 1982

KENNETH G WILSON

for his theory for critical phenomena in connection with phase transitions

THE NOBEL PRIZE FOR PHYSICS

Speech by Professor STIG LUNDQVIST of the Royal Academy of Sciences.
Translation from the Swedish text

Your Majesties, Your Royal Highnesses, Ladies and Gentlemen,

The development in physics is on the whole characterized by a close interaction between experiment and theory. New experimental discoveries lead often rapidly to the development of theoretical ideas and methods that predict new phenomena and thereby stimulate further important experimental progress. This close interaction between theory and experiment keeps the frontiers of physics moving forward very rapidly.

However, there have been a few important exceptions, where the experimental facts have been well known for a long time but where the fundamental theoretical understanding has been lacking and where the early theoretical models have been incomplete or even seriously in error. I mention here three classical examples from the physics of the twentieth century, namely superconductivity, critical phenomena and turbulence. Superconductivity was discovered in the beginning of this century, but in spite of great theoretical efforts by many famous physicists, it took about fifty years until a satisfactory theory was developed. The theory of superconductivity was awarded the Nobel Prize in physics exactly ten years ago. The critical phenomena occur at phase transitions, for example between liquid and gas. These phenomena were known even before the turn of the century, and some simple but incomplete theoretical models were developed at an early stage. In spite of considerable theoretical efforts over many decades, one had to wait until the early seventies for the solution. The problem was solved in an elegant and profound way by Kenneth Wilson, who developed the theory which has been awarded this year's Nobel Prize in physics. The third classical problem I mentioned, namely turbulence, has not yet been solved, and remains a challenge for the theoretical physicists.

From daily life we know that matter can exist in different phases and that transitions from one phase to another may occur if we change, for example, the temperature. A liquid goes over into gas phase when sufficiently heated, a metal melts at a certain temperature, a permanent magnet loses its magnetization above a certain critical temperature, just to give a few examples. Let us consider the transition between liquid and gas. When we come close to the critical point, there will appear fluctuations in the density of the liquid at all possible scales. These fluctuations take the forms of drops of liquids mixed with bubbles of gas. There will be drops and bubbles of all sizes from the size of a single molecule to the volume of the system. Exactly at the critical point the scale of the largest fluctuations becomes infinite, but the role of the smaller fluctuations can by no means be ignored. A proper theory for the critical phenomena must take into account the entire spectrum of length scales. In most problems in physics one has to deal with only one length scale. This problem required the development of a new type of theory capable of describing phenomena at all possible length scales, for example, from the order of a centimeter down to less than one millionth of a centimeter.

Wilson succeeded in an ingenious way to develop a method to solve the problem, published in two papers from 1971. A frontal attack on this problem is impossible, but he found a method to divide the problem into a sequence of simpler problems, in which each part can be solved. Wilson built his theory on an essential modification of a method in theoretical physics called renormalization group theory.

Wilson's theory gave a complete theoretical description of the behaviour close to the critical point and gave also methods to calculate numerically the crucial quantities. During the decade since he published his first papers we have seen a complete breakthrough of his ideas and methods. The Wilson theory is now also successfully applied to a variety of problems in other areas of physics.

Professor Wilson,

You are the first theoretical physicist to develop a general and tractable method, where widely different scales of length appear simultaneously. Your theory has given a complete solution to the classical problem of critical phenomena at phase transitions. Your new ideas and methods seem also to have a great potential to attack other important and up to now unsolved problems in physics.

I am very happy to have the privilege of expressing the warmest congratulations of the Royal Swedish Academy of Sciences. I now ask you to receive your Nobel Prize from the hands of His Majesty the King.

Kenneth G Wilson

KENNETH G. WILSON

I was born 1936 in Waltham, Massachusetts, the son of E. Bright Wilson Jr. and Emily Buckingham Wilson. My father was on the faculty in the Chemistry Department of Harvard University; my mother had one year of graduate work in physics before her marriage. My grandfather on my mother's side was a professor of mechanical engineering at the Massachusetts Institute of Technology; my other grandfather was a lawyer, and one time Speaker of the Tennessee House of Representatives.

My schooling took place in Wellesley, Woods Hole, Massachusetts (second, third/fourth grades in two years), Shady Hill School in Cambridge, Mass. (from fifth to eighth grade), ninth grade at the Magdalen College School in Oxford, England, and tenth and twelfth grades (skipping the eleventh) at the George School in eastern Pennsylvania. Before the year in England I had read about mathematics and physics in books supplied by my father and his friends. I learned the basic principle of calculus from *Mathematics and Imagination* by Kasner and Newman, and went of to work through a calculus text, until I got stuck in a chapter on involutes and evolutes. Around this time I decided to become a physicist. Later (before entering college) I remember working on symbolic logic with my father; he also tried, unsuccessfully, to teach me group theory. I found high school dull. In 1952 I entered Harvard. I majored in mathematics, but studied physics (both by intent), participated in the Putnam Mathematics competition, and ran the mile for the track team (and cross-country as well). I began research, working summers at the Woods Hole Oceanographic Institution, especially for Arnold Arons (then based at Amherst).

My graduate studies were carried out at the California Institute of Technology. I spent two years in the Kellogg Laboratory of nuclear physics, gaining experimental experience while taking theory courses; I then worked on a thesis for Murray Gell-Mann. While at Cal Tech I talked a lot with Jon Mathews, then a junior faculty member; he taught me how to use the Institute's computer; we also went on hikes together. I spent a summer at the General Atomic Company in San Diego working with Marshall Rosenbluth in plasma physics. Another summer Donald Groom (then a fellow graduate student) and I hiked the John Muir Trail in the Sierra Nevada from Yosemite Park to Mt. Whitney. After my third year I went off to Harvard to be a Junior Fellow while Gell-Mann went off to Paris. During the first year of the fellowship I went back to Cal Tech for a few months to finish my thesis. There was relatively little theoretical activity at Harvard at the time; I went often to M.I.T. to use their computer and eat lunch with the M. I. T. theory group, led by Francis Low.

In 1962 I went to CERN for a calendar year, first on my Junior Fellowship and then as a Ford Foundation fellow. Mostly, I worked but I found time to join Henry Kendall and James Bjorken on a climb of Mt. Blanc. I spent January through August of 1963 touring Europe.

In September of 1963 I came to Cornell as an Assistant Professor. I received tenure as an Associate Professor in 1965, became Full Professor in 1971 and the James A. Weeks Professor in 1974. I came to Cornell in response to an unsolicited offer I received while at CERN; I accepted the offer because Cornell was a good university, was out in the country and was reputed to have a good folk dancing group, folk-dancing being a hobby I had taken up as a graduate student.

I have remained at Cornell ever since, except for leaves and summer visits: I spent the 1969–1970 academic year at the Stanford Linear Accelerator Center, the spring of 1972 at the Institute for Advanced Study in Princeton, the fall of 1976 at the California Institute of Technology as a Fairchild Scholar, and the academic year 1979–80 at the IBM Zurich Laboratory.

In 1975 I met Alison Brown and in 1982 we were married. She works for Cornell Computer Services. Together with Douglas Von Houweling, then Director of Academic Computing and Geoffrey Chester of the Physics Department we initiated a computing support project based on a Floating Point Systems Array Processor. I helped write the initial Fortran Compiler for the Array Processor. Since that time I have (aside from using the array processor myself) been studying the role of large scale scientific computing in science and technology and the organizational problems connected with scientific computing. At the present time I am trying to win acceptance for a program of support for scientific computing in universities from industry and government.

I have benefitted enormously from the high quality and selfless cooperation of researchers at Cornell, in the elementary particle group and in materials research; for my research in the 1960's I was especially indebted to Michael Fisher and Ben Widom.

One other hobby of mine has been playing the oboe but I have not kept this up after 1969.

The home base for my research has been elementary particle theory, and I have made several contributions to this subject: a short distance expansion for operator products presented in an unpublished preprint in 1964 and a published paper in 1969; a discussion of how the renormalization group might apply to strong interactions, in which I discussed all possibilities except the one (asymptotic freedom) now believed to be correct; the formulation of the gauge theory in 1974 (discovered independently by Polyakov), and the discovery that the strong coupling limit of the lattice theory exhibits quark confinement. I am currently interested in trying to solve Quantum Chromodynamics (the theory of quarks) using a combination of renormalization group ideas and computer simulation.

I am also interested in trying to unlock the potential of the renormalization group approach in other areas of classical and modern physics. I have continued to work on statistical mechanics (specifically, the Monte Carlo Renormalization Group, applied to the three dimensional Ising model) as part of this effort.

(*added in 1991*): Wilson became the Director of the Center for Theory and Simulation in Science and Engineering (Cornell Theory Center) — one of five national supercomputer centers created by the National Science Foundation in 1985. In 1988, he moved to The Ohio State University's Department of Physics where he became the Hazel C. Youngberg Trustees Distinguished Professor. He is now heavily engaged in educational reform as a Co-Principal Investigator on Ohio's Project Discovery, one of the National Science Foundation's Statewide Systemic Initiatives.

He was elected to the National Academy of Sciences in 1975, the American Academy of Arts and Sciences in 1975, and the American Philosophical Society in 1984.

THE RENORMALIZATION GROUP AND CRITICAL PHENOMENA

Nobel lecture, 8 December 1982

by

KENNETH G. WILSON

Laboratory of Nuclear Studies, Cornell University, Ithaca, New York 14853

I. *Introduction*

This paper has three parts. The first part is a simplified presentation of the basic ideas of the renormalization group and the ε expansion applied to critical phenomena, following roughly a summary exposition given in 1972[1]. The second part is an account of the history (as I remember it) of work leading up to the papers in 1971—1972 on the renormalization group. Finally, some of the developments since 1971 will be summarized, and an assessment for the future given.

II. *Many Length Scales and the Renormalization Group*

There are a number of problems in science which have, as a common characteristic, that complex microscopic behavior underlies macroscopic effects.

In simple cases the microscopic fluctuations average out when larger scales are considered, and the averaged quantities satisfy classical continuum equations. Hydrodynamics is a standard example of this where atomic fluctuations average out and the classical hydrodynamic equations emerge. Unfortunately, there is a much more difficult class of problems where fluctuations persist out to macroscopic wavelengths, and fluctuations on all intermediate length scales are important too.

In this last category are the problems of fully developed turbulent fluid flow, critical phenomena, and elementary particle physics. The problem of magnetic impurities in non-magnetic metals (the Kondo problem) turns out also to be in this category.

In fully developed turbulence in the atmosphere, global air circulation becomes unstable, leading to eddies on a scale of thousands of miles. These eddies break down into smaller eddies, which in turn break down, until chaotic motions on all length scales down to millimeters have been excited. On the scale of millimeters, viscosity damps the turbulent fluctuations and no smaller scales are important until atomic scales are reached.[2]

In quantum field theory, "elementary" particles like electrons, photons, protons and neutrons turn out to have composite internal structure on all size scales down to 0. At least this is the prediction of quantum field theory. It is hard to make observations of this small distance structure directly; instead the particle scattering cross sections that experimentalists measure must be interpreted using quantum field theory. Without the internal structure that apperars in the theory, the predictions of quantum field theory would disagree with the experimental findings.[3]

A critical point is a special example of a phase transition. Consider, for example, the water-steam transition. Suppose the water and steam are placed under pressure, always at the boiling temperature. At the critical point: a pressure of 218 Atm and temperature of 374°C,[4] the distinction between water and steam disappears, and the whole boiling phenomenon vanishes. The principal distinction between water and steam is that they have different densities. As the pressure and temperature approach their critical values, the difference in density between water and steam goes to zero. At the critical point one finds bubbles of steam and drops of water intermixed at all size scales from macroscopic, visible sizes down to atomic scales. Away from the critical point, surface tension makes small drops or bubbles unstable; but as water and steam become indistinguishable at the critical point, the surface tension between the two phases vanishes. In particular, drops and bubbles near micron sizes cause strong light scattering, called "critical opalescence", and the water and steam become milky.

In the Kondo effect, electrons of all wavelengths from atomic wavelengths up to very much larger scales, all in the conduction band of a metal, interact with the magnetic moment of each impurity in the metal.[5]

Theorists have difficulties with these problems because they involve very many coupled degrees of freedom. It takes many variables to characterize a turbulent flow or the state of a fluid near the critical point. Analytic methods are most effective when functions of only one variable (one degree of freedom) are involved. Some extremely clever transformations have enabled special cases of the problems mentioned above to be rewritten in terms of independent degrees of freedom which could be solved analytically. These special examples include Onsager's solution of the two dimensional Ising model of a critical point,[6] the solution of Andrei and Wiegmann of the Kondo problem,[7] the solution of the Thirring model of a quantum field theory,[8] and the simple solutions of noninteracting quantum fields. These are however only special cases; the entire problem of fully developed turbulence, many problems in critical phenomena and virtually all examples of strongly coupled quantum fields have defeated analytic techniques up till now.

Computers can extend the capabilities of theorists, but even numerical computer methods are limited in the number of degrees of freedom that are practical. Normal methods of numerical integration fail beyond only 5 to 10 integration variables; partial differential equations likewise become extremely difficult beyond 3 or so independent variables. Monte Carlo and statistical averaging methods can treat some cases of thousands or even millions of variables but the slow convergence of these methods versus computing time used is a perpetual hassle. An atmospheric flow simulation covering all length scales of turbulence would require a grid with millimeter spacing covering thousands of miles horizontally and tens of miles vertically: the total number of grid points would be of order 10^{25}, far beyond the capabilities of any present or conceivable computer.

The "renormalization group" approach is a strategy for dealing with problems involving many length scales. The strategy is to tackle the problem in steps, one step for each length scale. In the case of critical phenomena, the problem, technically, is to carry out statistical averages over thermal fluctuations on all size scales. The renormalization group approach is to integrate out the fluctuations in sequence starting with fluctuations on an atomic scale and then moving to successively larger scales until fluctuations on all scales have been averaged out.

To illustrate the renormalization group ideas the case of critical phenomena will be discussed in more detail. First the mean field theory of Landau will be described, and important questions defined. The renormalization group will be presented as an improvment to Landau's theory.

The Curie point of a ferromagnet will be used as a specific example of a critical point. Below the Curie temperature, an ideal ferromagnet exhibits spontaneous magnetization in the absence of an external magnetic field; the direction of the magnetization depends on the history of the magnet. Above the Curie temperature T_c, there is no spontaneous magnetization. Figure 1 shows a typical plot of the spontaneous magnetization versus temperature. Just below the Curie temperature the magnetization is observed to behave as $(T_c-T)^\beta$, where β is an exponent somewhere near 1/3 (in three dimensions).[9, 10]

Magnetism is caused at the atomic level by unpaired electrons with magnetic moments, and in a ferromagnet, a pair of nearby electrons with moments aligned has a lower energy than if the moments are anti-aligned.[10] At high temperatures, thermal fluctuations prevent magnetic order. As the temperature is reduced towards the Curie temperature, alignment of one moment causes preferential alignment out to a considerable distance called the correlation length ξ. At the Curie temperature, the correlation length becomes infinite, marking the onset of preferential alignment of the entire system. Just above T_c the correlation length is found to behave as $(T-T_c)^{-\upsilon}$, where υ is about 2/3 (in three dimensions).[11]

A simple statistical mechanical model of a ferromagnet involves a Hamiltonian which is a sum over nearest neighbor moment pairs with different energies for the aligned and antialigned case. In the simplest case, the moments are allowed only to be positive or negative along a fixed spatial axis; the resulting model is called the Ising model.[12]

The formal prescription for determining the properties of this model is to compute the partition function Z, which is the sum of the Boltzmann factor $\exp(-H/kT)$ over all configurations of the magnetic moments, where k is Boltzmann's constant. The free energy F is proportional to the negative logarithm of Z.

The Boltzmann factor $\exp(-H/kT)$ is an analytic function of T near T_c, in fact for all T except T = 0. A sum of analytic functions is also analytic. Thus it is puzzling that magnets (including the Ising model) show complex non-analytic behavior at $T = T_c$. The true non-analytic behavior occurs only in the thermodynamic limit of a ferromagnet of infinite size; in this limit there are an infinite number of configurations and there are no analyticity theorems for the infinite sums appearing in this limit. However, it is difficult to understand how even an infinite sum can give highly non-analytic behavior. A major challenge has been to show how the non-analyticity develops.

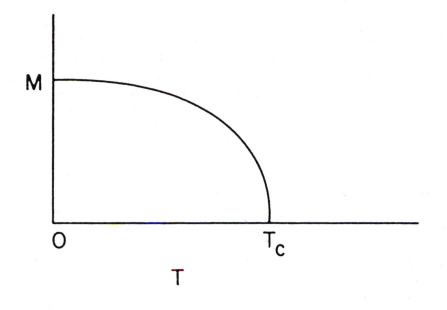

Landau's proposal[13] was that if only configurations with a given magnetization density M are considered then the free energy is analytic in M. For small
M, the form of the free energy (to fourth order in M) is (from the analyticity
assumption)

$$F = V\{RM^2+UM^4\} \tag{1}$$

where V is the volume of the magnet and R and U are temperature-dependent
constants. (A constant term independent of M has been omitted). In the absence
of an external magnetic field, the free energy cannot depend on the sign of M,
hence only even powers of M occur. The true free energy is the minimum of
F over all possible values of M. In Landau's theory, R is 0 at the critical
temperature, and U must be positive so that the minimum of F occurs at
M = 0 when at the critical temperature. The minimum of F continues to be at
M = 0 if R is positive: this corresponds to temperatures above critical. If R is
negative the minimum occurs for non-zero M, namely the M value satisfying

$$0 = \frac{\partial F}{\partial M} = (2RM+4UM^3) \tag{2}$$

or

$$M = \sqrt{-R/(2U)} \tag{3}$$

This corresponds to temperatures below critical.

   Along with the analyticity of the free energy in M, Landau assumed analyticity
in T, namely that R and U are analytic functions of T. Near T_c this means that

to a first approximation, U is a constant and R (which vanishes at T_c) is proportional to $T-T_c$. (It is assumed that dR/dT does not vanish at T_c). Then, below T_c, the magnetization behaves as

$$M \propto (T_c-T)^{1/2} \tag{4}$$

i.e. the exponent β is ½ which disagrees with the evidence, experimental and theoretical, that β is about $1/3$.[9]

Landau's theory allows for a slowly varying space-dependent magnetization. The free energy for this case takes the Landau-Ginzburg form[14]

$$F = \int d^3x\{[\nabla M(x)]^2+RM^2(x)+UM^4(x)-B(x)M(x)\} \tag{5}$$

where $B(x)$ is the external magnetic field. The gradient term is the leading term in an expansion involving arbitrarily many gradients as well as arbitrarily high powers of M. For slowly varying fields $M(x)$ higher powers of gradients are small and are neglected. (Normally the $\nabla M^2(x)$ term has a constant coefficient — in this paper this coefficient is arbitrarily set to 1). One use of this generalized free energy is to compute the correlation length ξ above T_c. For this purpose let $B(x)$ be very small δ function localized at $x = 0$. The U term in F can be neglected, and the magnetization which minimize the free energy satisfies

$$-\nabla^2 M(x)+RM(x) = B\delta^3(x) \tag{6}$$

The solution $M(x)$ is

$$M(x) \propto Be^{-\sqrt{R}|\vec{x}|}/|\vec{x}| \tag{7}$$

and the correlation length can be read off to be

$$\xi \propto 1/\sqrt{R} \tag{8}$$

Hence near T_c, ξ is predicted to behave as $(T-T_c)^{-1/2}$, which again disagrees with experimental and theoretical evidence.[11]

The Landau theory assumes implicitly that analyticity is maintained as all space-dependent fluctuations are averaged out. The loss of analyticity arises only when averaging over the values of the overall average magnetization M. It is this overall averaging, over $e^{-F/kT}$, which leads to the rule that F must be minimized over M, and the subsequent non-analytic formula (4) for M. To be precise, if the volume of the magnet is finite, $e^{-F/kT}$ must be integrated over M, with analytic results. It is only in the thermodynamic limit $V \rightarrow \infty$ that the average of $e^{-F/kT}$ is constructed by minimizing F with respect to M, and the nonanalyticity of Eqn. (4) occurs.

The Landau theory has the same physical motivation as hydrodynamics. Landau assumes that only fluctuations on an atomic scale matter. Once these have been averaged out the magnetization $M(x)$ becomes a continuum, continous function which fluctuates only in response to external space-dependent stimuli. $M(x)$ (or, if it is a constant, M) is then determined by a simple classical equation. Near the critical point the correlation function is itself the solution of the classical equation (6).

In a world with greater than four dimensions, the Landau picture is correct.[15] Four dimensions is the dividing line — below four dimensions, fluctuations on all scales up to the correlation length are important and Landau theory breaks down,[16] as will be shown below. An earlier criterion by Ginzburg[15] also would predict that four dimensions is the dividing line.

The role of long wavelength fluctuations is very much easier to work out near four dimensions where their effects are small. This is the only case that will be discussed here. Only the effects of wavelengths long compared to atomic scales will be discussed, and it will be assumed that only modest corrections to the Landau theory are required. For a more careful discussion see ref. 17.

Once the atomic scale fluctuations have been averaged out, the magnetization is a function $M(x)$ on a continuum, as in Landau theory. However, long wavelength fluctuations are still present in $M(x)$ — they have not been averaged out — and the allowed forms of $M(x)$ must be stated with care. To be precise, suppose fluctuations with wavelengths $< 2\pi L$ have been averaged out, where L is a length somewhat larger than atomic dimensions. Then $M(x)$ can contain only Fourier modes with wavelengths $> 2\pi L$. This requirement written out, means

$$M(\vec{x}) = \int_{\vec{k}} e^{i\vec{k} \cdot \vec{x}} M_{\vec{k}} \tag{9}$$

where the integral over $\vec{k}$ means $(2\pi)^{-d} \int d^d k$, d is the number of space dimensions, and the limit on wavelengths means that the integration over $\vec{k}$ is restricted to values of $\vec{k}$ with $|\vec{k}| < L^{-1}$.

Averaging over long wavelength fluctuations now reduces to integrating over the variables $M_{\vec{k}}$, for all $|\vec{k}| < L^{-1}$. There are many such variables; normally this would lead to many coupled integrals to carry out, a hopeless task. Considerable simplifications will be made below in order to carry out these integrations.

We need an integrand for these integrations. The integrand is a constrained sum of the Boltzmann factor over all atomic configurations. The constraints are that all $M_{\vec{k}}$ for $|\vec{k}| < L^{-1}$ are held fixed. This is a generalization of the constrained sum in the Landau theory; the difference is that in the Landau theory only the average magnetization is held fixed. The result of the constrained sum will be written e^{-F}, similarly to Landau theory except for convenience the exponent is written F rather than F/kT (i.e. the factor 1/kT is absorbed into an unconventional definition of F). The exponent F depends on the magnetization function $M(x)$ of Eq. (9). We shall assume Landau's analysis is still valid for the form of F, namely F is given by Eq. (5). However, the importance of long wavelength fluctuations means that the parameters R and U depend on L. Thus F should be denoted F_L:

$$F_L = \int d^d x \{ (\nabla M)^2 (x) + R_L M^2 (x) + U_L M^4 (x) \} \tag{10}$$

(in the absence of any external field) (in the simplified analysis presented here, the coefficient of $\nabla M^2 (x)$ is unchanged at 1). The assumption will be reviewed later.

The L dependence of R_L and U_L will be determined shortly. However, the breakdown of analyticity at the critical point is a simple consequence of this L dependence. The L dependence persists only out to the correlation length ξ: fluctuations with wavelengths $> \xi$ will be seen to be always negligible. Once all wavelengths of fluctuations out to $L \sim \xi$ have been integrated out, one can use the Landau theory; this means (roughly speaking) substituting R_ξ and U_ξ in the formulae (4) and (8) for the spontaneous magnetization and the correlation length. Since ξ is itself non-analytic in T at $T = T_c$ the dependence of R_ξ and U_ξ on ξ introduces new complexities at the critical point. Details will be discussed shortly.

In order to study the effects of fluctuations, only a single wavelength scale will be considered; this is the basic step in the renormalization group method. To be precise, consider only fluctuations with wavelengths lying in an infinitesimal interval L to $L+\delta L$. To average over these wavelengths of fluctuations one starts with the Boltzmann factor e^{-F_L} where the wavelengths between L and $L+\delta L$ are still present in $M(x)$, and then averages over fluctuations in $M(x)$ with wavelengths between L and $L+\delta L$. The result of these fluctuation averages is a free energy $F_{L+\delta L}$ for a magnetization function (which will be denoted $M_H(x)$) with wavelengths $> L+\delta L$ only. The Fourier components of $M_H(x)$ are the same $\vec{k}$ that appear in $M(x)$ except that $|\vec{k}|$ is now restricted to be less than $1/(L+\delta L)$.

The next step is to count the number of integration variables $M_{\vec{k}}$ with $|\vec{k}|$ lying between $1/L$ and $1/(L+\delta L)$. To make this count it is necessary to consider a finite system in a volume V. Then the number of degrees of freedom with wavelengths between $2\pi L$ and $2\pi(L+\delta L)$ is given by the corresponding phase space volume, namely the product of k space and position space volumes. This product is (apart from constant factors like π, etc.) $L^{-(d+1)}V\delta L$.

It is convenient to choose the integration variables not to be the $M_{\vec{k}}$ themselves but linear combinations which correspond to localized wave packets instead of plane waves. That is, the difference $M_H(x) - M(x)$ should be expanded in a set of wave packet functions $\psi_n(x)$, each of which has momenta only in the range $1/L$ to $1/(L+\delta L)$, but which is localized in x space as much as possible. Since each function $\psi_n(x)$ must (by the uncertainty principle) fill unit volume in phase space, the position space volume for each $\psi_n(x)$ is

$$\delta V = L^{d+1}/\delta L \tag{11}$$

and there are $V/\delta V$ wavefunctions $\psi_n(x)$. We can write

$$M(x) = M_H(x) + \sum_n m_n \psi_n(x) \tag{12}$$

and the integrations to be performed are integrations over the coefficients m_n.

Because of the local nature of the Landau-Ginzburg free energy, it will be assumed that the overlap of the different wavefunctions ψ_n can be neglected. Then each m_n integration can be treated separately, and only a single such integration will be discussed here. For this single integration, the form of $M(x)$ can be written

$$M(x) = M_H(x) + m\psi(x) \tag{13}$$

since only one term from the sum over n contributes within the spatial volume occupied by the wavefunction $\psi(x)$.

The other simplification that will be made is to treat $M_H(x)$ as if it were a constant over the volume occupied by $\psi(x)$. In other words the very long wavelengths in $M_H(x)$ are emphasized relative to wavelengths close to L.

The calculation to be performed is to compute

$$e^{-F_{L+\delta L}[M_H]} = \int_{-\infty}^{\infty} dm \; e^{-F_L[M_H + m\psi]} \tag{14}$$

where $F_{L+\delta L}$ and F_L involve integration only over the volume occupied by $\psi(x)$. In expanding out $F_L[M_H + m\psi]$ the following simplifications will be made. First, all terms linear in $\psi(x)$ are presumed to integrate to 0 in the x integration defining F_L. Terms of third order and higher in ψ are also neglected. The function $\psi(x)$ is presumed to be normalized so that

$$\int d^d x \; \psi^2(x) = 1 \tag{15}$$

and due to the limited range of wavelengths in $\psi(x)$, there results

$$\int [\nabla \psi(x)]^2 d^d x \simeq 1/L^2 \tag{16}$$

The result of these simplifications is that the integral becomes

$$e^{-F_{L+\delta L}[M_H]} = e^{-F_L[M_H]} \int_{-\infty}^{\infty} dm \; \exp\{(R_L + \frac{1}{L^2}) \; m^2 + 6U_L M_H^2 m^2\} \tag{17}$$

or

$$F_{L+\delta L}[M_H] = F_L\{M_H\} + \frac{1}{2}\ell n(\frac{1}{L^2} + R_L + 6U_L M_H^2) \tag{18}$$

The logarithm must be rewritten as an integral over the volume occupied by $\psi(x)$; this integral can then be extended to an integral over the entire volume V when the contributions from all other m_n integrations are included. Also the logarithm must be expanded in powers of M_H; only the M_H^2 and M_H^4 terms will be kept. Further it will be assumed that R_L changes slowly with L. When L is at the correlation length ξ, $1/L^2$ and R_L are equal (as already argued) so that for values of L intermediate between atomic sizes and the correlation length, R_L is small compared to $1/L^2$. Expanding the logarithm in powers of $R_L + 6U_L M_H^2$, to second order (to obtain an M_H^4 term) gives (cf. Eq. (11)):

$$\frac{1}{2}\ell n\left(\frac{1}{L^2} + R_L + 6U_L M_H^2\right) = \text{terms independent of } M_H$$
$$+ (\delta V)(\delta L)L^{-d-1}\{3U_L M_H^2 L^2 - 9U_L^2 M_H^4 L^4 - 3R_L U_L M_H^2 L^4\} \tag{19}$$

One can rewrite δV as an integral over the volume δV. There results the equations

$$R_{L+\delta L} = R_L + (3U_L L^{1-d} - 3R_L U_L L^{3-d})\delta L \tag{20}$$

$$U_{L+\delta L} = U_L - 9U_L^2 L^{3-d}\delta L \tag{21}$$

or

$$L\frac{dR_L}{dL} = 3L^{2-d}U_L - 3R_L U_L \cdot L^{4-d} \tag{22}$$

$$L\frac{dU_L}{dL} = -9U_L^2 L^{4-d} \tag{23}$$

These equations are valid only for $L<\xi$; for $L>\xi$ there is very little further change in R_L or U_L, due to the switchover in the logarithm caused by the dominance of R_L rather than $1/L^2$. If d is greater than 4, it can be seen that R_L and U_L are constant for large L, as expected in the Landau theory. For example, if one assumes R_L and U_L are constant for large L it is easily seen that integration of (22) and (23) only gives negative powers of L. For $d<4$ the solutions are not constant. Instead, U_L behaves for sufficiently large L as

$$U_L \simeq \frac{(4-d)}{9}L^{d-4} \tag{24}$$

(which is easily seen to be a solution of (23)), R_L satisfies the equation

$$\frac{dR_L}{dL}+\frac{(4-d)}{3L}R_L = \frac{(4-d)}{3}L^{-3} \tag{25}$$

whose solution is

$$R_L = cL^{(d-4)/3}-\frac{(4-d)}{3}\frac{1}{2-(4-d)/3}L^{-2} \tag{26}$$

where c is related to the value of R_L at some initial value of L. For large enough L, the L^{-2} term can be neglected.

The parameter c should be analytic in temperature, in fact proportional to $T-T_c$. Hence, for large L

$$R_L \propto L^{(d-4)/3}(T-T_c) \tag{27}$$

which is analytic in T for fixed L. However the equation for ξ is

$$\xi \propto R_\xi^{-1/2} = (T-T_c)^{-1/2}\xi^{(4-d)/6} \tag{28}$$

Let

$$\varepsilon = 4-d \tag{29}$$

then the correlation length exponent is

$$\upsilon = \frac{1}{2}\frac{1}{1-\varepsilon/6} \tag{30}$$

which gives $\upsilon = 0.6$ in 3 dimensions. Similarly, the spontaneous magnetization below T_c behaves as $(R_\xi/U_\xi)^{1/2}$ giving

$$\beta = \frac{1}{2}-\frac{\varepsilon}{3}\frac{1}{1-\varepsilon/6} \tag{31}$$

These computations give an indication of how non-trivial values can be obtained for β and υ. The formulae derived here are not exact, due to the severe simplifications made, but at least they show that β and υ do not have to be $\frac{1}{2}$ and in fact can have a complicated dependence on the dimension d.

A correct treatment is much more complex. Once $M_H(x)$ is not treated as a constant, one could imagine expanding $M_H(x)$ in a Taylor's series about its value at some central location x_0 relative to the location of the wavefunction $\psi(x)$, thus bringing in gradients of M_H. In addition, higher order terms in the expansion of the logarithm give higher powers of M_H. All this leads to a more complex form for the free energy functional F_L with more gradient terms and more powers of M_H. The whole idea of the expansion in powers of M_H and powers of gradients can in fact be called into question. The fluctuations have an intrinsic size (i.e., m^2 has a size $\sim L^2$ as a consequence of the form of the integrand in Eq. 17) and it is not obvious that in the presence of these fluctuations, M is small. Since arbitrary wavelengths of fluctuations are important the function M is not sufficiently slowly varying to justify an expansion in gradients either. This means that $F_L[M]$ could be an arbitrarily complicated function of M, an expression it is hard to write down, with thousands of parameters, instead of the simple Landau-Ginzburg form with only two parameters R_L and U_L.

Fortunately, the problem simplifies near 4 dimensions, due to the small magnitude of U_L, which is proportional to $\varepsilon = 4-d$. All the complications neglected above arise only to second order or higher in an expansion in U_L which means second order or higher in ε. The computations described here are exact to order ε. See Ref. 17.

The renormalization group approach that was defined in 1971 embraces both practical approximations leading to actual computations and a formalism.[17] The full formalism cannot be discussed here but the central idea of "fixed points" can be illustrated.

As the fluctuations on each length scale are integrated out a new free energy functional $F_{L+\delta L}$ is generated from the previous functional F_L. This process is repeated many times. If F_L and $F_{L+\delta L}$ are expressed in dimensionless form, then one finds that the transformation leading from F_L to $F_{L+\delta L}$ is repeated in identical form many times. (The transformation group thus generated is called the "renormalization group"). As L becomes large the free energy F_L approaches a fixed point of the transformation, and thereby becomes independent of details of the system at the atomic level. This leads to an explanation of the universality[18] of critical behavior for different kinds of systems at the atomic level. Liquid-gas transitions, magnetic transitions, alloy transitions, etc. all show the same critical exponents experimentally; theoretically this can be understood from the hypothesis that the same "fixed point" interaction describes all these systems.

To demonstrate the fixed point form of the free energy functional, it must be put into dimensionless form. Lengths need to be expressed in units of L, and M, R_L, and U_L rewritten in dimensionless form. These changes are easily determined: write

$$x = Ly \tag{32}$$

$$M(x) = L^{1-d/2}m(y) \tag{33}$$

$$R_L = 1/L^2 r_L \tag{34}$$

$$U_L = L^{d-4}u_L \tag{35}$$

$$F_L = \int d^d y \{(\nabla m)^2 + r_L m^2(y) + u_L m^4(y)\} \tag{36}$$

The asymptotic solution for the dimensionless parameters r_L and u_L is

$$r_L = cL^{2-\varepsilon/3} - \frac{\varepsilon}{3}\frac{1}{2-\varepsilon/3} \tag{37}$$

$$u_L = \frac{\varepsilon}{9} \tag{38}$$

Apart from the term in r_L, these dimensionless parameters are independent of L, denoting a free energy form which is also independent of L. The c term designates an instability of the fixed point, namely a departure from the fixed point which grows as L increases. The fixed point is reached only if the thermodynamic system is at the critical temperature for which c vanishes; any departure from the critical temperature triggers the instability.

For further analysis of the renormalization group formalism and its relation to general ideas about critical behavior, see e.g. ref. 17.

III. *Some History Prior to 1971*

The first description of a critical point was the description of the liquid-vapor critical point developed by Van der Waals,[19] developed over a century ago following experiments of Andrews.[19] Then Weiss provided a description of the Curie point in a magnet.[20] Both the Van der Waals and Weiss theories were special cases of Landau's mean field theory.[13] Even before 1900, experiments indicated discrepancies with mean field theory; in particular the experiments indicated that β was closer to 1/3 than 1/2.[19] In 1944, Onsager[6] published his famous solution to the two dimensional Ising model,[12] which explicity violated the mean field predictions. Onsager obtained $\upsilon = 1$ instead of the mean field prediction $\upsilon = 1/2$, for example. In the 1950's, Domb, Sykes, Fisher and others[21] studied simple models of critical phenomena in three dimensions with the help of high temperature series expansions carried to very high order, exacting critical point exponents by various extrapolation methods. They obtained exponents in disagreement with mean field theory but in reasonable agreement with experiment. Throughout the sixties a major experimental effort pinned down critical exponents and more generally provided a solid experimental basis for theoretical studies going beyond mean field theory. Experimentalists such as Voronel; Fairbanks, Buckingham, and Keller; Heller and Benedek; Ho and Litster, Kouvel and Rodbell, and Comly; Sengers; Lorentzen; Als-Nielsen and Dietrich; Birgeneau and Shirane; Rice; Chu; Teaney; Moldover; Wolf and Ahlers all contributed to this development, with M. Green, Fisher, Widom, and Kadanoff providing major coordination efforts.[22] Theoretically, Widom[23] proposed a scaling law for the equation of state near the critical point that accommodated non-mean field exponents and predicted relations among them. The full set of scaling hypotheses were developed by Essam and Fisher, Domb and Hunter, Kadanoff, and Patashinskii, and Pokrovskii.[24] See also the inequalities of Rushbrooke[25] and Griffith.[26]

My own work began in quantum field theory, not statistical mechanics. A convenient starting point is the development of renormalization theory by Bethe, Schwinger, Tomonaga, Feynman, Dyson and others[27] in the late 1940's. The first discussion of the "renormalization" group appeared in a paper by Stueckelberg and Petermann,[28] published in 1953.

In 1954 Murray Gell-Mann and Francis Low published a paper entitled "Quantum Electrodynamics at Small Distances"[29] which was the principal inspiration for my own work prior to Kadanoff's formulation[30] of the scaling hypothesis for critical phenomena in 1966.

Following the definition of Quantum Electrodynamics (QED) in the 1930's by Dirac, Fermi, Heisenberg, Pauli, Jordan, Wigner, et al.[27], the solution of QED was worked out as perturbation series in e_0, the "bare charge" of QED. The QED Lagrangian (or Hamiltonian) contains two parameters: e_0 and m_0, the latter being the "bare" mass of the electron. As stated in the introduction in QED the physical electron and photon have composite structure. In consequence of this structure the measured electric charge e and electron mass m are not identical to e_0 and m_0 but rather are given by perturbation expansions in powers of e_0. Only in lowest order does one find $e = e_0$ and $m = m_0$. Unfortunately, it was found in the 30's that higher order corrections in the series for e and m are all infinite, due to integrations over momentum that diverge in the large momentum (or small distance) limit.[27]

In the late 1940's renormalization theory was developed, which showed that the divergences of Quantum Electrodynamics could all be eliminated if a change of parametrization was made from the Lagrangian parameters e_0 and m_0 to the measurable quantities e and m, and if at the same time the electron and electromagnetic fields appearing in the Lagrangian were rescaled to insure that observable matrix elements (especially of the electromagnetic field) are finite.[27]

There are many reparametrizations of Quantum Electrodynamics that eliminate the divergences but use different finite quantities than e and m to replace e_0 and m_0. Stueckelberg and Petermann observed that transformation groups could be defined which relate different reparametrizations — they called these groups "groupes de normalization" which is translated "renormalization group". The Gell-Mann and Low paper,[29] one year later but independently, presented a much deeper study of the significance of the ambiguity in the choice of reparametrization and the renormalization group connecting the difference choices of reparametrization. Gell-Mann and Low emphasized that e, measured in classical experiments, is a property of the very long distance behavior of QED (for example it can be measured using pith balls separated by centimeters, whereas the natural scale of QED is the Compton wavelength of the electron, $\sim 10^{-11}$ cm). Gell-Mann and Low showed that a family of alternative parameters e_λ could be introduced, any one of which could be used in place of e to replace e_0. The parameter e_λ is related to the behavior of QED at an arbitrary momentum scale λ instead of at very low momenta for which e is appropriate.

The family of parameters e_λ introduced by Gell-Mann and Low interpolate between the physical charge e and the bare charge e_0, namely e is obtained as the low momentum ($\lambda \to 0$) limit of e_λ and e_0 is obtained as the high momentum ($\lambda \to \infty$) limit of e_λ.

Gell-Mann and Low found that e_λ^2 obeys a differential equation, of the form

$$\lambda^2 d(e_\lambda^2)/d(\lambda^2) = \psi\ (e_\lambda^2,\ m^2/\lambda^2) \tag{39}$$

where the ψ function has a simple power series expansion with non-divergent coefficients independently of the value of λ, in fact as $\lambda \to \infty$, ψ becomes a function of e_λ^2 alone. This equation is the forerunner of my own renormalization group equations such as (22) and (23).

The main observation of Gell-Mann and Low was that despite the ordinary nature of the differential equation, Eq. (38), the solution was *not* ordinary, and in fact predicts that the physical charge e has divergences when expanded in powers of e_0, or vice versa. More generally, if e_λ is expanded in powers of $e_{\lambda'}$, the higher order coefficients contain powers of $\ell n(\lambda^2/\lambda'^2)$, and these coefficients diverge if either λ or λ' go to infinity, and are very large if λ^2/λ'^2 is either very large or very small.

Furthermore, Gell-Mann and Low argued that, as a consequence of Eqn. (38), e_0 must have a fixed value independently of the value of e; the fixed value of e_0 could be either finite or infinite.

When I entered graduate school at California Institute of Technology, in 1956, the default for the most promising students was to enter elementary particle theory, the field in which Murray Gell-Mann, Richard Feynman, and Jon Mathews were all engaged. I rebelled briefly against this default, spending a summer at the General Atomic Corp. working for Marshall Rosenbluth on plasma physics and talking with S. Chandresekhar who was also at General Atomic for the summer. After about a month of work I was ordered to write up my results, as a result of which I swore to myself that I would choose a subject for research where it would take at least five years before I had anything worth writing about. Elementary particle theory seemed to offer the best prospects of meeting this criterion and I asked Murray for a problem to work on. He first suggested a topic in weak interactions of strongly interacting particles (K mesons, etc.) After a few months I got disgusted with trying to circumvent totally unknown consequences of strong interactions, and asked Murray to find me a problem dealing with strong interactions directly, since they seemed to be the bottleneck. Murray suggested I study K meson-nucleon scattering using the Low equation in the one meson approximation. I wasn't very impressed with the methods then in use to solve the Low equation, so I wound up fiddling with various methods to solve the simpler case of pion-nucleon scattering. Despite the fact that the one meson approximation was valid, if at all, only for low energies, I studied the high energy limit, and found that I could perform a "leading logarithms" sum very reminiscent of a very mysterious chapter in Bogoliubov and Shirkov's field theory text[31]; the chapter was on the renormalization group.

In 1960 I turned in a thesis to Cal Tech containing a mish-mash of curious calculations. I was already a Junior Fellow at Harvard. In 1962 I went to CERN for a year. During this period (1960—1963) I partly followed the fashions of the time. Fixed source meson theory (the basis for the Low equation) died, to be replaced by S matrix theory. I reinvented the "strip approximation" (Ter-Martirosyan had invented it first[32]) and studied the Amati-Fubini-Stanghellini theory of multiple production.[33] I was attentive at seminars (the only period of my life when I was willing to stay fully awake in them) and I also pursued back waters such as the strong coupling approximation to fixed source meson theory.[34]

By 1963 it was clear that the only subject I wanted to pursue was quantum field theory applied to strong interactions. I rejected S matrix theory because the equations of S matrix theory, even if one could write them down, were too complicated and inelegant to be a theory; in contrast the existence of a strong coupling approximation as well as a weak coupling approximation to fixed source meson theory helped me believe that quantum field theory might make sense. As far as strong interactions were concerned, all that one could say was that the theories one could write down, such as pseudoscalar meson theory, were obviously wrong. No one had any idea of a theory that could be correct. One could make these statements even though no one had the foggiest notion how to solve these theories in the strong coupling domain.

My very strong desire to work in quantum field did not seem likely to lead to quick publications; but I had already found out that I seemed to be able to get jobs even if I didn't publish anything so I did not worry about 'publish or perish' questions.

There was very little I could do in quantum field theory — there were very few people working in the subject, very few problems open for study. In the period 1963—1966 I had to clutch at straws. I thought about the "ξ-limiting" process of Lee and Yang.[35] I spent a major effort disproving Ken Johnson's claims[36] that he could define quantum electrodynamics for arbitrarily small e_0, in total contradiction to the result of Gell-Mann and Low. I listened to K. Hepp and others describe their results in axiomatic field theory[37]; I didn't understand what they said in detail but I got the message that I should think in position space rather than momentum space. I translated some of the work I had done on Feynman diagrams with some very large momenta (to disprove Ken Johnson's ideas) into position space and arrived at a short distance expansion for products of quantum field operators. I described a set of rules for this expansion in a preprint in 1964. I submitted the paper for publication; the referee suggested that the solution of the Thirring model might illustrate this expansion. Unfortunately, when I checked out the Thirring model, I found that while indeed there was a short distance expansion for the Thirring model,[38] my rules for how the coefficient functions behaved were all wrong, in the strong coupling domain. I put the preprint aside, awaiting resolution of the problem.

Having learned the fixed source meson theory as a graduate student, I continued to think about it. I applied my analysis of Feynman diagrams for some large momenta, to the fixed source model. I realized that the results I was getting became much clearer if I made a simplification of the fixed source model itself, in which the momentum space continuum was replaced by momentum slices.[39] That is, I rubbed out all momenta except well separated slices, e.g., $1 \leqslant |k| \leqslant 2$, $\Lambda \leqslant |k| \leqslant 2\Lambda$, $\Lambda^2 \leqslant |k| \leqslant 2\Lambda^2$, ..., $\Lambda^n \leqslant |k| \leqslant 2\Lambda^n$, etc. with Λ a large number.

This model could be solved by a perturbation theory very different from the methods previously used in field theory. The energy scales for each slice were very different, namely of order Λ^n for the nth slice. Hence the natural procedure was to treat the Hamiltonian for the largest momentum slice as the unperturbed Hamiltonian, and the terms for all lesser slices as the perturbation. In each slice the Hamiltonian contained both a free meson energy term and an interaction term, so this new perturbation method was neither a weak coupling nor a strong coupling perturbation.

I showed that the effect of this perturbation approach was that if one started with n momentum slices, and selected the ground state of the unperturbed Hamiltonian for the nth slice, one wound up with an effective Hamiltonian for the remaining n-1 slices. This new Hamiltonian was identical to the original Hamiltonian with only n-1 slices kept, except that the meson-nucleon coupling constant g was renormalized (i.e., modified): the modification was a factor involving a non-trivial matrix element of the ground state of the nth-slice Hamiltonian.[39]

This work was a real breakthrough for me. For the first time I had found a natural basis for renormalization group analysis: namely the solution and elimination of one momentum scale from the problem. There was still much to be done: but I was no longer grasping at straws. My ideas about renormalization were now reminiscent of Dyson's analysis of Quantum Electrodynamics.[40] Dyson argued that renormalization in Quantum Electrodynamics should be carried out by solving and eliminating high energies before solving low energies. I studied Dyson's papers carefully but was unable to make much use of his work.[41]

Following this development, I thought very hard about the question "what is a field theory", using the ϕ^4 interaction of a scalar field (identical with the Landau-Ginzburg model of a critical point[14] discussed in my 1971 papers) as an example. Thoughout the '60's I taught quantum mechanics frequently, and I was very impressed by one's ability to understand simple quantum mechanical systems. The first step is a qualitative analysis minimizing the energy (defined by the Hamiltonian) using the uncertainty principle; the second step might be a variational calculation with wavefunctions constructed using the qualitative information from the first step; the final stage (for high accuracy) would be a numerical computation with a computer helping to achieve high precision. I felt that one ought to be able to understand a field theory the same way.

I realized that I had to think about the degrees of freedom that make up a field theory. The problem of solving the ϕ^4 theory was that kinetic term in the Hamiltonian (involving $(\nabla\phi)^2$) was diagonal only in terms of the Fourier components ϕ_k of the field, whereas the ϕ^4 term was diagonal only in terms of the field $\phi(x)$ itself. Therefore I looked for a compromise representation in which both the kinetic term and the interaction term would be at least roughly diagonal. I needed to expand the field $\phi(x)$ in terms of wavefunctions that would have minimum extent in both position space and momentum space, in other words wavefunctions occupying the minimum amount of volume in phase space. The uncertainty principle defines the lower bound for this volume, namely 1, in suitable units. I thought of phase space being divided up into blocks of unit volume. The momentum slice analysis indicated that momentum space should be marked off on a logarithmic scale, i.e. each momentum space volume should correspond to a shell like the slices defined earlier, except that I couldn't leave out any momentum range so the shells had to be e.g. ..., $1 < |k| < 2$, $2 < |k| < 4$, etc. By translational invariance the position space blocks would all be the same size for a given momentum shell, and would define a simple lattice of blocks. The position space blocks would have different sizes for different momentum shells.

When I tried to study this Hamiltonian I didn't get very far. It was clear that the low momentum terms should be a perturbation relative to the high momentum terms but the details of the perturbative treatment became too complicated. Also my analysis was too crude to identify the physics of highly relativistic particles which should be contained in the Hamiltonian of the field theory.[42]

However, I learned from this picture of the Hamiltonian that the Hamiltonian would have to be cutoff at some large but finite value of momentum k in order to make any sense out of it, and that once it was cutoff, I basically had a lattice theory to deal with, the lattice corresponding roughly to the position space blocks for the largest momentum scale. More precisely, the sensible procedure for defining the lattice theory was to define phase space cells covering all of the cutoff momentum space, in which case there would be a single set of position space blocks, which in turn defined a position space lattice on which the field ϕ would be defined. I saw from this that to understand quantum field theories I would have to understand quantum field theories on a lattice.

In thinking and trying out ideas about "what is a field theory" I found it very helpful to demand that a correctly formulated field theory should be soluble by computer, the same way an ordinary differential equation can be solved on a computer, namely with arbitrary accuracy in return for sufficient computing power. It was clear, in the '60's, that no such computing power was available in practice; all that I was able to actually carry out were some simple exercises involving free fields on a finite lattice.

In the summer of 1966 I spent a long time at Aspen. While there I carried out a promise I had made to myself while a graduate student, namely I worked through Onsager's solution of the two dimensional Ising model. I read it in translation, studying the field theoretic form given in Lieb, Mattis and Schultz.[43]

When I entered graduate school, I had carried out the instructions given to me by my father and had knocked on both Murray Gell-Mann's and Feynman's doors, and asked them what they were currently doing. Murray wrote down the partition function for the three dimensional Ising model and said it would be nice if I could solve it (at least that is how I remember the conversation). Feynman's answer was "nothing". Later, Jon Mathews explained some of Feynman's tricks for reproducing the solution for the two dimensional Ising model. I didn't follow what Jon was saying, but that was when I made my promise. Sometime before going to Aspen, I was present when Ben Widom presented his scaling equation of state,[23] in a seminar at Cornell. I was puzzled by the absence of any theoretical basis for the form Widom wrote down; I was at that time completely ignorant of the background in critical phenomena that made Widom's work an important development.

As I worked through the paper of Mattis, Lieb, and Schultz, I realized there should be applications of my renormalization group ideas to critical phenomena, and discussed this with some of the solid state physicists also at Aspen. I was informed that I had been scooped by Leo Kadanoff and should look at his preprint.[30]

Kadanoff's idea was that near the critial point one could think of blocks of magnetic moments, for example containing $2 \times 2 \times 2$ atoms per block, which would act like a single effective moment, and these effective moments would have a simple nearest neighbor interaction like simple models of the original system. The only change would be that the system would have an effective temperature and external magnetic field that might be distinct from the original. More generally the effective moments would exist on a lattice of arbitrary spacing L times the original atomic spacing; Kadanoff's idea was that there would be L-dependent temperature and field variables T_L and h_L and that T_{2L} and h_{2L} would be analytic functions of T_L and h_L. At the critical point, T_L and h_L would have fixed values independent of L. From this hypothesis Kadanoff was able to derive the scaling laws of Widom,[23] Fisher, etc.[24]

I now amalgamated my thinking about field theories on a lattice and critical phenomena. I learned about Euclidean (imaginary time) quantum field theory and the "transfer matrix" method for statistical mechanical models and found there was a close analogy between the two (see Ref. 17). I learned that for a field theory to be relativistic, the corresponding statistical mechanical theory had to have a large correlation length, i.e., be near a critical point. I studied Schiff's strong coupling approximation to the ϕ^4 theory,[44] and found that he had ignored renormalization effects; when these were taken into account the strong coupling expansion was no longer so easy as he claimed. I thought about the implications of the scaling theory of Kadanoff, Widom et al. applied to quantum field theory, along with the scale invariance of the solution of the Thirring model[8] and the discussion of Kastrup and Mack of scale invariance in quantum field theory.[45] These ideas suggested that scale invariance would apply, at least at short distances, but that field operators would have non-trivial scale dimensions corresponding to the non-trivial exponents in critical phenomena. I redid my theory of short distance expansions based on these scaling ideas and published the result.[46] My theory did not seem to fit the main experimental ideas about short distance behavior (coming from Bjorken's and Feynman's analysis[47] of deep inelastic electron scattering) but I only felt confused about this problem and did not worry about it.

I returned to the fixed source theory and the momentum slice approximation. I made further simplifications on the model. Then I did the perturbative analysis more carefully. Since in real life the momentum slice separation factor Λ would be 2 instead of very large, the ratio $1/\Lambda$ of successive energy scales would be $1/2$ rather than very small, and an all orders perturbative treatment was required in $1/\Lambda$. When the lower energy scales were treated to all orders relative to the highest energy scale, an infinitely complicated effective Hamiltonian was generated, with an infinite set of coupling constants. Each time an energy scale was eliminated through a perturbative treatment, a new infinitely complicated Hamiltonian was generated. Nevertheless, I found that for sufficiently large Λ I could mathematically control rigorously the effective Hamiltonians that were generated; despite the infinite number of couplings I was able to prove that the higher orders of perturbation theory had only a small and boundable impact on the effective Hamiltonians, even after arbitrarily many iterations.[48]

This work showed me that a renormalization group transformation, whose purpose was to eliminate an energy scale or a length scale or whatever from a problem, could produce an effective interaction with arbitrarily many coupling constants, without being a disaster. The renormalization group formalism based on fixed points could still be correct, and furthermore one could hope that only a small finite number of these couplings would be important for the qualitative behavior of the transformations, with the remaining couplings being important only for quantitative computations. In other words the couplings should have an order of importance, and for any desired but given degree of accuracy only a finite subset of the couplings would be needed. In my model the order of importance was determined by orders in the expansion in powers of $1/\Lambda$. I realized however that in the framework of an interaction on a lattice, especially for Ising-type models, locality would provide a natural order of importance — in any finite lattice volume there are only a finite mumber of Ising spin interactions that can be defined. I decided that Kadanoff's emphasis on the nearest neighbor coupling of the Ising model[30] should be restated: the nearest neighbor coupling would be the most important coupling because it is the most localized coupling one can define, but other couplings would be present also in Kadanoff's effective "block spin" Hamiltonians. A reasonable truncation procedure on these couplings would be to consider a finite region, say 3^3 or 4^3 lattice sites in size, and consider only multispin couplings that could fit into these regions (plus translations and rotations of these couplings).

Previously all the renormalization group transformations I was familiar with involved a fixed number of couplings: in the Gell-Mann-Low case just the electric charge c_λ, in Kadanoff's case an effective temperature and external field. I had tried many ways to try to derive transformations just for these fixed number of couplings, without success. Liberated from this restriction, it turned out to be easy to define renormalization group transformations; the hard problem was to find approximations to these transformations which would be computable in practice. Indeed a number renormalization group transformations now exist (see Section IV and its references).

In the fall of 1970 Ben Widom asked me to address his statistical mechanics seminar on the renormalization group. He was particularly interested because Di Castro and Jona-Lisinio had proposed applying the field theoretic renormalization group formalism to critical phenomena,[49] but no one in Widom's group could understand Di Castro and Jona-Lasinio's paper. In the course of lecturing on the general ideas of fixed points and the like I realized I would have to provide a computable example, even if it was not accurate or reliable. I applied the phase space cell analysis to the Landau-Ginzburg model of the critical point and tried to simplify it to the point of a calculable equation, making no demands for accuracy but simply trying to preserve the essence of the phase space cell picture. The result was a recursion formula in the form of a nonlinear integral transformation on a function of one variable, which I was able to solve by iterating the transformation on a computer.[50] I was able to compute numbers for exponents from the recursion formula at the same time that I could show (at least in part) that it had a fixed point and that the scaling theory of critical phenomena of Widom et al. followed from the fixed point formalism. Two papers of 1971 on the renormalization group presented this work.[50]

Some months later I was showing Michael Fisher some numerical results from the recursion formula, when we realized, together, that the nontrivial fixed point I was studying became trivial at four dimensions and ought to be easy to study in the vicinity of four dimensions. The dimension d appeared in a simple way as a parameter in the recursion formula and working out the details was straightforward; Michael and I published a letter[51] with the results. It was almost immediately evident that the same analysis could be applied to the full Landau-Ginzberg model without the approximations that went into the recursion formula. Since the simplifying principle was the presence of a small coefficient of the ϕ^4 term, a Feynman diagram expansion was in order. I used my field theoretic training to crank out the diagrams and my understanding of the renormalization group fixed point formalism to determine how to make use of the diagrams I computed. The results were published in a second letter in early 1972.[52] The consequent explosion of research is discussed in Part IV.

There were independent efforts on the same area taking place while I completed my work. The connection between critical phenomena and quantum field theory was recognized by Gribov and Migdal and Polyakov[53] and by axiomatic field theorists such as Symanzik.[54] T. T. Wu[55] worked on both field theory and the Ising model. Larkin and Khmelnitskii applied the field theoretic renormalization group of Gell-Mann and Low to critical phenomena in four dimensions and to the special case of uniaxial ferromagnets in three dimensions,[56] in both cases deriving logarithmic corrections to Landau's theory. Dyson formulated a somewhat artificial "hierarchical" model of a phase transition which was exactly solved by a one dimensional integral recursion formula.[54] This formula was almost identical to the one I wrote down later, in the 1971 paper. Anderson[5] worked out a simple but approximate procedure for eliminating momentum scales in the Kondo problem, anticipating my own work in the Kondo problem (see Sec. IV). Many solid state theorists were trying to apply diagrammatic expansions to critical phenomena, and Abe[58] and Scalapino and Ferrell[59] laid the basis for a diagrammatic treatment of models with a large number of degrees of freedom, for any dimension. (The limit of an infinite number of degrees of freedom had already been solved by Stanley[60]). Kadanoff was making extensive studies of the Ising model,[61] and discovered a short distance expansion for it similar to my own expansion for field theories. Fractional dimensions had been thought about before in critical phenomena.[62] Continuation of Feynman diagrams to non-integer dimensions was introduced into quantum field theory in order to provide a gauge invariant regularization procedure for non-abelian gauge theories:[63] this was done about simultaneously with its use to develop the ε expansion.

In the late '60's, Migdal and Polyakov[64] developed a "bootstrap" formulation of critical phenomena based on a skeleton Feynman graph expansion, in which all parameters including the expansion parameter inself would be determined self-consistently. They were unable to solve the bootstrap equations because of their complexity, although after the ε expansion about four dimensions was discovered, Mack showed that the bootstrap could be solved to lowest order in ε^{65}. If the 1971 renormalization group ideas had not been developed, the Migdal-Polyakov bootstrap would have been the most promising framework of its time for trying to further understand critical phenomena. However, the renormalization group methods have proved both easier to use and more versatile, and the bootstrap receives very little attention today.

In retrospect the bootstrap solved a problem I tried and failed to solve; namely how to derive the Gell-Mann-Low and Kadanoff dream of a fixed point involving only one or two couplings — there was only one coupling constant to be determined in the Migdal-Polyakov bootstrap. However, I found the bootstrap approach unacceptable because prior to the discovery of the ε expansion no formal argument was available to justify truncating the skeleton expansion to a finite number of terms. Also the skeleton diagrams were too complicated to test the truncation in practice by means of brute force computation of a large number of diagrams. Even today, as I review the problems that remain unsolved either by ε expansion or renormalization group methods, the problem of convergence of the skeleton expansion leaves me unenthusiastic about pursuing the bootstrap approach, although its convergence has never actually been tested. In the meantime, the Monte Carlo Renormalization group[66] has recently provided a framework for using small number of couplings in a reasonably effective and nonperturbative way: see Section IV.

I am not aware of any other independent work trying to understand the renormalization group from first principles as a means to solve field theory or critical phenomena one length scale at a time, or suggesting that the renormalization group should be formulated to allow arbitrarily many couplings to appear at intermediate stages of the analysis.

IV. *Results after 1971*

There was an explosion of activity after 1972 in both renormalization group and ε expansion studies. To review everything that has taken place since 1972 would be hopeless. I have listed a number of review papers and books which provide more detailed information at the end of this paper. Some principal results and some thoughts for the future will be outlined here. The "ε expansion" about four dimensions gave reasonable qualitative results for three dimensional systems. It enabled a much greater variety of details of critical behavior to be studied than was previously possible beyond the mean field level. The principal critical point is characterized by two parameters: the dimension d and the number of internal components n. Great efforts were made to map out critical behavior as a function of d and n. ε expansion and related small coupling expansions were carried to very high orders by Brézin, Le Guillou, Zinn-Justin,[67] and Nickel[68] led to precise results for d = 3.[69,70] The large n limit and 1/n expansion was pursued further.[71] A new expansion in 2+ε dimensions was developed for $n > 2$ by Polyakov.[72] For $n = 1$ there is an expansion in $1 + \varepsilon$ dimensions.[73] The full equation of state in the critical region was worked out in the ε expansion[74] and $1/n$ expansion.[75] The special case $n = 0$ was shown by De Gennes to describe the excluded volume problem in polymer configuraton problems and random walks.[76] Corrections to scaling were first considered by Wegner[77]. A recent reference is Aharony and Ahlers.[78]

Besides the careful study of the principal critical point other types of critical points and critical behavior were pursued. Tricritical phenomena were investigated by Riedel and Wegner,[79] where Landau theory was found to breakdown starting in three dimensions instead of four.[80] More general multicritical points have been analyzed.[81] Effects of dipolar forces,[82] other long range forces,[83] cubic perturbations and anisotropies[84,85] were pursued. The problems of dynamics of critical behavior were extensively studied.[86] Liquid crystal transitions were studied by Halperin, Lubensky, and Ma.[87]

Great progress has been made in understanding special features of two dimensional critical points, even though two dimensions is too far from four for the ε expansion to be practical. The Mermin-Wagner theorem[88] foreshadowed the complex character of two dimensional order in the presence of continuous symmetries. The number of exactly soluble models generalizing the Ising model steadily increases.[89] Kosterlitz and Thouless[90] blazed the way for renormalization group applications in two dimensional systems, following earlier work by Berezinskii.[91] They analyzed the transition to topological order in the 2-dimensional xy model with its peculiar critical point adjoining a critical line at lower temperatures; for further work see José et al[92] and Fröhlich and Spencer[93,94]. Kadanoff and Brown have given an overview of how a number of the two-dimensional models interrelate.[95] A subject of burning recent interest is the two-dimensional melting transition.[96] Among generalizations of the Ising model, the 3 and 4 state Potts model have received special attention. The three-state Potts model has only a first order transition in mean field theory and an expansion in $6 - \varepsilon$ dimensions but has a second order transition in two dimensions.[97] The four state Potts model has exceptional behavior in two dimensions (due to a "marginal variable"), which provides a severe challenge to approximate renormalization methods. Notable progress on this model has been made recently.[98]

A whole vast area of study concerns critical behavior or ordering in random systems, such as dilute magnets, spin glasses, and systems with random external fields. Random systems have qualitative characteristics of a normal system in two higher dimensions as was discovered by Lacour-Gayet and Toulouse[99] Imry, Ma, Grinstein, Aharony,[100] and Young[101] and confirmed by Parisi and Sourlas[102] in a remarkable paper applying 'supersymmetry' ideas from quantum field theory.[103] The "replica method" heavily used in the study of random systems[104] involves an $n \to 0$ limit, where n is the number of replicas similar to the De Gennes $n \to 0$ limit defining random walks.[76] There are serious unanswered questions surrounding this limiting process. Another curious discovery is the existence of an $\varepsilon^{1/2}$ expansion found by Khmelnitskii and Grinstein and Luther.[105]

Further major areas for renormalization group applications have been in percolation,[106] electron localization or conduction in random media,[107] the problems of structural transitions and "Lifshitz" critical points,[108] and the problem of interfaces between two phases.[109]

Much of the work on the ε expansion involved purely Feynman graph techniques; the high order computations involved the Callan-Symanzik formulation[110] of Gell-Mann Low theory. The computations also depended on the special diagram computation techniques of Nickel[68] and approximate formulae for very large orders of perturbation theory first discussed by Lipatov.[111] In lowest order other diagrammatic techniques also worked, for example the Migdal-Polyakov bootstrap was solved to order ε by Mack.[65]

The modern renormalization group has also developed considerably, Wegner[77,112,113] strengthened the renormalization group formalism considerably. A number of studies, practical and formal[114] were based on the approximate recursion formula introduced in 1971. Migdal and Kadanoff[115] developed an alternative approximate recursion formula (based on "bond moving" techniques). Real space renormalization group methods were initiated by Niemeijer and Van Leeuwen[116] and have been extensively developed since.[117, 118] The simplest real space transformation is Kadanoff's "spin decimation" transformation[119,120] where roughly speaking some spins are held fixed while other spins are summed over, producing an effective interaction on the fixed spins.

The decimation method was very successful in two dimensions where the spins on alternative diagonals of a square lattice were held fixed.[120] Other real space formulations[116, 117] involved kernels defining block spin variables related to sums of spins in a block (the block could be a triangle, square, cube, a lattice site plus all its nearest neighbors, or whatever).

Many of the early applications of real space renormalization group methods gave haphazard results — sometimes spectacularly good, sometimes useless. One could not apply these methods to a totally new problem with any confidence of success. The trouble was the severe truncations usually applied to set up a practical calculation; interactions which in principle contained thousands of parameters were truncated to a handful of parameters. In addition, where hundreds of degrees of freedom should be summed over (or integrated over) to execute the real space transformation, a very much simplified computation would be substituted. A notable exception is the *exactly* soluble differential renormalization group transformation of Hilhorst, Schick and Van Leeuwen, which unfortunately can be derived only for a few two dimensional models.[121, 122]

Two general methods have emerged which do not involve severe truncations and the related unreliability. First of all, I carried out a brute force calculation for the two dimensional Ising model using the Kadanoff decimation approach[119, 120] (as generalized by Kadanoff). Many interaction parameters (418) were kept and the spin sums were carried out over a very large finite lattice. The results were very accurate and completely confirmed my hypothesis that the local couplings of the shortest range were the most important. Most importantly the results could be an *optimization principle*. The fixed point of Kadanoff's decimation transformation depends on a single arbitrary parameter; it was possible to determine a best value for this parameter from internal consistency considerations. Complex calculations with potentially serious errors always are most effective when an optimization principle is available and parameters exist to optimize on.[123] This research has never been followed up, as is often the case when large scale computing is involved. More recently, the Monte Carlo Renormalization group method,[66] developed by Swendsen, myself, Shenker, and Tobochnik (see also Hilhorst and Van Leeuwen)[121] has proved very accurate and may shortly overtake both the high temperature expansions and the ε expansion as the most accurate source of data on the three dimensional Ising model. The Monte Carlo Renormalization Group is currently most successful on two dimensional problems where computing requirements are less severe: it has been applied successfully to tricritical models and the four-state Potts model.[124] In contrast, the ε expansion is all but useless for two dimensional problems. Unfortunately, none of the real space methods as yet provide the detailed information about correlation functions and the like that are easily derived in the ε expansion.

A serious problem with the renormalization group transformations (real space or otherwise) is that there is no guarantee that they will exhibit fixed points. Bell and myself[125] and Wegner in a more general and elegant way[113] have shown that for some renormalization group transformations, iteration of a critical point does not lead to a fixed point, presumably yielding instead interactions with increasingly long range forces. There is no known principle for avoiding this possibility, and as Kadanoff has showed using his decimation procedure,[120] a simple approximation to a transformation can misleadingly give a fixed point even when the full transformation cannot. The treatment that I gave of the two dimensional Ising model has self consistency checks that signal immediately when long range forces outside the 418 interactions kept are becoming important. Nothing is known yet about how the absence of a fixed point would be manifested in the Monte Carlo renormalization group computations. Cautions about real space renormalization group methods have also been advanced by Griffiths et al.[126]

There is a murky connection between scaling ideas in critical phenomena and Mandelbrot's "fractals" theory — a theory of scaling of irregular geometrical structures (such as coastlines).[127]

Renormalization group methods have been applied to areas other than critical phenomena. The Kondo problem is one example. Early renormalization group work was by Anderson[5] and Fowler and Zawadowski.[128] I then carried out a very careful renormalization group analysis of the Kondo Hamiltonian,[129] producing effective Hamiltonians with many couplings for progressively smaller energy scales, following almost exactly the prescription I learned for fixed source meson theory. The result was the zero-temperature susceptibility to about 1 % accuracy, which was subsequently confirmed by Andrei and Wiegmann's[7] exact solution. Renormalization group methods have been applied to other Hamiltonian problems, mostly one dimensional.[130] In multidimensional systems and in many one dimensional systems, the effective Hamiltonians presently involve too many states to be manageable.

The renormalization group has played a key role in the development of Quantum Chromodynamics — the current theory of quarks and nuclear forces. The original Gell-Mann Low theory[29] and the variant due to Callan and Symanzik[110] was used by Politzer, Gross and Wilczek[131] to show that nonabelian gauge theories are asymptotically free. This means that the short distance couplings are weak but increase as the length scale increases; it is now clear that this is the only sensible framework which can explain, qualitatively, the weak coupling that is evident in the analysis of deep inelastic electron scattering results (off protons and neutrons) and the strong coupling which is evident in the binding of quarks to form protons, neutrons, mesons, etc.[132] I should have anticipated the idea of asymptotic freedom[133] but did not do so. Unfortunately, it has been hard to study quantum chromodynamics in detail because of the effects of the strong binding of quarks at nuclear distances, which cannot be treated by diagrammatic methods. The development of the lattice gauge theory by Polyakov and myself[134] following pioneering work of Wegner[135] has made possible the use of a variety of lattice methods on the problems of quantum chromodynamics,[136] including strong coupling expansions, Monte Carlo simulations, and the Monte Carlo renormalization group methods.[67, 137] As computers become more powerful I expect there will be more emphasis on various modern renormalization group methods in these lattice studies, in order to take accurately into account the crossover from weak coupling at short distances to strong coupling at nuclear distances.

The study of unified theories of strong, weak and electromagnetic interactions makes heavy use of the renormalization group viewpoint. At laboratory energies the coupling strengths of the strong and electromagnetic interactions are too disparate to be unified easily. Instead, a unification energy scale is postulated at roughly 10^{15} GeV; in between renormalization group equations cause the strong and electroweak couplings to approach each other, making unification possible. Many grand unified theories posit important energy scales in the region between 1 and 10^{15} GeV. It is essential to think about these theories one energy scale at a time to help sort out the wide range of phenomena that are predicted in these theories. See Langacker[138] for a review. The study of grand unification has made it clear that Lagrangians describing laboratory energies are phenomenological rather than fundamental, and this continues to be the case through the grand unification scale, until scales are reached where quantum gravity is important. It has been evident for a long time that there should be applications of the renormalization group to turbulence, but not much success has been achieved yet. Feigenbaum[138] developed a renormalization group-like treatment of the conversion from order to chaos in some simple dynamical systems,[140] and this work may have applications to the onset of turbulence. Feigenbaum's method is probably too specialized to be of broader use, but dynamical systems may be a good starting point for developing more broadly based renormalization group methods applicable to classical partial differential equations.[141]

In my view the extensive research that has already been carried out using the renormalization group and the ε expansion is only the beginning of the study of a much larger range of applications that will be discovered over the next twenty years (or perhaps the next century will be required). The quick successes of the ε expansion are now past, and I believe progress now will depend rather on the more difficult, more painful exercises such as my own computations on the two dimensional Ising model and the Kondo problem,[120] or the Monte Carlo Renormalization group[66] computations. Often these highly quantitative, demanding computations will have to precede simpler qualitative analysis in order to be certain the many traps potentially awaiting any renormalization group analysis have been avoided.

Important potential areas of application include the theory of the chemical bond, where an effective interaction describing molecules at the bond level is desperately needed to replace current ab initio computations starting at the individual electron level.[142] A method for understanding high energy or large momentum transfer Quantum Chromodynamics (QCD) cross sections (including non-perturbative effects) is needed which will enable large QCD backgrounds to be computed accurately and subtracted away from experimental results intended to reveal smaller non-QCD effects. Practical areas like percolation, frost heaving, crack propagation in metals, and the metallurgical quench all involve very complex microscopic physics underlying macroscopic effects, and most likely yield a mixture of some problems exhibiting fluctuations on all length scales and other problems which become simpler classical problems without fluctuations in larger scales.

I conclude with some general references. Two semi-popular articles on the renormalization group are Wilson (1979) and Wilson (1975). Books include Domb and Green (1976); Pfeuty and Toulouse (1977); Ma (1976); Amit (1978); Patashinskii and Pokrovskii (1979)[143]; and Stanley (1971)[144]. Review articles and conference proceedings include Widom (1975)[145]; Wilson and Kogut (1974); Wilson (1975); Fisher (1974)[146]; Wallace and Zia (1978); Greer and Moldover (1981); aand Lévy et al. (1980).[147]

I thank the National Science Foundation for providing funding to me, first as a graduate student, then throughout most of my research career. The generous and long term commitment of the United States to basic research was essential to my own success. I thank my many colleagues at Cornell, especially Michael Fisher and Ben Widom, for encouragement and support. I am grateful for the opportunity to be a member of the international science community during two decades of extraordinary discoveries.

REFERENCES

1. Wilson, K. G. Physica *73*, 119 (1974).
2. See, e.g. Rose H. A. and Sulem, P. L. J. Physique *39*, 441 (1978).
3. See, e.g. Criegee L. and Knies, G. Phys. Repts. *83*, 151 (1982).
4. See, CRC Handbook of Chemistry and Physics (62nd Edition), Weast, R. C. ed. (CRC Press, Boca Raton, Florida, 1981), p. F-76.
5. See, e.g., Anderson, P. W. J. Phys. *C3*, 2346 (1970).
6. Onsager, L. Phys. Rev. *65*, 117 (1944).
7. Andrei, N. Phys. Rev. Lett. *45*, 379 (1980; Phys. Lett. *87A*, 299 (1982); Andrei, N. and Lowenstein, J. H. Phys. Rev. Lett. *46*, 356 (1981); Wiegmann, P. B. Phys. Lett. *80A*, 163 (1980); J. Phys. *C14*, 1463 (1981); Filyov, V. M. Tzvelik, A. M. and Weigmann, P. B. Phys. Lett. *81A*, 175 (1981).
8. Johnson, K. Nuovo Cimento *20*, 773 (1961).
9. The experimental measurements on fluids (e.g., SF_6, He^3, and various organic fluids,) give $\beta = 0.32 \pm 0.02$, while current theoretical computations give $\beta = 0.325 \pm 0.005$; see Greer S. C. and Moldover, M. R. Annual Review of Physical Chemistry *32*, 233 (1981) for data and caveats.
10. For experimental reviews, see e. g. Heller, P. Repts. Prog. Phys. *30*, 731 (1967); Kadanoff, L. Götze, W. Hamblen, D. Hecht, R. Lewis, E. A. S. Palciauskas, V. V. Rayl, M. Swift, J. Aspres, D. Kane, J. *Static Phenomena near Critical Points: Theory and Experiment*, Rev. Mod. Phys. *39*, 395 (1967).
11. For the alloy transition in β brass see Als-Nielsen, J. in Domb, C. and Green, M. S. eds., *Phase Transitions and Critical Phenomena*, Vol. 5a (Academic Press, New York, 1978) p.87. See Ref. 10 for earlier reviews of other systems.
12. For a history of the Lenz-Ising model see Brush, S. G. Rev. Mod. Phys. *39*, 883 (1967).
13. Landau, L. D. Phys. Zurn. Sowjetunion *11*, 545 (1937); English translation: ter Haar, D. *Men of Physics: L. D. Landau*, Vol. II (Pergamon, Oxford, 1969).
14. Ginzburg, V. L. and Landau, L. D. Zh. Eksperim, i Teor. Fiz. *20*, 1064 (1950). See also Schrieffer, J. *Superconductivity* (Benjamin, New York, 1964) p.19.
15. Ginzburg, V. L. Fiz. Tverd. Tela *2* 2031 (1960). English translation: Soviet Physics: Solid St. *2* 1824 (1960).
16. Wilson, K. G. and Fisher, M. E. Phys. Rev. Letters *28*, 240 (1972).
17. Wilson, K. G. and Kogut, J. Phys. Repts. *12C*, 75 (1974).
18. See, e. g., Guggenheim, E. A. J. Chem. Phys. *13*, 253 (1945); Griffiths, R. B. Phys. Rev. Letts. *24*, 1479 (1970); Griffiths R. B. and Wheeler, J. B. Phys. Rev. *A2* 1047 (1970); Kadanoff L. in *Phase Transitions and Critical Phenomena*, Domb C. and Green, M. S. Eds. Vol. *5a* (Academic, New York, 1978, p.1.).
19. See, e.g., DeBoer, J. Physica *73*, 1 (1974); Klein, M. D. Physica *73*, 28 (1974); Levelt. J. M. H. Sengers, Physica *73*, 73 (1974).
20. Weiss, P. J. Phys. *6*, 661 (1907).
21. See, Domb, C. Proc. Roy. Soc. *A199*, 199 (1949) and, e.g. Fisher, M. E. Reports. Prog. Phys. *30*, 615 (1967).
22. See, e.g., ref. 10 and Ahlers, G. Revs. Mod. Phys. *52*, 489 (1980).
23. Widom, B. J. Chem. Phys. *43*, 3898 (1965).
24. Essam J. W. and Fisher, M. E. J. Chem. Phys. *38*, 802 (1963); Domb C. and Hunter, D. L. Proc. Phys. Soc. *86*, 1147 (1965); Kadanoff, L. P. Physics, *2*, 263 (1966); Patashiniskii A. Z. and Prokrovskii, V. L. Zh. Eksper. Teor. Fiz. *50*, 439 (1966); English translation: Soviet Physics—JETP *23*, 292 (1966).
25. Rushbrooke, G. S., 1963, J. Chem. Phys. *39*, 842.
26. Griffiths, R. B., 1965, Phys. Rev. Lett. *14*, 623.
27. See the reprint collection Schwinger, J. Ed., *Quantum Electrodynamics* (Dover, New York, 1958).
28. Stueckelberg E. C. G. and Petermann, A. Helv. Phys. Acta *26*, 499 (1953); see also Petermann, A. Phys. Repts. *53*, 157 (1979).
29. Gell-Mann M. and Low, F. E. Phys. Rev. *95*, 1300 (1954).

30. Kadanoff, L. P. Physics *2*, 263 (1966).
31. Bogoliubov N. N. and Shirkov, D. V. *Introduction to the Theory of Quantized Fields* (Interscience; New York, 1959), Ch. VIII.
32. Ter-Martirosyan, K. A. Zh. Eksperim. Teor. Fiz. *39*, 827 (1960); Soviet Physics—JETP *12*, 575 (1961).
33. See, e.g. Wilson, K. G. Acta Physica Austriaca *17*, 37 (1963—64) and references cited therein.
34. Wenzel, G. Helv. Phys. Acta *13*, 269 (1940); *14*, 633 (1941); see also Henley E. M. and Thirring, W. *Elementary Quantum Field Theory* (McGraw Hill, New York, 1962).
35. Lee T. D. and Yang, C. N. Phys. Rev. *128*, 885 (1962).
36. Johnson, K. Baker, M. and Willey, R. S. Phys. Rev. Lett. *11*, 518 (1963). For a subsequent view see Baker, M. and Johnson, K., Phys. Rev. *183*, 1292 (1969); ibid *D3*, 2516, 2541 (1971).
37. See e.g. Hepp, K. Acta Physica Austriaca *17*, 85 (1963—64).
38. See e.g. Lowenstein, J. H. Commun. Math. Phys. *16*, 265 (1970); Wilson, K. G. Phys. Rev. *D2*, 1473 (1970).
39. Wilson, K. G. Phys. Rev. *140*, B445 (1965).
40. Dyson, F. J. Phys. Rev. *83*, 608, 1207 (1951).
41. Mitter, P. K., and G. Valent, 1977, Phys. Lett. *B70*, 65.
42. See, e.g. Kogut J. and Susskind, L. Phys. Repts. *8C*, 75 (1973) and references cited therein.
43. Schultz, T. D. Mattis, D. C. and Lieb, E. H. Rev. Mod. Phys. *36*, 856 (1964).
44. Schiff, L. I. Phys. Rev. *92*, 766 (1953).
45. See, e.g. Mack, G. Nucl. Phys. *B5*, 499 (1968) and references cited therein.
46. Wilson, K. G. Phys. Rev. *179*, 1499 (1969).
47. Bjorken, J. Phys. Rev. *179*, 1547 (1969); Feynman, R. P. Phys. Rev. Lett. *23*, 1415 (1969); for a review see Yan, T. M. Annual Review of Nuclear Science, *26*, 199 (1976) and Feynman, R. P. *Photon-Hadron Interactions*, Benjamin, W. A. (Reading, Mass, 1972).
48. Wilson, K. G. Phys. Rev. *D2*, 1438 (1970).
49. Di Castro, C. and Jona-Lasinio, G. Phys. Letters *29A*, 322 (1969).
50. Wilson, K. G. Phys. Rev. *B4*, 3174, 3184 (1971).
51. Wilson, K. G. and Fisher, M. E. Phys. Rev. Lett. *28*, 240 (1972).
52. Wilson, K. G. Phys. Rev. Lett. *28*, 548 (1972).
53. Gribov, V. N. and Migdal, A. A. Zh. Eksp. Teor. Fiz. *55*, 1498 (1968); English Translation: Soviet Physics—JETP *28*, 784 (1968); Migdal, A. A. ibid. *59*, 1015 (1970); JETP *32*, 552 (1971); Polyakov, A. M. ibid. *55*, 1026 (1968); JETP *28*, 533 (1969); Polyakov, A. M. ibid. *57*, 271 (1969); JETP *30*, 151 (1970); Polyakov, A. M. ibid. *59*, 542 (1970); JETP *32*, 296 (1971).
54. Symanzik, K. J. Math. Phys. *7*, 510 (1966).
55. Wu, T. T. Phys. Rev. *149*, 380 (1966); McCoy, B. M. and Wu, T. T. *The Two Dimensional Ising Model* (Harvard University Press, Cambridge, 1973); McCoy, B. M. Tracy, C. and Wu, T. T. Phys. Lett. *61A*, 283 (1977).
56. Larkin, A. I. and Khmelnitskii, D. E. Zh. Eksp. Teor. Fiz. *56*, 2087 (1969); English Translation: Soviet Physics JETP *29*, 1123 (1969).
57. Dyson, F. J. Commun. Math. Phys. *12*, 91 (1969); see also Baker, G. A. Jr., Phys. Rev. *B5*, 2622 (1972).
58. Abe, R. Progr. Theor. Phys. (Kyoto) *48*, 1414 (1972); ibid. *49*, 113 (1973); Abe, R. and Hikami, S. Phys. Letters *45A*, 11 (1973); Hikami, S. Progr. Theor. Phys. (Kyoto) *49*, 1096 (1973).
59. Ferrell, R. A. and Scalapino, D. J. Phys. Rev. Letters *29*, 413 (1972); Phys. Letters *41A*, 371 (1972).
60. Stanley, H. E. Phys. Rev. *176*, 718 (1968).
61. Kadanoff, L. P. Phys. Rev. Lett. *23*, 1430 (1969); Phys. Rev. *188*, 859 (1969).
62. See, e.g. Fisher, M. E. and Gaunt, D. S. Phys. Rev. *133*, A224 (1964); Widom, B. Mol. Phys. *25*, 657 (1973).
63. 't Hooft, G. and Veltman, M. Nucl. Phys. *B44*, 189 (1972); Bollini, C. G. and Giambiagi, J. J. Phys. Letts. *40B*, 566 (1972); Ashmore, J. F. Lettere Nuovo Cimento *4*, 289 (1972).
64. Patashinskii, A. Z. and Pokrovskii, V. L. Zh. Eksp. Teor. Fiz. *46*, 994 (1964); English Translation: Soviet Physics JETP *19*, 677 (1964); Polyakov, A. M. Zh. ETF. Pis. Red. *12*, 538

(1970); English Translation: JETP Letters *12*, 381 (1970); Migdal, A. A. Phys. Letts. *37B*, 386 (1971); Mack, G. and Symanzik, K. Common Math Phys. *27*, 247 (1972) and references cited therein.

65. Mack, G. in Strong Interaction Physics, Rühl, W. and Vanira, A. eds. (Springer-Verlag, Berlin, 1973), p.300.

66. Swendsen, R. H. Phys. Rev. Lett. *42*, 859 (1979); Phys. Rev. *B20*, 2080 (1979); Wilson, K. G. Monte Carlo Calculation for the Lattice Gauge Theory, in *Recent Developments in Gauge Theories*, 't Hooft, G. et al. Eds. (Plenum Press, New York, 1980); Shenker, S. and Tobochnik, J. Phys. Rev. *B22*, 4462 (1980).

67. See, e.g. the review of Zinn-Justin, J. Phys. Repts. *70*, 109 (1981); see further Nickel, B. Physica *106A*, 48 (1981).

68. Nickel, B. (unpublished), cited in Ref. 64 and prior papers.

69. Parisi, G., 1980, J. Stat. Phys. *23*, 49.

70. Vladimirov, A. A., D. I. Kazakov, and O. Tarasov, 1979, Sov. Phys. JETP *50*, 521.

71. See, e.g. Ma, S. K. in Domb, C. and Green, M. S. eds. *Phase Transitions and Critical Phenomena*, Vol. 6 (Academic Press, London, 1976), p.250, and references cited therein.

72. Polyakov, A. M. Phys. Lett. *59B*, 79 (1975); Migdal, A. A., Zh. Eksp. Teor. Fiz. *69*,. 1457 (1975); English translation: Soviet Physics–JETP *42*, 743 (1976); Brézin, E. Zinn-Justin, J. and LeGuillou, J. C. Phys. Rev. *B14*, 4976, (1976); Bhattacharjee, J. K. Cardy, J. L. and Scalapino, D. J. Phys. Rev. *B25*, 1681 (1982).

73. Wallace, D. J. and Zia, R. K. Phys. Rev. Lett. *35*, 1399 (1979); Widom, B. (ref. 59).

74. Brézin, E. Wallace, D. and Wilson, K. G. Phys. Rev. Letters, *29*, 591 (1972); Phys. Rev. *B7*, 232 (1973); Avdeeva, G. M. and Migdal, A. A. Zh. ETF. Pis. Red. *16*, 253 (1972); English Translation: Soviet Physics–JETP Letters *16*, 178 (1972); Avdeeva, G. M. Zh. Eksp. Teor. Fiz. *64*, 741 (1973), English Translation: Soviet Physics–JETP *37*, 377 (1973).

75. Brézin, E. and Wallace, D. J. Phys. Rev. *B7*, 1967 (1973); for a review see Ma S.-K. in Domb, C. and Green, M. S. Eds., Phase Transitions and Critical Phenomena, Vol. 6 (Academic Press, London, 1976) p.250.

76. DeGennes, P. G. Phys. Letters *38A*, 339 (1972); des Cloiseaux, J. J. Phys. (Paris) *36*, 281 (1975).

77. Wegner, F. J., 1972a, Phys. Rev. *B5*, 4529.

78. Aharony, A. and G. Ahlers, 1980, Phys. Rev. Lett. *44*, 782.

79. Riedel, E. K. and Wegner, F. J. Phys. Rev. Lett. *29*, 349 (1972); Phys. Rev. *B7*, 248 (1973).

80. Stephen, M. J., E. Abrahams, and J. P. Straley, 1975, Phys. Rev. *B12*, 256.

81. See, e.g., Fisher, M. E. in A.I.P. Conference Proceedings #24, *Magnetism and Magnetic Materials* 1974, p. 273.

82. Fisher, M. E. and Aharony, A. Phys. Rev. Lett. *30*, 559 (1973); Aharony, A. and Fisher, M. E. Phys. Rev. *B8*, 3323 (1973); Aharony, A. Phys. Rev. *B8*, 3342, 3349, 3358, 3363 (1973); Aharony, A. Phys. Letters *44A*, 313 (1973).

83. Suzuki, M. Phys. Letters, *42A*, 5 (1972); Suzuki, M. Progr. Theor. Phys. (Kyoto) *49*, 424, 1106, 1440 (1973); Fisher, M. E. Ma, S. and Nickel, B. G. Phys. Rev. Letters *29*, 917 (1972); Baker Jr., G. A. and Golner, G. R. Phys. Rev. Letters *31*, 22 (1973); Suzuki, M. Yamazaki, Y. and Igarishi, G. Phys. Letters *42A*, 313 (1972); Sak, J. Phys. Rev. *B8*, 281 (1973).

84. Pfeuty, P. and Fisher, M. E. Phys. Rev. *B6*, 1889 (1972); Wallace, D. J. J. Phys. *C6*, 1390 (1973); Ketley, L. J. and Wallace, D. J. J. Phys. *A6*, 1667 (1973); Aharony, A. Phys. Rev. Letters *31*, 1494 (1973); Suzuki, M. Progr. Theor. Phys. (Kyoto) *49*, 1451 (1973); Liu, L. Phys. Rev. Letters *31*, 459 (1973); Grover, M. K. Phys. Letters *44A*, 253 (1973); Chang, T. S. and Stanley, H. E. Phys. Rev. *B8*, 4435 (1973); Aharony, A. Phys. Rev. *B8*, 4270 (1973); A recent reference is Blankschtein, D. and Mukamel, D. Phys. Rev. *B25*, 6939 (1982).

85. Wegner, F. J., 1972b, Phys. Rev. *B6*, 1891.

86. Halperin, B. I. Hohenberg, P. C. and Ma, S.-K. Phys. Rev. Letters *29*, 1548 (1972); Suzuki, M. and Igarishi, G. Progr. Theor. Phys. (Kyoto) *49*, 1070 (1973); Suzuki, M. Phys. Letters *43A*, 245 (1973); Suzuki, M. Progr. Theor. Phys. (Kyoto) *50*, 1767 (1973); for a review see Hohenberg, P. C. and Halperin, B. I. Revs. Mod. Phys. *49*, 435 (1977); for a recent Monte Carlo Renormalization Group method see Tobochnik, J. Sarker, S. and Cordery, R. Phys.

Rev. Lett, *46*, 1417, (1981). Other recent references are Ahlers, G. Hohenberg, P. C. and Kornblit, A. Phys. Rev. *B25*, 3136 (1982); Heilig, S. J. Luscombe, J. Mazenko, G. F. Oguz, E. and Valls, O. T. Phys. Rev. *B25*, 7003 (1982).

87. Halperin B. I., T. C. Lubensky, and S.-k. Ma, 1974, Phys. Rev. Letters, *32*, 242.

88. Mermin, N. D. and Wagner, H. Phys. Rev. Lett. *17*, 1133 (1966); Mermin, N. D. J. Math. Phys. *8*, 1061 (1967); Hohenberg, P. C. Phys. Rev. *158*, 383 (1967).

89. See, e.g., Baxter, R. *Exactly Solved Models in Statistical Mechanics*, (Academic Press, New York, 1982).

90. Kosterlitz, J. M. and Thouless, D. J. J. Phys. *C6, 1*181 (1973); Kosterlitz, J. M. J. Phys. *C7*, 1046 (1974); for a review see Kosterlitz, J. M. and Thouless, D. J. Prog. Low Temp. Phys. *7*, 373 (1978).

91. Berezinskii, V. L. Zh. Eksp. Teor. Fiz. *59*, 907 (1970); (Soviet Physics−JETP *32*, 493 (1971)); Berezinskii, V. L. Zh. Eksp. Teor. Fiz. *61*, 1144 (1971) (Soviet Physics−JETP *34*, 610 (1972)).

92. José, J. V., L. P. Kadanoff, S. Kirkpatrick, and D. R. Nelson, 1977, Phys. Rev. *B16*, 1217.

93. Fröhlich, J., and T. Spencer, 1981, Phys. Rev. Lett. *46*, 1006.

94. Fröhlich, J., and T. Spencer, 1981, Commun. Math. Phys. *81*, 527.

95. Kadanoff, L. P. and Brown, A. C. Annals Phys. (N.Y.) *121*, 318 (1979).

96. Nelson, D. R. and Halperin, B. I. Phys. Rev. *B19*, 2457 (1979); Young, A. P. Phys. Rev. *B19*, 1855 (1979).

97. Baxter, R. J. J. Phys. *C6*, 2445 (1973) (rigorous 2-d solution); see e.g., Banavar, J. R. Grest, G. S and Jasnow, D. Phys. Rev. *B25*, 4639 (1982) and references cited therin (d>2).

98. Nienhuis, B. Berker, A. N. Riedel, E. K. and Schick, M. Phys. Rev. Lett. *43*, 737 (1979); Swendsen, R. H. Andelman, D. and Berker, A. N. Phys. Rev. *B24*, 6732 (1982) and references cited therein.

99. Lacour-Gayet, P., and G. Toulouse, 1974, J. Phys. *35*, 425.

100. Imry, Y. and Ma, S.-K. Phys. Rev. Lett. *35*, 1399 (1975); Grinstein, G. Phys. Rev. Lett. *37*, 944 (1976); Aharony, A. Imry, Y. and Ma, S.-K. Phys. Rev. Lett. *37*, 1367 (1976); a recent reference is: Mukamel, D. and Grinstein, G. Phys. Rev. *B25*, 381 (1982).

101. Young, A. P., 1977, J. Phys. *C10*, L257.

102. Parisi, G. and Sourlas, N. Phys. Rev. Lett. *43*, 744 (1979).

103. Parisi, G., and N. Sourlas, 1981, Phys. Rev. Lett. *46*, 871.

104. Edwards, S. F. and Anderson, P. W. J. Phys. F5, 965 (1975).

105. Khmelnitskii, D. E. Zh. Eksp. Teor. Fiz. *68*, 1960 (1975), (Sov. Phys.−JETP *41*, 981 (1976)); Grinstein, G. and Luther, A. Phys. Rev. *B13*, 1329 (1976).

106. See, e.g., the review of Essam, J. W. Repts. Prog. Phys. *43*, 833 (1980); a recent reference is Lobb, C. J. and Karasek, K. R. Phys. Rev. *B25*, 492 (1982).

107. Nagaoka, Y., and H. Fukuyama, editors, 1982, *Anderson Localization* Springer Series in Solid-State Science (Springer-Berlin), Vol. 39.

108. See, e.g., the review of Bruce, A. D. Adv. Phys. *29*, 111 (1980); a recent paper is Grinstein, G. and Jayaprakash, C. Phys. Rev. *B25*, 523 (1982).

109. See Ref. 68 and references cited therein.

110. Callan, C. G. Phys. Rev. *D2*, 1541 (1970); Symanzik, K. Commun. Math. Phys. *18*, 227 (1970); Symanzik, K. Springer Tracts in Modern Physics *57*, 222 (1971).

111. Lipatov, L. N. Zh. Eksp. Teor. Fiz. *72*, 411 (1977); English translation, Soviet Physics−JETP *45*, 216 (1977).
    Brezin, E. Le Guillou, J. C. and Zinn-Justin, J. Phys. Rev. *D15*, 1544, 1558 (1977).

112. Wegner, F. J., 1974, J. Phys. *C7*, 2098.

113. See, e.g. F. Wegner, in C. Domb and M. S. Green, eds., *Phase Transitions and Critical Phenomena*, (Academic Press, London, 1976) p. 7.

114. See, e.g., Golner, G. Phys. Rev. *B8*, 3419 (1973); Langer, J. S. and Bar-on, M. Ann Phys. (N.Y.) *78*, 421 (1973); Bleher, P. M. and Ya. Ya. G. Sinai, Commun. Math. Phys. *45*, 247 (1975).

115. Migdal, A. A. Zh. Eksp. Teor. Fiz. *69*, 810, 1457 (1975); English Translations: Soviet Physics−JETP *42*, 413, 743 (1975), for a review and more references, see Kadanoff, L. P. Revs. Mod. Phys. *49*, 267 (1977).

116. See, e.g., Niemeijer, Th. and Van Leeuwen, J. M. J. in Domb, C. and Green, M. S. eds., Phase Transitions and Critical Phenomena, Vol. 6 (Academic Press, London, 1976), p. 425.
117. See Ref. 80 and e.g., Riedel, E. K. Physica *106A*, 110 (1981) and references cited therein.
118. Burkhardt, T. W., and J. M. J. Van Leeuwen, 1982, *Real-Space Renormalization* (Springer, Berlin)
119. Kadanoff, L. P., and A. Houghton, 1975, Phys. Rev. *B11*, 377.
120. See Wilson, K. G. Rev. Mod. Phys. *47*, 773 (1975) esp. Sect. IV.
121. See Hilhorst H. J. and Van Leeuwen, J. M. J. Physica *106A*, 301 (1981).
122. Hilhorst, H. J., M. Schick, and J. M. J. Van Leeuwen, 1978, Phys. Rev. Lett. *40*, 1605.
123. I thank Yves Parlange for reminding me of this.
124. See, e.g. Swendsen, R. H. Andelman, D. and Berker, A. N. Phys. Rev. *B24*, 6732 (1982); Landau, D. P. and Swendsen R. H. Phys. Rev. Lett. *46*, 1437 (1981).
125. Bell, T. L. and Wilson, K. G. Phys., Rev. *B10*, 3935 (1974).
126. See, e.g. Griffiths, R. B. Physica *106A*, 59 (1981).
127. See, e.g., Mandelbrot, Benoit B. *The Fractal Geometry of Nature* (Freeman, W. H. San Francisco, 1982);
128. Fowler, M. and Zawadowski, A. Solid State Commun. *9*, 471 (1971).
129. Wilson, K. G. Revs. Mod. Phys. *47*, 773 (1975); Krishna-Murthy, H. R. Wilkins, J. N. and Wilson, K. G. Phys. Rev. Letters *35*, 1101 (1975); Phys. Rev. *B21*, 1044, 1104 (1980).
130. See, e.g. Drell, S. D. Weinstein, M. and Yankielowicz, S. Phys. Rev. *D17*, 1769 (1977); Jullien, R. Fields, J. N. and Doniach, S. Phys. Rev. *B16*, 4889 (1977); recent references are Hanke, W. and Hirsch, J. E. Phys. Rev. *B25*, 6748 (1982); Penson, K. A. Jullien, R. and Pfeuty, P. Phys. Rev. *B25*, 1837 (1982) and references cited therein.
131. Politzer, H. Phys. Rev. Lett. *30*, 1346 (1973); Gross, D. and Wilczek, F. Phys. Rev. Lett. *30*, 1343 (1973).
132. See, e.g. Altarelli, G. Phys. Repts. *81*, 1 (1982).
133. Wilson, K. G. Phys. Rev. *D3*, 1818 (1971).
134. Polyakov A. M. (unpublished); Wilson, K. G. Phys. Rev. *D10*, 2455 (1974).
135. Wegner, F. J. Math. Phys. *12*, 2259 (1971).
136. See the review of Bander, M. Phys. Repts. *75*, 205 (1981).
137. Swendsen, R. H., 1982, in *Real-Space Renormalization*, edited by J. W. Burkhardt and J. M. J. Van Leeuwen, (Springer, Berlin).
138. Langacker P., 1981, Phys. Rep. *72*, 185.
139. Feigenbaum, M. J. J. Stat. Phys. *19*, 25 (1978).
140. See, e.g. Eckmann, J. P. Rev. Mod. Phys. *53*, 643 (1981); Ott, E. Rev. Mod. Phys. *53*, 655 (1981).
141. See e.g. Coppersmith, S. and Fisher, D. Bell Laboratories preprint.
142. See, e.g., Mulliken, R. S. Ann. Rev. Phys. Chem. *29*, 1 (1978); Löwdin, P. O. Adv. Quantum Chem. *12*, 263 (1980); Hirst, D. M. Adv. Chem. Phys. *50*, 517 (1982); Bartlett, R. J. Ann. Rev. Phys. Chem. *32*, 359 (1981); Case, D. A. Ann. Rev. Phys. Chem. *33*, 151 (1982).
143. Patashinskii, A. Z., and V. L. Pokrovskii, 1979, *Fluctuating Theory of Phase Transitions* (Pergamon)
144. Stanley, H. E., 1972, *Introduction to Phase Transitions and Critical Phenomena* (Oxford University, London)
145. Widom, B., 1975 in *Fundamental Problems in Statistical Mechanics*, edited by E. C. G. Cohen (North-Holland) Vol. III, p. l.
146. Fisher, M. E., 1964, J. Math. Phys. *5*, 944.
147. Levy, M., J. C. Le Guillou, and J. Zinn-Justin, editors, 1980, *Phase Transitions* (Plenum, New York).

Books Related to the Renormalization Group

Domb, C. and Green, M. S. Eds., *Phase Transitions and Critical Phenomena* (Academic Press, London) esp. Vol. 6 (1976);
Pfeuty, P. and Toulouse, G. *Introduction to the Renormalization Group and to Critical Phenomena* (Wiley, New York, 1977);

Ma, S.-K. *Modern Theory of Critical Phenomena* (Benjamin/Cummings, Reading, Mass., 1976);
Amit, D. J. *Field Theory, the Renormalization Group, and Critical Phenomena* (McGraw-Hill, New York, 1978).

Popular Articles (sort of)
Wilson, K. G. Sci. Am. *241*, 158 (August, 1979);
Wilson, K. G. Adv. Math *16*, 176 (1975).

Review Articles
Greer, S. C. and Moldover, M. R. Ann. Rev. Phys. Chem. *32*, 233 (1981): Critical Behavior of Fluids and many references to prior reviews.
Bander, M. ref. 116: Quark Confinement
Zinn-Justin, J. (Ref. 67): Precise Computation of Critical Exponents from the ε expansion;
Baret, J. F. Prog. Surf. Membrane Sci. *14*, 292 (1981);
Cadenhead, D. A. and Danielli, J. F. Eds., (Academic, New York, 1981): A practical area involving phase transitions.
Wallace, D. J. and Zia, R. K. Repts. Progr. Phys. *41*, 1 (1978).

Physics 1983

SUBRAHMANYAN CHANDRASEKHAR

*for his theoretical studies of the physical processes of importance to the
structure and evolution of the stars*

and

WILLIAM A FOWLER

*for his theoretical and experimental studies of the nuclear reactions of
importance in the formation of the chemical elements in the universe*

THE NOBEL PRIZE FOR PHYSICS

Speech by Professor SVEN JOHANSSON of the Royal Academy of Sciences
Translation from the Swedish text

Your Majesties, Your Royal Highnesses, Ladies and Gentlemen,

Astrophysics is one of the areas in physics which has developed most rapidly during recent years. Through satellite technology it has become possible to study the different physical processes which are taking place in stars and other astronomical objects. Space has become a new and exciting laboratory for the physicist. It is true that experiments, in the proper sense of the word, cannot be carried out, but one may observe phenomena which can never be observed in terrestrial laboratories. In space we find matter in the most extreme forms; stars at immensely high temperatures and with enormously high densities, and particles and radiation with an energy which we cannot reach, even with our largest accelerators.

The common theme for this year's prize in physics is the evolution of the stars. From the moment of their birth out of interstellar matter until their extinction, the stars exhibit many physical processes of great interest. In order to put this year's prize in perspective, it is perhaps appropriate to give a short description of the evolution of the stars.

Stars are formed from the gas and dust clouds which are present in galaxies. Under the influence of gravity, this matter condenses and contracts to form a star. During these processes energy is released which leads to a rise in the temperature of the newly formed star. Eventually, the temperature becomes so high that nuclear reactions are initiated inside the star. Hydrogen, which is the primary constituent, burns to form helium. During this process pressure builds up which prevents further contraction, the star stabilizes, and may continue to exist for millions or thousands of millions of years. When the supply of hydrogen has been used up, other nuclear reactions come into play, especially in more massive stars, and heavier elements are thus formed. A particularly effective type of nuclear reaction is the successive addition of neutrons. Finally, the star is, to a large extent, composed of heavier elements, mainly iron and neighbouring elements, and the supply of nuclear fuel is exhausted. When the star has evolved this far it can no longer withstand the pressure of its own gravitational force and collapses, the product of collapse depending on the mass of the star.

For lighter stars with a mass roughly equal to that of the Sun, the collapse results in a so-called white dwarf. The star is so named because of its reduction in size, leading to an increase in its density to about 10 tons per cubic centimetre. The mechanism for the collapse is that the electron shell structure is crushed, so that the star consists of atomic nuclei in an electron gas.

For somewhat heavier stars, the collapse can lead to an explosion, the visible result being a supernova. This is accompanied by a short-lived but intense neutron flux which leads to the formation of the heaviest elements. In these heavy stars the collapse can go even further, the atomic nuclei and the electrons combining to form neutrons. This results in a so-called neutron star which has the enormously high density of 100 million tons per cubic centimetre. A star with a mass of 1 to 2 times that of the Sun may be compressed so that the radius is only about 10 km. A neutron star is essentially a sphere of neutrons in a fluid form surrounded by a solid crust which is very much harder than steel.

The collapse of still heavier stars can lead to an even more exotic object, a black hole. Here the gravitational force is so strong that all matter which is sucked into the hole loses its identity, and is compressed into an infinitely small volume, i.e. a mathematical point. Not even light, emitted from within the black hole, may escape into the outside world, hence the name, black hole. The existence of a black hole may be revealed through the radiation which is emitted by matter which, when being sucked into it, undergoes a considerable increase in temperature before finally disappearing. Certain strange objects called quasars may possibly be a black hole in the centre of a galaxy.

It should now be clear that during their evolution stars exhibit many different physical processes of fundamental importance. Many scientists have studied the problems involved with these processes, but especially important contributions have been made by Subrahmanyan Chandrasekhar and William Fowler.

Chandrasekhar's work is particularly many-sided and covers many aspects of the evolution of stars. An important part of his work is a study concerning the problems of stability in different phases of their evolution. In recent years he has studied relativistic effects, which become important because of the extreme conditions which arise during the later stages of the star's development. One of Chandrasekhar's most well known contributions is his study of the structure of white dwarfs. Even if some of these studies are from his earlier years, they have become topical again through advances in the fields of astronomy and space research.

Fowler's work deals with the nuclear reactions which take place during the evolution of stars. Apart from generating the energy which is emitted, they are important because they lead to the production of the chemical elements from the starting material, which mainly consists of the lightest element, hydrogen. Not only has Fowler carried out a great deal of experimental work on nuclear reactions of interest in the astrophysical context, but has also worked on this problem from a theoretical point of view. In the 1950's, together with a number of colleagues, he developed a complete theory for the formation of the chemical elements in the Universe. This theory is still the basis of our knowledge in this area, and the latest advances in nuclear physics and space research have further shown this theory to be correct.

Professor Chandrasekhar and Professor Fowler,

Your pioneering work has laid the foundation for important developments in astrophysics and you have both been the source of inspiration for other scientists working in this field. The remarkable achievements of astronomy and space research in recent years have vindicated your ideas and demonstrated their importance.

It is my privilege and pleasure to convey to you the warmest congratulations of the Royal Swedish Academy of Sciences. May I now ask you to come forward and receive your prize from the hands of His Majesty the King.

S. Chandrasekhar.

SUBRAHMANYAN CHANDRASEKHAR

I was born in Lahore (then a part of British India) on the 19th of October 1910, as the first son and the third child of a family of four sons and six daughters. My father, Chandrasekhara Subrahmanya Ayyar, an officer in Government Service in the Indian Audits and Accounts Department, was then in Lahore as the Deputy Auditor General of the Northwestern Railways. My mother, Sita (neé Balakrishnan) was a woman of high intellectual attainments (she translated into Tamil, for example, Henrik Ibsen's *A Doll House*), was passionately devoted to her children, and was intensely ambitious for them.

My early education, till I was twelve, was at home by my parents and by private tuition. In 1918, my father was transferred to Madras where the family was permanently established at that time.

In Madras, I attended the Hindu High School, Triplicane, during the years 1922—25. My university education (1925—30) was at the Presidency College. I took my bachelor's degree, B.Sc. (Hon.), in physics in June 1930. In July of that year, I was awarded a Government of India scholarship for graduate studies in Cambridge, England. In Cambridge, I became a research student under the supervision of Professor R. H. Fowler (who was also responsible for my admission to Trinity College). On the advice of Professor P. A. M. Dirac, I spent the third of my three undergraduate years at the Institut for Teoretisk Fysik in Copenhagen.

I took my Ph.D. degree at Cambridge in the summer of 1933. In the following October, I was elected to a Prize Fellowship at Trinity College for the period 1933—37. During my Fellowship years at Trinity, I formed lasting friendships with several, including Sir Arthur Eddington and Professor E. A. Milne.

While on a short visit to Harvard University (in Cambridge, Massachusetts), at the invitation of the then Director, Dr. Harlow Shapley, during the winter months (January—March) of 1936, I was offered a position as a Research Associate at the University of Chicago by Dr. Otto Struve and President Robert Maynard Hutchins. I joined the faculty of the University of Chicago in January 1937. And I have remained at this University ever since.

During my last two years (1928—30) at the Presidency College in Madras, I formed a friendship with Lalitha Doraiswamy, one year my junior. This friendship matured; and we were married (in India) in September 1936 prior to my joining the University of Chicago. In the sharing of our lives during the past forty-seven years, Lalitha's patient understanding, support, and encouragement have been the central facts of my life.

After the early preparatory years, my scientific work has followed a certain pattern motivated, principally, by a quest after perspectives. In practise, this quest has consisted in my choosing (after some trials and tribulations) a certain area which appears amenable to cultivation and compatible with my taste, abilities, and temperament. And when after some years of study, I feel that I have accumulated a sufficient body of knowledge and achieved a view of my own, I have the urge to present my point of view, ab initio, in a coherent account with order, form, and structure.

There have been seven such periods in my life: stellar structure, including the theory of white dwarfs (1929–1939); stellar dynamics, including the theory of Brownian motion (1938–1943); the theory of radiative transfer, including the theory of stellar atmospheres and the quantum theory of the negative ion of hydrogen and the theory of planetary atmospheres, including the theory of the illumination and the polarization of the sunlit sky (1943–1950); hydrodynamic and hydromagnetic stability, including the theory of the Rayleigh-Bernard convection (1952–1961); the equilibrium and the stability of ellipsoidal figures of equilibrium, partly in collaboration with Norman R. Lebovitz (1961–1968); the general theory of relativity and relativistic astrophysics (1962–1971); and the mathematical theory of black holes (1974–1983). The monographs which resulted from these several periods are:

1. An Introduction to the Study of Stellar Structure (1939, University of Chicago Press; reprinted by Dover Publications, Inc., 1967).

2a. Principles of Stellar Dynamics (1943, University of Chicago Press; reprinted by Dover Publications, Inc., 1960).

2b. 'Stochastic Problems in Physics and Astronomy', *Reviews of Modern Physics*, 15, 1–89 (1943); reprinted in *Selected Papers on Noise and Stochastic Processes* by Nelson Wax, Dover Publications, Inc., 1954.

3. Radiative Transfer (1950, Clarendon Press, Oxford; reprinted by Dover Publications, Inc., 1960).

4. Hydrodynamic and Hydromagnetic Stability (1961, Clarendon Press, Oxford; reprinted by Dover Publications, Inc., 1981).

5. Ellipsoidal Figures of Equilibrium (1968; Yale University Press).

6. The Mathematical Theory of Black Holes (1983, Clarendon Press, Oxford).

However, the work which appears to be singled out in the citation for the award of the Nobel Prize is included in the following papers:

'The highly collapsed configurations of a stellar mass', *Mon. Not. Roy. Astron. Soc.*, 91, 456–66 (1931).

'The maximum mass of ideal white dwarfs', *Astrophys. J.*, 74, 81–2 (1931).

'The density of white dwarf stars', *Phil. Mag.*, 11, 592–96 (1931).

'Some remarks on the state of matter in the interior of stars', *Z. f. Astrophysik*, 5, 321–27 (1932).

'The physical state of matter in the interior of stars', *Observatory*, 57, 93–9 (1934).

'Stellar configurations with degenerate cores', *Observatory*, 57, 373–77 (1934).

'The highly collapsed configurations of a stellar mass' (second paper), *Mon. Not. Roy. Astron. Soc.*, 95, 207—25 (1935).

'Stellar configurations with degenerate cores', *Mon. Not. Roy. Astron. Soc.*, 95, 226—60 (1935).

'Stellar configurations with degenerate cores' (second paper), *Mon. Not. Roy. Astron. Soc.*, 95, 676—93 (1935).

'The pressure in the interior of a star', *Mon. Not. Roy. Astron. Soc.*, 96, 644—47 (1936).

'On the maximum possible central radiation pressure in a star of a given mass', *Observatory*, 59, 47—8 (1936).

'Dynamical instability of gaseous masses approaching the Schwarzschild limit in general relativity', *Phys. Rev. Lett.*, 12, 114—16 (1964); Erratum, *Phys. Rev. Lett.*, 12, 437—38 (1964).

'The dynamical instability of the white-dwarf configurations approaching the limiting mass' (with Robert F. Tooper), *Astrophys. J.*, 139, 1396—98 (1964).

'The dynamical instability of gaseous masses approaching the Schwarzschild limit in general relativity', *Astrophys. J.*, 140, 417—33 (1964).

'Solutions of two problems in the theory of gravitational radiation', *Phys. Rev. Lett.*, 24, 611—15 (1970); Erratum, *Phys. Rev. Lett.*, 24, 762 (1970).

'The effect of graviational radiation on the secular stability of the Maclaurin spheroid', *Astrophys. J.*, 161, 561—69 (1970).

ON STARS, THEIR EVOLUTION
AND THEIR STABILITY

Nobel lecture, 8 December, 1983

by

SUBRAHMANYAN CHANDRASEKHAR

The University of Chicago, Chicago, Illinois 60637, USA

1. *Introduction*

When we think of atoms, we have a clear picture in our minds: a central nucleus and a swarm of electrons surrounding it. We conceive them as small objects of sizes measured in Ängstroms ($\sim 10^{-8}$ cm); and we know that some hundred different species of them exist. This picture is, of course, quantified and made precise in modern quantum theory. And the success of the entire theory may be traced to two basic facts: *first*, the Bohr radius of the ground state of the hydrogen atom, namely,

$$\frac{h^2}{4\pi^2 m e^2} \sim 0.5 \times 10^{-8} \text{ cm,} \tag{1}$$

where h is Planck's constant, m is the mass of the electron and e is its charge, provides a correct measure of atomic dimensions; and *second*, the reciprocal of *Sommerfeld's fine-structure constant*,

$$\frac{hc}{2\pi e^2} \sim 137, \tag{2}$$

gives the maximum positive charge of the central nucleus that will allow a stable electron-orbit around it. This maximum charge for the central nucleus arises from the effects of special relativity on the motions of the orbiting electrons.

We now ask: can we understand the basic facts concerning stars as simply as we understand atoms in terms of the two combinations of natural constants (1) and (2). In this lecture, I shall attempt to show that in a limited sense we can.

The most important fact concerning a star is its mass. It is measured in units of the mass of the sun, $\odot$, which is 2×10^{33} gm: stars with masses very much less than, or very much more than, the mass of the sun are relatively infrequent. The current theories of stellar structure and stellar evolution derive their successes largely from the fact that the following combination of the dimensions of a mass provides a correct measure of stellar masses:

$$\left(\frac{hc}{G} \right)^{3/2} \frac{1}{H^2} \simeq 29.2 \odot, \tag{3}$$

where G is the constant of gravitation and H is the mass of the hydrogen atom. In the first half of the lecture, I shall essentially be concerned with the question: how does this come about?

2. *The role of radiation pressure*

A central fact concerning normal stars is the role which radiation pressure plays as a factor in their hydrostatic equilibrium. Precisely the equation governing the hydrostatic equilibrium of a star is

$$\frac{dP}{dr} = -\frac{GM(r)}{r^2}\rho, \tag{4}$$

where P denotes the total pressure, ρ the density, and $M(r)$ is the mass interior to a sphere of radius r. There are two contributions to the total pressure P: that due to the material and that due to the radiation. On the assumption that the matter is in the state of a perfect gas in the classical Maxwellian sense, the material or the gas pressure is given by

$$p_{gas} = \frac{k}{\mu H}\rho T, \tag{5}$$

where T is the absolute temperature, k is the Boltzmann constant, and μ is the mean molecular weight (which under normal stellar conditions is ~ 1.0). The pressure due to radiation is given by

$$p_{rad} = \frac{1}{3}\alpha T^4, \tag{6}$$

where α denotes Stefan's radiation-constant. Consequently, if radiation contributes a fraction $(1-\beta)$ to the total pressure, we may write

$$P = \frac{1}{1-\beta}\frac{1}{3}\alpha T^4 = \frac{1}{\beta}\frac{k}{\mu H}\rho T. \tag{7}$$

To bring out explicitly the role of the radiation pressure in the equilibrium of a star, we may eliminate the temperature, T, from the foregoing equations and express P in terms of ρ and β instead of in terms of ρ and T. We find:

$$T = \left(\frac{k}{\mu H}\frac{3}{\alpha}\frac{1-\beta}{\beta}\right)^{1/3}\rho^{1/3} \tag{8}$$

and

$$P = \left[\left(\frac{k}{\mu H}\right)^4\frac{3}{\alpha}\frac{1-\beta}{\beta^4}\right]^{1/3}\rho^{4/3} = C(\beta)\rho^{4/3} \text{ (say)}. \tag{9}$$

The importance of this ratio, $(1-\beta)$, for the theory of stellar structure was first emphasized by Eddington. Indeed, he related it, in a famous passage in his book on *The Internal Constitution of the Stars*, to the 'happening of the stars'.[1] A more rational version of Eddington's argument which, at the same time, isolates the combination (3) of the natural constants is the following:

There is a general theorem[2] which states that the pressure, P_c, at the centre of a star of a mass M in hydrostatic equilibrium in which the density, $\rho(r)$, at a point at a radial distance, r, from the centre does not exceed the mean density, $\bar{\rho}(r)$, interior to the same point r, must satisfy the inequality,

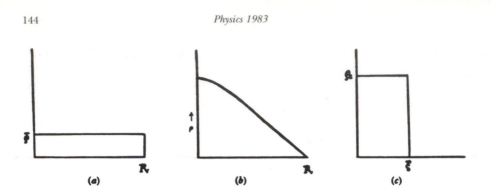

Fig. 1. A comparison of an inhomogeneous distribution of density in a star (*b*) with the two homogeneous configurations with the constant density equal to the mean density (*a*) and equal to the density at the centre (*c*).

$$\frac{1}{2}\,G\left(\frac{4}{3}\pi\right)^{1/3}\bar{\rho}^{4/3}\,M^{2/3} \leqslant \rho_c \leqslant \frac{1}{2}\,G\left(\frac{4}{3}\pi\right)^{1/3}\rho_c^{4/3}\,M^{2/3}, \tag{10}$$

where $\bar{\rho}$ denotes the mean density of the star and ρ_c its density at the centre. The content of the theorem is no more than the assertion that the actual pressure at the centre of a star must be intermediate between those at the centres of the two configurations of uniform density, one at a density equal to the mean density of the star, and the other at a density equal to the density ρ_c at the centre (see Fig. 1). If the inequality (10) should be violated then there must, in general, be some regions in which adverse density gradients must prevail; and this implies instability. In other words, we may consider conformity with the inequality (10) as equivalent to the condition for the stable existence of stars.

The right-hand side of the inequality (10) together with P given by equation (9), yields, for the stable existence of stars, the condition,

$$\left[\left(\frac{k}{\mu H}\right)^4 \frac{3}{a} \frac{1-\beta_c}{\beta_c^4}\right]^{1/3} \leqslant \left(\frac{\pi}{6}\right)^{1/3} G\,M^{2/3}, \tag{11}$$

or, equivalently,

$$M \geqslant \left(\frac{6}{\pi}\right)^{1/2}\left[\left(\frac{k}{\mu H}\right)^4 \frac{3}{a} \frac{1-\beta_c}{\beta_c^4}\right]^{1/2} \frac{1}{G^{3/2}}, \tag{12}$$

where in the foregoing inequalities, β_c is a value of β at the centre of the star. Now Stefan's constant, a, by virtue of Planck's law, has the value

$$a = \frac{8\pi^5 k^4}{15h^3 c^3}. \tag{13}$$

Inserting this value a in the inequality (12) we obtain

$$\mu^2 M\left(\frac{\beta_c^4}{1-\beta_c}\right)^{1/2} \geqslant \frac{(135)^{1/2}}{2\pi^3}\left(\frac{hc}{G}\right)^{3/2}\frac{1}{H^2} = 0.1873\left(\frac{hc}{G}\right)^{3/2}\frac{1}{H^2}. \tag{14}$$

We observe that the inequality (14) has isolated the combination (3) of natural constants of the dimensions of a mass; by inserting its numerical value given in equation (3), we obtain the inequality,

$$\mu^2 M \left(\frac{\beta_c^4}{1-\beta_c}\right)^{1/2} \geqslant 5.48 \odot. \tag{15}$$

This inequality provides an upper limit to $(1-\beta_c)$ for a star of a given mass. Thus,

$$1-\beta_c \leqslant 1-\beta_*, \tag{16}$$

where $(1-\beta_*)$ is uniquely determined by the mass M of the star and the mean molecular weight, μ, by the quartic equation,

$$\mu^2 M = 5.48 \left(\frac{1-\beta_*}{\beta_*^4}\right)^{1/2} \odot. \tag{17}$$

In Table 1, we list the values of $1-\beta_*$ for several values of $\mu^2 M$. From this table it follows in particular, that for a star of solar mass with a mean molecular weight equal to 1, the radiation pressure at the centre cannot exceed 3 percent of the total pressure.

Table 1
The maximum radiation pressure, $(1-\beta_*)$,
at the centre of a star of a given mass, M.

$1-\beta_*$	$M\mu^2/\odot$	$1-\beta_*$	$M\mu^2/\odot$
0.01	0.56	0.50	15.49
.03	1.01	.60	26.52
.10	2.14	.70	50.92
.20	3.83	.80	122.5
.30	6.12	.85	224.4
0.40	9.62	0.90	519.6

What do we conclude from the foregoing calculation? We conclude that to the extent equation (17) is at the base of the equilibrium of actual stars, to that extent the combination of natural constants (3), providing a mass of proper magnitude for the measurement of stellar masses, is at the base of a physical theory of stellar structure.

3. *Do stars have enough energy to cool?*

The same combination of natural constants (3) emerged soon afterward in a much more fundamental context of resolving a paradox Eddington had formulated in the form of an aphorism: 'a star will need energy to cool.' The paradox arose while considering the ultimate fate of a gaseous star in the light of the then new knowledge that white-dwarf stars, such as the companion of Sirius, exist, which have mean densities in the range 10^5-10^7 gm cm^{-3}. As Eddington stated[3]

> I do not see how a star which has once got into
> this compressed state is ever going to get out of it....
> It would seem that the star will be in an awkward pre-
> dicament when its supply of subatomic energy fails.

The paradox posed by Eddington was reformulated in clearer physical terms by R. H. Fowler.[4] His formulation was the following:

> The stellar material, in the white-dwarf state,
> will have radiated so much energy that it has less en-
> ergy than the same matter in normal atoms expanded at
> the absolute zero of temperature. If part of it were
> removed from the star and the pressure taken off, what
> could it do?

Quantitatively, Fowler's question arises in this way.

An estimate of the electrostatic energy, E_V, per unit volume of an assembly of atoms, of atomic number Z, ionized down to bare nuclei, is given by

$$E_V = 1.32 \times 10^{11} Z^2 \rho^{4/3}, \tag{18}$$

while the kinetic energy of thermal motions, E_{kin}, per unit volume of free particles in the form of a perfect gas of density, ρ, and temperature, T, is given by

$$E_{kin} = \frac{3}{2} \frac{k}{\mu H} \rho\, T = \frac{1.24 \times 10^8}{\mu} \rho\, T. \tag{19}$$

Now if such matter were released of the pressure to which it is subject, it can resume a state of ordinary normal atoms only if

$$E_{kin} > E_V, \tag{20}$$

or, according to equations (18) and (19), only if

$$\rho < \left(0.94 \times 10^{-3} \frac{T}{\mu Z^2} \right)^3. \tag{21}$$

This inequality will be clearly violated if the density is sufficiently high. This is the essence of Eddington's paradox as formulated by Fowler. And Fowler resolved this paradox in 1926 in a paper[4] entitled 'Dense Matter' — one of the great landmark papers in the realm of stellar structure: in it the notions of Fermi statistics and of electron degeneracy are introduced for the first time.

4. *Fowler's resolution of Eddington's paradox; the degeneracy of the electrons in white-dwarf stars*

In a completely degenerate electron gas all available parts of the phase space, with momenta less than a certain 'threshold' value p_o — the Fermi 'threshold' — are occupied consistently with the Pauli exclusion-principle i.e., with two electrons per 'cell' of volume h^3 of the six-dimensional phase space. Therefore, if $n(p) \, dp$ denotes the number of electrons, per unit volume, between p and $p+dp$, then the assumption of complete degeneracy is equivalent to the assertion,

$$
\begin{aligned}
n(p) &= \frac{8\pi}{h^3}p^2 \quad (p \leqslant p_o), \\
&= 0 \quad\quad (p > p_o).
\end{aligned}
\Bigg\}
\tag{22}
$$

The value of the threshold momentum p_o, is determined by the normalization condition

$$
n = \int_0^{p_o} n(p)dp = \frac{8\pi}{3h^3}p_o^3,
\tag{23}
$$

where n denotes the total number of electrons per unit volume.

For the distribution given by (22), the pressure P and the kinetic energy E_{kin} of the electrons (per unit volume), are given by

$$
P = \frac{8\pi}{3h^3}\int_0^{p_o} p^3 v_p \, dp
\tag{24}
$$

and

$$
E_{\text{kin}} = \frac{8\pi}{h^3}\int_0^{p_o} p^2 \, T_p \, dp,
\tag{25}
$$

where v_p and T_p are the velocity and the kinetic energy of an electron having a momentum p.

If we set

$$
v_p = p/m \text{ and } T_p = p^2/2m,
\tag{26}
$$

appropriate for non-relativistic mechanics, in equations (24) and (25), we find

$$
P = \frac{8\pi}{15h^3m}p_o^5 = \frac{1}{20}\left(\frac{3}{\pi}\right)^{2/3}\frac{h^2}{m}n^{5/3}
\tag{27}
$$

and

$$
E_{\text{kin}} = \frac{8\pi}{10h^3m}p_o^5 = \frac{3}{40}\left(\frac{3}{\pi}\right)^{2/3}\frac{h^2}{m}n^{5/3}.
\tag{28}
$$

Fowler's resolution of Eddington's paradox consists in this: at the temperatures and densities that may be expected to prevail in the interiors of the white-dwarf stars, the electrons will be highly degenerate and E_{kin} must be evaluated in accordance with equation (28) and *not* in accordance with equation (19); and equation (28) gives,

$$
E_{\text{kin}} = 1.39 \times 10^{13} \, (\rho/\mu)^{5/3}.
\tag{29}
$$

Comparing now the two estimates (18) and (29), we see that, for matter of the density occurring in the white dwarfs, namely $\rho \approx 10^5$ gm cm^{-3}, the total kinetic energy is about two to four times the negative potential-energy; and Eddington's paradox does not arise. Fowler concluded his paper with the following highly perceptive statement:

> The black-dwarf material is best likened to a single gigantic molecule in its lowest quantum state. On the Fermi-Dirac statistics, its high density can be achieved in one and only one way, in virtue of a correspondingly great energy content. But this energy can no more be expended in radiation than the energy of a normal atom or molecule. The only difference between black-dwarf matter and a normal molecule is that the molecule can exist in a free state while the black-dwarf matter can only so exist under very high external pressure.

5. *The theory of the white-dwarf stars; the limiting mass*

The internal energy ($= 3P/2$) of a degenerate electron gas that is associated with a pressure P is *zero-point energy;* and the essential content of Fowler's paper is that this zero-point energy is so great that we may expect a star to eventually settle down to a state in which all of its energy is of this kind. Fowler's argument can be more explicitly formulated in the following manner.[5]

According to the expression for the pressure given by equation (27), we have the relation,

$$P = K_1\rho^{5/3} \text{ where } K_1 = \frac{1}{20}\left(\frac{3}{\pi}\right)^{2/3}\frac{h^2}{m(\mu_e H)^{5/3}}, \tag{30}$$

where μ_e is the mean molecular weight per electron. An equilibrium configuration in which the pressure, P, and the density ρ, are related in the manner,

$$P = K\rho^{1+1/n}, \tag{31}$$

is an *Emden polytrope* of index n. The degenerate configurations built on the equation of state (30) are therefore polytropes of index $3/2$; and the theory of polytropes immediately provides the relation,

$$K_1 = 0.4242\,(GM^{1/3}\,R) \tag{32}$$

or, numerically, for K_1 given by equation (30),

$$\log_{10}(R/R_\odot) = -\frac{1}{3}\log_{10}(M/\odot) - \frac{5}{3}\log_{10}\mu_e - 1.397. \tag{33}$$

For a mass equal to the solar mass and $\mu_e = 2$, the relation (33) predicts $R = 1.26 \times 10^{-2} R_\odot$ and a mean density of 7.0×10^5 gm/cm^3. These values are precisely of the order of the radii and mean densities encountered in white-dwarf stars. Moreover, according to equations (32) and (33), the radius of the white-dwarf configuration is inversely proportional to the cube root of the mass. On this account, finite equilibrium configurations are predicted for all masses. And it came to be accepted that the white-dwarfs represent the last stages in the evolution of all stars.

But it soon became clear that the foregoing simple theory based on Fowler's premises required modifications. For the electrons, at their threshold energies at the centres of the degenerate stars, begin to have velocities comparable to that of light as the mass increases. Thus, already for a degenerate star of solar mass (with $\mu_e = 2$) the central density (which is about six times the mean density) is 4.19×10^6 gm/cm^3; and this density corresponds to a threshold momentum $p_o = 1.29 \, mc$ and a velocity which is $0.63 \, c$. Consequently, the equation of state must be modified to take into account the effects of special relativity. And this is easily done by inserting in equations (24) and (25) the relations,

$$v_p = \frac{p}{m \, (1+p^2/m^2c^2)^{1/2}} \text{ and } T_p = mc^2\left[(1+p^2/m^2c^2)^{1/2}-1\right], \tag{34}$$

in place of the non-relativistic relations (26). We find that the resulting equation of state can be expressed, parametrically, in the form

$$P = A f(x) \text{ and } \rho = Bx^3, \tag{35}$$

where

$$A = \frac{\pi m^4 c^5}{3h^3}, \; B = \frac{8\pi m^3 c^3 \mu_e \, H}{3h^3} \tag{36}$$

and

$$f(x) = x(x^2+1)^{1/2} \, (2x^2-3)+3 \sinh^{-1} x. \tag{37}$$

And similarly

$$E_{\text{kin}} = Ag(x), \tag{38}$$

where

$$g(x) = 8x^3\left[(x^2+1)^{1/2}-1\right]-f(x). \tag{39}$$

According to equations (35) and (36), the pressure approximates the relation (30) for low enough electron concentrations ($x \ll 1$); but for increasing electron concentrations ($x \gg 1$), the pressure tends to[6]

$$P = \frac{1}{8}\left(\frac{3}{\pi}\right)^{1/3} hc \, n^{4/3}. \tag{40}$$

This limiting form of relation can be obtained very simply by setting $v_p = c$ in equation (24); then

$$P = \frac{8\pi c}{3h^3}\int_0^{p_o} p^3 \, dp = \frac{2\pi c}{3h^3}p_o^4; \tag{41}$$

and the elimination of p_o with the aid of equation (23) directly leads to equation (40).

    While the modification of the equation of state required by the special theory of relativity appears harmless enough, it has, as we shall presently show, a dramatic effect on the predicted mass-radius relation for degenerate configurations.

The relation between P and ρ corresponding to the limiting form (41) is

$$P = K_2\,\rho^{4/3} \text{ where } K_2 = \frac{1}{8}\left(\frac{3}{\pi}\right)^{1/3}\frac{hc}{(\mu_e H)^{4/3}}. \tag{42}$$

In this limit, the configuration is an Emden polytrope of index 3. And it is well known that when the polytropic index is 3, the mass of the resulting equilibrium configuration is uniquely determined by the constant of proportionality, K_2, in the pressure-density relation. We have accordingly,

$$M_{\text{limit}} = 4\pi\left(\frac{K_2}{\pi G}\right)^{3/2}(2.018) = 0.197\left(\frac{hc}{G}\right)^{3/2}\frac{1}{(\mu_e H)^2} = 5.76\mu_e^{-2}\,\odot. \tag{43}$$

(In equation (43), 2.018 is a numerical constant derived from the explicit solution of the Lane-Emden equation for $n = 3$.)

It is clear from general considerations[7] that *the exact mass-radius relation for the degenerate configurations must provide an upper limit to the mass of such configurations given by equation* (43); *and further, that the mean density of the configuration must tend to infinity*, while the radius tends to zero, and $M \rightarrow M_{\text{limit}}$. These conditions, straightforward as they are, can be established directly by considering the equilibrium of configurations built on the exact equation of state given by equations (35)−(37). It is found that the equation governing the equilibrium of such configurations can be reduced to the form[8,9]

$$\frac{1}{\eta^2}\frac{d}{d\eta}\left(\eta^2\frac{d\phi}{d\eta}\right) = -\left(\phi^2 - \frac{1}{y_0^2}\right)^{3/2}, \tag{44}$$

where

$$y_0^2 = x_0^2 + 1, \tag{45}$$

and mcx_0 denotes the threshold momentum of the electrons at the centre of the configuration and η measures the radial distance in the unit

$$\left(\frac{2A}{\pi G}\right)^{1/2}\frac{1}{By_0} = l_1\,y_0^{-1}\,(\text{say}). \tag{46}$$

By integrating equation (44), with suitable boundary conditions and for various initially prescribed values of y_0, we can derive the exact mass-radius relation, as well as the other equilibrium properties, of the degenerate configurations. The principal results of such calculations are illustrated in Figures 2 and 3.

The important conclusions which follow from the foregoing considerations are: *first*, there is an upper limit, M_{limit}, to the mass of stars which can become degenerate configurations, as the last stage in their evolution; and *second*, that stars with $M > M_{\text{limit}}$ must have end states which cannot be predicted from the considerations we have presented so far. And finally, we observe that the

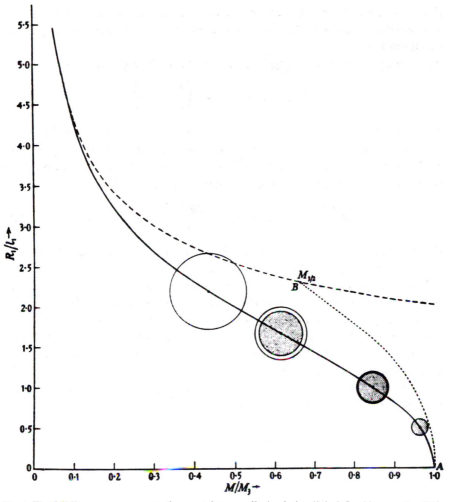

Fig. 2. The full-line curve represents the exact (mass-radius)-relation (l_1 is defined in equation (46) and M_3 denotes the limiting mass). This curve tends asymptotically to the - - - - curve appropriate to the low-mass degenerate configurations, approximated by polytropes of index 3/2. The regions of the configurations which may be considered as relativistic ($\varrho > (K_1/K_2)^3$) are shown shaded. (From Chandrasekhar, S., *Mon. Not. Roy. Astr. Soc.*, 95, 207 (1935).)

combination of the natural constant (3) now emerges in the fundamental context of M_{limit} given by equation (43): its significance for the theory of stellar structure and stellar evolution can no longer be doubted.

6. *Under what conditions can normal stars develop degenerate cores?*

Once the upper limit to the mass of completely degenerate configurations had been established, the question that required to be resolved was how to relate its existence to the evolution of stars from their gaseous state. If a star has a mass less than M_{limit}, the assumption that it will eventually evolve towards the completely degenerate state appears reasonable. But what if its mass is greater than M_{limit}?

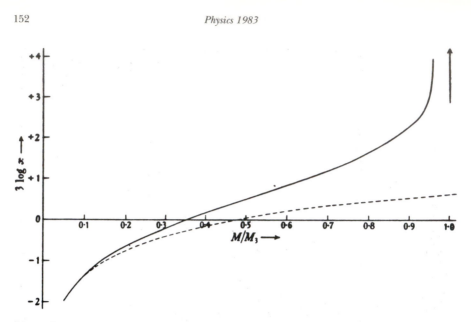

Fig. 3. The full-line curve represents the exact (mass-density)-relation for the highly collapsed configurations. This curve tends asymptotically to the dotted curve as $M \rightarrow 0$. (From Chandrasekhar, S., *Mon. Not. Roy. Astr. Soc.*, 95, 207 (1935).)

Clues as to what might ensue were sought in terms of the equations and inequalities of §§2 and 3.[10, 11]

The first question that had to be resolved concerns the circumstances under which a star, initially gaseous, will develop degenerate cores. From the physical side, the question, when departures from the perfect-gas equation of state (5) will set in and the effects of electron degeneracy will be manifested, can be readily answered.

Suppose, for example, that we continually and steadily increase the density, at constant temperature, of an assembly of free electrons and atomic nuclei, in a highly ionized state and initially in the form of a perfect gas governed by the equation of state (5). At first the electron pressure will increase linearly with ρ; but soon departures will set in and eventually the density will increase in accordance with the equation of state that describes the fully degenerate electron-gas (see Fig. 4). The remarkable fact is that this limiting form of the equation of state is independent of temperature.

However, to examine the circumstances when, during the course of evolution, a star will develop degenerate cores, it is more convenient to express the electron pressure (as given by the classical perfect-gas equation of state) in terms of ρ and β_e defined in the manner (cf. equation (7)).

$$p_e = \frac{k}{\mu_e H} \rho T = \frac{\beta_e}{1-\beta_e} \frac{1}{3} \alpha T^4, \tag{47}$$

where p_e now denotes the electron pressure. Then, analogous to equation (9), we can write

$$p_e = \left[\left(\frac{k}{\mu_e H} \right)^4 \frac{3}{\alpha} \frac{1-\beta_e}{\beta_e} \right]^{1/3} \rho^{4/3}. \tag{48}$$

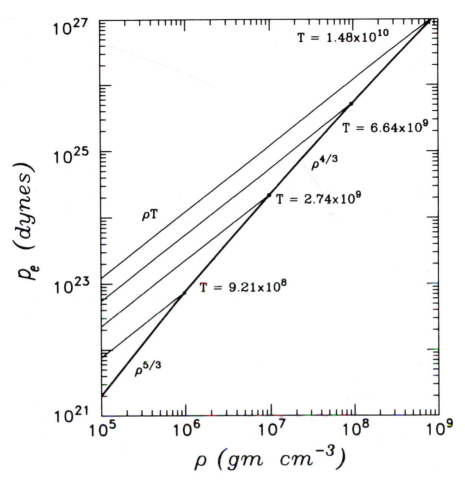

Fig. 4. Illustrating how by increasing the density at constant temperature degeneracy always sets in.

Comparing this with equation (42), we conclude that if

$$\left[\left(\frac{k}{\mu_e H}\right)^4 \frac{3}{\alpha} \frac{1-\beta_e}{\beta_e}\right]^{1/3} > K_2 = \frac{1}{8}\left(\frac{3}{\pi}\right)^{1/3} \frac{hc}{(\mu_e H)^{4/3}}, \tag{49}$$

the pressure p_e given by the classical perfect-gas equation of state will be greater than that given by the equation if degeneracy were to prevail, not only for the prescribed ρ and T, but for *all p and T having the same β_e.*

Inserting for α its value given in equation (13), we find that the inequality (49) reduces to

$$\frac{960}{\pi^4} \frac{1-\beta_e}{\beta_e} > 1, \tag{50}$$

or equivalently,

$$1-\beta_e > 0.0921 = 1-\beta_\omega \text{ (say).} \tag{51}$$

(See Fig. 5)

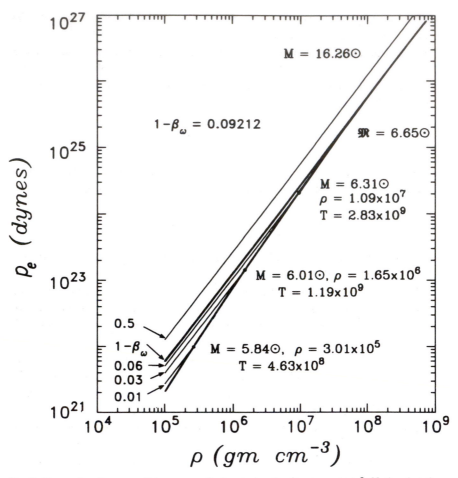

Fig. 5. Illustrating the onset of degeneracy for increasing density at constant β. Notice that there are no intersections for β> 0.09212. In the figure, 1-β is converted into mass of a star built on the standard model.

For our present purposes, the principal content of the inequality (51) is the criterion that for a star to develop degeneracy, it is necessary that the radiation pressure be less than 9.2 percent of (p_e+p_{rad}). This last inference is so central to all current schemes of stellar evolution that the directness and the simplicity of the early arguments are worth repeating.

The two principal elements of the early arguments were these: *first*, that radiation pressure becomes increasingly dominant as the mass of the star increases; and *second*, that the degeneracy of electrons is possible only so long as the radiation pressure is not a significant fraction of the total pressure — indeed, as we have seen, it must not exceed 9.2 percent of (p_e+p_{rad}). The second of these elements in the arguments is a direct and an elementary consequence of the physics of degeneracy; but the first requires some amplification.

That radiation pressure must play an increasingly dominant role as the mass of the star increases is one of the earliest results in the study of stellar structure that was established by Eddington. A quantitative expression for this fact is given by Eddington's *standard model* which lay at the base of early studies summarized in his *The Internal Constitution of the Stars*.

On the standard model, the fraction β (= gas pressure/total pressure) is a constant through a star. On this assumption, the star is a polytrope of index 3 as is apparent from equation (9); and, in consequence, we have the relation (cf. equation (43))

$$M = 4\pi \left[\frac{C(\beta)}{\pi G}\right]^{3/2} (2.018) \qquad (52)$$

where $C(\beta)$ is defined in equation (9). Equation (52) provides a quartic equation for β analogous to equation (17) for β^*. Equation (52) for $\beta = \beta_\omega$ gives

$$M = 0.197 \beta_\omega^{-3/2} \left(\frac{hc}{G}\right)^{3/2} \frac{1}{(\mu H)^2} = 6.65 \mu^{-2} \odot = \boldsymbol{M} \text{ (say)}. \qquad (53)$$

On the standard model, then stars with masses exceeding $\boldsymbol{M}$ will have radiation pressures which exceed 9.2 percent of the total pressure. Consequently stars with $M > \boldsymbol{M}$ cannot, at any stage during the course of their evolution, develop degeneracy in their interiors. Therefore, for such stars an eventual white-dwarf state is not possible unless they are able to eject a substantial fraction of their mass.

The standard model is, of course, only a model. Nevertheless, except under special circumstances, briefly noted below, experience has confirmed the standard model, namely that the evolution of stars of masses exceeding $7-8 \odot$ must proceed along lines very different from those of less massive stars. These conclusions, which were arrived at some fifty years ago, appeared then so convincing that assertions such as these were made with confidence:

> Given an enclosure containing electrons and atomic nuclei (total charge zero) what happens if we go on compressing the material indefinitely? $(1932)^{10}$

> The life history of a star of small mass must be essentially different from the life history of a star of large mass. For a star of small mass the natural white-dwarf stage is an initial step towards complete extinction. A star of large mass cannot pass into the white-dwarf stage and one is left speculating on other possibilities. $(1934)^8$

And these statements have retained their validity.

While the evolution of the massive stars was thus left uncertain, there was no such uncertainty regarding the final states of stars of sufficiently low mass.[11] The reason is that by virtue, again, of the inequality (10), the maximum central pressure attainable in a star must be less than that provided by the degenerate equation of state, so long as

$$\frac{1}{2} G \left(\frac{4}{3}\pi\right)^{1/3} M^{2/3} < K_2 = \frac{1}{8} \left(\frac{3}{\pi}\right)^{1/3} \frac{hc}{(\mu_e H)^{4/3}} \qquad (54)$$

or, equivalently

$$M < \frac{3}{16\pi} \left(\frac{hc}{G}\right)^{3/2} \frac{1}{(\mu_e H)^2} = 1.74 \mu_e^{-2} \odot. \qquad (55)$$

We conclude that there can be no surprises in the evolution of stars of mass less than 0.43 $\odot$ (if $\mu_e = 2$). The end stage in the evolution of such stars can only be that of the white dwarfs. (Parenthetically, we may note here that the inequality (55) implies that the so-called 'mini' black-holes of mass $\sim 10^{15}$ gm cannot naturally be formed in the present astronomical universe.)

7. Some brief remarks on recent progress in the evolution of massive stars and the onset of gravitational collapse

It became clear, already from the early considerations, that the inability of the massive stars to became white dwarfs must result in the development of much more extreme conditions in their interiors and, eventually, in the onset of gravitational collapse attended by the super-nova phenomenon. But the precise manner in which all this will happen has been difficult to ascertain in spite of great effort by several competent groups of investigators. The facts which must be taken into account appear to be the following.[*]

In the first instance, the density and the temperature will steadily increase without the inhibiting effect of degeneracy since for the massive stars considered $1-\beta_e > 1-\beta_\omega$. On this account, 'nuclear ignition' of carbon, say, will take place which will be attended by the emission of neutrinos. This emission of neutrinos will effect a cooling and a lowering of $(1-\beta_e)$; but it will still be in excess of $1-\beta_\omega$. The important point here is that the emission of neutrinos acts selectively in the central regions and is the cause of the lowering of $(1-\beta_e)$ in these regions. The density and the temperature will continue to increase till the next ignition of neon takes place followed by further emission of neutrinos and a further lowering of $(1-\beta_e)$. This succession of nuclear ignitions and lowering of $(1-\beta_e)$ will continue till $1-\beta_e < 1-\beta_\omega$ and a relativistically degenerate core with a mass approximately that of the limiting mass ($=1.4 \odot$ for $\mu_e = 2$) forms at the centre. By this stage, or soon afterwards, instability of some sort is expected to set in (see following §8) followed by gravitational collapse and the phenomenon of the super-nova (of type II). In some instances, what was originally the highly relativistic degenerate core of approximately 1.4 $\odot$, will be left behind as a neutron star. That this happens sometimes is confirmed by the fact that in those cases for which reliable estimates of the masses of pulsars exist, they are consistently close to 1.4 $\odot$. However, in other instances — perhaps, in the majority of the instances — what is left behind, after all 'the dust has settled', will have masses in excess of that allowed for stable neutron stars; and in these instances black holes will form.

In the case of less massive stars ($M \sim 6-8 \odot$) the degenerate cores, which are initially formed, are not highly relativistic. But the mass of core increases with the further burning of the nuclear fuel at the interface of the core and the mantle; and when the core reaches the limiting mass, an explosion occurs following instability; and it is believed that this is the cause underlying super-nova phenomenon of type I.

---

[*] I am grateful to Professor D. Arnett for guiding me through the recent literature and giving me advice in the writing of this section.

From the foregoing brief description of what may happen during the late stages in the evolution of massive stars, it is clear that the problems one encounters are of exceptional complexity, in which a great variety of physical factors compete. This is clearly not the occasion for me to enter into a detailed discussion of these various questions. Besides, Professor Fowler may address himself to some of these matters in his lecture that is to follow.

8. *Instabilities of relativistic origin: (1) The vibrational instability of spherical stars*

I now turn to the consideration of certain types of stellar instabilities which are derived from the effects of general relativity and which have no counterparts in the Newtonian framework. It will appear that these new types of instabilities of relativistic origin may have essential roles to play in discussions pertaining to gravitational collapse and the late stages in the evolution of massive stars.

We shall consider first the stability of spherical stars for purely radial perturbations. The criterion for such stability follows directly from the linearized equations governing the spherically symmetric radial oscillations of stars. In the framework of the Newtonian theory of gravitation, the stability for radial perturbations depends only on an average value of the adiabatic exponent, Γ_1, which is the ratio of the fractional Lagrangian changes in the pressure and in the density experienced by a fluid element following the motion; thus,

$$\Delta P/P = \Gamma_1 \Delta\rho/\rho. \tag{56}$$

And the Newtonian criterion for stability is

$$\overline{\Gamma}_1 = \int_0^M \Gamma_1(r)\, P(r)\, d\, M(r) \div \int_0^M P(r)\, d\, M(r) > \frac{4}{3}. \tag{57}$$

If $\overline{\Gamma}_1 < 4/3$, *dynamical instability* of a global character will ensue with an *e*-folding time measured by the time taken by a sound wave to travel from the centre to the surface.

When one examines the same problem in the framework of the general theory of relativity, one finds [12] that, again, the stability depends on an average value of Γ_1; but contrary to the Newtonian result, the stability now depends on the radius of the star as well. Thus, one finds that no matter how high $\overline{\Gamma}_1$ may be, instability will set in provided the radius is less than a certain determinate multiple of the *Schwarzschild radius*,

$$R_s = 2\, GM/c^2. \tag{58}$$

Thus, if for the sake of simplicity, we assume that Γ_1 is a constant through the star and equal to 5/3, then the star will become dynamically unstable for radial perturbations, if $R_1 < 2.4\, R_s$. And further, if $\Gamma_1 \to \infty$, instability will set in for all $R < (9/8)\, R_s$. The radius $(9/8)\, R_s$ defines, in fact, *the minimum radius which any gravitating mass, in hydrostatic equilibrium, can have in the framework of general relativity.* This important result is implicit in a fundamental paper by Karl Schwarzschild published in 1916. (Schwarzschild actually proved that for a star in which the energy density is uniform, $R > (9/8)\, R_s$.)

In one sense, the most important consequence of this instability of relativistic origin is that if Γ_1 (again assumed to be a constant for the sake of simplicity) differs from and is greater than $4/3$ only by a small positive constant, then the instability will set in for a radius R which is a large multiple of R_s; and, therefore, under circumstances when the effects of general relativity, on the structure of the equilibrium configuation itself, are hardly relevant. Indeed, it follows[13] from the equations governing radial oscillations of a star, in a first post-Newtonian approximation to the general theory of relativity, that instability for radial perturbations will set in for all

$$R < \frac{K}{\Gamma_1 - 4/3} \frac{2GM}{c^2}, \tag{59}$$

where K is a constant which depends on the *entire** march of density and pressure in the equilibrium configuration in the Newtonian frame-work. Thus, for a polytrope of index n, the value of the constant is given by

$$K = \frac{5-n}{18} \left[\frac{2(11-n)}{(n+1) \xi_1^4 |\theta_1'|^3} \int_0^{\xi_1} \theta \left(\frac{d\theta}{d\xi}\right)^2 \xi^2 \, d\xi + 1 \right], \tag{60}$$

where θ is the Lane-Emden function in its standard normalization ($\theta = 1$ at $\xi = 0$), ξ is the dimensionless radial coordinate, ξ_1 defines the boundary of the polytrope (where $\theta = 0$) and θ_1' is the derivative of θ at ξ_1.

Table 2
Values of the constant K in the inequality (59)
for various polytropic indices, n.

n	K	n	K
0	0.452381	3.25	1.28503
1.0	.565382	3.5	1.49953
1.5	.645063	4.0	2.25338
2.0	.751296	4.5	4.5303
2.5	.900302	4.9	22.906
3.0	1.12447	4.95	45.94

In Table 2, we list the values of K for different polytropic indices. It should be particularly noted that K increases without limit for $n \to 5$ and the configuration becomes increasingly centrally condensed.** Thus, already for $n = 4.95$ (for which polytropic index $\rho_c = 8.09 \times 10^6 \bar{\rho}$), $K \sim 46$. In other words, for the highly centrally condensed massive stars (for which Γ_1 may differ from $4/3$ by as little

---

* It is for this reason that we describe the instability as *global*.
** Since this was written, it has been possible to show (Chandrasekhar and Lebovitz 13a) that for $n \to 5$, the asymptotic behaviour of K is given by

$$K \to 2.3056/(5-n).$$

and, further, that along the polytropic sequence, the criterion for instability (59) can be expressed alternatively in the form

$$R < 0.2264 \frac{2GM}{c^2} \frac{1}{\Gamma_1 - 4/3} \left(\frac{\rho_c}{\bar{\rho}}\right)^{1/3} \qquad (\rho_c > 10^6 \bar{\rho})$$

as 0.01),* the instability of relativistic origin will set in, already, when its radius falls below $5 \times 10^3\ R_s$. Clearly this relativistic instability must be considered in the contexts of these problems.

A further application of the result described in the preceding paragraph is to degenerate configurations near the limiting mass[14]. Since the electrons in these highly relativistic configurations have velocities close to the velocity of light, the effective value of Γ_1 will be very close to 4/3 and the post-Newtonian relativistic instability will set in for a mass slightly less than that of the limiting mass. On account of the instability for radial oscillations setting in for a mass less than M_{limit}, the period of oscillation, along the sequence of the degenerate configurations, must have a minimum. This minimum can be estimated to be about two seconds (see Fig. 6). Since pulsars, when they were discovered, were known to have periods much less than this minimum value, the possibility of their being degenerate configurations near the limiting mass was ruled out; and this was one of the deciding factors in favour of the pulsars being neutron stars. (But by a strange irony, for reasons we have briefly explained in § 7, pulsars which have resulted from super-nova explosions have masses close to 1.4 $\odot$!)

Finally, we may note that the radial instability of relativistic origin is the underlying cause for the *existence* of a maximum mass for stability: it is a direct consequence of the equations governing hydrostatic equilibrium in general relativity. (For a complete investigation on the periods of radial oscillation of neutron stars for various admissible equations of state, see a recent paper by Detweiler and Lindblom[15].)

9. Instabilities of relativistic origin: (2) The secular instability of rotating stars derived from the emission of gravitational radiation by non-axisymmetric modes of oscillation

I now turn to a different type of instability which the general theory of relativity predicts for rotating configurations. This new type of instability[16] has its origin in the fact that the general theory of relativity builds into rotating masses a dissipative mechanism derived from the possibility of the emission of gravitational radiation by non-axisymmetric modes of oscillation. It appears that this instability limits the periods of rotation of pulsars. But first, I shall explain the nature and the origin of this type of instability.

It is well known that a possible sequence of equilibrium figures of rotating homogeneous masses is the Maclaurin sequence of oblate spheroids[17]. When one examines the second harmonic oscillations of the Maclaurin spheroid, in a frame of reference rotating with its angular velocity, one finds that for two of these modes, whose dependence on the azimuthal angle is given by $e^{2i\varphi}$, the characteristic frequencies of oscillation, σ, depend on the eccentricity e in the manner illustrated in Figure 7. It will be observed that one of these modes becomes neutral (i.e., $\sigma = 0$) when $e = 0.813$ and that the two modes coalesce when $e = 0.953$ and become complex conjugates of one another beyond this

---

* By reason of the dominance of the radiation pressure in these massive stars and of β being very close to zero.

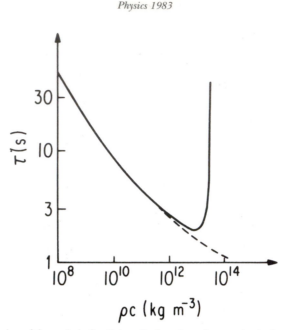

Fig. 6. The variation of the period of radial oscillation along the completely degenerate configurations. Notice that the period tends to infinity for a mass close to the limiting mass. There is consequently a minimum period of oscillation along these configurations; and the minimum period is approximately 2 seconds. (From J. Skilling, *Pulsating Stars* (Plenum Press, New York, 1968), p. 59.)

point. Accordingly, the Maclaurin spheroid becomes *dynamically unstable* at the latter point (first isolated by Riemann). On the other hand, the origin of the neutral mode at $e = 0.813$ is that at this point a new equilibrium sequence of triaxial ellipsoids—the ellipsoids of Jacobi—bifurcate. On this latter account, Lord Kelvin conjectured in 1883 that

> if there be any viscosity, however slight ... the equilibrium beyond $e = 0.81$ cannot be secularly stable.

Kelvin's reasoning was this: viscosity dissipates energy but not angular momentum. And since for equal angular momenta, the Jacobi ellipsoid has a lower energy content than the Maclaurin spheroid, one may expect that the action of viscosity will be to dissipate the excess energy of the Maclaurin spheroid and transform it into the Jacobi ellipsoid with the lower energy. A detailed calculation[18] of the effect of viscous dissipation on the two modes of oscillation, illustrated in Figure 7, does confirm Lord Kelvin's conjecture. It is found that viscous dissipation makes the mode, which becomes neutral at $e = 0.813$, unstable beyond this point with an e-folding time which depends inversely on the magnitude of the kinematic viscosity and which further decreases monotonically to zero at the point, $e = 0.953$ where the dynamical instability sets in.

Since the emission of gravitational radiation dissipates *both* energy and angular momentum, it does *not* induce instability in the Jacobi mode; instead it

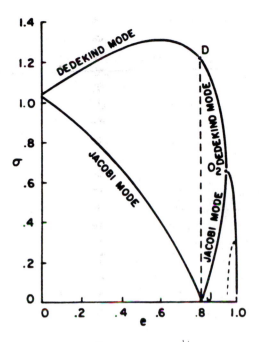

Fig. 7. The characteristic frequencies (in the unit $(\pi G \varrho)^{1/2}$) of the two even modes of second-harmonic oscillation of the Maclaurin spheriod. The Jacobi sequence bifurcates from the Maclaurin sequence by the mode that is neutral ($\sigma = 0$) at $e = 0.813$; and the Dedekind sequence bifurcates by the alternative mode at D. At O_2 ($e = 0.9529$) the Maclaurin spheroid becomes dynamically unstable. The real and the imaginary parts of the frequency, beyond O_2, are shown by the full line and the dashed curves, respectively. Viscous dissipation induces instability in the branch of the Jacobi mode; and radiation-reaction induces instability in the branch DO_2 of the Dedekind mode.

induces instability in the *alternative* mode at the same eccentricity. In the first instance this may appear surprising; but the situation we encounter here clarifies some important issues.

If instead of analyzing the normal modes in the rotating frame, we had analyzed them in the inertial frame, we should have found that the mode which becomes unstable by radiation reaction at $e = 0.813$, is in fact neutral at this point. And the neutrality of *this* mode in the inertial frame corresponds to the fact that the neutral deformation at this point is associated with the bifurcation (at this point) of a new triaxial sequence – the sequence of the Dedekind ellipsoids. These Dedekind ellipsoids, while they are congruent to the Jacobi ellipsoids, they differ from them in that they are at rest in the inertial frame and owe their triaxial figures to internal vortical motions. An important conclusion that would appear to follow from these facts is that in the framework of general relativity we can expect secular instability, derived from radiation reaction, to arise from a Dedekind mode of deformation (which is quasi-stationary in the inertial frame) rather than the Jacobi mode (which is quasi-stationary in the rotating frame).

A further fact concerning the secular instability induced by radiation reaction, discovered subsequently by Friedman[19] and by Comins[20], is that the modes belonging to higher values of m (= 3, 4, , ,) become unstable at smaller eccentricities though the e-folding times for the instability becomes rapidly longer. Nevertheless it appears from some preliminary calculations of Friedman[21] that it is the secular instability derived from modes belonging to $m = 3$ (or 4) that limit the periods of rotation of the pulsars.

It is clear from the foregoing discussions that the two types of instabilities of relativistic origin we have considered are destined to play significant roles in the contexts we have considered.

10. *The mathematical theory of black holes*

So far, I have considered only the restrictions on the last stages of stellar evolution that follow from the existence of an upper limit to the mass of completely degenerate configurations and from the instabilities of relativistic origin. From these and related considerations, the conclusion is inescapable that black holes will form as one of the natural end products of stellar evolution of massive stars; and further that they must exist in large numbers in the present astronomical universe. In this last section I want to consider very briefly what the general theory of relativity has to say about them. But first, I must define precisely what a black hole is.

A black hole partitions the three-dimensional space into two regions: an inner region which is bounded by a smooth two-dimensional surface called the *event horizon;* and an outer region, external to the event horizon, which is asymptotically flat; and it is required (as a part of the definition) that no point in the inner region can communicate with any point of the outer region. This incommunicability is guaranteed by the impossibility of any light signal, originating in the inner region, crossing the event horizon. The requirement of asymptotic flatness of the outer region is equivalent to the requirement that the black hole is isolated in space and that far from the event horizon the space-time approaches the customary space-time of terrestrial physics.

In the general theory of relativity, we must seek solutions of Einstein's vacuum equations compatible with the two requirements I have stated. It is a startling fact that compatible with these very simple and necessary requirements, the general theory of relativity allows for stationary (i.e., time-independent) black-holes exactly a single, unique, two-parameter family of solutions. This is the Kerr family, in which the two parameters are the mass of the black hole and the angular momentum of the black hole. What is even more remarkable, the metric describing these solutions is simple and can be explicitly written down.

I do not know if the full import of what I have said is clear. Let me explain.

Black holes are macroscopic objects with masses varying from a few solar masses to millions of solar masses. To the extent they may be considered as stationary and isolated, to that extent, they are all, every single one of them, described *exactly* by the Kerr solution. This is the only instance we have of an exact description of a macroscopic object. Macroscopic objects, as we see them all around us, are governed by a variety of forces, derived from a variety of approximations to a variety of physical theories. In contrast, the only elements in the construction of black holes are our basic concepts of space and time. They are, thus, almost by definition, the most perfect macroscopic objects there are in the universe. And since the general theory of relativity provides a single unique two-parameter family of solutions for their descriptions, they are the simplest objects as well.

Turning to the physical properties of the black holes, we can study them best by examining their reaction to external perturbations such as the incidence of waves of different sorts. Such studies reveal an analytic richness of the Kerr space-time which one could hardly have expected. This is not the occasion to elaborate on these technical matters[22]. Let it suffice to say that contrary to every prior expectation, all the standard equations of mathematical physics can be solved exactly in the Kerr space-time. And the solutions predict a variety and range of physical phenomena which black holes must exhibit in their interaction with the world outside.

The mathematical theory of black holes is a subject of immense complexity; but its study has convinced me of the basic truth of the ancient mottoes,

The simple is the seal of the true

and

Beauty is the splendour of truth.

REFERENCES

1. Eddington, A. S., *The Internal Constitution of the Stars* (Cambridge University Press, England, 1926), p. 16.
2. Chandrasekhar, S., *Mon. Not. Roy. Astr. Soc.*, 96, 644 (1936).
3. Eddington, A. S., *The Internal Constitution of the Stars* (Cambridge University Press, England, 1926), p. 172.
4. Fowler, R. H., *Mon. Not. Roy. Astr. Soc.*, 87, 114 (1926).
5. Chandrasekhar, S., *Phil. Mag.*, 11, 592 (1931).
6. Chandrasekhar, S., *Astrophys. J.*, 74, 81 (1931).
7. Chandrasekhar, S., *Mon. Not. Roy. Astr. Soc.*, 91, 456 (1931).
8. Chandrasekhar, S., *Observatory*, 57, 373 (1934).
9. Chandrasekhar, S., *Mon. Not. Roy. Astr. Soc.*, 95, 207 (1935).
10. Chandrasekhar, S., *Z. f. Astrophysik*, 5, 321 (1932).
11. Chandrasekhar, S., *Observatory*, 57, 93 (1934).
12. Chandrasekhar, S., *Astrophys. J.*, 140, 417 (1964); see also *Phys. Rev. Lett.*, 12, 114 and 437 (1964).
13. Chandrasekhar, S., *Astrophys. J.*, 142, 1519 (1965).
13 a. Chandrasekhar, S. and Lebovitz, N. R., *Mon. Not. Roy. Astr. Soc.*, 207, 13 P (1984).
14. Chandrasekhar, S. and Tooper, R. F., *Astrophys. J.*, 139, 1396 (1964).
15. Detweiler, S. and Lindblom, L., *Astrophys. J. Supp.*, 53, 93 (1983).
16. Chandrasekhar, S., *Astrophys. J.*, 161, 561 (1970); see also *Phys. Rev. Lett.*, 24, 611 and 762 (1970).
17. For an account of these matters pertaining to the classical ellipsoids see Chandrasekhar, S., *Ellipsoidal Figures of Equilibrium* (Yale University Press, New Haven, 1968).
18. Chandrasekhar, S., *Ellipsoidal Figures of Equilibrium* (Yale University Press, New Haven, 1968), Chap. 5, § 37.
19. Friedman, J. L., *Comm. Math. Phys.*, 62, 247 (1978); see also Friedman, J. L. and Schutz, B. F., *Astrophys. J.*, 222, 281 (1977).
20. Comins, N., *Mon. Not. Roy. Astr. Soc.*, 189, 233 and 255 (1979).
21. Friedman, J. L., *Phys. Rev. Lett.*, 51, 11 (1983).
22. The author's investigations on the mathematical theory of black holes, continued over the years 1974–1983, are summarized in his last book *The Mathematical Theory of Black Holes* (Clarendon Press, Oxford, 1983).

The reader may wish to consult the following additional references:
1. Chandrasekhar, S., 'Edward Arthur Milne: his part in the development of modern astrophysics,' *Quart. J. Roy. Astr. Soc.*, 21, 93–107 (1980).
2. Chandrasekhar, S., *Eddington: The Most Distinguished Astrophysicist of His Time* (Cambridge University Press, 1983).

WILLIAM ALFRED FOWLER

I was born in 1911 in Pittsburgh, Pennsylvania, the son of John MacLeod Fowler and Jennie Summers Watson Fowler. My parents had two other children, my younger brother, Arthur Watson Fowler and my still younger sister, Nelda Fowler Wood. My paternal grandfather, William Fowler, was a coal miner in Slammannan, near Falkirk, Scotland who emigrated to Pittsburgh to find work as a coal miner around 1880. My maternal grandfather, Alfred Watson, was a grocer. He emigrated to Pittsburgh, also around 1880, from Taniokey, near Clare in County Armagh, Northern Ireland. His parents taught in the National School, the local grammar school for children, in Taniokey, for sixty years. The family lived in the central part of the school building; my great grandfather taught the boys in one wing of the building and my great grandmother taught the girls in the other wing. The school is still there and I have been to see it.

I was raised in Lima, Ohio, from the age of two when my father, an accountant, was transferred to Lima from Pittsburgh. Each summer during my childhood the family went back to Pittsburgh during my father's vacation from work. He was an ardent sportsman and through him I became (and still am) a loyal fan of the Pittsburgh Pirates in the National Baseball League and of the Pittsburgh Steelers in the National Football League.

Lima was a railroad center served by the Pennsylvania, Erie, Nickel Plate and Baltimore & Ohio railroads. It was also the home of the Lima Locomotive Works which built steam locomotives. My brother, Arthur Watson Fowler, a mechanical engineer, worked for Lima Locomotive all his life until his retirement. After 1960 the company produced power shovels and construction cranes. As a boy I spent many hours in the switch yards of the Pennsylvania Railroad not far from my family home. It is no wonder that I go around the world seeking passenger trains still pulled by steam locomotives. In 1973 I travelled the Trans Siberian Railroad from Khabarovsk to Moscow because, among other reasons, the train was powered by steam for almost 2 500 kilometers from Khabarovsk to Chita. It's not powered by steam but now I can afford to ride on the new Orient Express. It is also no wonder that on my 60th birthday my colleagues and former students presented me in Cambridge, England, with a working model, 3 1/4″ gauge (1/16 standard size) British Tank Engine. I operated it frequently on the elevated track of the Cambridge and District Model Engineering Society. It is my pride and joy. I have named it *Prince Hal*.

I attended Horace Mann Grade School and Lima Central High School. A few of my high school teachers are still alive and I met them at my 50th class reunion in 1979. I was President of the Senior Class of 1929. My teachers encouraged and fostered my interest in engineering and science but also insisted that I take four years of Latin rather than French or German. My family home was located across the street from the extensive playgrounds of Horace Mann School. There were baseball diamonds, tennis courts, a running track and a football field. During my high school days I played on the Central High School football team and won my letter as a senior. Horace Mann was Central's home football field. During my college days I served as Recreational Director of the Horace Mann playground during the summer. Not far from my home was Baxter's Woods with a running creek and swimming hole. What a wonderful environment it all was for my boyhood!

On graduation from school I enrolled at the Ohio State University in Columbus, Ohio, in ceramic engineering. I had won a prize for an essay on the production of Portland cement and ceramic engineering seemed a natural choice for me. Fortunately all engineering students took the same courses including physics and mathematics. I became fascinated with physics and when I learned from Professor Alpheus Smith, head of the Physics Department, that there was a new degree offered in Engineering Physics I enrolled in that option at the start of my sophomore year. So also did Leonard I. Schiff, who became a very great theoretical physicist. We were lifelong friends until his death a few years ago.

My parents were not affluent and my summer salary as recreation director did not cover my expenses at Ohio State. For my meals I waited table, washed the dishes and stoked the furnaces at the Phi Sigma Sigma Sorority. I worked Saturdays cutting and selling ham and cheese in an outside stall at the Central Market in Columbus. Early in the morning we put up the stall and unloaded the hams and cheeses from the wholesaler's truck; late at night we cleaned up and took down the stall. For eighteen hours work I was paid five dollars. I did scrape enough money together to join a social fraternity, Tau Kappa Epsilon. In my junior year I was elected to the engineering honorary society, Tau Beta Pi, and in my senior year I was elected President of the Ohio State Chapter.

My professors at Ohio State solidified my interest in experimental physics. Willard Bennett permitted me to do an undergraduate thesis on the "Focussing of Electron Beams" in his laboratory. From him I learned how different a working laboratory is from a student laboratory. The answers are not known! John Byrne permitted me to work after school hours in the electronic laboratory of the Electrical Engineering Department. I studied the characteristics of the Pentode! It was the best of worlds—the thrills of making real measurements in physics along with practical training in engineering.

On graduation from Ohio State I came to Caltech and became a graduate student under Charles Christian Lauritsen—physicist, engineer, architect and violinist—in the W. K. Kellogg Radiation Laboratory. Kellogg was constructed to Lauritsen's architectural plans by funds obtained from the American corn flakes king by Robert Andrews Millikan. Lauritsen was a native of Denmark and in common with many Scandinavians he loved the songs of Carl Michael Bellman, the 18th century Swedish poet-musician. He tried to teach me to sing

Bellman's drinking songs with a good Swedish accent but I failed miserably except in spirit or should I say spirits. 'Del Delsasso dubbed me Willy and it stuck'.

Charlie Lauritsen was the greatest influence in my life. He supervised my doctoral thesis on "Radioactive Elements of Low Atomic Number" in which we discovered *mirror nuclei* and showed that the nuclear forces are charge symmetric—the same between two protons as between two neutrons when charged particle Coulomb forces are excluded. He taught me many practical things—how to repair motors, plumbing, and electrical wiring. Most of all he taught me how to *do* physics and how to *enjoy* it. I also learned from my fellow graduate students Richard Crane and Lewis Delsasso. Charlie's son, Tommy Lauritsen, did his doctoral work under us and the three of us worked together as a team for over thirty-five years. We were primarily experimentalists. In the early days Robert Oppenheimer taught us the theoretical implications of our results. Richard Tolman taught us not to rush into the publication of premature results in those days of intense competition between nuclear laboratories.

Hans Bethe's announcement of the CN-cycle in 1939 changed our lives. We were studying the nuclear reactions of protons with the isotopes of carbon and nitrogen in the laboratory, the very reactions in the CN-cycle. World War II intervened. The Kellogg Laboratory was engaged in defense research throughout the war. I spent three months in the South Pacific during 1944 as a civilian with simulated military rank. I saw at first hand the heroism of soldiers and seamen and the horrors they endured.

Just before the war I married Ardiane Foy Olmsted whose family came to California over the plains and mountains of the western United States in the Gold Rush around 1850. We are the parents of two daughters, Mary Emily and Martha Summers, whom we refer to as our biblical characters. Martha and her husband, Robert Schoenemann, are the parents of our grandson, Spruce William Schoenemann. They live in Pawlet, a small village in Vermont-the Green Mountain State.

After the war the Lauritsens and I restored Kellogg as a nuclear laboratory and decided to concentrate on nuclear reactions which take place in stars. We called it Nuclear Astrophysics. Before the war Hans Staub and William Stephens had confirmed that there was no stable nucleus at mass 5. After the war Alvin Tollestrup, Charlie Lauritsen and I confirmed that there was no stable nucleus at mass 8. These mass gaps spelled the doom of George Gamow's brilliant idea that all nuclei heavier than helium (mass 4) could be built by neutron addition one mass unit at a time in his *big bang*. Edwin Salpeter of Cornell came to Kellogg in the summer of 1951 and showed that the fusion of three helium nuclei of mass four into the carbon nucleus of mass twelve could probably occur in Red Giant stars but not in the *big bang*. In 1953 Fred Hoyle induced Ward Whaling in Kellogg to perform an experiment which quantitatively confirmed the fusion process under the temperature and density conditions which Hoyle, Martin Schwarzschild and Allan Sandage had shown occur in Red Giants.

Fred Hoyle was the second great influence in my life. The grand concept of nucleosynthesis in stars was first definitely established by Hoyle in 1946. After Whaling's confirmation of Hoyle's ideas I became a believer and in 1954/1955 spent a sabbatical year in Cambridge, England, as a Fulbright Scholar in order to work with Hoyle. There Geoffrey and Margaret Burbidge joined us. In 1956 the Burbidges and Hoyle came to Kellogg and in 1957 our joint efforts culminated in the publication of "Synthesis of the Elements in Stars" in which we showed that all of the elements from carbon to uranium could be produced by nuclear processes in stars starting with the hydrogen and helium produced in the *big bang*. This paper has come to be known from the last initials of the authors as B^2FH. A. G. W. Cameron single-handedly came forward with the same broad ideas at the same time.

Fred Hoyle became the Plumian Professor at Cambridge, was knighted by the Queen and founded the Institute of Theoretical Astronomy in Cambridge in 1966. I spent many happy summers at the Institute until Hoyle's retirement to Cumbria in the Lake District of England. Fred taught me more than astrophysics. He introduced me to English cricket, rugby and association football (we call it soccer). He took me to the Scottish Highlands and taught me how to read an ordnance map as well as how to enjoy climbing the 3000 ft peaks called Munros. I still go climbing somewhere in the British Isles every summer. It keeps me fit and renews my soul.

If has been a long row to hoe. Experimental measurements of the cross section of hundreds of nuclear reactions and their conversion into stellar reaction rates are essential if nucleosynthesis in stars is to be quantitatively confirmed. The Kellogg Laboratory has played a leading role for many years in this effort. I am fortunate that the Nobel Prize was awarded from team work. It is impossible to credit all my colleagues. In experimental nuclear astrophysics Charles Barnes and Ralph Kavanagh have played leading roles. So did Thomas Tombrello and Ward Whaling until they found other fields of interest and promise. In addition Robert Christy and Steven Koonin in theoretical nuclear physics, Jesse Greenstein in observational and theoretical astronomy and Gerald Wasserburg in precision geochemistry on meteoritic and lunar samples have played essential roles. Of my 50 graduate students who have contributed to the field I must single out Donald D. Clayton. His graduate student Stanford Woosley is my grand student and his student Rick Wallace is my great grand student. Nuclear Astrophysics continues to be an active and exciting field. This is clearly evident in my 70th birthday festschrift, "Essays in Nuclear Astrophysics" in which the Cambridge University Press presents the research studies of my colleagues and former students around the world as of 1982.

It is appropriate to conclude, without elaboration, with some details of my life outside the laboratory:

Awarded Medal for Merit by President Harry Truman, 1948
Elected member of the National Academy of Sciences, 1956
Awarded Barnard Medal for Meritorious Service to Science, 1965
Member of the National Science Board, 1968–74
Member of the Space Science Board, 1970–73, 1977–80
Designated Benjamin Franklin Fellow of the Royal Society of Arts, 1970

Awarded the G. Unger Vetlesen Prize, 1973
Awarded National Medal of Science by President Gerald Ford, 1974
Designated Associate of the Royal Astronomical Society, 1975
Elected President of the American Physical Society, 1976
Designated an Honorary Member of the Mark Twain Society, 1976
Awarded Eddington Medal of the Royal Astronomical Society, 1978
Awarded Bruce Gold Medal, Astronomical Society of the Pacific, 1979
Elected to the Society of American Baseball Research, 1980–
Honorary degrees from University of Chicago, 1976, Ohio State University, 1978, University of Liege 1981, Observatory of Paris 1981 and Denison University 1982.

(*added in 1991*): My 80th birthday celebration was held August 11 to 14, 1991 as a Nuclear Astrophysics Symposium, which was one part of the Caltech Centennial Year events. Again my colleagues and former students participated along with other experts in the field of nuclear astrophysics.

Ardiane Fowler died in May 1988. In December 1989 I married Mary Dutcher, a descendant of the Dutch founders of New Amsterdam, now New York. She had taught grade school for many years on Long Island and had not previously been married. We reside in the two-story, New England style white frame house, which I purchased in 1958. It is only a ten-minute walk from Caltech. I am retired from teaching so my only routine trips to the Insitute are on Wednesdays for the Astronomy Seminar, Thursdays for the Physics Colloquium and Fridays for the Kellogg Nuclear Physics Seminar. Mary Dutcher Fowler has painted all her life and she now attends a painting school in Pasadena. We keep busy by taking long walks on many weekends and in general try to stay out of trouble.

Honorary degrees
Arizona State University, 1985
Georgetown University, 1986
University of Massachusetts, 1987
Williams College, 1988
Gustavus Adolphus College, 1991

Honours
Nobel Prize for Physics, 1983
Sullivant Medal, The Ohio State University, 1985
First recipient of the William A. Fowler Award for Excellence and Distinguished Accomplishments in Physics, Ohio Section, American Physical Society, 1986
Legion d'Honneur awarded by President Mitterand of France, 1989
Member of Lima City Schools Distinguished Alumni Hall of Fame, 1990
Member of Ohio Sci. & Tech. Hall of Fame, 1991

EXPERIMENTAL AND THEORETICAL NUCLEAR ASTROPHYSICS; THE QUEST FOR THE ORIGIN OF THE ELEMENTS

Nobel lecture, 8 December, 1983

by

WILLIAM A. FOWLER

W. K. Kellogg Radiation Laboratory
California Institute of Technology, Pasadena, California 91125

Ad astra per aspera et per ludum

I. *Introduction*

We live on planet Earth warmed by the rays of a nearby star we call the Sun. The energy in those rays of sunlight comes initially from the nuclear fusion of hydrogen into helium deep in the solar interior. Eddington told us this in 1920 and Hans Bethe developed the detailed nuclear processes involved in the fusion in 1939. For this he was awarded the Nobel Prize in Physics in 1967.

All life on earth, including our own, depends on sunlight and thus on nuclear processes in the solar interior. But the sun did not produce the chemical elements which are found in the earth and in our bodies. The first two elements and their stable isotopes, hydrogen and helium, emerged from the first few minutes of the early high temperature, high density stage of the expanding Universe, the so-called "big bang". A small amount of lithium, the third element in the periodic table, was also produced in the big bang, but the remainder of the lithium and all of beryllium, element four, and boron, element five, are thought to have been produced by the spallation of still heavier elements by the cosmic radiation in the interstellar medium between stars. These elements are in general very rare in keeping with this explanation of their origin as reviewed in detail by Audouze and Reeves (1).

Where did the heavier elements originate? The generally accepted answer is that all of the heavier elements from carbon, element six, up to long-lived radioactive uranium, element ninety-two, were produced by nuclear processes in the interior of stars in our own Galaxy. The stars we see at the present time in what we call the *Milky Way* are located in a spiral arm of our Galaxy. In Sweden you call it *Vintergatan,* the Winter Street. We see with our eyes only a small fraction of the one hundred billion stars in the Galaxy. Astronomers cover almost the full range of the electromagnetic spectrum and thus can observe many more Galactic stars and even individual stars in other galaxies.

The stars which synthesized the heavy elements in the solar system were

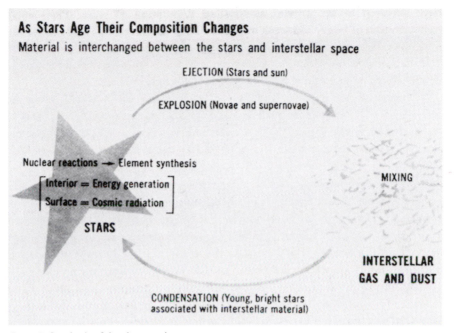

Figure 1. Synthesis of the elements in stars.

formed or born, evolved or aged, and eventually ejected the ashes of their nuclear fires into the interstellar medium over the lifetime of the Galaxy *before* the solar system itself formed four and one-half billion years ago.

The lifetime of the Galaxy is thought to be more than ten billion years but less than twenty billion years. In any case the Galaxy is much older than the solar system. The ejection of the nuclear ashes or newly formed elements took place by slow mass loss during the old age of the star, called the *giant* stage of stellar evolution, or during the relatively frequent outbursts which astronomers call *novae,* or during the final spectacular stellar explosions called *supernovae.* Supernovae can be considered to be the death of stars. White dwarfs or neutron stars or black holes which result from stellar evolution may represent a form of stellar purgatory.

In any case the sun and the earth and all the other planets in the solar system condensed under gravitational and rotational forces from a gaseous solar nebula in the interstellar medium consisting of "big bang" hydrogen and helium mixed with the heavier elements synthesized in earlier generations of Galactic stars. All of this is illustrated in Figure 1.

This idea can be generalized to successive generations of stars in the Galaxy with the result that the heavy element content of the interstellar medium and the stars which form from it increases with time. The oldest stars in the Galactic halo, that is, those we believe to have formed first, are found to have heavy element abundances less than one percent of the heavy element abundance of the solar system. The oldest stars in the Galactic disk have approximately ten percent. Only the less massive stars among those first formed can have survived to the present as so-called Population II stars. Their small concentration of heavy elements may have been produced in a still earlier but more massive generation of stars, Population III, which rapidly exhausted their fuels and survived for only a very short lifetime. Stars formed in the disk of the Galaxy over its lifetime are referred to as Population I stars.

We speak of this element building as nucleosynthesis in stars. It can be generalized to other galaxies such as our twin, the Andromeda Nebula, and so this mechanism can be said to be a universal one. Astronomical observations on other galaxies have contributed much to our understanding of nucleosynthesis in stars.

We refer to the basic physics of energy generation and element synthesis in stars as Nuclear Astrophysics. It is a benign application of nuclear physics in contrast to military reactors and bombs. For the nuclear physicist this contrast is a personal and professional paradox. However, there is one thing of which I am certain. The science which explains the origin of sunlight must not be used to raise a dust cloud which will black out that sunlight from our planet.

As for all physics the field of Nuclear Astrophysics involves experimental and theoretical activities on the part of its practitioners and hence the first part of the title of this lecture. This lecture will emphasize nuclear experimental results and the theoretical analysis of those results almost but not entirely to the exclusion of other theoretical aspects. It will not in any way do justice to the observational activities of astronomers and cosmochemists which are necessary to complete the cycle: experiment, theory, observation. Nor will it do justice to the calculations by many theoretical astrophysicists of the results of nucleosynthesis of the elements and their isotopes under astrophysical conditions during the many stages of stellar evolution.

My deepest personal interest is in experimental data, in the analysis of the data and in the proper use of the data in theoretical stellar models. I continue to be encouraged in this regard by this one-hundred and nine year old quotation from Mark Twain:

> *There is something fascinating about science. One gets such wholesale returns of conjecture out of such a trifling investment of fact.*
>
> — Life on the Mississippi 1874

For me Twain's remark is a challenge to the experimentalist. The experimentalist must try to eliminate the word "trifling" through his endeavors in uncovering the facts of nature.

Experimental research and theoretical research are often very hard work. Fortunately this is lightened by the fun of doing physics and in obtaining results which bring a personal feeling of intellectual satisfaction. To my mind the hard work and the resulting intellectual fun transcend in a way the benefits which may accrue to society through subsequent technological applications. Please understand — I do not belittle these applications but I am unable to overlook the fact that they are a two-edged sword. My subject matter resulted from the hard work of a nuclear astrophysicist which when successful brought him joy and satisfaction. It was hard work but it was fun. Thus I have chosen the subtitle for this lecture — "Ad astra per aspera et per ludum" which can be freely translated — "To the stars through hard work and fun." This is in keeping with my paraphrase of the biblical quotation from Matthew "Man shall not live by work alone."

With that in the record let us next ask what are the goals of Nuclear Astrophysics? First of all, Nuclear Astrophysics attempts to understand energy generation in the sun and other stars at all stages of stellar evolution. Energy generation by nuclear processes requires the transmutation of nuclei into new nuclei with lower mass. The small decrease in mass is multiplied by the velocity of light squared as Einstein taught us and a relatively large amount of energy is released.

Thus the first goal is closely related to the second goal that attempts to understand the nuclear processes which produced under various astrophysical circumstances the relative abundances of the elements and their isotopes in nature; whence the second part of the little of this lecture. Figure 2 shows a schematic curve of atomic abundances as a function of atomic weight. The data for this curve was first systemized from a plethora of terrestrial, meteoritic, solar and stellar data by Hans Suess and Harold Urey (2) and the available data has been periodically updated by A. G. W. Cameron (3). Major contributions to the experimental measurement of atomic transition rates needed to determine solar and stellar abundances have been made by my colleague, Ward Whaling (4). References (3) and (4) occur in a book *Essays in Nuclear Astrophysics* which reviews the field up to 1982. In the words of one of America's baseball immortals, Casey Stengel, "You can always look it up."

The curve in Figure 2 is frequently referred to as "universal" or "cosmic" but in reality it primarily represents relative atomic abundances in the solar system and in Main Sequence stars similar in mass and age to the sun. In current usage the curve is described succinctly as "solar". It is beyond the scope of this lecture to elaborate on the difficult, beautiful research in astronomy and cosmochemistry which determined this curve. How this curve serves as a goal can be simply put. In the sequel it will be noticed that calculations of atomic abundances produced under astronomical circumstances at various postulated stellar sites are almost invariably reduced to ratios relative to "solar" abundances.

II. *Early Research on Element Synthesis*

George Gamow and his collaborators, R. A. Alpher and R. C. Herman (5), attempted to synthesize all of the elements during the big bang using a nonequilibrium theory involving neutron (n) capture with gamma-ray (γ) emission and electron (e) beta-decay by successively heavier nuclei. The synthesis proceeded in steps of one mass unit at a time since the neutron has approximately unit mass on the mass scale used in all the physical sciences. As they emphasized, this theory meets grave difficulties beyond mass 4 (^{4}He) because no stable nuclei exist at atomic mass 5 and 8. Enrico Fermi and Anthony Turkevich attempted valiantly to bridge these "mass gaps" without success and permitted Alpher and Herman to publish the results of their

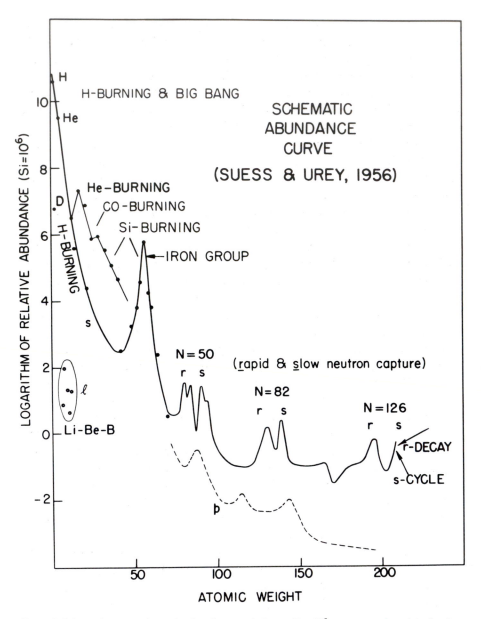

Figure 2. Schematic curve of atomic abundances relative to Si $= 10^6$ versus atomic weight for the sun and similar Main Sequence stars.

attempts. Seventeen years later Wagoner, Fowler, and Hoyle (6) armed with nuclear reaction data accumulated over the intervening years succeeded only in producing ^{7}Li at a mass fraction of at most 10^{-8} compared to hydrogen plus helium for acceptable model universes. All heavier elements totaled less than 10^{-11} by mass. Wagoner, Fowler, and Hoyle (6) did succeed in producing ^{2}D, ^{3}He, ^{4}He, and ^{7}Li in amounts in reasonable agreement with observations at the time. More recent observations and calculations are frequently used to place constraints on models of the expanding universe and in general favor open models in which the expansion continues indefinitely. In other words there is not enough ordinary matter to close the universe. However, if neutrinos have only 10^{-5} the mass of the electron, they close the universe.

It was in connection with the "mass gaps" that the W. K. Kellogg Radiation Laboratory first became involved, albeit unwittingly, in astrophysical and cosmological phenomena. Before proceeding it is appropriate at this point to discuss briefly the origins of the Kellogg Radiation Laboratory where I have worked for 50 years. The laboratory was designed and the construction supervised by Charles Christian Lauritsen in 1930 through 1931. Robert Andrews Millikan, the head of Calteach, acquired the necessary funds from Will Keith Kellogg, the American "corn flakes king." The Laboratory was built to study the physics of 1 MeV X-rays and the application of those X-rays in the treatment of cancer. In 1932 Cockcroft and Walton discovered that nuclei could be disintegrated by protons (p), the nuclei of the light hydrogen atom ^{1}H, accelerated to energies well under 1 MeV. Lauritsen immediately converted one of his X-ray tubes into a positive ion accelerator (they were powered by alternating current transformers!) and began research in nuclear physics. Robert Oppenheimer and Richard Tolman were instrumental in convincing Millikan that Lauritsen was doing the right thing. Oppenheimer played an active role in the theoretical interpretation of the experimental results obtained in the Kellogg Laboratory in the early crucial years.

Lauritsen supervised my doctoral research from 1933−1936 and I worked closely with him until his death. It was he who taught me that physics was both hard work and fun. He was a native of Denmark and was an accomplished violinist as well as physicist, architect and engineer. He loved the works of Carl Michael Bellman, the famous Swedish poet-musician of the 18th century, and played and sang Bellman for his students. It is well known that many of Bellman's works were drinking songs. That made it all the better.

We must now return to the first involvement of the Kellogg Radiation Laboratory in the mass gap at mass 5. In 1939, in Kellogg, Hans Staub and William Stephens (7) detected resonance scattering by ^{4}He of neutrons with orbital angular momentum equal to one in units of $\hbar$ (p-wave) and energy somewhat less than 1 MeV as shown in Figure 3. This confirmed previous reaction studies by Williams, Shepherd, and Haxby (8) and showed that the ground state of ^{5}He is unstable. As fast as ^{5}He is made it disintegrates! The same was later shown to be true for ^{5}Li, the other candidate nucleus at mass 5. The Pauli exclusion principle dictates for fermions that the third neutron in ^{5}He must have at least unit angular momentum and not zero as permitted for the first two neutrons with antiparallel spins. The attractive nuclear force cannot match the outward centrifugal force in classical terminology. Still later, in the Kellogg Radiation Laboratory, Tollestrup, Fowler, and Lauritsen (9) confirmed, with improved precision, the discovery of Hemmendinger (10) that the ground state of ^{8}Be is unstable. They (9) found the energy of the ^{8}Be break-up to be 89 ± 5 keV compared to the currently accepted value of 91.89 ± 0.05 keV! The Pauli exclusion principle is again at work in the instability of ^{8}Be. As

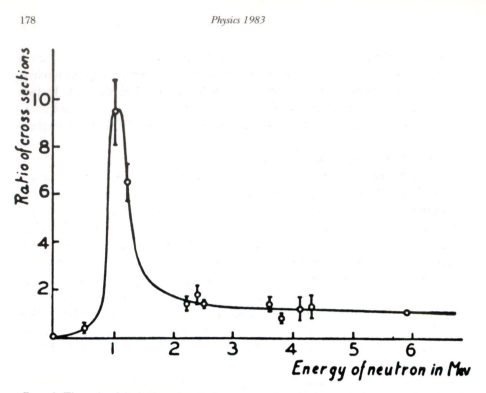

Figure 3. The ratio of the backward scattering cross section of helium to hydrogen as a function of the laboratory energy in MeV of the incident neutron.

fast as ^{8}Be is made it disintegrates into two ^{4}He-nuclei. The latter may be bosons but they consist of fermions. The mass gaps at 5 and 8 spelled the doom of Gamow's hopes that all nuclear species could be produced in the big bang one unit of mass at a time.

The eventual commitment of the Kellogg Radiation Laboratory to Nuclear Astrophysics came about in 1939 when Bethe (11) brought forward the operation of the CN-cycle as one mode of the fusion of hydrogen into helium in stars (since oxygen has been found to be involved the cycle is now known as the CNO-cycle). Charles Lauritsen, his son Thomas Lauritsen, and I were measuring the cross sections of the proton bombardment of the isotopes of carbon and nitrogen which constitute the CN-cycle. Bethe's paper (11) told us that we were studying in the laboratory processes which are occurring in the sun and other stars. It made a lasting impression on us. World War II intervened but in 1946 on returning the laboratory to nuclear experimental research, Lauritsen decided to continue in low-energy, *classical* nuclear physics with emphasis in the study of nuclear reactions thought to take place in stars. In this he was strongly supported by Ira Bowen, a Caltech Professor of Physics who had just been appointed Director of the Mt. Wilson Observatory, by Lee DuBridge, the new President of Caltech, by Carl Anderson, Nobel Prize winner 1936, and by Jesse Greenstein, newly appointed to establish research in astronomy at Caltech. In Kellogg, Lauritsen did not follow the fashionable trend to higher and higher energies which has continued to this day. He did support Robert Bacher and others in establishing high energy physics at Caltech.

Although Bethe (11) in 1939 and others still earlier had previously discussed energy generation by nuclear processes in stars the grand concept of nucleosynthesis in stars was first definitely established by Fred Hoyle (12). In two classic papers the basic ideas of the concept were presented within the framework of stellar structure and evolution with the use of the then known nuclear data.

Again the Kellogg Laboratory played a role. Before his second paper Hoyle was puzzled by the slow rate of the formation of ^{12}C-nuclei from the fusion $(3\alpha \rightarrow {}^{12}C)$ of three alpha-particles (α) or ^{4}He-nuclei in Red Giant Stars. Hoyle was puzzled because his own work with Schwarzschild (13) and previous work of Sandage and Schwarzschild (14) had convinced him that helium burning through $3\alpha \rightarrow {}^{12}C$ should commence in Red Giants just above 10^8 K rather than at 2×10^8 K as required by the reaction rate calculation of Salpeter (15). Salpeter made his calculation while a visitor at the Kellogg Laboratory during the summer of 1951 and used the Kellogg value (9) for the energy of ^{8}Be in excess of two ^{4}He to determine the resonant rate for the process $(2\alpha \leftrightarrow {}^8Be)$ which takes into account both the formation and decay of the ^{8}Be. However, in calculating the next step, $^8Be + \alpha \rightarrow {}^{12}C + \gamma$, Salpeter had treated the radiative fusion as nonresonant.

Hoyle realized that this step would be speeded up by many orders of magnitude, thus reducing the temperatures for its onset, if there existed an excited state of ^{12}C with energy 0.3 MeV in excess of $^8Be + \alpha$ at rest and with the angular momentum and parity $(0^+, 1^-, 2^+, 3^-, ...)$ dictated by the selection rules for these quantities. Hoyle came to the Kellogg Laboratory early in 1953 and questioned the staff about the possible existence of his proposed excited state. To make a long story short Ward Whaling and his visiting associates and graduate students (16) decided to go into the laboratory and search for the state using the $^{14}N(d, \alpha)^{12}$C-reaction. They found it to be located almost exactly where Hoyle had predicted. It is now known to be at 7.654 MeV excitation in ^{12}C or 0.2875 MeV above $^8Be + \alpha$ and 0.3794 MeV above 3α. Cook, Fowler, Lauritsen, and Lauritsen (17) then produced the state in the decay of radioactive ^{12}B and showed it could break up into 3α and thus by reciprocity could be formed from 3α. They argued that the spin and parity of the state must be 0^+ as is now known to be the case.

The $3\alpha \rightarrow {}^{12}C$ fusion in Red Giants jumps the mass gaps at 5 and 8. This process could never occur under big bang conditions. By the time the ^{4}He was produced in the early expanding Universe the subsequent density and temperature were too low for the helium fusion to carbon to occur. In contrast, in Red Giants, after hydrogen conversion to helium during the Main Sequence stage, gravitational contraction of the helium core raises the density and temperature to values where helium fusion is ignited. Hoyle and Whaling showed that conditions in Red Giant stars are just right.

Fusion processes can be referred to as nuclear burning in the same way we speak of chemical burning. Helium burning in Red Giants succeeds hydrogen burning in Main Sequence stars and is in turn succeeded by carbon, neon, oxygen, and silicon burning to reach to the elements near iron and somewhat beyond in the periodic table. With these nuclei of intermediate mass as seeds, subsequent processes similar to Gamow's involving neutron capture at a slow rate (s-process) or at a rapid rate (r-process) continued the synthesis beyond ^{209}Bi, the last stable nucleus, up through short lived radioactive nuclei to long lived ^{232}Th, ^{235}U, and ^{238}U, the parents of the natural radioactive series. This last requires the r-process which actually builds beyond mass 238 to radioactive nuclei which decay back to ^{232}Th, ^{235}U, and ^{238}U rapidly at the cessation of the process.

The need for two neutron capture processes was provided by Suess and Urey (2). With the adroit use of relative isotopic abundances for elements with several isotopes they demonstrated the existence of the double peaks (r and s) in Figure 2. It was immediately clear that these peaks were associated with neutron shell filling at the magic neutron numbers $N = 50,82$, and 126 in the nuclear shell model of Hans Jensen and Maria Goeppert-Mayer who won the Nobel Prize in Physics just twenty years ago.

In the s-process the nuclei involved have low capture cross-sections at shell closure and thus large abundances to maintain the s-process flow. In the r-process it is the proton-deficient radioactive progenitors of the stable nuclei which are involved. Low capture cross-sections and small beta-decay rates at shell closure lead to large abundances but after subsequent radioactive decay these large abundances appear at lower A values than for the s-process since Z is less and thus $A = N + Z$ is less. In Hoyle's classic papers (12) stellar nucleosynthesis up to the iron group elements was attained by charged particle reactions. Rapidly rising Coulomb barriers for charged particles curtailed further synthesis. Suess and Urey (2) made the breakthrough which led to the extension of nucleosynthesis in stars by neutrons unhindered by Coulomb barriers all the way to ^{238}U.

The complete run of the synthesis of the elements in stars was incorporated into a paper by Burbidge, Burbidge, Fowler, and Hoyle (18), commonly referred to as B^{2}FH, and was independently developed by Cameron (19). Notable contributions to the astronomical aspects of the problem were made by Jesse Greenstein (20) and by many other observational astronomers. Since that time Nuclear Astrophysics has developed into a full-fledged scientific activity including the exciting discoveries of isotopic anomalies in meteorites by my colleagues Gerald Wasserburg, Dimitri Papanastassiou and Samuel Epstein and many other cosmochemists. What follows will highlight a few of the many experiments and theoretical researches under way at the present time or carried out in the past few years. This account will emphasize research activities in the Kellogg Laboratory because they are closest to my interest and knowledge. However, copious references to the work of other laboratories and institutions are cited in the hope that the reader will obtain a broad view of current experimental and theoretical studies in Nuclear Astrophysics.

This account cannot discuss the details of the nucleosynthesis of all the elements and their isotopes which would, for a given nuclear species, involve discussing all the reactions producing that nucleus and all those which destroy it. The reader will find some of these details for ^{12}C, ^{16}O and ^{55}Mn.

It will be noted that the measured cross sections for the reactions are customarily very small at the lowest energies of measurement, for $^{12}C(\alpha,\gamma)^{16}O$ even less than one nanobarn $(10^{-33}cm^2)$ near 1.4 MeV. This means that experimental Nuclear Astrophysics requires accelerators with large currents of well focussed, monoenergic ion beams, thin targets of high purity and stability, detectors of high sensitivity and energy resolution and experimentalists with great tolerance for the long running times required and with patience in accumulating data of statistical significance. Classical Rutherfordian measurements of nuclear cross sections are required in experimental nuclear astrophysics and the results are in turn essential to our understanding of the physics of nuclei.

A comment on nuclear reaction notation is necessary at this point. In the reaction $^{12}C(\alpha,\gamma)^{16}O$ discussed in the previous paragraph ^{12}C is the laboratory target nucleus, α is the incident nucleus (^4He) accelerated in the laboratory, γ is the photon produced and detected in the laboratory, and ^{16}O is the residual nucleus which can also be detected if it is desireable to do so. If ^{12}C is accelerated against a gas target of ^4He and the ^{16}O-products are detected but not the gamma rays then the laboratory notation is $^4He(^{12}C,^{16}O)\gamma$. The stars could not care less. In stars all the particles are moving and only the center-of-momentum system is important for the determination of stellar reaction rates. In $^{12}C(\alpha,n)^{15}O(e^+\nu)^{15}N$, n is the neutron promptly produced and detected and e^+ is the beta-delayed positron which can also be detected. The neutrino emitted with the position is designated by ν.

As an aside at this point I am proud to recall that I first spoke to the Royal Swedish Academy of Sciences on "Nuclear Reactions in Stars" on January 26, 1955. It does not seem so long ago and some of you in the audience heard that talk!

III. Stellar Reaction Rates from Laboratory Cross Sections

Thermonuclear reaction rates in stars are customarily expressed as $N_A <\sigma v>$ reactions per second per (mole cm^{-3}) where $N_A = 6.022 \times 10^{23}$mole^{-1} is Avogadro's number and $<\sigma v>$ is the Maxwell-Boltzmann average as a function of temperature for the product of the reaction cross section, σ, in cm^2, and the relative velocity of the reactants, v in cm sec^{-1}. Multiplication of $<\sigma v>$ by the product of the number densities per cm^3 of the two reactants is necessary to obtain rates in reactions per second per cm^3. N_A is incorporated so that mass fractions can be used as described in detail in Fowler, Caughlan and Zimmerman (21). These authors also describe procedures for reactions involving more than two reactants and give analytical expressions for reactions mainly involving γ, e, n, p and α with nuclei having atomic mass number $A \lesssim 30$. Bose-Einstein statistics for γ have been necessarily incorporated but the extension to Fermi-Dirac statistics for degenerate e, n and p and the extension to Bose-Einstein statistics for α are not included. Factors for calculating reverse reaction rates are given.

Early work on the evaluation of stellar reaction rates from experimental laboratory cross sections was reviewed in Bethe's Nobel Lecture (11). Fowler,

DEFINITION OF THE S-FACTOR (BETHE 1967)
AS A FUNCTION OF REACTION ENERGY(E)

$\sigma(E) = \pi\lambda^2 \times P \times$ INTRINSIC NUCLEAR FACTOR

$\pi\lambda^2 \propto E^{-1}$ $\qquad \lambda =$ DE BROGLIE WAVE LENGTH$/2\pi$

$P(E) =$ GAMOW PENETRATION FACTOR

$\qquad \propto \exp(-E_G^{\frac{1}{2}}/E^{\frac{1}{2}})$ $\qquad E_G \approx Z_0^2 Z_1^2 A$ MeV

$S(E) \equiv E\sigma(E) \exp(+E_G^{\frac{1}{2}}/E^{\frac{1}{2}})$

$S(E)$ $\left\{\begin{array}{l}\text{PERMITS MORE PRECISE EXTRAPOLATION FROM} \\ \text{LOWEST ENERGY MEASUREMENTS IN LABORATORY} \\ \text{TO VERY LOW EFFECTIVE STELLAR ENERGIES}\end{array}\right.$

Table 1.

Caughlan and Zimmerman (21) have provided detailed numerical and analytical procedures for converting laboratory cross sections into stellar reaction rates. It is first of all necessary to accommodate the rapid variation of the nuclear cross sections at low energies which are relevant in astrophysical circumstances. For neutron induced reactions this is accomplished by defining a cross-section S-factor equal to the cross section (σ) multiplied by the interaction velocity (v) in order to eliminate the usual v^{-1} singularity in the cross section at low velocities and low energies.

For reactions induced by charged particles such as protons, alpha particles or the heavier ^{12}C, ^{16}O... nuclei it is necessary to accommodate the decrease by many orders of magnitude from the lowest laboratory measurements to the energies of astrophysical relevance. This is done in the way first suggested by E. E. Salpeter (22) and emphasized by the second of references Bethe (11). Table 1 shows how a relatively slowly varying S-factor can be defined by eliminating the rapidly varying term in the Gamow penetration factor governing transmission through the Coulomb barrier. The cross section is usually expressed in barns ($10^{-24} cm^2$) and the energy in MeV (1.602×10^{-6} erg) so the S-factor is expressed in MeV-barns although keV-barns is sometimes used. In Table 1, the two charge numbers and the reduced mass in atomic mass units of the interacting nuclei are designated by Z_0, Z_1, and A. Table 2 then shows how stellar reaction rates can be calculated as an average over the Maxwell-Boltzmann distribution for both nonresonant and resonant cross sections. In Table 2 the effective stellar reaction energy is given numerically by

STELLAR REACTION RATES AS
FUNCTIONS OF TEMPERATURE(T)

$$\langle \sigma v \rangle_{MB} = f(T) \propto T^{-3/2} \int S(E) \exp(-E_G^{1/2}/E^{1/2} - E/kT) dE$$

MB $\equiv$ AVERAGE OVER MAXWELL–BOLTZMANN DISTRIBUTION

MAXIMUM IN INTEGRAND OCCURS AT E_r AND AT

$E_0 \equiv$ EFFECTIVE STELLAR REACTION ENERGY $\propto E_G^{1/3} T^{2/3}$

NONRESONANT RATE

$$\langle \sigma v \rangle_{nr} \propto S(E_0) T^{-2/3} \exp(-3E_0/kT) \qquad E_0/kT \propto T^{-1/3}$$

RESONANT RATE

$$\langle \sigma v \rangle_r \propto S(E_r) T^{-3/2} \exp(-E_r/kT)$$

E_r = ENERGY AT RESONANCE

Table 2.

$E_0 = 0.122(Z_0^2 Z_1^2 A)^{1/3} T_9^{2/3}$ MeV where T_9 is the temperature in units of 10^9K. Expressions for reaction rates derived from theoretical statistical model calculations are given by Woosley, Fowler, Holmes, and Zimmerman (23).

It is true that the extrapolation from the cross sections measured at the lowest laboratory energies to the cross sections at the effective stellar energy can often involve a decrease by many orders of magnitude. However the elimination of the Gamow penetration factor, which causes this decrease, is based on the solution of the Schroedinger equation for the Coulomb wave functions in which one can have considerable confidence. The main uncertainty lies in the variation of the S-factor with energy which depends primarily on the value chosen for the radius at which formation of a compound nucleus between two interacting nuclei or nucleons occurs as discussed long ago in reference (18). The radii used by my colleagues and me in recent work are given in reference (23). There is, in addition, the uncertainty in the *intrinsic nuclear factor* of Table 1 which can only be eliminated by recourse to laboratory experiments. The effect of a resonance in the compound nucleus just below or just above the threshold for a given reaction can often be ascertained by determination of the properties of the resonance in other reactions in which it is involved and which are easier to study.

IV. *Hydrogen Burning in Main Sequence Stars and the Solar Neutrino Problem*
Hydrogen burning in Main Sequence stars has contributed at the present time only about 20 percent more helium than that which resulted from the big bang.

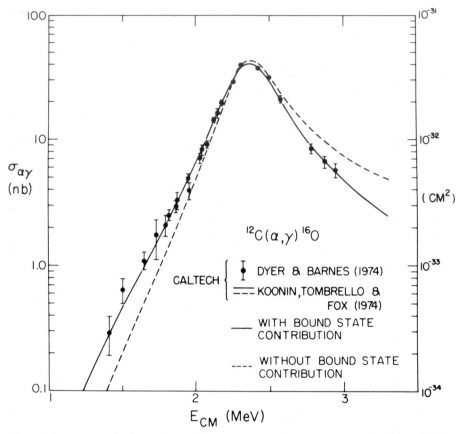

Figure 4. The cross section in nanobarns (*nb*) versus center-of-momentum energy in Mev for ^{12}C(α, γ)^{16}O measured by Dyer and Barnes (35) and compared with theoretical calculations by Koonin, Tombrello and Fox (see 35).

However, hydrogen burning in the sun has posed a problem for many years. In 1938 Bethe and Critchfield (24) proposed the proton-proton or pp-chain as one mechanism for hydrogen burning in stars. From many cross-section measurements in Kellogg and elsewhere it is now known to be the mechanism which operates in the sun rather than the CNO-cycle.

Our knowledge of the weak nuclear interaction (beta decay, neutrino emission and absorption, etc.) tells us that two neutrinos are emitted when four hydrogen nuclei are converted into helium nuclei. Detailed elaboration of the pp-chain by Fowler (25) and Cameron (26) showed that a small fraction of these neutrinos, those from the decay of ^{7}Be and ^{8}B, should be energetic enough to be detectable through interaction with the nucleus ^{37}Cl to form radioactive ^{37}Ar, a method of neutrino detection suggested by Pontecorvo (27) and Alvarez (28). Raymond Davis (29) and his collaborators have attempted for more than 25 years to detect these energetic neutrinos employing a 380,000 liter tank of perchloroethylene ($C_2{}^{35}Cl_3{}^{37}Cl_1$) located one mile deep in the Homestake Gold Mine in Lead, South Dakota. They find only about one quarter of the number expected on the basis of the model dependent calculations of Bahcall *et al.* (30).

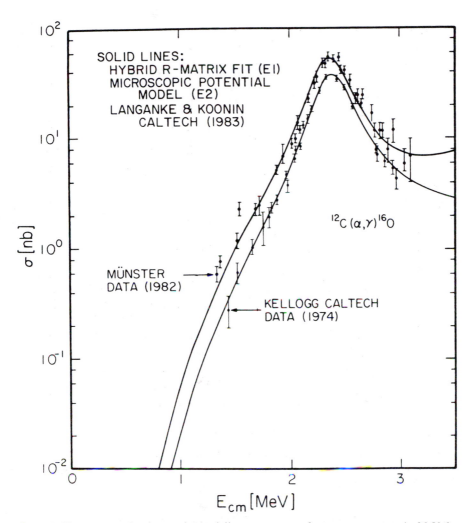

Figure 5. The cross section in nanobarns (*nb*) versus center-of-momentum energy in MeV for $^{12}C(\alpha, \gamma)^{16}O$. The Münster data was obtained by Kettner *et al.* (36) and the Kellogg Caltech data was obtained by Dyer and Barnes (35). The solid lines are theoretical calculations made by Langanke and Koonin (34).

Something is wrong—either the standard solar models are incorrect, the relevant nuclear cross sections are in error, or the electron-type neutrinos produced in the sun are converted in part into undetectable muon neutrinos or tauon neutrinos on the way from the sun to the earth. There indeed have been controversies about the nuclear cross sections which have been for the most part resolved as reviewed in Robertson *et al.* and Osborne *et al.* (31) and Skelton and Kavanagh (32).

It is generally agreed that the next step is to build a detector which will detect the much larger and model independent flux of low energy neutrinos from the sun through neutrino absorption by the nucleus ^{71}Ga to form radioactive ^{71}Ge. This will require 30 to 50 tons of gallium at a cost (for 50 tons) of approximately 25 million dollars or 200 million Swedish crowns. An interna-

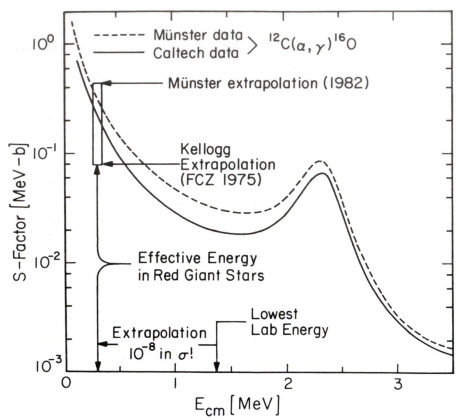

Figure 6. The cross section factor, S in MeV-barns, versus center-of-momentum energy in MeV for $^{12}C(\alpha, \gamma)^{16}O$. The dashed and solid curves are the theoretical extrapolations of the Münster and Kellogg Caltech data by Langanke and Koonin (34).

tional effort is being made to obtain the necessary amount of gallium. We are back at square one in Nuclear Astrophysics. Until the solar neutrino problem is resolved the basic principles underlying the operation of nuclear processes in stars are in question. A gallium detector should go a long way toward resolving the problem.

The Homestake detector must be maintained in low level operation until the chlorine and gallium detectors can be operated at full level simultaneously. Otherwise endless conjecture concerning time variations in the solar neutrino flux will ensue. Morever the results of the gallium observations may uncover information that has been overlooked in the past chlorine observations. In the meantime bromine could be profitably substituted for chlorine in the Homestake detector. The chlorine could eventually be resubstituted.

The CNO-cycle operates at the higher temperatures which occur during hydrogen burning in Main Sequence stars somewhat more massive than the sun. This is the case because the CNO-cycle reaction rates rise more rapidly with temperature than do those of the pp-chain. The cycle is important because ^{13}C, ^{14}N, ^{15}N, ^{17}O, and ^{18}O are produced from ^{12}C and ^{16}O as seeds. The role of these nuclei as sources of neutrons during helium burning is discussed in Section V.

V. *The Synthesis of* ^{12}*C and* ^{16}*O and Neutron Production in Helium Burning*

The human body is 65 % oxygen by mass and 18 % carbon with the remainder mostly hydrogen. Oxygen (0.85 %) and carbon (0.39 %) are the most abundant elements heavier than helium in the sun and similar Main Sequence stars. It is little wonder that the determination of the ratio ^{12}C/^{16}O produced in helium burning is a problem of paramount importance in Nuclear Astrophysics. This ratio depends in a fairly complicated manner on the density, temperature and duration of helium burning but it depends directly on the relative rates of the $3\alpha \rightarrow {}^{12}$C process and the ^{12}C$(\alpha,\gamma)^{16}$O process. If $3\alpha \rightarrow {}^{12}$C is much faster than ^{12}C$(\alpha,\gamma)^{16}$O then no ^{16}O is produced in helium burning. If the reverse is true then no ^{12}C is produced. For the most part the subsequent reaction ^{16}O$(\alpha,\gamma)^{20}$Ne is slow enough to be neglected.

There is general agreement about the rate of the $3\alpha \rightarrow {}^{12}C$ process as reviewed by Barnes (33). However, there is a lively controversy at the present time about the laboratory cross section for ^{12}C$(\alpha,\gamma)^{16}$O and about its theoretical extrapolation to the low energies at which the reaction effectively operates. The situation is depicted in Figures 4, 5 and 6 taken with some modification from Langanke and Koonin (34), Dyer and Barnes (35) and Kettner *et al.* (36). The Caltech data obtained in the Kellogg Laboratory is shown as the experimental points in Figure 4 taken from Dyer and Barnes (35) who compared their results with theoretical calculations by Koonin, Tombrello and Fox (see 35). The Münster data is shown as the experimental points in Figure 5 taken from Kettner *et al.* (36) in comparison with the data of Dyer and Barnes (35). The theoretical curves which yield the best fit to the two sets of data are from Langanke and Koonin (34).

The crux of the situation is made evident in Figure 6 which shows the extrapolations of the Caltech and Münster cross section factors from the lowest measured laboratory energies ($\sim$1.4MeV) to the effective energy $\sim$0.3 MeV, at $T = 1.8 \times 10^8$ K, a representative temperature for helium burning in Red Giant stars. The extrapolation in cross sections covers a range of 10^{-8}! The rise in the cross section factor is due to the contributions of two bound states in the ^{16}O nucleus just below the ^{12}C$(\alpha,\gamma)^{16}$O threshold as clearly indicated in Figure 4. It is these contributions plus differences in the laboratory data which produce the current uncertainty in the extrapolated S-factor. Note that Langanke and Koonin (34) increase the 1975 extrapolation of the Caltech data by Fowler, Caughlan, and Zimmerman (21) by a factor of 2.7 and lower the 1982 extrapolation of the Münster data by 23 %. There remains a factor of 1.6 between their extrapolation of the Münster data and of the Caltech data. There is a lesson in all of this. The semiempirical extrapolation of their data by the experimentalists, Dyer and Barnes (35), was only 30 % lower than that of Langanke and Koonin (34) and their quoted uncertainty extended to the value of Langanke and Koonin (34). Caughlan *et al.* (21) will tabulate the analysis of the Caltech data by Langanke and Koonin (34).

With so much riding on the outcome it will come as no surprise that both laboratories are engaged in extending their measurements to lower energies with higher precision. In the discussion of quasistatic silicon burning in what follows it will be found that the abundances produced in that stage of nucleo-synthesis depend in part on the ratio of ^{12}C to ^{16}O produced in helium burning and that the different extrapolations shown in Figure 6 are in the range crucial to the ultimate outcome of silicon burning. These remarks do not apply to explosive nucleosynthesis.

Recently the ratio of ^{12}C to ^{16}O produced under the special conditions of helium flashes during the asymptotic giant phase of evolution has become of great interest. The hot blue star PG 1159-035 has been found to undergo nonradial pulsations with periods of 460 and 540 seconds and others not yet accurately determined. The star is obviously highly evolved having lost its hydrogen atmosphere, leaving only a hot dwarf of about 0.6 solar masses behind. Theoretical analysis of the pulsations by Starrfield *et al.* and Becker (37) requires substantial amounts of oxygen in the pulsation-driving regions where the oxygen is alternately ionized and deionized. Carbon is completely ionized in these regions and only diminishes the pulsation amplitude. It is not yet clear that sufficient oxygen is produced in helium flashes which certainly involve $3\alpha \rightarrow {}^{12}$C but may not last long enough for ^{12}C$(\alpha,\gamma)^{16}$O to be involved. The problem may not lie in the nuclear reaction rates according to references (37). We shall see!

In what follows in this paper β^+-decay is designated by $(e^+\nu)$ since both a positron (e^+) and a neutrino (ν) are *emitted*. Similarly β^--decay will be desig-nated by $(e^-\bar{\nu})$ since both an electron (e^-) and antineutrino $(\bar{\nu})$ are *emitted*. Electron capture (often indicated by ε) will be designated by (e^-,ν), the comma indicating that an electron is captured and a neutrino emitted. The notations $(e^+,\bar{\nu}),(\nu,e^-)$ and $(\bar{\nu},e^+)$ should now be obvious.

Neutrons are produced when helium burning occurs under circumstances in which the CNO-cycle has been operative in the previous hydrogen burning. When the cycle does not go to completion copious quantities of ^{13}C are produced in the sequence of reactions ^{12}C$(p,\gamma)^{13}$Ne$(e^+\nu)^{13}$C. In subsequent helium burning, neutrons are produced by ^{13}C$(\alpha,n)^{16}$O. When the cycle goes to completion the main product ($>95\%$) is ^{14}N. In subsequent helium burning, ^{18}O and ^{22}Ne are produced in the sequence of reactions ^{14}N$(\alpha,\gamma)^{18}$F$(e^+\nu)^{18}$O$(\alpha,\gamma)^{22}$Ne and these nuclei in turn produce neutrons through ^{18}O(α,n) ^{21}Ne$(\alpha,n)^{24}$Mg and ^{22}Ne$(\alpha,n)^{25}$Mg. However, the astrophysical circumstances and sites under which the neutrons produce heavy elements through the s-process and the r-process are, even today, matters of some controversy and much study (See Section XI).

VI. *Carbon, Neon, Oxygen, and Silicon Burning*

The advanced burning processes discussed in this section involve the network of reactions shown in Figure 7. Because of the high temperature at which this network can operate, radioactive nuclei can live long enough to serve as live reaction targets. In addition excited states of even the stable nuclei are populat-ed and also serve as targets. The determination of the nuclear cross sections and stellar rates of the approximately 1000 reactions in the network has

Reaction Network

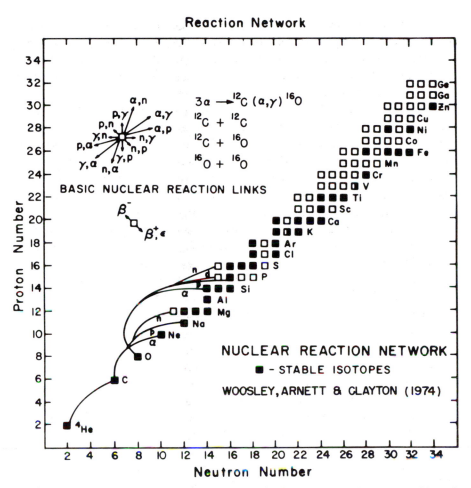

Figure 7. The reaction network for nucleosynthesis involving the most important stable and radioactive nuclei with $N = 2$ to 34 and $Z = 2$ to 32. Stable nuclei are indicated by solid squares. Radioactive nuclei are indicated by open squares. Excited states of both are involved in the reaction network.

involved and will continue to involve extensive experimental and theoretical effort.

The following discussion applies to massive enough stars such that electron degeneracy does not set in as nuclear evolution proceeds through the various burning stages discussed in this section. In less massive stars electron-degeneracy can terminate further nuclear evolution at certain stages with catastrophic results leading to the disruption of the stellar system. The reader will find Figure 8, especially 8(a), instructive in following the discussion in this section. Figure 8 is taken from Woosley and Weaver (38) and a much more detailed recent version is shown in Figure 9 from Weaver, Woosley and Fuller (39). Figure 8(a) applies to the preexplosive stage of a young (Population I) star of 25 solar masses and shows the result of various nuclear burnings in the following mass zones: (1)$>10M_\odot$, convective envelope with the results of some CNO-burning; (2) $7-10\ M_\odot$, products mainly of H-burning; (3)$6.5-7M_\odot$,

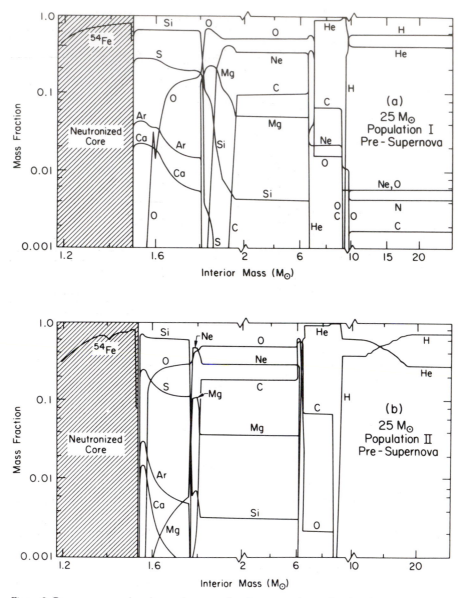

Figure 8. Pre-supernova abundances by mass fraction versus increasing interior mass in solar masses, $M_\odot$, measured from zero at the stellar center to 25 $M_\odot$, the total stellar mass from Woosley and Weaver (38). (a) Population I star. (b) Population II star.

products of He-burning; (4)$1.9-6.5M_\odot$ products of C-burning; (5)$1.8-1.9M_\odot$ products of Ne-burning; (6)$1.5-1.8M_\odot$, products of O-burning; (7)$<1.5M_\odot$, the products of Si-burning in the partially neutronized core are not shown in detail but consist mainly of ^{54}Fe as well as substantial amounts of other neutron-rich nuclei such as ^{48}Ca, ^{50}Ti, ^{54}Cr and ^{58}Fe. ^{54}Fe, ^{48}Ca and ^{50}Ti have $N = 28$, for which a neutron subshell is closed. Both Figures 8(a) and 8(b) have been evaluated shortly after photodisintegration has initiated core collapse which will then be subsequently sustained by the reduction of the outward

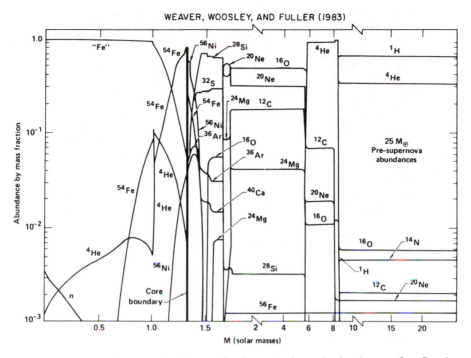

Figure 9. Pre-supernova abundances by mass fraction versus increasing interior mass for a Population I star with total mass equal to 25 $M_\odot$ from Weaver, Woosley and Fuller (39).

pressure through electron-capture and the resulting almost complete neutronization of the core.

It must be realized that the various burning stages took place originally over the central regions of the star and finally in a shell surrounding that region. Subsequent stages modify the inner part of the previous burning stage. For example, in the 25 solar mass Population I star of Figure 8(a), C-burning took place in the central 6.5 solar masses of the star but the inner 1.9 solar masses were modified by subsequent Ne-, O- and Si-burning.

Helium burning produces a stellar core consisting mainly of ^{12}C and ^{16}O. After core contraction the temperature and density rise until carbon burning through ^{12}C+^{12}C fusion is ignited. The S-factor for the total reaction rate shown in Figure 10 has been taken from page 213 of reference (33) and is based on measurements in a number of laboratories. The extrapolation to the low energies of astrophysical relevance is uncertain as Figure 10 makes clear and more experimental and theoretical studies are urgently needed. At the lowest bombarding energy, 2.4 MeV, the cross section is ~10^{-8} barns. For a representative burning temperature of 6×10^8 K the effective energy is $E_0 = 1.7$ MeV and the extrapolated cross section is ~10^{-13} barns. The main product of carbon burning is ^{20}Ne produced primarily in the ^{12}C(^{12}C,α)^{20}Ne reaction. The reactions ^{12}C(^{12}C,p)^{23}Na and ^{12}C(^{12}C,n)^{23}Mg(e^+v)^{23}Na also occur as well as many secondary reactions such as ^{23}Na(p,α)^{20}Ne. When the ^{12}C is exhausted, ^{20}Ne and ^{16}O are the major remaining constituents. As the temperature rises,

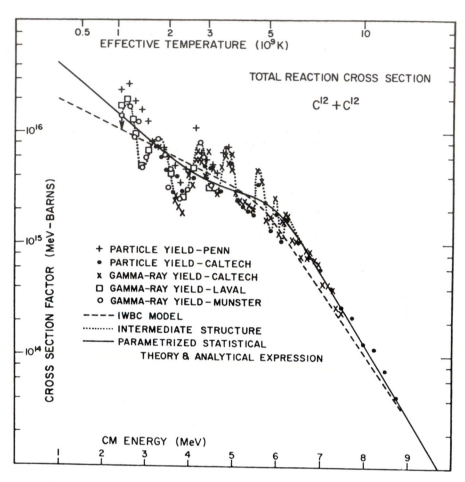

Figure 10. The total cross-section factor in MeV-barns versus center-of-momentum energy in MeV for the fusion of ^{12}C and ^{12}C. The experimental data from several laboratories are shown along with schematic intermediate structure in the dotted curve. Two parametrized adjustments to the data, ignoring intermediate structure, are shown in the dashed and solid curves.

from further gravitational contraction, the ^{20}Ne is destroyed by photodisintegration, ^{20}Ne$(\gamma,\alpha)^{16}$O. This occurs because the alpha-particle in ^{20}Ne is bound to its closed-shell partner, ^{16}O, by only 4.731 MeV. In ^{16}O, for example, the binding of an alpha-particle is 7.162 MeV.

The next stage is oxygen burning through ^{16}O$+^{16}$O fusion. The S-factor for the total reaction rate is shown in Figure 11 and is based entirely on data obtained in the Kellogg Laboratory at Caltech. The work of Hulke, Rolfs, and Trautvetter (40) using gamma-ray detection is in fair agreement with the gamma-ray measurements at Caltech. As in the case of ^{12}C$+^{12}$C the extrapolation to the low energies of astrophysical relevance is uncertain although only one of many possible extrapolations is shown in Figure 11. The main product of oxygen burning is ^{28}Si through the primary reaction ^{16}O$(^{16}$O$,\alpha)^{28}$Si and a number of secondary reactions. Under some conditions neutron induced reactions lead to the synthesis of significant quantities of ^{30}Si. Oxygen burning can result in nuclei with a small but important excess of neutrons over protons.

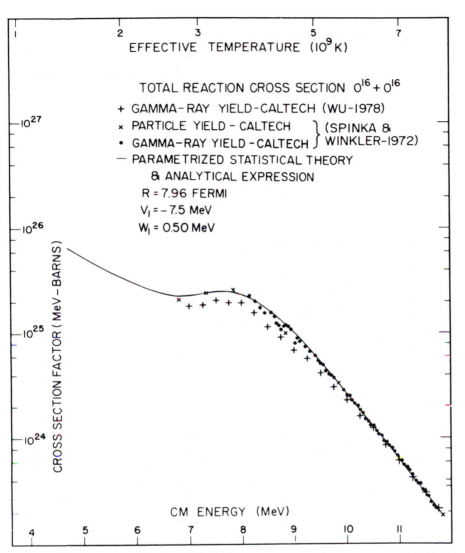

Figure 11. The total cross-section factor in MeV-barns versus center-of-momentum energy in MeV for the fusion of $^{16}O + {}^{16}O$. The experimental data from several measurements at Caltech are shown and compared with a parametrized theoretical adjustment in the solid curve.

The onset of Si-burning signals a marked change in the nature of the *fusion process*. The Coulomb barrier between two ^{28}Si nuclei is too great for fusion to produce the compound nucleus, ^{56}Ni, directly at the ambient temperatures ($T_9 = 3$ to 5) and densities ($\varrho = 10^5$ to 10^9 g cm^{+3}). However, the ^{28}Si and subsequent products are easily photodisintegrated by $(\gamma,\alpha),(\gamma,n)$ and (γ,p)-reactions. As Si-burning proceeds more and more ^{28}Si is reduced to nucleons and alpha particles which can be captured by the remaining ^{28}Si nuclei to build through the network in Figure 7 up to the iron group nuclei. The main product in explosive Si-burning is ^{56}Ni which transforms eventually through two beta-decays to ^{56}Fe.

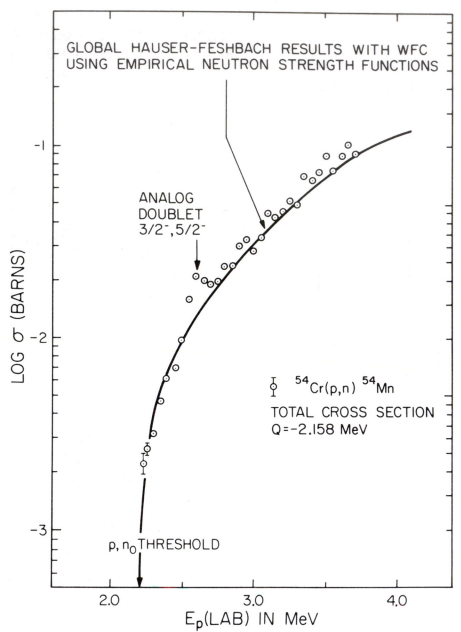

Figure 12. The total cross section in barns integrated over all outgoing angles versus laboratory proton energy in MeV for the reaction $^{54}Cr(p, n)^{54}Mn$. The data of Zysking *et al.* (42) are compared with unnormalized global Hauser-Feshbach calculations made by Woosley *et al.* (23).

In quasistatic Si-burning the weak interactions are fast enough that ^{54}Fe, with two more neutrons than protons, is the main product. Because of the important role played by alpha particles (α) and because of the inexorable trend to equilibrium (e) involving nuclei near mass 56, which have the largest binding energies per nucleon of all nuclear species, B^{2}FH (18) broke down, what is now called Si-burning, into their α-process and e-process. Quasi-

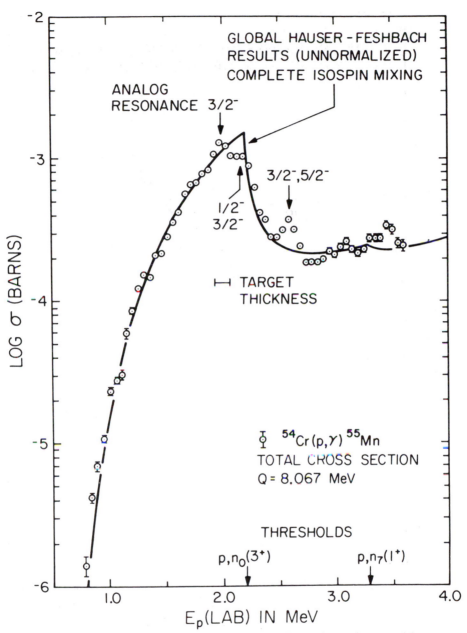

Figure 13. The total cross section in barns integrated over all outgoing angles versus laboratory proton energy in MeV for the reaction $^{54}\text{Cr}(p, \gamma)^{55}\text{Mn}$. The data of Zyskind *et al.* (42) are compared with unnormalized global Hauser-Feshbach calculations made by Woosley *et al.* (23).

equilibrium calculations for Si-burning were made by Bodansky, Clayton and Fowler (41) who cite the original papers in which the basic ideas of Si-burning were developed. Modern computers permit detailed network flow calculations to be made as discussed in references (38) and (39).

The extensive laboratory studies of Si-burning reactions are reviewed in reference (33). Figures 12 and 13 adapted from Zyskind *et al.* (42) show the

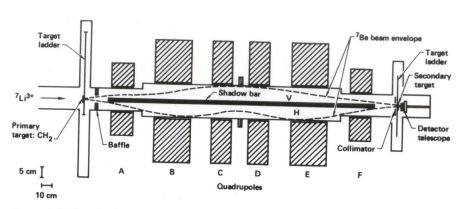

Figure 14. Radioactive beam transport system developed by Haight *et al.* (44).

laboratory excitation curves for $^{54}Cr(p,n)^{54}$Mn and $^{54}Cr(p,\gamma)^{55}Mn$ as examples. The neutrons produced in the first of these reactions will increase the number of neutrons available in Si-burning but will not contribute directly to the synthesis of ^{55}Mn as does the second reaction. In fact, above its threshold at 2.158 MeV the (p,n)-reaction competes strongly with the (p,γ)-reaction, which is of primary interest, and produces the pronounced *competition cusp* in the excitation curve in Figure 13. Competition in the disintegration of the compound nucleus produced in nuclear reactions was stressed very early by Niels Bohr so perhaps the cusps should be called *Bohr Cusps*. They arise from the same basic cause but are not the long known *Wigner Cusps*. It will be clear from Figure 13 that the rate of the $^{54}Cr(p,\gamma)^{55}$Mn reaction at very high temperatures will be an order of magnitude lower because of the cusp than would otherwise be the case.

The element manganese has only one isotope, ^{55}Mn. The manganese in nature is produced in quasistatic Si-burning most probably through the $^{54}Cr(p,\gamma)^{55}$Mn-reaction just discussed in the previous paragraph. The reaction network extends to ^{54}Cr and then on through ^{55}Mn. $^{51}V(\alpha,\gamma)^{55}$Mn and $^{52}V(\alpha,n)^{55}$Mn may also contribute especially in explosive Si-burning. The overall synthesis of ^{55}Mn involves a balance in its production and destruction. In quasistatic Si-burning the reactions which destroy ^{55}Mn are most probably $^{55}Mn(p,\gamma)^{56}$Fe and $^{55}Mn(p,n)^{55}$Fe, which are discussed and illustrated in Mitchell and Sargood (43). $^{55}Mn(\alpha,\gamma)^{59}$Co, $^{55}Mn(\alpha,p)^{58}$Fe, and $^{55}Mn(\alpha,n)^{58}$Co may also destroy some ^{55}Mn in explosive Si-burning. In the figures discussed in Section VIII it will be noted that calculations of the overall synthesis of ^{55}Mn yield values in fairly close agreement with the abundance of this nucleus in the solar system. Unfortunately the same can not be said about many other nuclei.

The laboratory measurements on Si-burning reactions have covered only about 20% of the reactions in the network of Figure 7 involving stable nuclei as targets. Direct measurements on short lived radioactive nuclei and the excited states of all nuclei are impossible at the present time. In this connection the production of radioactive ion-beams holds great promise for the future. Richard Boyd and Haight *et al.* (44) have pioneered in the development of this

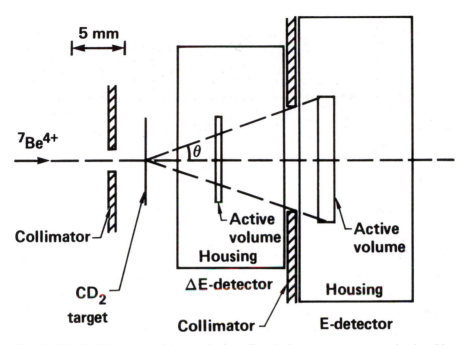

Figure 15. Detail of the target and detector in the radioactive beam transport system developed by Haight *et al.* (44).

technique. It will also be possible to study with this technique the reaction rates of the fairly long-lived isomeric excited states of stable nuclei. Figures 14 and 15 show the beam transport system developed by Haight *et al.* (44) which has produced accelerated beams of ^{7}Be and ^{13}N and successfully determined the cross section of the reaction ^{2}II(^{7}Be,^{8}B)n to be 59 ± 11 millibarns for 16.9 MeV ^{7}Be-ions. The equivalent center-of-momentum energy for the ^{7}Be(d,n)^{8}B reaction is 3.8 MeV. It is my view that continued development and application of radioactive ion-beam techniques could bring the most exciting results in laboratory Nuclear Astrophysics in the next decade. For example the rate of the ^{13}N(p,γ)^{14}O reaction, which will be studied as ^{1}H(^{13}N,γ)^{14}O, is crucial to the operation of the so-called fast CN-cycle.

In any case it has been clear for some time that experimental results on Si-burning reactions must be systematized and supplemented by comprehensive theory. Fortunately theoretical average cross sections will suffice in many cases. This is because the stellar reaction rates integrate the cross sections over the Maxwell−Boltzmann distribution. For most Si-burning reactions resonances in the cross section are closely spaced and even overlapping and the integration covers a wide enough range of energies that the detailed structure in the cross sections is automatically averaged out. The statistical model of nuclear reactions developed by Hauser and Feshbach (45), which yields average cross sections, is ideal for the purpose. Accordingly Holmes, Woosley, Fowler and Zimmerman (46) undertook the task of developing a *global, parametrized* Hauser-Feshbach theory and computer program for use in Nuclear Astrophysics. Reference (23) is an extension of this work. The free parameters are the

STATISTICAL MODEL CALCULATIONS VS MEASUREMENTS (I)
RATIO OF REACTION RATE (GROUND STATE OF TARGET) FROM WOOSLEY, FOWLER, HOLMES
& ZIMMERMAN (AD & ND TABLES 22, 371, 1978) TO REACTION RATES FROM
EXPERIMENTAL YIELD MEASUREMENTS (1970-1982) AT BOMBAY,
CALTECH, COLORADO, KENTUCKY, MELBOURNE & TORONTO

REACTION	$T_9 = T/10^9$ K				
	1	2	3	4	5
^{23}Na(p,n)^{23}Mg	1.4	1.2	1.1	1.1	1.0
^{25}Mg(p,γ)^{26}Al	1.2	1.1	1.0	0.9	0.8
^{25}Mg(p,n)^{25}Al	1.1	1.0	0.9	0.8	0.8
^{27}Al(p,γ)^{28}Si	3.7	2.1	1.5	1.3	1.1
^{27}Al(p,n)^{27}Si	1.8	1.4	1.3	1.3	1.2
	0.9	0.9	0.9	1.0	1.0
^{28}Si(p,γ)^{29}P		1.2	1.3	1.2	0.9
^{29}Si(p,γ)^{30}P		1.0	1.6	1.6	1.5
^{39}K(p,γ)^{40}Ca	15	4.5	3.0	2.6	2.5
^{41}K(p,γ)^{42}Ca	0.5	0.5	0.5	0.4	0.4
^{41}K(p,n)^{41}Ca	0.8	1.0	1.1	1.2	1.3
^{40}Ca(p,γ)^{41}Sc				0.1	0.2
^{42}Ca(p,γ)^{43}Sc	1.3	1.4	1.4	1.4	1.3
	0.8	1.1	1.3	1.4	1.4

Table 3.

radius, depth and compensating reflection factor of the black-body, square-well
equivalent of the Woods–Saxon potential characteristic of the interaction
between n, p and α with nuclei having $Z \geqslant 8$. Two free parameters must also be
incorporated to adjust the intensity of electric and magnetic dipole transitions
for gamma radiation. Weak interaction rates must also be specified and ways
and means for doing this will be discussed later in Section VII.

The parameters originally chosen for n, p and α-reactions were taken from
earlier work of Michaud and Fowler (47) who depended heavily on studies by
Vogt (see 47). These parameters and those chosen for electromagnetic and
weak interactions have survived comparison of the theory with a plethora of
laboratory measurements. More sophisticated programs have been developed
which use experimental neutron strength functions instead of that from the
equivalent square well or which use realistic Woods–Saxon potentials for all
interactions as done by Mann (48). In addition marked improvement in the

correspondence between theory and experiment is found when width-fluctu-
ation corrections are made as described in Zyskind *et al.* (49).

It is well known that the free parameters can always be adjusted to fit the
cross sections and reaction rates of any one particular nuclear reaction. This is
not done in a *global* program. The parameters are in principle determined by
the best least squares fit to all reactions for which experimental results are
available. For example see the figure, p. 307, in reference (46). It is on this
basis that some confidence can be had in predictions in those cases where
experimental results are unavailable.

The original program, references (46) and (23), has produced reaction rates
either in numerical or analytical form as a function of temperature. Ready
comparison with integrations of laboratory cross sections for target ground
states are possible. Using the same *global* parameters which apply to reactions
involving the ground states of stable nuclei the theoretical program calculates
rates for the ground states of radioactive nuclei and for the excited states of
both stable and radioactive nuclei. Summing over the statistically weighted
contributions of the ground and known excited states or theoretical level
density functions yields the stellar reaction rate for the equilibrated statistical
population of the nuclear states. After summing, division by the partition
function of the target nucleus is necessary. Analytical parametrized expressions
for the partition functions of nuclei with $8 \leqslant Z \leqslant 36$ are given in Table IIA of
reference (23) as a function of temperature over the range $0 \leqslant T \leqslant 10^{10}$K.

Sargood (50) has compared experimental results from a number of laboratories for protons and alpha particles reacting with 80 target nuclei which are, of
course, in their ground states with the theoretical predictions of reference (23).
Ratios of statistical model calculations to laboratory measurements for 12 cases
are shown in Table 3 for temperatures in the range from 1 to 5×10^9K. The
double entry for ^{27}Al$(p,n)^{27}$Si signifies ratios of theory to measurements made in
two different laboratories. It is fair to note that the theoretical calculations
match the experimental results within 50% with a few marked exceptions. In
American vernacular "You win some and you lose some". For the rather light
targets in Table 3, especially at low temperature, the global mean rates can be
in error whenever more and stronger resonances or fewer and weaker resonances than expected on average occur in the excitation curve of the reaction at
low energies.

Sargood (50) has also compared the ratio of the stellar rate of a reaction with
target nuclei in a thermal distribution of ground and excited states with the rate
for all target nuclei in their ground state. The latter is of course determined
from laboratory measurements. A number of cases are tabulated for
$T = 5 \times 10^9$K in Table 4. In many cases, notably for reactions producing gamma rays, the ratio of stellar to laboratory rates is close to unity. In other cases
the ratios can be high by several orders of magnitude. This can occur for a
number of reasons. It frequently occurs when the ground state can interact only
through partial waves of high angular momentum resulting in small penetration factors and thus small cross sections and rates. This makes clear a basic
assumption in the prediction of stellar rates: a statistical theory which does well

STELLAR/LABORATORY REACTION RATES

$$\langle \sigma v \rangle^* / \langle \sigma v \rangle^0$$

TEMPERATURE = 5×10^9 K

D. G. Sargood, Australian Journal of Physics (1983)

Woosley, Fowler, Holmes & Zimmerman, At. Data & Nucl. Data Tables, 22, 371 (1978)

Target nucleus	Reaction								
	(n,γ)	(n,p)	(n,α)	(p,γ)	(p,n)	(p,α)	(α,γ)	(α,n)	(α,p)
^{20}Ne	0.959	12.2	4.98	0.954	34.1	6.86	0.907	4.90	1.29
^{21}Ne	0.808	6.15	1.13	0.818	1.78	1.95	0.943	0.985	1.37
^{22}Ne	0.917	159	22.1	0.895	5.11	2.72	0.968	0.996	2.46
^{23}Na	0.897	4.95	9.70	0.890	2.27	0.944	0.826	1.30	0.918
^{24}Mg	0.939	20.4	7.30	0.924	120	15.0	0.835	4.70	1.04
^{25}Mg	0.905	5.05	3.18	0.862	3.48	5.02	0.958	0.973	1.10
^{26}Mg	0.968	71.4	53.8	0.958	8.05	4.92	0.974	1.00	1.41
^{27}Aℓ	0.934	4.12	10.9	0.913	3.22	1.14	0.905	1.13	0.972
^{28}Si	0.976	6.51	7.26	0.950	140	23.5	0.933	3.55	1.02
^{29}Si	0.943	8.67	3.34	0.907	3.18	50.1	0.927	0.964	1.18
^{30}Si	0.989	89.4	28.6	0.982	2.99	6.63	0.973	1.01	1.09
^{31}P	0.972	2.63	18.4	0.901	3.77	1.11	0.969	1.70	0.978
^{32}S	0.988	2.33	1.57	0.980	90.1	7.35	0.975	3.79	1.00
^{33}S	0.943	1.46	1.06	0.920	4.73	3.24	0.916	0.995	1.01
^{34}S	1.00	25.8	13.1	0.979	8.02	2.02	0.964	1.05	1.02
^{36}S	0.998	428	95.9	1.00	1.00	1.02	0.995	1.00	1.68
^{35}Cℓ	0.972	1.19	3.06	0.948	4.48	1.05	0.945	1.23	0.992
^{37}Cℓ	0.994	26.0	13.7	0.987	1.00	1.00	0.985	1.00	0.995

Table 4.

predicting ground state results is assumed to do equally well in predicting excited state results. This assumption is frequently not valid. Bahcall and Fowler (51) have shown that in a few cases laboratory measurements on inelastic scattering involving excited states can be used indirectly to determine reaction cross sections for those states.

Ward and Fowler (52) have investigated in detail the circumstances under which long lived isomeric states do not come into equilibrium with ground states. When this occurs it is necessary to incorporate into network calculations the stellar rates for both the isomeric and ground state. An example of great interest is the nucleus ^{26}Al. The ground state has spin and parity, $J^\pi = 5^+$ and isospin, $T = 0$, and has a mean lifetime for positron emission to ^{26}Mg of 10^6 years. The isomeric state at 0.228 MeV has $J^\pi = 0$, $T = 1$ and mean lifetime 9.2 seconds. Ward and Fowler (52) show that the isomeric state effectively does not come into equilibrium with the ground state for $T < 4 \times 10^8$K. At these low temperatures both the isomeric state and the ground state of ^{28}Al must be included in the network of Figure 7.

VII. *Astrophysical Weak-Interaction Rates*

Weak nuclear interactions play an important role in astrophysical processes in conjunction with the strong nuclear interactions as indicated in Figure 7. Only through the weak interaction can the overall proton number and neutron number of nuclear matter change during stellar evolution, collapse, and explosion. The formation of a neutron star requires that protons in ordinary stellar matter capture electrons. Gravitational collapse of a Type II supernova core is retarded as long as electrons remain to exert outward pressure.

Many years of theoretical and experimental work on weak-interaction rates in the Kellogg Laboratory and elsewhere have culminated in the calculation and tabulation by Fuller, Fowler and Newman (53) of electron and positron emission rates and continuum electron and positron capture rates, as well as the associated neutrino energy loss rates for free nucleons and 226 nuclei with mass numbers between $A = 21$ and 60. Extension to higher and lower values for A is now underway.

These calculations depended heavily on experimental determinations in Kellogg by Wilson, Kavanagh and Mann (54) of Gamow-Teller elements for 87 discrete transitions in intermediate-mass nuclei. The majority of the experimental matrix elements for both Fermi and Gamow-Teller discrete transitions as well as nuclear level data were taken from the exhaustive tabulation of Lederer and Shirley (55). Unmeasured matrix elements for allowed transitions were assigned a mean value as described in the second of references (53). These mean values were $|M_F|^2 = .062$ and $|M_{GT}|^2 = .039$ corresponding to $\log ft = 5$, where f is the phase space factor and t is the half-life for the transition. Nuclear physicists traditionally think in terms of log ft-values in connection with weak interaction rates.

Simple shell model arguments were employed to estimate Gamow–Teller sum rules and collective state resonance excitation energies. These estimates have been shown to be high by $\sim$50% fair approximations for $T^<$-nuclei and $T^>$-nuclei by recent high resolution measurements on p,n-reactions and $^3T, ^3$He-reactions by Goodman *et al.* and Ajzenberg-Selove *et al.* respectively (56). Here $T^<$, with $T = |N-Z|$ represents, for example, ^{56}Fe with $T = 2$ in ^{56}Fe$(e^-,\nu)^{56}$Mn or ^{56}Fe$(n,p)^{56}$Mn. Similarly $T^>$ designates ^{56}Mn with $T = 3$. The work described in references (53) emphasizes the great need for additional results for $T^>$-nuclei using the n,p-reaction as well as the $^3T, ^3$He-reaction from which matrix elements for electron capture can be obtained.

Moment method shell model calculations of Gamow–Teller strength functions have been performed by S. D. Bloom and G. M. Fuller (57) with the Lawrence Livermore National Laboratory's vector shell model code for the ground states and first excited states of ^{56}Fe, ^{60}Fe, and ^{64}Fe. These detailed calculations confirm the general trends in Gamow–Teller strength distributions used in the approximations of references (53).

The discrete state contribution to the rates, dominated by experimental information and the Fermi transitions, determines the weak nuclear rates in the regime of temperature and densities characteristic of the quasistatic phases of presupernova stellar evolution. At the higher temperatures and densities characteristic of the supernova collapse phase, which is of such great current interest as discussed in detail in Brown, Bethe and Baym (58), the electron-capture rates are dominated by the Gamow-Teller collective resonance contribution.

The detailed nature and the difficulty of the theoretical aspects of the combined atomic, nuclear, plasma, and hydrodynamic physics problems in Type II supernova implosion and explosion were brought home to us by Hans Bethe during his stay in our laboratory as a Caltech Fairchild Scholar early in 1982. His visit plus long-distance interaction with his collaborators resulted in the preparation of two seminal papers, Bethe, Yahil, and Brown (59) and Bethe, Brown, Cooperstein, and Wilson (60).

Current ideas on the nuclear equation of state predict that early in the collapse of the iron core of a massive star the nuclei present will become so neutron rich that allowed electron capture on protons in the nuclei is blocked. Allowed electron capture, for which $\Delta l = 0$, is not permitted when neutrons have filled the subshells having orbital angular momentum, l, equal to that of the subshells occupied by the protons.

This neutron shell blocking phenomenon, and several unblocking mechanisms operative at high temperature and density, including forbidden electron capture, have been studied in terms of the simple shell model by Fuller (61). Though the unblocking mechanisms are sensitive to details of the equation of state, typical conditions result in a considerable reduction of the electron capture rates on heavy nuclei leading to significant dependence on electron capture by the small number of free protons and a decrease in the overall neutronization rate.

The results of one-zone collapse calculations which have been made by Fuller (61) suggest that the effect of neutron shell blocking is to produce a larger core lepton fraction (leptons per baryon) at neutrino trapping. In keeping with the Chandrasekhar relation that core mass is proportional to the square of the lepton fraction this leads to a larger *final*-core mass and hence a stronger post-bounce shock. On the other hand the incorporation of the new electron capture rates during precollapse Si-burning reduces the lepton fraction and leads to a smaller *initial*-core mass and thus to a smaller amount of material (*initial*-core mass minus *final*-core mass) in which the post-bounce shock can be dissipated. The dissipation of the shock is thus reduced. This is discussed in detail in reference (39).

Recent work on the weak-interaction has concentrated on making the previously calculated reaction rates as efficient as possible for users of the published tables and the computer tapes which are made available on request. The stellar weak interaction rates of nuclei are in general very sensitive functions of temperature and density. Their temperature dependence arises from thermal excitation of parent excited states and from the lepton distribution functions in the integrands of the decay and continuum capture phase space factors.

For electron and positron emission, most of the temperature dependence is due to thermal population of parent excited states at all but the lowest temperatures and highest densities. In general, only a few transitions will contribute to these decay rates and hence the variation of the rates with temperature is usually not so large that rates cannot be accurately interpolated in temperature and density with the standard grids provided in references (53). The density dependence of these decay rates is minimal. In the case of electron emission, however, there may be considerable density dependence due to Pauli blocking for electrons where the density is high and the temperature is low. This does not present much of a problem for practical interpolation since the electron-emission rate is usually very small under these conditions.

The temperature and density dependence of continuum electron and positron capture is much more serious problem. In addition to temperature sensitivity introduced through thermal population of parent excited states, there are considerable effects from the lepton distribution functions in the integrands of the continuum-capture phase-space factors. This sensitivity of the capture rates means that interpolation in temperature and density on the standard grid to obtain a rate can be difficult, requiring a high-order interpolation routine and a relatively large amount of computer time for an accurate value. This is especially true for electron capture processes with threshold above zero energy.

We have found that the interpolation problem can be greatly eased by defining a simple continuum-capture phase-space integral, based on the parent-ground-state to daughter-ground-state transition Q-value, and then dividing this by the tabulated rates (53) at each temperature and density grid point to obtain a table of effective ft-values; these turn out to be much less dependent on temperature and density. This procedure requires a formulation of the capture phase-space factors which is simple enough to use many times in the inner loop of stellar evolution nucleosynthesis computer programs. Such a formulation in terms of standard Fermi integrals has been found, along with approximations for the requisite Fermi integrals. When the chemical potential (Fermi energy) which appears in the Fermi integrals goes through zero these approximations have continuous values and continuous derivatives.

We have recently found expressions for the reverse reactions to e^-, e^+-capture, (i.e., $\nu,\bar\nu$-capture) and for $\nu,\bar\nu$ blocking of the direct reactions when $\nu,\bar\nu$-states are partially or completely filled. These reverse reactions and the blocking are important during supernova core collapse when neutrinos and antineutrinos eventually become trapped, leading to equilibrium between the two directions of capture. General analytic expressions have been derived and approximated with computer-usable equations. All of these new results described in the previous paragraphs will be published in Fuller, Fowler, and Newman (62) and new tapes including $\nu,\bar\nu$-capture will be made available to users on request.

VIII. *Calculated abundances for A $\lesssim$ 60 with Brief Comments on Explosive Nucleosynthesis*

Armed with the available strong and weak nuclear reaction rates which apply to the advanced stages of stellar evolution, theoretical astrophysicists have attempted to derive the elemental and isotopic abundances produced in quasi-static presupernova nucleosynthesis and in explosive nucleosynthesis occurring during supernova outbursts.

The various stages of preexplosive nucleosynthesis have been discussed in Sections IV, V and VI and it is fair to say that there is reasonably general agreement on nucleosynthesis during these stages. On the other hand explosive nucleosynthesis is still an unsettled matter, subject to intensive study at the present time as reviewed for example, in Woosley, Axelrod, and Weaver (63).

The abundance produced in explosive nucleosynthesis must of necessity depend on the detailed nature of supernova explosions. Ideas concerning the nature of Type I and Type II supernova explosions were published many years ago by Hoyle and Fowler (64) and Fowler and Hoyle (65). It was suggested that Type I supernovae of small mass were precipitated by the onset of explosive carbon burning under conditions of electron-degeneracy where pressure is approximately independent of temperature. Carbon burning raises the temperature to the point where the electrons are no longer degenerate and explosive disruption of the star results. For Type II supernovae of larger mass it was suggested that Si-burning produced iron-group nuclei which have the maximum binding energies of all nuclei so that nuclear energy is no longer available. Subsequent photodisintegration and electron-capture in the stellar core leads to core implosion and ignition of explosive nucleosynthesis in the infalling inner mantle which still contains nuclear fuel. These ideas have "survived" but, to say the least, with considerable modification over the years as indicated in the excellent review by Wheeler (66). Modern views on Type II supernovae are given in references (38), (39), (58), (59), and (60) and on Type I supernovae in Nomoto (67).

We can return to the nuclear abundance problem by reference to Figure 16 taken from reference (38), which shows the distribution of the final abundances by mass fraction in the supernova ejecta of a 25 $M_\odot$ Population I star. The presupernova distribution is that shown in Figure 8(a). The modification in the abundances for the mass zones interior to $2.2M_\odot$ is very apparent. The mass exterior to $2.2M_\odot$ is ejected with little or no modification in nuclear abundances. The supernova explosion was simulated by arbitrarily assuming that the order of 10^{51} ergs was delivered to the ejected material by the shock generated in the bounce or rebound of the collapsing and hardening core.

Integration over the mass zones of Figure 16 for $1.5M_\odot < M < 2.2M_\odot$ and over those of Figure 8(a) for $M > 2.2M_\odot$ enabled Woosley and Weaver (38) to calculate the isotopic abundances ejected into the interstellar medium by their $25M_\odot$ Population I simulated supernova. The results relative to solar abundances (the reader should refer to the last paragraph of Section I) are shown in Figure 17 taken from reference (38). The relative ratios are normalized to unity for ^{16}O for which the overproduction ratio was 14, that is, for each gram of ^{16}O originally in the star, 14 grams were ejected. This overproduction in a single supernova can be expected to have produced the heavy element abundances in the interstellar medium just prior to formation of the solar system given the fact that supernovae occur approximately every one hundred years in the Galaxy. The ultimate theoretical calculations will yield a constant overproduction factor of the order of 10.

The results shown in Figure 17 are disappointing if one expects the ejecta of $25M_\odot$ Population I supernovae to match solar system abundances with a relatively constant overproduction factor. The dip in abundances from sulfur to chromium is readily apparent. Woosley and Weaver (38) point out that calculations must be made for other stellar masses and properly integrated over the

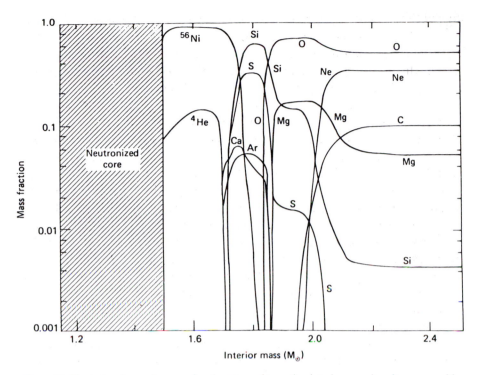

Figure 16. Final abundances by mass fraction versus increasing interior mass in solar masses, $M_\odot$, in Type II supernova ejecta from a Population I star with total mass equal to $25M_\odot$ from Woosley and Weaver (38).

mass distribution for stellar formation which varies roughly inversely proportional to mass. Woosley, Axelrod and Weaver (63) discuss their expectations of the abundances produced in stellar explosions for stars in the mass range $10M_\odot$ to $10^6 M_\odot$ They show that a $200M_\odot$ Population III star produces abundant quantities of sulfur, argon, and calcium which possibly compensate for the dip in figure 17. Population III stars are massive stars in the range $100M_\odot < M < 300M_\odot$ which are thought to have formed from hydrogen and helium early in the history of the Galaxy and evolved very rapidly. Since their heavy element abundance was zero they have no counterparts in presently forming Population I stars as well as no counterparts among old, low mass Population II stars.

Other authors have suggested a number of solutions to the problem depicted in Figure 17. Nomoto, Thielemann and Wheeler (68) have calculated the abundances produced in carbon deflagration models of Type I supernovae. By adding equal contributions from Type I and Type II supernovae they obtain Figure 18 which can be considered somewhat more satisfactory than Figure 17. On the other hand Arnett and Thielemann (69) have recalculated quasistatic presupernova nucleosynthesis for $M \approx 20M_\odot$ using a value for the $^{12}C(\alpha, \gamma)^{16}O$ rate equal to three times that given in references listed under (21). This would seem to be justified by the recent analysis of $^{12}C(\alpha, \gamma)^{16}O$ data in reference (34) as discussed in Section V. They then assume that explosive nucleosynthesis will

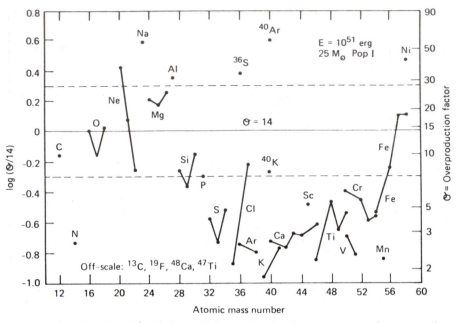

Figure 17. Overabundance (ϑ) relative to 14 times solar abundances versus atomic mass number for nucleosynthesis resulting from a Type II, Population I supernova with total mass equal to 25 $M_\odot$ from Woosley and Weaver (38).

not substantially modify their quasistatic abundances and obtain the results shown in Figure 19. The average overproduction ratio is roughly 14 and deviations are in general within a factor of two of this value. However, their assumption of minor modification during explosion and ejection is questionable.

I feel that the results discussed in this section and those obtained by numerous other authors show promise of an eventual satisfactory answer to the question where and how did the elements from carbon to nickel originate. We shall see!

IX. *Isotopic Anomalies in Meteorites and Observational Evidence for Ongoing Nucleosynthesis*

Almost a decade ago it became clear that nucleosynthesis occurred in the Galaxy up to the time of formation of the solar system or at least up until several million years before the formation. For slightly over one year it has been clear that nucleosynthesis continues up to the present time or at least within several million years of the present. The decay of radioactive ^{26}Al($\bar{\tau} = 1.04 \times 10^6$ years) is the key to these statements which bring great satisfaction to most experimentalists, theorists, and observers in Nuclear Astrophysics. For the record it must be admitted that the word "clear" is subject to certain reservations in the minds of some investigators but as a believer, "clear" is clear to me.

Isotopic anomalies in meteorites produced by the decay of shortlived radioactive nuclei were first demonstrated in 1960 by Reynolds (70) who found large enrichments of ^{129}Xe in the Richardson meteorite. Jeffery and Reynolds (71)

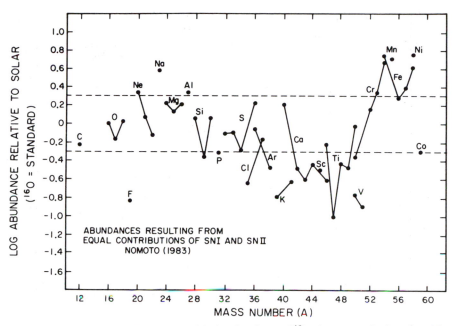

Figure 18. Abundances relative to solar with the abundance of ^{16}O taken as standard produced by equal contributions from typical Type I and Type II supernovae from Nomoto, Thielemann and Wheeler (68).

demonstrated in 1961 that the excess ^{129}Xe correlated with ^{127}I in the meteorite and thus showed that the ^{129}Xe resulted from the decay *in situ* of ^{129}I($\bar{\tau} = 23 \times 10^6$ years). Quantitative results indicated that ^{129}I/^{127}I$\approx 10^{-4}$ at the time of meteorite formation. On the assumption that ^{129}I and ^{127}I are produced in roughly equal abundances in nucleosynthesis (most probably in the r-process) over a period of $\sim 10^{10}$ years in the Galaxy prior to formation of the solar system and taking into account that only the ^{129}I produced over a period of the order of its lifetime survives, Wasserburg, Fowler, and Hoyle (72) suggested that a period of free decay of the order of 10^8 years or more occurred between the last nucleosynthetic event which produced ^{129}I and its incorporation in meteorites in the solar system. There remains evidence for such a period in some cases, notably ^{244}Pu, but probably not in the history of the nucleosynthetic events which produced ^{129}I and other "short"-lived radioactive nuclei such as ^{26}Al and ^{107}Pd($\bar{\tau} = 9.4 \times 10^6$ years).

The substantiated meteoritic anomalies in ^{26}Mg from ^{26}Al, in ^{107}Ag from ^{107}Pd, in ^{129}Xe from ^{129}I, and in the heavy isotopes of Xe from the fission of ^{244}Pu($\bar{\tau} = 117 \times 10^6$ years; fission tracks also observed) as well as searches in the future for anomalies in ^{41}K from ^{41}Ca($\bar{\tau} = 0.14 \times 10^6$ years), in ^{60}Ni from ^{60}Fe($\bar{\tau} = 0.43 \times 10^6$ years), in ^{53}Cr from ^{53}Mn($\bar{\tau} = 5.3 \times 10^6$ years), and in ^{142}Nd from ^{146}Sm($\bar{\tau} = 149 \times 10^6$ years; α-decay) are discussed exhaustively by my colleagues Wasserburg and Papanastassiou (73). They espouse *in situ* decay for the observations to date but my former student D. D. Clayton (74) argues that the anomalies occur in interstellar grains preserved in the meteorites and originally produced by condensation in the expanding and cooling envelopes of

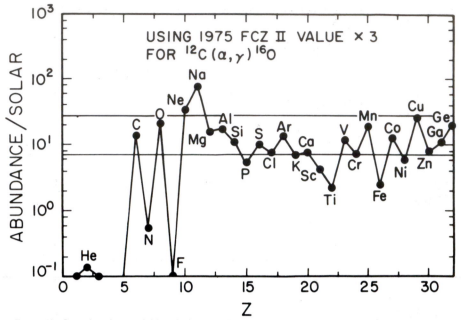

Figure 19. Overabundance yields relative to solar versus atomic number, Z, resulting from the explosion of a Type II supernova with mass approximately equal to $20M_\odot$ from Arnett and Thielemann (69). The horizontal lines are a factor two higher and lower than the average overabundance equal to 14. It is assumed that the pre-supernova abundances were not modified during the supernova explosion. The reaction rate for $^{12}C(\alpha, \gamma)^{16}0$ of Fowler, Caughlan and Zimmerman (21) was multiplied by a factor of 3 in accordance with the theoretical analysis by Langanke and Koonin (34).

supernovae and novae. Wasserburg and Papanastassiou (73) write on p. 90 "There is, as yet, no compelling evidence for the presence of preserved presolar grains in the solar system. All of the samples so far investigated appear to have melted or condensed from a gas, and to have chemically reacted to form new phases." With mixed emotions I accept this.

Before turning to some elaboration of the $^{26}Al/^{26}Mg$ case it is appropriate to return to a discussion of the free decay interval mentioned above. It is the lack of a detectable anomalies in ^{235}U from the decay of $^{247}Cm(\bar{\tau} = 23 \times 10^6$ years) in meteorites as shown by Chen and Wasserburg (75) coupled with the demonstrated occurrence of heavy Xe anomalies from the fission of $^{244}Pu(\bar{\tau} = 117 \times 10^6$ years) as discussed for example by Burnett, Stapanian and Jones (76) which demands a free decay interval of the order of several times 10^8 years. This interval is measured from the "last" r-process nucleosynthesis event (supernova?) which produced the actinides, Th, U, Pu, Cm, and beyond, up to the "last" nucleosynthesis events (novae?, supernovae with short-run r-processes?) which produced the short-lived nuclei ^{26}Al, ^{107}Pd, and ^{129}I before the formation of the solar system. The fact that the anomalies produced by these short-lived nuclei relative to normal abundances all are of the order of 10^{-4} despite a wide range in their mean lifetimes (1.04 to 23×10^6 years) indicates that this anomaly range must be the result of inhomogeneous mixing of exotic materials with much larger quantities of normal solar system materials over a short time

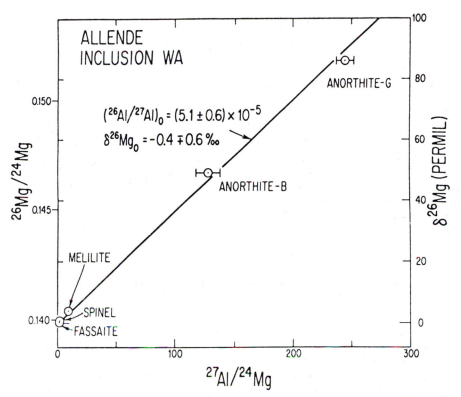

Figure 20. Evidence for the *in situ* decay of ^{26}Al in various minerals in inclusion WA of the Allende meteorite from Lee, Papanastassiou and Wasserburg (77). The linear relation between $^{26}Mg/^{24}Mg$ and $^{27}Al/^{24}Mg$ implies that $^{26}Al/^{27}Al = (5.1 \pm 0.6) \times 10^{-5}$ at the time of information of the inclusion with ^{26}Al considered to react chemically in the same manner as ^{27}Al.

rather than the result of free decay. The challenges presented by this conclusion are manifold. Figure 14 of reference (73) shows the time scale for the formation of dust, rain, and hailstones in the early solar system and for the aggregation into chunks and eventually the terrestrial planets. The solar nebula was almost but not completely mixed when it collapsed to form the solar system. From ^{26}Al it becomes clear that the mixing time down to an inhomogenity of only one part in 10^3 (see what follows) was the order of 10^6 years.

Evidence that ^{26}Al was alive in interstellar material in the solar nebula which condensed and aggregated to form the parent body (planet in the asteroid belt?) of the Allende meteorite is shown in Figure 20 taken with some modification from Lee, Papanastassiou, and Wasserburg (77). The Allende meteorite fell near Pueblito de Allende in Mexico on February 8, 1969 and is a carbonaceous chondrite, a type of meteorite thought to contain the most primitive material in the solar system *unaltered since its original solidification*.

Figure 20 depicts the results for $^{26}Mg/^{24}Mg$ versus $^{27}Al/^{24}Mg$ in different mineral phases (spinel, etc.) from a Ca-Al-rich inclusion called WA obtained from a chondrule found in Allende. It will be clear that excess ^{26}Mg correlates linearly with the amount of ^{27}Al in the mineral phases. Since ^{26}Al is chemically identical with ^{27}Al, it can be inferred that phases rich in ^{27}Al were initially rich

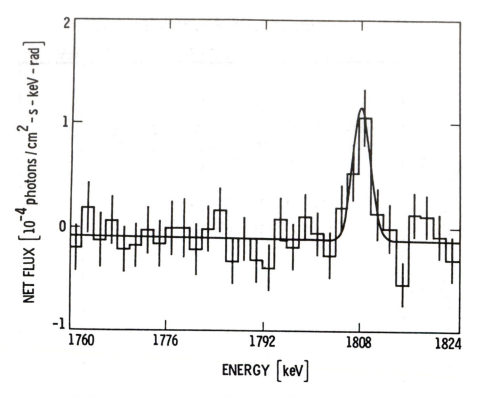

Figure 21. The High Energy Astrophysical Observatory (HEAO 3) data on gamma rays in the energy range 1760 to 1824 keV emitted from the Galactic equatorial plane from Mahoney *et al.* (78). The line at 1809 keV is attributed to the decay of radioactive ^{26}Al($\bar{\tau} = 1.04 \times 10^6$ years) to the excited state of ^{26}Mg at this energy.

in ^{26}Al which subsequently decayed *in situ* to produce excess ^{26}Mg. ^{26}Al was alive with abundance 5×10^{-5} that of ^{27}Al in one part of the solar nebula when the WA inclusion aggregated during the earliest stages of the formation of the solar system. The unaltered inclusion survived for 4.5 billion years to tell its story. Other inclusions in Allende and other meteorites yield ^{26}Al/^{27}Al from zero up to $\sim 10^{-3}$ with 10^{-4} a representative value. The reader is referred to reference (73) for the rich details of the story and the important and significance of non-accelerator-based contributions to Nuclear Astrophysics.

Evidence that ^{26}Al is alive in the interstellar medium today is shown in Figure 21 from Mahoney, Ling, Wheaton and Jacobson (78), my colleagues at Caltech's Jet Propulsion Laboratory (JPL). Figure 21 shows the gamma-ray spectrum observed in the range 1760 to 1824 keV by instruments aboard the High Energy Astronomical Observatory, HEAO 3, which searched for diffuse gamma-ray emission from the Galactic equatorial plane.

The discrete line at 1809 keV, detected with a significance of nearly five standard deviations, is without doubt due to the transition from the first excited state at 1809 keV in ^{26}Mg to its ground state. Radioactive ^{26}Al decays by ^{26}Al$(e^+v)^{26}$Mg $(\gamma)^{26}$Mg to this state and thence to the ground state of ^{25}Mg. This gamma-ray transition shows clearly that ^{26}Al is alive in the interstellar medium in the Galactic equatorial plane today. Given the mean life-time (1.04

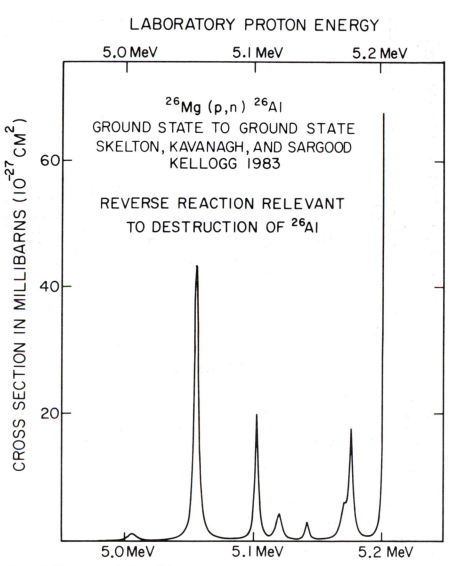

Figure 22. The cross section in millibarns versus laboratory proton energy for the ground-state to ground-state reaction $^{26}Mg(p, n)^{26}Al$ from Skelton, Kavanagh and Sargood (79).

$\times 10^6$ years) of ^{26}Al, this shows that ^{26}Al has been produced no longer than several million years ago and is probably being produced continuously. It is no great extrapolation to argue that nucleosynthesis in general continues in the Galaxy at the present time. Quantitatively the observations indicate that $^{26}Al/^{27}Al \sim 10^{-5}$ in the interstellar medium. This value averages over the Galactic plane interior to the sun at the present time. This average value was probably much the same when the solar system formed but the variations in $^{26}Al/^{27}Al$ for various meteoritic inclusions show that there were wide variations in the solar nebula about this value ranging from zero to 10^{-3}.

The question immediately arises, what is the site of the synthesis of the ^{26}Al? Since the preparation of reference (52) I have been convinced that ^{26}Al could not be by synthesized in supernovae at high temperatures where neutrons are copiously produced because of the expectation of a large cross section for $^{26}Al(n, p)^{26}Mg$. This expectation has been borne out by the measurements on the reverse reaction $^{26}Mg(p,n)^{26}Al$ in the Kellogg Laboratory by Skelton, Kavanagh and Sargood (79). Figure 22 is taken from Figure 1a of these authors and shows the great beauty of high resolution measurements in experimental Nuclear Astrophysics. Until the ^{26}Al-targets just recently available can be bombarded with neutrons it is necessary to supplement the laboratory measurements on $^{26}Mg(p,n)^{26}Al$, perforce involving the ground state of ^{26}Mg, with theoretical calculations involving excited states, reference (23), in order to calculate the stellar rate for $^{26}Al(n,p)^{26}Mg$. There is little doubt that this rate is very large indeed.

In references (74) and (78) and in Arnould *et al.* (80) it is suggested that ^{26}Al is produced in novae. This is quite reasonable on the basis of nucleosynthesis in novae as discussed in Truran (81). In current models for novae hydrogen from a binary companion is accreted by a white dwarf until a thermal runaway involving the fast CN cycle occurs. Similarly a fast MgAl cycle may occur with production of $^{26}Al/^{27}Al \geq 1$ as shown in Figure 9 of reference (52). The recent experimental measurements cited in reference (52) substantiate this conclusion. Clayton (74) argues that the estimated 40 novae occurring annually in the Galactic disk can produce the observed $^{26}Al/^{27}Al$ ratio of the order of 10^{-5} on average. He assumes that each nova ejects $10^{-4}M_\odot$ of material containing a mass fraction of ^{26}Al equal to 3×10^{-4}.

Another possible source of ^{26}Al is spallation induced by irradiation of proto-planetary material by high energy protons from the young sun as it settled on the main sequence. This possibility was discussed very early by Fowler, Greenstein, and Hoyle (82) who also attempted to produce D, Li, Be, and B in this way, requiring such large primary proton and secondary neutron fluxes that many features of the abundance curve in the solar system would have been changed substantially. A more reasonable version of the scenario was presented by Lee (83) but without notable success. I find it difficult to believe that an early irradiation produced the anomalies in meteorites. The ^{26}Al in the interstellar medium today certainly cannot have been produced in this way.

Anomalies have been found in meteorites in the abundances compared with normal solar system material of the stable isotopes of may elements: O, Ne, Mg, Ca, Ti, Kr, Sr, Xe, Ba, Nd, and Sm. The possibility that the oxygen anomalies are non-nuclear in origin has been raised by Thiemens and Heidenreich (84) but the anomalies in the remaining elements are generally attributed to nuclear processes.

One example is a neutron-capture/beta-decay (nβ) process studied by Sandler, Koonin, and Fowler (85). The seed nuclei consisted of all of the elements from Si to Cr with normal solar system abundances. When this process operates at neutron densities $\approx 10^7$ mole cm^{-3} and exposure times of $\approx 10^3$ s, small admixtures ($\leq 10^{-4}$) of the exotic material produced are sufficient to account for most of the Ca and Ti isotopic anomalies found in the Allende meteorite inclusion EK-1-4-1 by Niederer, Papanastassiou, and Wasserburg

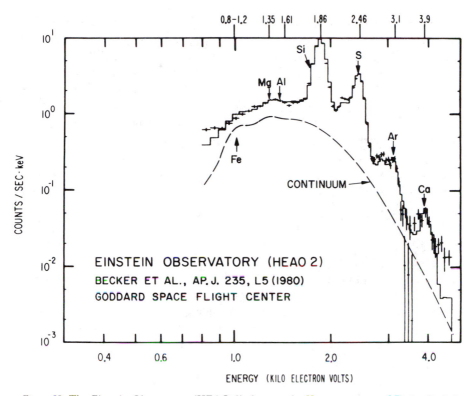

Figure 23. The Einstein Observatory (HEAO 2) data on the X-ray spectrum of Tycho Brahe's Supernova Remnant from Becker *et al.* (89).

(86). The anomalies in stable isotope abundances are of the same order as those for short lived radioactive nuclei and strongly support the view that the solar nebula was inhomogeneous and not completely mixed with regions containing exotic materials up to 10^{-4} or more of normal material.

Agreement for the ^{46}Ca and ^{49}Ti anomalies in EK-1-4-1 was obtained only by increasing the theoretical Hauser-Feshbach cross sections for ^{46}K(n,γ) and ^{49}Ca(n,γ) by a factor of 10 on the basis of probable thermal resonances just above threshold in the compound nuclei ^{47}K and ^{50}Ca respectively. In a CERN report which subsequently became available Huck *et al.* (87) reported an excited state in ^{50}Ca just 0.16 MeV above the ^{49}Ca(n,γ) threshold which can be produced by s-wave capture and fulfills the requirements of reference (85).

Reference (85) suggests that the $\approx 10^3$s exposure time scale is determined by the mean life-time of ^{13}N$(\bar{\tau} = 862$ s), produced through ^{12}C$(p,\gamma)^{13}$N by a jet of hydrogen suddenly introduced into the helium burning shell of a Red Giant star where a substantial amount of ^{12}C has been produced by the 3 $\alpha \rightarrow {}^{12}$C process. The beta decay ^{13}N$(e^+v)^{13}$C is followed by ^{13}C$(\alpha,n)^{16}$O as the source of the neutrons. All of this is very interesting, if true. More to the point reference (85) predicts the anomalies to be expected in the isotopes of chromium. Attempts to measure these anomalies are underway now by Wasserburg and his colleagues. Again, we shall see!

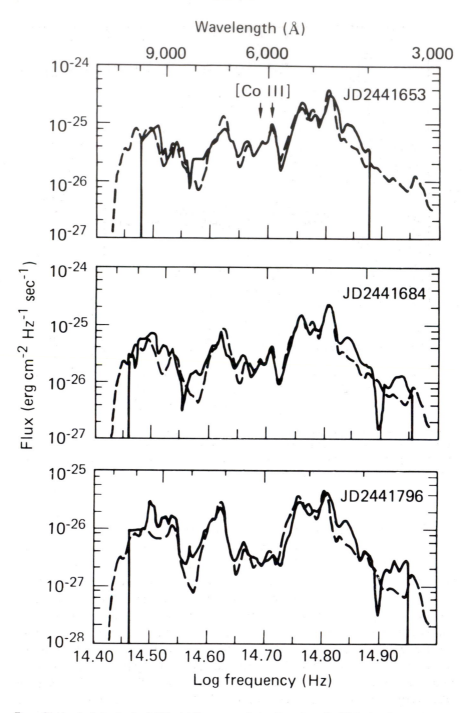

Figure 24. Analysis by Axelrod (91) yielding two emission lines from Co III in the observations on SN 1972e obtained by Kirshner and Oke (92). The observations were made 233, 264 and 376 days after JD2441420, assigned as the initial day of the supernova explosion. The mean lifetime of ^{56}Co is 114 days; the Co III lines appear to decay in keeping with their emission from radioactive ^{56}Co.

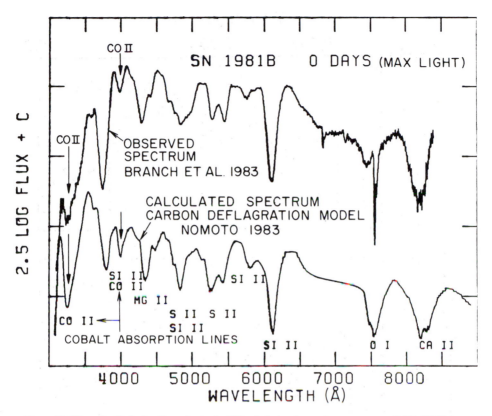

Figure 25. Top: Analysis by Branch *et al.* (93) of their absorption spectrum of SN 1981b at maximum light showing evidence for Co II absorption features. Bottom: Comparison with the calculated spectrum expected from the carbon deflagration model for Type I supernova according to the calculations of Nomoto (67).

X. *Observational Evidence for Nucleosynthesis in Supernovae*

Over the years there has been considerable controversy concerning elemental abundance observations in the optical wave-length range on Galactic supernovae remnants. To my mind the most convincing evidence for nucleosynthesis in supernovae has been provided by Chevalier and Kirshner (88) who obtained quantitative spectral information for several of the fast-moving knots in the supernova remnant Cassiopeia A (approximately dated 1659 but a supernova event was not observed at that time). The knots are considered to be material ejected from various layers of the original star in a highly asymmetric, nonspherical explosion. In one knot, labelled KB33, the following ratios *relative to solar, designated by brackets* were observed: [S/O] = 61, [Ar/O] = 55, [Ca/O] = 59. It is abundantly clear that oxygen burning to the silicon group elements in the layer in which KB33 originated has depleted oxygen and enhanced the silicon group elements. Other knots and other features designated as filaments show different abundance patterns, albeit, not so easily interpreted. The moral for supernova modelers is that spherically symmetric supernova explosions may be the easiest to calculate but are not to be taken as realistic. Admittedly they have a good answer: it is expensive enough to compute spherically symmetric models. OK, OK!

Most striking of all has been the payoff from the NASA investment in the High Energy Astronomy Observatory (HEAO 2), now called the Einstein Observatory. From this satellite Becker *et al.* (89) observed the X-ray spectrum in the range 1 to 4 keV of Tycho Brahe's supernova remnant (1572) shown in Figure 23. An X-ray spectrum is much simpler than an optical spectrum. For me it is wonderful that satellite observations show the K-level X-rays expected from Si, S, Ar, and Ca just where the Handbook of Chemistry and Physics says they ought to be! Such observations are not all that easy in a terrestrial laboratory. Shull (90) has used a single-velocity, non-ionization-equilibrium model of a supernova blast wave to calculate abundances in Tycho's remnant *relative to solar, designated by brackets* and finds: [Si] = 7.6, [S] = 6.5, [Ar] = 3.2 and [Ca] = 2.6. With considerably greater uncertainty he gives [Mg] = 2.0 and [Fe] = 2.1. He finds different enhancements in Kepler's remnant (1604) and in Cassiopeia A. One more lesson for the modelers: no two supernovae are alike. Nucleosynthesis in supernovae depends on their initial mass, rotation, mass loss during the Red Giant stage, the degree of symmetry during explosion, initial heavy element content, and probably other factors. These details aside it seems clear that supernovae produce enhancements in elemental abundances up to iron and probably beyond. Detection of the much rarer elements beyond iron will require more sensitive X-ray detectors operating at higher energies. The nuclear debris of supernovae eventually enriches the interstellar medium from which succeeding generations of stars are formed. It becomes increasingly clear that novae also enrich the interstellar medium. Sorting out these two contributions poses interesting problems in ongoing research in all aspects of Nuclear Astrophysics.

Explosive silicon burning in the shell just outside a collapsing supernova core primarily produces ^{56}Ni as shown in Figure 16. It is generally believed that the initial energy source for the light curves of Type I supernovae is electron capture by ^{56}Ni ($\bar{\tau}$ = 8.80 days) to the excited state of ^{56}Co at 1.720 MeV with subsequent gamma ray cascades to the ground state. These gamma rays are absorbed and provide energy to the ejected envelope. The subsequent source of energy is the electron capture and positron emission by ^{56}Co ($\bar{\tau}$ = 114 days) to a number of excited states of ^{56}Fe with gamma ray cascades to the stable ground state of ^{56}Fe. Both the positrons and the gamma rays heat the ejected material. If ^{56}Co is an energy source there should be spectral evidence for cobalt in newly discovered Type I supernovae since its lifetime is long enough for detailed observations to be possible after the initial discovery.

The cobalt has been observed! Axelrod (91) studied the optical spectra of SN1972e obtained by Kirshner and Oke (92). The spectra obtained at 233, 264 and 376 days after Julian day 2441420, assigned as the initial day of the explosive event, are shown in Figure 24. Axelrod assigned the two emission lines near 6000Å (*log ν* = 14.7) to Co III. They are clearly evident at 233 and 264 days, but are only marginally evident at 376 days (~ τ) later. The lines decay in reasonable agreement with the mean lifetime of ^{56}Co.

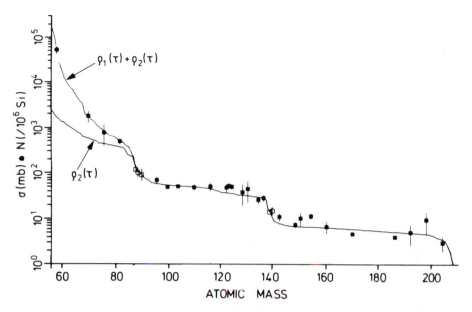

Figure 26. Neutron capture cross section at 30 keV in millibarns multiplied by solar system abundances relative to $Si = 10^6$ versus atomic mass for nuclei produced in the s-process from Almeida and Käppeler (99). Theoretical calculations are shown for a single exponential distribution $p_2(\tau)$ in neutron exposure, τ, and for two such distributions, $p_1(\tau) + p_2(\tau)$.

Branch *et al.* (93) have studied absorption spectra during the first hundred days of SN1981b. Their results at maximum light are shown in the top curve of Figure 25. Using the carbon deflagration model for Type I supernovae of Nomoto (67), Branch (94) has calculated the spectrum shown in the lower curve of Figure 25. Deep absorption lines of Co II are clearly evident near 3300Å and 4000Å.

It is my conclusion that there is substantial evidence for nucleosynthesis in supernovae of elements produced in oxygen and silicon burning. The role of neutron capture processes in supernovae will be discussed in the next section.

XI. *Neutron capture processes in nucleosynthesis*
In Section I the need for two neutron capture processes for nucleosynthesis beyond $A \gtrsim 60$ was discussed in terms of early historical developments in Nuclear Astrophysics. These two processes were designated the s-process for neutron capture slow (s) compared to electron beta-decay and the r-process for neutron capture rapid (r) compared to electron beta-decay in the process networks.

For a given element the heavier isotopes are frequently bypassed in the s-process and are produced only in the r-process; thus the designation r-only. Lighter isotopes are frequently shielded by more neutron rich stable isobars in the r-process and are produced only in the s-process; thus the designation s-only. The lightest isotopes are frequently very rare because they are not produced in either the s-process or the r-process and are thought to be produced in what is called the p-process. The p-process involves positron production and capture, proton-capture, neutron-photoproduction and/or (p, n)-reactions and will not be discussed further. The reader is referred to Audouze and Vauclair (95). The results of the s-process, the r-process and the p-process are frequently illustrated by reference to the ten stable isotopes of tin. The reader is referred to Figures 10 and 11 of the first reference in Fowler (96).

It is fair to say that the s-process has the clearest phenomenological basis of all processes of nucleosynthesis. This is primarily the result of the correlation of s-process abundances first delineated by Seeger, Fowler and Clayton (97) with the beautiful series of measurements on netron capture cross sections in the 1 to 100 keV range by the Oak Ridge National Laboratory group under Macklin and Gibbons (98).

This correlation is illustrated in Figure 26 which shows the product of neutron capture cross sections (σ) at 30 KeV multiplied by s-process abundances (N) as a function of atomic mass for s-only nuclei and those produced predominately by the s-process. It is not difficult to understand in first order approximation that the product σN should be constant in the s-process synthesis. A nucleus with a small (large) neutron capture cross section must have a large (small) abundance to maintain continuity in the s-capture path. Figure 26 demonstrates this in the *plateaus* shown from $A = 90$ to 140 and from $A = 140$ to 206. The anomalous behavior below $A = 80$ is discussed in Almeida and Käppeler (99) from which Figure 26 is taken.

Nuclear shell structure introduces the *precipices* shown in Figure 26 at $A{\sim}84$, $A{\sim}138$ and $A{\sim}208$ which correspond to the s-process abundance peaks in Figure 2. At these values for A the neutron numbers are "magic," $N = 50$, 82, and 126. The cross sections for neutron capture into new neutron shells are very small at these magic numbers. With a finite supply of neutrons it follows that the σN product must drop to a new plateau just as observed. Quantitative explanations of this effect have been given by Ulrich (100) and by Clayton and Ward (101).

What is the site of the s-process and what is the source of the neutrons? A very convincing answer has been given by Iben (102) that the site is the He burning shell of a pulsating Red Giant with the $^{22}\text{Ne}(\alpha,n)^{25}\text{Mg}$ reaction as the neutron source.Critical discussions have been given by Almeida and Käppeler (99) and by Truran (103). The latter reference reserves the possibility that the $^{13}\text{C}(\alpha,n)^{16}\text{O}$ reaction is the neutron source.

We turn now to the r-process. This process has been customarily treated by the *waiting point* method of B²FH (18). Under explosive conditions a large flux of neutrons drives nuclear seeds to the neutron rich side of the valley of stability where, depending on the temperature, the (n,γ)-reaction and the (γ,n)-reaction reach equality. The nuclei wait at this point until electron beta-decay transforms neutrons in the nuclei into protons whence further neutron capture can occur. At the cessation of the r-process the neutron rich nuclei decay to their stable isobars. In first order this means that the abundance of an r-process nucleus multiplied by the electron beta-decay rate of its neutron rich r-process isobar progenitor will be roughly constant. At magic neutron numbers in the neutron rich progenitors, beta-decay must perforce open the closed neutron

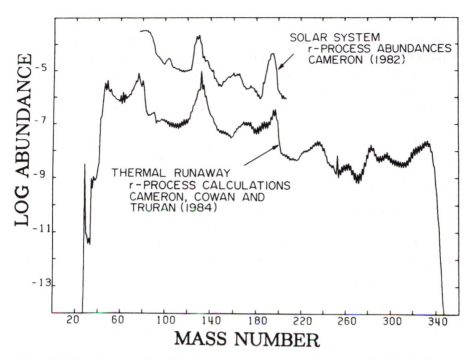

Figure 27. Abundances produced in the r-process versus atomic mass number in the thermal runaway model (lower curve) of Cameron, Cowan and Truran (107) compared with the solar system r-process abundances (upper curve) of Cameron (3).

shell in transforming a neutron into a proton and there the rate will be relatively small. Accordingly the abundance of progenitors with $N = 50, 82$ and 126 will be large. The associated number of protons will be less than in the corresponding s-process nuclei with a magic number of neutrons. It follows then that the stable daughter isobars will have smaller mass numbers and this is indeed the case, the r-process abundance peaks occurring at $A \sim 80$, $A \sim 130$ and $A \sim 195$, all below the corresponding s-process peaks as illustrated in Figure 2.

A phenomenological correlation of r-process abundances with beta-decay rates made by Becker and Fowler (104) and a detailed illustration of this correlation between solar system r-process abundances and theory is given in Figure 13 of the first of references (96). It is too phenomenological to satisfy critical nuclear astrophysicists. They wish to know the site of the high neutron fluxes demanded for r-process nucleosynthesis and the details of the r-process path through nuclei far from the line of beta-stability.

There is also general belief at the present time that the waiting point approximation is a poor one and must be replaced by dynamical r-process flow calculations taking into account explicit (n,γ), (γ,n) and beta-decay rates with time varying temperature and neutron flux. Schramm (105) has discussed such calculations in some detail and has emphasized that nonequilibrium effects are particularly important during the freeze-out at the end of the r-process when the temperature drops and the neutron flux falls to zero. Simple dynamical calculations have been made by Blake and Schramm (106) for a process they designated as the n-process and Sandler, Fowler, and Koonin (85) for their $n\beta$-process discussed in Section IX. The most ambitious calculations have been made by Cameron, Cowan, and Truran (107). This paper gives references to their previous herculean efforts in dynamical r-process calculations. An example of their results are shown in Figure 27. They emphasize that they have not been able to find a plausible astrophysical scenario for the initial ambient conditions required for Figure 27. In spite of this I am convinced that they are on the right track to an eventual understanding of the dynamics and site of the r-process.

Many suggestions have been made for possible sites of the r-process almost all in supernovae explosions where the basic requirement of a large neutron flux of short duration is met. These suggestions are reviewed in Schramm (105) and Truran (103). To my mind the helium core thermal runaway r-process of Cameron, Cowan, and Truran (107) is the most promising. These authors do not rule out the $^{22}Ne(\alpha,n)^{25}Mg$ reactions as the source of the neutrons but their detailed results shown in Figure 27 are based on the $^{13}C(\alpha,n)^{16}O$ reaction as the source. They start with a star formed from material with the same heavy element abundance distribution as in the solar system but with smaller total amount. They assume a significant amount of ^{13}C in the helium core of the star after hydrogen burning. This ^{13}C was produced previously by the introduction of hydrogen into the core which had already burned half of its helium into ^{12}C. For Figure 27 they assume a ^{13}C abundance of 14.3 percent by mass, density equal to $10^6 gm\ cm^{-3}$, and an initial temperature of 1.6×10^8 K which is raised by the initial slow ^{13}C burning to an eventual maximum of 3.6×10^8 K. The electrons in the core are initially degenerate but the rise in temperature lifts the degeneracy producing a thermal runaway with expansion and subsequent cooling of the core. This event is the second helium-flash episode in the history of the core and, if it occurs, only a small amount of the r-process material produced need escape into the interstellar medium to contribute the r-process abundance in solar system material. It is my belief that a realistic astrophysical site for the thermal runaway, perhaps with different initial conditions, will be found. I rest the case.

XII. *Nucleocosmochronology*

Armed with their r-process calculations of the abundances for the long lived parents of the natural radioactive series ^{232}Th, ^{235}U, and ^{238}U and with the then current solar system abundances of these nuclei B^2FH(18) were able to determine the duration of r-process nucleosynthesis from its beginning in the first stars in the Galaxy up to the last events before the formation of the solar system. The general idea was originally suggested by Rutherford (108). B^2FH (18) made a major advance in taking into account the contributions to the abundance of the long lived *eon glasses* from the decay of their short lived progenitors also produced in the r-process. The parents of the natural radioactive series are indeed excellent *eon glasses* with their mean lifetimes: ^{232}Th, 20.0×10^9 years; ^{238}U, 6.51×10^9 years; ^{235}U, 1.03×10^9 years. The analogy with *hour glasses* is fairly good; the sand in the top of the *hour glass* is the radioactive parent, that in the bottom is the daughter. The analogy fails in that in the *eon glasses* "sand" is being added or removed, top and bottom, by nucleosynthesis (production in stars) and astration (destruction in stars). Properly expressed differential equations can compensate for this failure.

The abundances used were those observed in meteorites assumed to be closed systems since their formation, taken to have occurred 4.55 billion years ago. It was necessary to correct for free decay during this period in order to obtain abundances for comparison with calculations based on r-process production plus decay over the duration Galactic nucleosynthesis before the meteorites became closed systems. Fortunately ratios of abundances, ^{232}Th/^{238}U and ^{235}U/^{238}U, sufficed since absolute abundances could not, and still cannot, be calculated with the necessary precision. The calculations required only the elemental ratio, Th/U, in meteorites since the isotopic ratio, ^{235}U/^{238}U, was assumed to be the same for meteoritic and terrestrial samples. The Apollo Program has added lunar data to the meteoritic and terrestrial in recent years.

B^{2}FH (18) considered a number of possible models, one of which assumed r-process nucleosynthesis *uniform* in time and an arbitrary time interval between the last r-process contribution to the material of the solar nebula and the closure of the meteorite systems. A zero value for this time interval indicated that the production of uranium started 18 billion years ago. When this time interval was taken to be 0.5 billion years, the production started 11.5 billion years ago. These values are in remarkable, if coincidental, concordance with current values.

It is appropriate to point out at this point that nucleocosmochronology yields, with additional assumptions, an estimate for the age of the expanding Universe completely independent of red shift-distance observations in astronomy on distant galaxies. The assumptions referred to in the previous sentence are that the r-process started soon, less than a billion years, after the formation of the Galaxy and that the Galaxy formed soon, less than a billion years, after the "big bang" origin of the Universe. Adding a billion years or so to the start of r-process nucleosynthesis yields an independent value, based on radioactivity, for the age or time back to the origin of the expanding Universe.

Much has transpired over the recent years in the field of nucleocosmochronology. I have kept my hand in most recently in Fowler (109). Exponentially decreasing nucleosynthesis with the time constant in the negative exponent a free parameter to be determined by the observed abundance ratios along with the duration of nucleosynthesis was introduced by Fowler and Hoyle (110). For the time constant in the denominator of the exponent set equal to infinity, uniform synthesis results. When it is set equal to zero, a single spike of synthesis results. With two observed ratios, two free parameters in a model can be determined. As time went on the ratios ^{129}I/^{127}I and ^{244}Pu/^{238}U with τ (^{129}I) $= 0.023 \times 10^9$ years and τ (^{244}Pu) $= 0.117 \times 10^9$ years were added to nucleocosmochronology to permit the determination of two additional free parameters, the arbitrary time interval of B^{2}FH (18) previously discussed and the

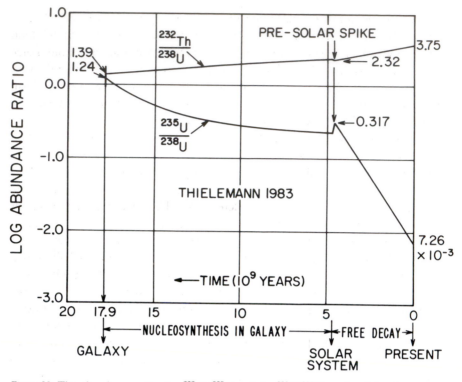

Figure 28. The abundance ratios for ^{232}Th/^{238}U and for ^{235}U/^{238}U produced by theoretical r-process nucleosynthesis over the lifetime of the Galaxy prior to the information of the solar system from Thielemann, Metzinger and Klapdor (113). The free decays over the lifetime of the solar system to reach the present values for these ratios, $(^{232}$Th/^{238}U$)_0 = 3.75$ and $(^{235}$U/^{238}U$)_0 = 7.26 \times 10^{-3}$ are also shown. The production ratios in each r-process event was theoretically calculated to be 1.39 for ^{232}Th/^{238}U and 1.24 for ^{235}U/^{238}U. Compare with Figure 10 in first reference under Fowler (109).

fraction of r-process nucleosynthesis produced in a last gasp "spike" at the end of the exponential time dependence.

Sophisticated models of Galactic evolution were introduced by Tinsley (111). A method for model independent determinations of the mean age of nuclear chronometers at the time of solar system formation was developed by Schramm and Wasserburg (112). In this method the mean age is one-half the duration for uniform synthesis and is equal to the actual time of single spike nucleosynthesis. This indicates that one can expect no more than a range of a factor of two in the time back to the beginning of nucleosynthesis in widely different models for its time variation. These developments are reviewed in Schramm (105).

The most recent calculations are those of Thielemann, Metzinger and Klapdor (113). Their results, revised by his own most recent calculations, are shown in Figure 28 prepared by F.-K. Thielemann. The pre-solar spike and its time of occurrence before the meteorites became closed systems depend primarily on the *minute glasses*, ^{129}I and ^{244}Pu. The *eon glasses*, ^{232}Th/^{238}U and ^{235}U/^{238}U, indicate that r-process nucleosynthesis in the Galaxy started 17.9 billion years ago with uncertainties of $+2$ billion years and -4 billion years according to reference (113). This value is to be compared with my value of 10.5 ± 2.3 billion years ago given in Fowler (109). Inputs of production and final abundance ratios have changed in (113)! Thielemann and I are now recomputing the new value for the duration using an initial spike in Galactic synthesis plus uniform synthesis thereafter. It should be noted that 1 to 2 billion years must be added to the age of the Galaxy to obtain the age of the Universe.

Reference (113) indicates that the age of the expanding Universe is 19 billion years give or take several billion years. This is to be compared to the Hubble time or reciprocal of Hubble's constant given by Sandage and Tammann (114) as 19.5 ± 3 billion years. However, the Hubble time is equal to the age of the expanding Universe only for a completely open Universe with mean matter density much less than the critical density for closure which can be calculated from the value for the Hubble time just given to be 5×10^{-30}gm cm^{-3}. The observed visible matter in galaxies is estimated to be ten percent of this which reduces the age of the Universe to 16.5 billions years. Invisible matter, neutrinos, black holes, etc. may add to the gravitational forces which decrease the velocity of expansion and may thus decrease the age to that corresponding to critical density which is 13.0 billion years. The new concept of the inflationary universe yields exactly the critical density and thus support the value of 13 billion years. If the expansion velocity was greater in the past, the time to the present radius of the Universe is correspondingly less. Moreover, there are those who obtain results for the Hubble time equal to about one-half that of Sandage and Tammann (114) as reviewed in van den Bergh (115). There is much to be done on all fronts!

A completely independent nuclear chronology involving the radiogenic ^{187}Os produced during Galactic nucleosynthesis by the decay of ^{187}Re$(\tau = 65 \times 10^9$ years) was suggested by Clayton (116). Schramm (105) discusses still other chronometric pairs. Clayton's suggestion involves the s-process even though the ^{187}Re is produced in the r-process. It requires that the abundance of ^{187}Re, the parent, be compared to that of its daughter, ^{187}Os, when the s-only production of this daughter nucleus is subtracted from its total solar system abundance. This was to be done by comparing the neutron capture cross section of ^{187}Os with that of its neighboring s-only isotope ^{186}Os which does not have a longlived radioactive parent and using the $N\sigma = constant$ rule for the s-process.

Fowler (117) threw a monkey wrench into the works by pointing out that ^{187}Os has a low-lying excited state at 9.75 keV which is practically fully populated at $kT = 30$ keV corresponding to the temperature $T = 3.5 \times 10^8$K, at which the s-process is customarily assumed to occur. Moreover with spin, $J = 3/2$, this state has twice the statistical weight $(2J + 1)$ of the ground state with spin, $J = 1/2$. Measurements of the ground state neutron capture cross section yields only one-third of what one needs to know.

All of this has led to a series of beautiful and difficult measurements for neutron induced reactions on the isotopes of osmium. Winters and Macklin (118) found the Maxwell−Boltzmann average ground state (laboratory) cross-section ratio for ^{186}Os(n, γ) relative to that for ^{187}Os(n, γ) to be 0.478 ± 0.022 at $kT = 30$ keV with a slow dependence on temperature. This ratio must be multiplied by a theoretical factor to correct the ^{187}Os cross section in the denominator of the cross-section ratio for that of its excited state. The larger the theoretical ^{187}Os excited state capture, the smaller this factor. Woosley and Fowler (119) used Hauser-Feshbach theory to give an estimate for this factor in the range 0.8 to 1.10 which is little comfort in view of the fact that it multiplies one number comparable to the number from which it must be subtracted. These factors translated into a time for the beginning of the r-process in the Galaxy in the range 14 to 19 billion years. In desperation I suggested privately that inelastic neutron scattering off the ground state of ^{187}Os to its excited state at 9.75 keV might yield information on the properties of the excited state. Measurements by Macklin *et al.* (120) and Hershberger *et al.* (121) determined these inelastic neutron scattering cross sections which yielded inherent support of the lower value of the Woosley and Fowler (119) factor and thus a greater value for the time back to the beginning of r-process nucleosynthesis in the 18 to 20 billion year range. It has to be admitted that this is concordant with the latest value from the Th/U-nucleocosmochronology.

Once again in desperation I privately suggested that measurement of the neutron capture cross section on the ground state of ^{189}Os might be helpful. In ^{189}Os the ground state has the same spin and Nilsson numbers as the excited state of ^{187}Os and the excited state corresponds to the ground state of ^{187}Os. Measurements by Browne and Berman (122) are available but are now being checked by an Oak Ridge National Laboratory, Denison University, and University of Kentucky consortium.

It will be clear that the lifetime of ^{187}Re comes directly into the calculations under discussion. There has been some discrepancy in the past between lifetimes measured geochemically and those measured directly by counting the electrons emitted in the 2.6 keV decay ^{187}Re$(e^-\nu)$ ^{187}Os. Direct measurement yields only the lifetime for electron-emission to the continuum while geochemistry yields the lifetime for electron-emission both to the continuum and to bound states in ^{187}Os. The entire matter is treated in considerable theoretical detail by Williams, Fowler, and Koonin (123) who found that the bound-state decay is negligible and that the direct measurements by Payne and Drever (124), which agree with the geochemical measurements of Hirt *et al.* (125), are correct.

There is also the vexing problem of the possible decrease in the effective lifetime of ^{187}Re in the Galactic environment. The ^{187}Re included in the material of the interstellar medium which forms new stars is subject to destruction by the s-process (astration) as well as being produced by the r-process. This decreases the effective lifetime of the ^{187}Re and all chronometric times based on the Re/Os chronology. This problem is discussed in elaborate detail by Yokoi, Takahashi and Arnould (126). The time back to the beginning of r-process nucleosynthesis could be as low as 12 billion years. It appropriates to end this last section before the concluding section with considerable uncertainity in nucleocosmochronology indicating that, as in all nuclear astrophysics, there is much exciting experimental and theoretical work to be done for many years to come. Amen!

XIII. *Conclusion*

In spite of the past and current researches in experimental and theoretical Nuclear Astrophysics, illustrated in what I have just shown you, the ultimate goal of the field has not been attained. Hoyle's grand concept of element synthesis in the stars will not be truly established until we attain a deeper and more precise understanding of many nuclear processes operating in astrophysical environments. Hard work must continue on all aspects of the cycle: experiment, theory, observation. It is not just a matter of filling in the details. There are puzzles and problems in each part of the cycle which challenge the basic ideas underlying nucleosynthesis in stars. Not to worry—that is what makes the field active, exciting and fun! It is a great source of satisfaction to me that the Kellogg Laboratory continues to play a leading role in experimental and theoretical Nuclear Astrophysics.

And now permit me to pass along one final thought in concluding my lecture. My major theme has been that all of the heavy elements from carbon to uranium have been synthesized in stars. Let me remind you that your bodies consist for the most part of these heavy elements. Apart from hydrogen you are 65 percent oxygen and 18 percent carbon with smaller percentages of nitrogen, sodium, magnesium, phosphorus, sulphur, chlorine, potassium, and traces of still heavier elements. Thus it is possible to say that you and your neighbor and I, each one of us and all of us, are truly and literally a little bit of stardust.

Charles Christian Lauritsen taught me a Swedish toast. I conclude with this toast to my Swedish friends: *"Din skål, min skål, alla vackra flickors skål. Skål!"*

ACKNOWLEDGEMENTS

My work in Nuclear Astrophysics has involved collaborative team work with many people and I am especially grateful to Fay Ajzenberg-Selove, Jean Audouze, C. A. Barnes, E. M. Burbidge, G. R. Burbidge, G. R. Caughlan, R. F. Christy, D. D. Clayton, G. M. Fuller, J. L. Greenstein, Fred Hoyle, Jean Humblet, R. W. Kavanagh, S. E. Koonin, C. C. Lauritsen, Thomas Lauritsen, D. N. Schramm, T. A. Tombrello, R. V. Wagoner, G. J. Wasserburg, Ward Whaling, S. E. Woosley, and B. A. Zimmerman.

For aid and helpful cooperation in all aspects of my scientific work, especially in the preparation of publications, I am grateful to Evaline Gibbs, Jan Rasmussen, Kim Stapp, Marty Watson and Elisabeth Wood. I acknowledge support for my research over the years by the Office of Naval Research (1946 to 1970) and by the National Science Foundation (1968 to present).

REFERENCES

1. J. Audouze and H. Reeves, p. 355, *Essays in Nuclear Astrophysics* edited by C. A. Barnes, D. D. Clayton, and D. N. Schramm (Cambridge University Press 1983).

2. H. E. Suess and H. C. Urey, Revs. Mod. Phys. 28, 53 (1956).

3. A. G. W. Cameron, p. 23, *Essays in Nuclear Astrophysics,* edited by C. A. Barnes, D. D. Clayton and D. N. Schramm (Cambridge University Press 1982).

4. W. Whaling, p. 65, *Essays in Nuclear Astrophysics,* edited by C. A. Barnes, D. D. Clayton, and D. N. Schramm (Cambridge University Press 1982).

5. R. A. Alpher and R. C. Herman, Revs. Mod. Phys. 22, 153 (1950).

6. R. V. Wagoner, W. A. Fowler, and F. Hoyle, Ap. J. 148, 3 (1967).

7. H. Staub and W. E. Stephens, Phys. Rev 55, 131 (1939).

8. J. H. Williams, W. G. Shepherd, and R. O. Haxby, Phys. Rev. 52, 390 (1937).

9. A. V. Tollestrup, W. A. Fowler, and C. C. Lauritsen, Phys. Rev 76, 428 (1949).

10. A. Hemmendinger, Phys. Rev. 73, 806 (1948); Phys. Rev. 74, 1267 (1949).

11. H. A. Bethe, Phys. Rev. 55, 434 (1939); H. A. Bethe, *Les Prix Nobel 1967* (Almquist & Wiksell International, Stockholm).

12. F. Hoyle, Mon. Not. R. Astron. Soc. 106, 343 (1946); Ap. J. Suppl. 1, 121 (1954). For a discussion of earlier ideas, including suggestions and retractions, see Chap. XII in S. Chandrasekhar, *Stellar Structure* 1939 (University of Chicago Press, Chicago).

13. F. Hoyle and M. Schwarzschild, Ap. J. Suppl. 2, 1 (1955).

14. A. R. Sandage and M. Schwarzschild, Ap. J. 116, 463 (1952). In particular see last paragraph on p. 475.

15. E. E. Salpeter, Ap. J. 115, 326 (1952).

16. D. N. F. Dunbar, R. E. Pixley, W. A. Wenzel, and W. Whaling, Phys. Rev. 92, 649 (1953).

17. C. W. Cook, W. A. Fowler, C. C. Lauritsen, and T. Lauritsen, Phys. Rev. 107, 508 (1957).

18. E. M. Burbidge, G. R. Burbidge, W. A. Fowler, and F. Hoyle, Rev. Mod. Phys. 29, 547 (1957). Hereafter referred to as B²FH (18). Also see F. Hoyle, W. A. Fowler, E. M. Burbidge, and G. R. Burbidge, Science 124, 611 (1956).

19. A. G. W. Cameron, Publ. Astron. Soc. of Pac., 69, 201 (1957).

20. J. L. Greenstein, Chapter 10, *Modern Physics for the Engineer* edited by L. N. Ridenour (McGraw-Hill 1954); p. 45, *Essays in Nuclear Astrophysics,* edited by C. A. Barnes, D. D. Clayton, and D. N. Schramm (Cambridge University Press 1982). Of key importance was the discovery of technicium lines in S-stars by P.W. Merrill, Science, 115, 484 (1952).

21. W. A. Fowler, G. R. Caughlan, and B. A. Zimmerman, Ann. Rev. Astron. and Astroph. 5, 525 (1967); 13, 69 (1975). Also see M. J. Harris *et al. ibid.* 21, 165 (1983) and G. R. Caughlan *et al.* Atomic Data and Nuclear Data Tables (1984), accepted for publication.

22. E. E. Salpeter, Phys. Rev. 88, 547 (1957); 97, 1237 (1955).

23. S. E. Woosley, W. A. Fowler, J. A. Holmes, and B. A. Zimmerman, Atomic Data and Nuclear Data Tables 22, 371 (1978).

24. H. A. Bethe and C. L. Critchfield, Phys. Rev. 54, 248 (1938).

25. W. A. Fowler, Ap. J. 127, 551 (1958).

26. A. G. W. Cameron, Ann. Rev. Nucl. Sci. 8, 249 (1958).

27. B. Pontecorvo, Chalk River Laboratory Report PD-205 (1946).

28. L. W. Alvarez, University of California Radiation Laboratory Report UCRL-328 (1949).

29. R. Davis, Jr., p. 2, *Science Underground,* edited by M. M. Nieto *et al.* (American Institute of Physics, New York 1983).

30. J. N. Bahcall, W. F. Huebner. S. H. Lubow, P. D. Parker, and R. K. Ulrich, Rev. Mod. Phys. 54, 767 (1982).

31. R. G. H. Robertson, P. Dyer, T. J. Bowles, Ronald E. Brown, Nelson Jarmie, C. J. Maggiore, and S. M. Austin, Phys. Rev. C27, 11 (1983); J. L. Osborne, C. A. Barnes, R. W. Kavanagh, R. M. Kremer, G. J. Mathews and J. L. Zyskind, Phys, Rev. Letters 48, 1664 (1982).

32. R. T. Skelton and R. W. Kavanagh, Nuclear Physics A414, 141 (1984).

33. C. A. Barnes, p. 193, *Essays in Nuclear Astrophysics,* edited by C. A. Barnes, D. P. Clayton, and D. N. Schramm (Cambridge University Press 1982).

34. K. Langanke and S. E. Koonin, Nucl. Phys. A410, 334 (1983) and private communication (1983).

35. P. Dyer and C. A. Barnes, Nucl. Phys. A233, 495 (1974); S. E. Koonin, T. A. Tombrello, and G. Fox, Nucl. Phys. A220, 221 (1974).

36. K U. Kettner, H. W. Becker, L. Buchmann, J. Gorres, H. Kräwinkel, C. Rolfs, P. Schmalbrock, H. P. Trautvetter, and A. Vlieks, Zeits. f. Physik A308, 73 (1982).

37. S. G. Starrfield, A. N. Cox, S. W. Hodson, and W. D. Pesnell, Ap. J. 268, L27 (1983); S. A. Becker, private communication (1983).

38. S. E. Woosley and T. A. Weaver, p. 381, *Essays in Nuclear Astrophysics*, edited by C. A. Barnes, D. D. Clayton, and D. N. Schramm (Cambridge University Press 1982).

39. T. A. Weaver, S. E. Woosley and G. M. Fuller, *Proceedings of the Conference on Numerical Astrophysics*, edited by R. Bowers, J. Centrella, J. LeBlanc, and M. LeBlanc (Science Books International 1983).

40. G. Hulke, C. Rolfs, and H. P. Trautvetter, Zeits, f. Physik A297, 161 (1980).

41. D. Bodansky, D. D. Clayton, and W. A. Fowler, Ap. J. Suppl. 16, 299 (1968).

42. J. L. Zyskind, J. M. Davidson, M. T. Esat, M. H. Shapiro, and R. H. Spear, Nucl. Phys. A301, 179 (1978).

43. L. W. Mitchell and D. G. Sargood, Aust. J. Phys. 36, 1 (1983).

44. R. N. Boyd, *Proceedings of the Workshop on Radioactive Ion Beams and Small Cross Section Measurements*, (The Ohio State University Press, Columbus 1981); R. C. Haight, G. J. Mathews, R. M. White, L. A. Avilés, and S. E. Woodward, Nuclear Instruments and Methods 212, 245 (1983).

45. W. Hauser and H. Feshbach, Phys. Rev. 78, 366 (1952).

46. J. A. Holmes, S. E. Woosley, W. A. Fowler, and B. A. Zimmerman, Atomic Data and Nuclear Data Tables 18, 305 (1976).

47. G. Michaud and W. A. Fowler, Phys. Rev. C2, 2041 (1970); also see E. W. Vogt, *Advances in Nuclear Physics*, 1, 261 (1969).

48. F.M. Mann, Hanford Engineering and Development Laboratory internal report HEDL-TME-7680 (1976, unpublished).

49. J. L. Zyskind, C. A. Barnes, J. M. Davidson, W. A. Fowler, R. E. Marrs, and M. H. Shapiro, Nucl. Phys. A343,295 (1980).

50. D. G. Sargood, Physics Reports 93, 61 (1982); Aust. J. Phys. 36, 583 (1983).

51. N. A. Bahcall and W. A. Fowler, Ap. J. 157, 645 (1969).

52. R. A. Ward and W. A. Fowler, Ap. J. 238, 266 (1980). For recent experimental data on the production of ^{26}Al, through ^{25}Mg(p, γ) ^{26}Al, see A. E. Champagne, A. J. Howard, and P. D. Parker, Ap. J. 269, 686 (1983). For recent experimental data on the destruction of ^{26}Al, through ^{26}Al (p, γ) ^{27}Si, see L. Buchmann, M. Hilgemaier, A. Krauss, A. Redder, C. Rolfs, and H. P. Trautvetter (in publication, 1984).

53. G. M. Fuller, W. A. Fowler, and M. J. Newman, Ap. J. Suppl. 42, 447 (1980); Ap. J. 252, 715 (1982); Ap. J. Suppl. 48, 279 (1982).

54. H. S. Wilson, R. W. Kavanagh, and F. M. Mann, Phys. Rev. C22, 1696 (1980).

55. C. M. Lederer and V. S. Shirley, Editors, *Table of Isotopes: Seventh Edition* (John Wiley & Sons, Inc., New York 1978).

56. C. D. Goodman, C. A. Goulding, M. B. Greenfield, J. Rapaport, D. E. Bainum, C. C. Foster, W. G. Love, and F. Petrovich, Phys. Rev. Lett. 44, 1755 (1980); F. Ajzenberg-Selove, R. E. Brown, E. R. Flynn, and J. W. Sunier, Phys. Rev. Lett. (1984), in publication.

57. S. D. Bloom and G. M. Fuller (in preparation, 1984).

58. G. E. Brown, H. A. Bethe and G. Baym, Nucl. Phys. A375, 481 (1982).

59. H. A. Bethe, A. Yahil, and G. E. Brown, Ap. J. Letters 262, L7 (1982).

60. H. A. Bethe, G. E. Brown, J. Cooperstein and J. R. Wilson, Nucl. Phys. A403, 625 (1983).

61. G.M. Fuller, Ap. J. 252, 741 (1982).

62. G. M. Fuller, W. A. Fowler, and M. J. Newman (in preparation, 1984).

63. S. E. Woosley, T. S. Axelrod, and T. A. Weaver, *Stellar Nucleosynthesis*, edited by C. Chiosi and A. Renzini (Dordrecht: Reidel 1984).

64. F. Hoyle and W. A. Fowler, Ap. J. 132, 565 (1960).

65. W. A. Fowler and F. Hoyle, Ap. J. Suppl. 9, 201 (1964).

66. J. C. Wheeler, Rep. Prog. Phys. 44, 85 (1981).

67. K. Nomoto, A. J. 253, 798 (1982); Ap. J. 257, 780 (1982); *Stellar Nucleosynthesis,* edited by C. Chiosi and A. Renzini (Dordrecht: Reidel 1984).

68. K. Nomoto, F.-K. Thielemann, and J. C. Wheeler, Ap. J. (1984), in publication.

69. W. D. Arnett and F.-K. Thielemann, *Stellar Nucleosynthesis,* edited by C. Chiosi and A. Renzini (Dordrecht: Reidel 1984).

70. J. H. Reynolds, Phys. Rev. Lett. 4, 8 (1960).

71. P. M. Jeffery and J. H. Reynolds, J. Geophys. Res. 66, 3582 (1961).

72. G. J. Wasserburg, W. A. Fowler, and F. Hoyle, Phys. Rev. Lett. 4, 112 (1960).

73. G. J. Wasserburg and D. A. Papanastassiou, p. 77, *Essays in Nuclear Astrophysics,* edited by C. A. Barnes, D. D. Clayton, and D. N. Schramm (Cambridge University Press 1982).

74. D. D. Clayton, Ap. J. 199, 765 (1975); Space Sci. Rev. 24, 147 (1979); Ap. J. 268, 381 (1983); Ap. J. (1984), in publication; also see D. D. Clayton and F. Hoyle, Ap. J. Lett. 187, L101 (1974); Ap. J. 203, 490 (1976).

75. J. H. Chen and G. J. Wasserburg, Earth Planet Sci. Lett. 52, 1 (1981).

76. D. S. Burnett, M. I. Stapanian, and J. H. Jones, p. 144, *Essays in Nuclear Astrophysics,* edited by C. A. Barnes, D. D. Clayton and D. N. Schramm (Cambridge University Press 1982).

77. Typhoon Lee, D. A. Papanastassiou and G. J. Wasserburg, Ap. J. Lett. 211, L107 (1977).

78. W. A. Mahoney, J. C. Ling, W. A. Wheaton, and A. S. Jacobson, Ap. J. (1984), in publication; also see W. A. Mahony, J. C. Ling, A. S. Jacobson, and R. E. Lingenfelter, Ap. J. 262, 742 (1982).

79. R. T. Skelton, R. W. Kavanagh, and D. G. Sargood, Ap. J. 271, 404 (1983).

80. M. Arnould, H. Nørgaard, F.-K. Thielemann, and W. Hillebrandt, Ap. J. 237, 931 (1980).

81. J. W. Truran, p. 467, *Essays in Nuclear Astrophysics,* edited by C. A. Barnes, D. D. Clayton and D. N. Schramm (Cambridge University Press 1982).

82. W. A. Fowler, J. L. Greenstein, and F. Hoyle, Geophys. J. 6, 148 (1962).

83. Typhoon Lee, Ap. J. 224, 217 (1978).

84. M. H. Thiemens and J. E. Heidenreich, Science 219, 1073 (1983).

85. D. G. Sandler, S. E. Koonin, and W. A. Fowler, Ap. J. 259, 908 (1982).

86. F. R. Niederer, D. A. Papanastassiou and G. J. Wasserburg, Ap. J. Lett. 228, L93 (1979).

87. A. Huck, G. Klotz, A. Knipper, C. Miéhé, C. Richard-Serre, and G. Walter, CERN 81-09, 378 (1981).

88. R. A. Chevalier and R. P. Kirshner, Ap. J. 233, 154 (1979).

89. R. H. Becker, S. S. Holt, B. W. Smith, N. E. White, E. A. Boldt, R. F. Mushotsky, and P. J. Serlemitsos, Ap. J. Letters 234, L73 (1979). For overabundances in oxygen and neon in the x-ray spectra of Puppis A see C. R. Canizares and P. F. Winkler, Ap. J. Letters 246, L33 (1981).

90. J. M. Shull, Ap. J. 262, 308 (1982) and private communication (1983).

91. T. S. Axelrod, *Late Time Optical Spectra from the* ^{56}Ni *Model for Type I Supernovae,* Thesis, University of California, Berkeley, UCRL-52994, (1980).

92. R. P. Kirshner and J. B. Oke, Ap. J. 200, 574 (1975).

93. D. Branch, C. H. Lacy, M. L. McCall, P. G. Sutherland, A. Uomoto, J. C. Wheeler, and B. J. Wills, Ap. J. 270, 123 (1983).

94. D. Branch, Proceedings of Yerkes Observatory Conference on "Challenges and New Developments in Nucleosynthesis," edited by W. D. Arnett (University of Chicago Press 1984).

95. J. Audouze and S. Vauclair, p. 92, *An Introduction to Nuclear Astrophysics,* (D. Reidel, Dordrecht 1980).

96. W. A. Fowler, Proc. Nat. Acad. Sciences 52, 524 (1964); *Nuclear Astrophysics,* (American Philosophical Society, Philadelphia 1967).

97. P. A. Seeger, W. A. Fowler, and D. D. Clayton, AP. J. Suppl. 11, 121 (1965).

98. R. L. Macklin and J. H. Gibbons, Rev. Mod Phys. 37, 166 (1965). Also see B. J. Allen, R. L. Macklin, and J. H. Gibbons, Adv. Nucl. Phys. 4, 205 (1971).

99. J. Almeida and F. Käppeler, Ap. J. 265, 417 (1983).

100. R. K. Ulrich, p. 139, *Explosive Nucleosynthesis,* edited by D. N. Schramm and W. D. Arnett (University of Texas Press, Austin 1973).

101. D. D. Clayton and R. A. Ward, AP. J. 193, 397 (1974).

102. I. Iben, Jr., Ap. J. 196, 525 (1975).

103. J. W. Truran, p. 95. *International Physics Conference Series No. 64,* (The Institute of Physics, London 1983).

104. R. A. Becker and W. A. Fowler, Phys. Rev. 115, 1410 (1959).

105. D. N. Schramm, p. 325, *Essays in Nuclear Astrophysics,* edited by C. A. Barnes, D. D. Clayton, and D. N. Schramm (Cambridge University Press 1982).

106. J. B Blake and D. N. Schramm, Ap. J. 209, 846 (1976).

107. A. G. W. Cameron, J. J. Cowan, and J. W. Truran, Proceedings of Yerkes Observatory Conference on "Challenges, and New Developments in Nucleosynthesis," edited by W. D. Arnett (University of Chicago Press 1984).

108. E. Rutherford, Nature 123, 313 (1929).

109. W. A. Fowler, p. 61, *Proceedings of the Welch Foundation Conferences on Chemical Research, XXI, Cosmochemistry,* edited by W. D. Milligan (Robert A. Welch Foundation, Houston 1977); also see W. A. Fowler, p. 67, *Cosmology, Fusion and Other Matters,* edited by F. Reines (Colorado Associated University Press 1972).

110. W. A. Fowler and F. Hoyle, Ann. Phys. 10, 280 (1960).

111. B. M. Tinsley, Ap. J. 198, 145 (1975).

112. D. N. Schramm and G. J. Wasserburg, Ap. J. 163, 75 (1970).

113. F.-K. Thielemann, J. Metzinger, and H. V. Klapdor Z. Phys. A309, 301 (1983) and private communication.

114. A. Sandage and G. A. Tammann, Ap. J. 256, 339 (1982).

115. S. van den Bergh, Nature 229, 297 (1982).

116. D. D. Clayton, Ap. J. 139, 637 (1964).

117. W. A. Fowler, Bulletin American Astronomical Society, 4, 412 (1972).

118. R. R. Winters and R. L. Macklin, Phys. Rev. C25, 208 (1982).

119. S. E. Woosley and W. A. Fowler, Ap. J. 233, 411 (1979).

120. R. L. Macklin, R. R. Winters, N. W. Hill, and J. A. Harvey, Ap. J. 274, 408 (1983).

121. R. L. Hershberger, R. L. Macklin, M. Balakrishnan, N. W. Hill and M. T. McEllistrem, Phys. Rev, C28, 2249 (1983).

122. J. C. Browne and B. L. Berman, Phys. Rev. C23, 1434 (1981).

123. R. D. Williams, W. A. Fowler, and S. E. Koonin, Ap. J. (1984), in publication.

124. J. A. Payne, Thesis: *An Investigation of the Beta Decay of Rhenium to Osmium, Using High Temperature Proportional Counters,* University of Glasgow (1965); R. W. P. Drever, private communication (1983).

125. B. Hirt, G. R. Tilton, W. Herr, and W. Hoffmeister, *Earth Sciences and Meteorites,* edited by J. Geiss and E. D. Goldberg (North Holland Press, Amsterdam 1963).

126. K. Yokoi, K. Takahashi, and M. Arnould, Astron. Astrophys. 117, 65 (1983).

Physics 1984

CARLO RUBBIA and SIMON VAN DER MEER

for their decisive contributions to the large project, which led to the discovery of the field particles W and Z, communicators of weak interaction

THE NOBEL PRIZE FOR PHYSICS

Speech by Professor GÖSTA EKSPONG of the Royal Academy of Sciences.
Translation from the Swedish text

Your Majesties, Your Royal Highnesses, Ladies and Gentlemen,

This year's Nobel Prize for Physics has been awarded to Professor CARLO RUBBIA and Dr. SIMON VAN DER MEER. According to the decision of the Royal Swedish Academy of Sciences the prize is given "for their decisive contributions to the large project, which led to the discovery of the field particles W and Z, communicators of the weak interaction".

The large project mentioned in the citation is the antiproton project at CERN, the international centre for research devoted to the study of elementary particles, which has 13 European states as members. CERN straddles, in a uniqe way, the border between two countries, Switzerland and France, and has grown progressively in importance over the 30 years of its life. The international character is underlined by the fact that Carlo Rubbia is Italian, Simon van der Meer is Dutch and the collaborators in the various phases of the project are scientists, engineers, and technicians of many nationalities, either employed by CERN or in one of the many universities or research institutes involved in the experiments. The project has been made possible by collaboration, by the pooling of financial resources and of scientific and technical skill. When the antiproton project was proposed eight years ago the CERN ship had two captains—two Directors General, Professor Leon van Hove from Belgium and Sir John Adams, from the United Kingdom. Navigating through the high waves generated by the convincing enthusiasm of Rubbia but having van der Meer on board as pilot to steer through the more difficult waters, they directed their ship towards new challenging frontiers. The late Sir John Adams had been responsible for the construction of the two outstanding proton accelerators, which were called into action in new roles for the new project.

A former Nobel Laureate expressed his opinion about the CERN project with the following words: van der Meer made it possible, Rubbia made it happen. Looking closer one finds that two conditions had to be fulfilled in order to produce the W and Z in particle collisions. The first is that the particles must collide at sufficiently high energy so that the conversion of energy into mass could create the heavy W and Z particles. The second is that the number of collisions must be large enough to give a chance of seeing the rare creation process taking place. The name of Rubbia is connected with the first condition, that of van der Meer with the second. Rubbia's proposal was to use the largest accelerator at CERN, the SPS, as a storage ring for circulating antiprotons as well as for protons, circulating in the opposite direction. The particles in the two beams would cross the French-Swiss border about 100,000 times every second for a whole day or more, to be repeated with new beams during months of operation. Antiprotons cannot be found in nature, in any case not on Earth.

But they can be created at CERN where sufficient energy is available at the other accelerator, the PS. The antiprotons are accumulated in a special storage ring, built by a team led by van der Meer.

It is here that his ingenious method, called stochastic cooling, enables an intense antiproton beam to be built up. The signals from produced particles are recorded in huge detector systems set up around two collision points along the periphery of the SPS storage ring. The largest of these detectors was designed, built and operated by a team led by Rubbia. A second large detector was built by another team, operating it in parallel with the first one, nicely confirming the extremely important results.

An old dream was fulfilled last year when the discoveries of the W and Z were made at CERN—the dream of better understanding the weak interaction, which turns out to be weak just because the W and Z are so very heavy. The weak interaction is unique in that it can change the nature of a particle, for example transforming a neutron into a proton or vice versa. Such transformations are crucial for the sun and it is the weakness of the interaction which leads to the very slow burning of the nuclear fuel of the sun and thus creates the conditions on earth which can support life.

At first radioactive decays were the only weak interaction phenomenon available for study. Nowadays thanks to accelerators and storage rings this field of research is quite large. The theory which synthesizes a vast amount of knowledge and combines our understanding of the weak and electromagnetic interactions was honoured by the award of the Nobel Prize for physics in 1979 to Sheldon Glashow, Abdus Salam and Steven Weinberg. It also predicted new phenomena caused by the invented particle Z, introduced to make the theory consistent. Such phenomena were first observed in a CERN experiment about ten years ago. The only historical parallel goes back 120 years to Maxwells theory for electric and magnetic phenomena. In that case the theory was made consistent by a new ingredient, which contained the seed for the prediction of radio waves, discovered by Heinrich Herz almost 100 years ago. The modern electroweak theory contains not only the electromagnetic photons as communicators of force but also the communicators W and Z which act as a kind of shock-absorber, especially noticeable in hard collisions—such as those which must have occurred frequently during the Big Bang era at the early stage of the evolution of our universe. The collisions in the CERN collider may be hard enough to break loose the communicators, the shock-absorbers, for a short moment. The resulting fireworks of newly produced particles have been observed in the detectors, and the signs showing the presence of the W and Z have been seen and a start has been made on measuring their properties.

Professor Rubbia and Dr. van der Meer,

Your achievements in recent years, leading to the successful operation of the CERN proton-antiproton collider, have been widely admired in the whole world. The discovery of the W and Z particles will go down in the history of physics like the discovery of radio waves and the photons of light, the communicators of electromagnetism.

I know that you share your joy with many collaborators at CERN and in the participating universities. I also know that they congratulate you in many ways, also by setting new records for energy and for the rate of collisions, and by finding new interesting phenomena produced in the collisions. The discovery of the W and Z is not the end—it is the beginning.

On behalf of the Royal Swedish Academy of Sciences, I have the pleasure and the honour of extending to you our warmest congratulations. I now invite you to receive your prizes from the hands of His Majesty the King.

Carlo Rubbia

CARLO RUBBIA

I was born in the small town of Gorizia, Italy, on 31 March, 1934. My father was an electrical engineer at the local telephone company and my mother an elementary school teacher. At the end of the World War II most of the province of Gorizia was overtaken by Yugoslavia and my family fled to Venice first and then to Udine.

As a boy, I was deeply interested in scientific ideas, electrical and mechanical, and I read almost everything I could find on the subject. I was attracted more by the hardware and construction aspects than by the scientific issues. At that time I could not decide if science or technology were more relevant for me.

After completing High School, I applied to the Faculty of Physics at the rather exclusive Scuola Normale in Pisa. My previous education had been seriously affected by the disasters of the war and the subsequent unrest. I badly failed the admission tests and my application was turned down. I forgot about physics and I started engineering at the University of Milan (Politecnico). To my great surprise and joy a few months later I was offered the possiblity of entering the Scuola Normale. One of the people who had won the admission contest had resigned! I am recollecting this apparently insignificant fact since it has determined and almost completely by accident my career of physicist. I moved to Pisa, where I completed the University education with at thesis on cosmic ray experiments. They have been very tough years, since I had to greatly improve my education, which was very deficient in a number of fundamental disciplines. At that time I also participated under my thesis advisor Marcello Conversi to new instrumentation developments and to the realization of the first pulsed gas particle detectors.

Soon after my degree, in 1958 I went to the United States to enlarge my experience and to familiarize myself with particle accelerators. I spent about one and a half years at Columbia University. Together with W. Baker, we measured at the Nevis Syncro-cyclotron the angular assymmetry in the capture of polarized muons, demonstrating the presence of parity violation in this fundamental process. This was his first of a long serie of experiments on Weak Interactions, which ever since has become my main field of interest. Of course at that time it would have been quite unthinkable for me to imagine to be one day amongst the people discovering the quanta of the weak field!

Around 1960 I moved back to Europe, attracted by the newly founded European Organization for Nuclear Research, where for the first time the idea of a joint European effort in a field of pure Science was to be tried in practice. The Syncro-cyclotron at CERN had a performance significantly superior to the one of the machine in Nevis and we succeeded in a number of very exciting experiments on the structure of weak interactions, amongst which I would like to mention the discovery of the beta decay process of the positive pion, $\pi^+ = \pi^0 + e + \nu$ and the first observation of the muon capture by free hydrogen, $\mu^- + p = n + \nu$.

In the early sixties John Adams brought to operation the CERN Proton-Syncrotron. I moved to the larger machine where I continued to do some weak interaction experiments, like for instance the determination of the parity violation in the beta decay of the lambda hyperon.

During the Summer of 1964 Fitch and Cronin announced the discovery of CP violation. This has been for me a tremendously important result and I abandoned all current work to start a long series of observations on CP-violation in K^0 decay and on the K_L-K_S mass difference. Unfortunately the subject did not turn out to be as prolific as in the case of the previous discovery of parity violation and even today, some thirty years afterwards we do not know much more about the origin of CP-violation than right after the announcement of the discovery.

I returned again to more ortodox weak interactions a few years later, when together with David Cline and Alfred Mann we proposed a major neutrino experiment at the newly started US laboratory of Fermilab. The operational problems associated with a limping accelerator and a new laboratory made very difficult, albeit impossible for us during the Summer of 1973 to settle definitively the question of the existence of neutral currents in neutrino interactions, when competing with the much more avanced instrumentation of Gargamelle at CERN. Instead, about one year later we could cleanly observe the presence of di-muons events in neutrino interactions and to confirm in this way one of the crucial predictions of the GIM mechanism, hinting at the existence of charm, glamorously settled only few months later with the observation of the Ψ/J particle.

In the meantime and under the impulse of Vicky Weisskopf a new, fascinating adventure had just started at CERN with a new type of colliding beams machine, the Intersecting Storage Rings, in which counter-rotating beams of protons collide against each other. This novel technique offered a much more efficient use of the accelerator energy than the traditional method of collisions against a fixed target. From the very first operation of this new type of accelerator, I have participated to a long series of experiments. They have been crucial to perfect the detection techniques with colliding beams of protons and antiprotons needed later on for the discovery of the Intermediate Bosons.

By that time it was quite clear that Unified Theories of the type $SU(2) \times U(1)$ had a very good chance of predicting the existence and the masses of the triplet of intermediate vector bosons. The problem of course was the one of finding a practical way of discovering them. To achieve energies high enough to create the intermediate vector bosons (roughly 100 times as heavy as the proton) together with David Cline and Peter Mc Intyre we proposed in 1976 a radically new approach. Along the lines discussed about ten years earlier by the russian physicist Budker, we suggested to transform an existing high energy accelerator in a colliding beam device in which a beam of protons and of antiprotons, their antimatter twins, are counter-rotating and colliding head-on. To this effect we had to develop a number of techniques for creating antiprotons, confining them in a concentrated beam and colliding them with an intense proton beam. These techniques were developed at CERN with the help of many people and in particular of Guido Petrucci, Jacques Gareyte and Simon van der Meer.

In view of the size and of the complexity of the detector, physics experiments at the proton-antiproton collider have required rather unusual techniques. Equally unusual has been the number and variety of different talents needed to reach the goal of observing the W and Z particles. International cooperation between many people from very different countries has been proven to be a very successful way of acheiving such goals.

(*added in 1991*): For eighteen years, I have dedicated one semester per year to teaching at Harvard University in Cambridge, Mass., where I have been appointed professor in 1970, spending the rest of my time mostly in Geneva, where I was conducting various experiments, especially the UA-1 Collaboration at the proton-antiproton collider until 1988.

On 17 December 1987, the Council of CERN decided to appoint me Director-General of the Organization as from 1st January 1989, for a mandate of five years.

My wife, Marisa, teaches Physics at High School, and we have two children, a married daughter Laura, medical doctor, and a son, André, student in high energy physics.

EXPERIMENTAL OBSERVATION OF THE INTERMEDIATE VECTOR BOSONS W+, W− and Z0.

Nobel lecture, 8 December, 1984

by

CARLO RUBBIA

CERN, CH-1211 GENEVA 23, Switzerland

1. *Introduction*

In this lecture I shall describe the discovery of the triplet of elementary particles W^+, W^-, and Z^0—by far the most massive elementary particles produced with accelerators up to now. They are also believed to be the propagators of the weak interaction phenomena.

On a cosmological scale, weak interactions play an absolutely fundamental role. For example, it is the weak process

$$p+p\rightarrow {}^2H+e^++\nu_e$$

that controls the main burning reactions in the sun. The most striking feature of these phenomena is their small rate of occurrence: at the temperature and density at the centre of the sun, this burning process produces a heat release per unit of mass which is only 1/100 that of the natural metabolism of the human body. It is indeed this slowness that makes them so precious, ensuring, for instance, the appropriate thermal conditions that are necessary for life on earth. This property is directly related to the very large mass of the W-field quanta.

Since the fundamental discoveries of Henri Becquerel and of Pierre and Marie Curie at the end of the last century, a large number of beta-decay phenomena have been observed in nuclei. They all appear to be related to a pair of fundamental reactions involving transformations between protons and neutrons:

$$n\rightarrow p+e^-+\bar\nu_e, \qquad p\rightarrow n+e^++\nu_e. \tag{1}$$

Following Fermi [1], these processes can be described perturbatively as a point interaction involving the product of the four participating fields.

High-energy collisions have led to the observation of many hundreds of new hadronic particle states. These new particles, which are generally unstable, appear to be just as fundamental as the neutron and the proton. Most of these new particle states exhibit weak interaction properties which are similar to those of the nucleons. The spectroscopy of these states can be described with the help of fundamental, point-like, spin-½ fermions, the quarks, with fractional electric charges $+\tfrac{2}{3}e$ and $-\tfrac{1}{3}e$ and three different colour states. The universality of the weak phenomena is then well interpreted as a Fermi

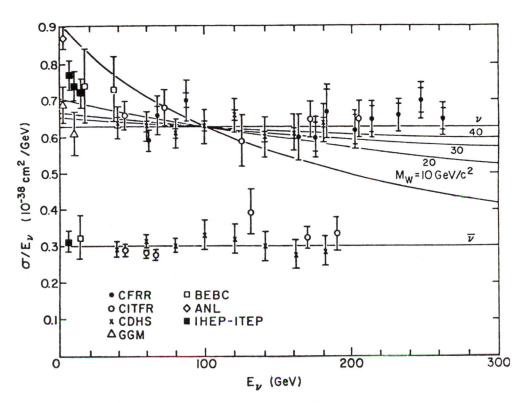

Fig. 1. The muon neutrino and antineutrino charged-current total cross-section as a function of the neutrino energy. Data are from the Particle Data Group (Rev. Mod. Phys. *56*, No. 2, Part 2, April 1984) reprinted at CERN. The lines represent the effects of the W propagator.

coupling occurring at the quark level [2]. For instance, reactions (1) are actually due to the processes

$$(d) \rightarrow (u) + e^- + \bar{v}_e, \qquad (u) \rightarrow (d) + e^+ + v_e, \qquad (2)$$

where (u) is a $+\frac{2}{3}e$ quark and (d) a $-\frac{1}{3}e$ quark. (The brackets indicate that particles are bound.) Cabibbo has shown that universality of the weak coupling to the quark families is well understood, assuming that significant mixing occurs in the $+\frac{1}{3}e$ quark states [3]. Likewise, the three leptonic families —namely (e, v_e), (μ, v_μ), and (τ, v_τ)—exhibit identical weak interaction behaviour, once the differences in masses are taken into account. It is not known if, in analogy to the Cabibbo phenomenon, mixing occurs also amongst the neutrino states (neutrino oscillations).

This has led to a very simple perturbative model in which there are three quark currents, built up from the (u, d_c), (c, s_c), and (t, b_c) pairs (the subscript C indicates Cabibbo mixing), and three lepton currents from (e, v_e), (μ, v_μ), and (τ, v_τ) pairs. Each of these currents has the standard vector form [4] $J_\mu = \bar{f}_1 \gamma_\mu (1 - \gamma_5) f_2$. Any of the pair products of currents J_μ, j_μ will relate to a basic four-fermion interaction occurring at a strength determined by the universal Fermi constant G_F:

$$\mathcal{L}(x) = (G_F/\sqrt{2}) J_\mu^*(x) j^\mu(x) + \text{c.c.},$$

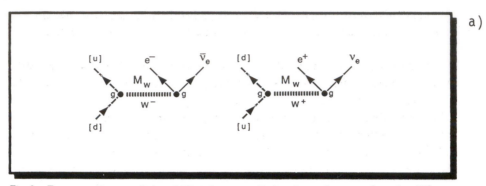

Fig. 2a. Feynman diagram of virtual W exchange mediating the weak process [reaction (2)].

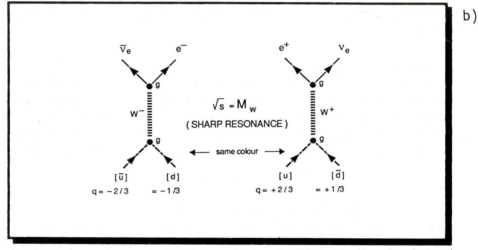

Fig. 2b. Feynman diagram for the direct production of a W particle. Note that the quark transformation has been replaced by a quark–antiquark annihilation.

where $G_F = 1.16632 \times 10^{-5} \text{GeV}^{-2}$ ($\hbar = c = 1$).

This perturbative, point-like description of weak processes is in excellent agreement with experiments, up to the highest q^2 experiments performed with the high-energy neutrino beams (Fig. 1). We know, however, that such a perturbative calculation is incomplete and unsatisfactory. According to quantum mechanics, all higher-order terms must also be included: they appear, however, as quadratically divergent. Furthermore, at centre-of-mass energies greater than about 300 GeV, the first-order cross-section violates conservation of probability.

It was Oskar Klein [5] who, in 1938, first suggested that the weak interactions could be mediated by massive, charged fields. Although he made use of Yukawa's idea of constructing a short-range force with the help of massive field quanta, Klein's theory established also a close connection between electromagnetism and weak interactions. We now know that his premonitory vision is embodied in the electroweak theory of Glashow, Weinberg and Salam [6], which will be discussed in detail later in this lecture. It is worth quoting Klein's view directly:

> *'The role of these particles, and their properties, being similar to those of the photons, we may perhaps call them "electro-photons" (namely electrically charged photons).'*

In the present lecture I shall follow today's prevalent notation of W^+ and W^- for these particles—from 'weak' [7]—although one must recognize that Klein's definition is now much more pertinent.

The basic Feynman diagrams of reaction (2) are the ones shown in Fig. 2 a.

The new, dimensionless coupling constant g is then introduced, related to $G_F/\sqrt{2} \cong g^2/m_W^2$, for $q^2 \ll m_W^2$. The V−A nature of the Fermi interaction requires that the spin J of the W particle be 1. It is worth remarking that in Klein's paper, in analogy to the photon, $J=1$ and $g=\alpha$. The apparently excellent fit of the neutrino data to the four-fermion point-like interaction (Fig. 1) indicates that m_W is very large ($\geqslant 60$ GeV/c^2) and is compatible with $m_W = \infty$.

2. Production of W particles

Direct production of W particles followed by their decay into the electron-neutrino is shown in Fig. 2 b. The centre-of-mass energy in the quark–antiquark collision must be large enough, namely $\sqrt{s} \simeq m_W$. The cross-section around the resonance will follow a characteristic Breit–Wigner shape, reminiscent of nuclear physics experiments. The cross-section is easily calculated:

$$\sigma(q\bar{q} \to W) = \tfrac{3}{4}\pi\lambda^2\Gamma_i\,\Gamma/[(E-m_W)^2+\Gamma^2\!/4],$$

where λ is the reduced quark wavelength in the centre of mass. Quark and antiquark must have identical colours. The initial-state width $\Gamma_i \equiv \Gamma_{q\bar{q}} \simeq 4.5 \times 10^{-7}$ m^3 (GeV) calculated from G_F is surprisingly wide: namely, for $m_W \simeq 82$ GeV/c^2 as predicted by SU(2)×U(1) theory, $\Gamma_{q\bar{q}} \cong 450$ MeV. The total width Γ depends on the number of quark and lepton generations. Taking $N_q = 3$ and $N_\ell = 3$, again for $m_W \simeq 100$ GeV, we find $\Gamma = 4 \times \Gamma_{q\bar{q}} = 2$ GeV.

At the peak of the resonance,

$$\sigma(q\bar{q} \to W, \sqrt{s} = m_W) = 3\pi\lambda_i^2\,B_i,$$

where $B_i = \Gamma_i/\Gamma$ is the branching ratio for the incoming channel.

Of course quark–antiquark collisions cannot be realized directly since free quarks are not available. The closest substitute is to use collisions between protons and antiprotons. The fraction of nucleon momentum carried by the quarks and antiquarks in a proton is shown in Fig. 3. Because of the presence of antiquarks, proton–proton collisions also can be efficiently used to produce W particles. However, a significantly greater beam energy is needed and there is no way of identifying the directions of the incoming quark and antiquark. As we shall see, this ambiguity will prevent the observation of important asymmetries associated with parity (P) and charge (C) violation of weak interactions. The centre-of-mass energy in the quark–antiquark collision $s_{q\bar{q}}$ is related to $S_{p\bar{p}}$ by the well-known formula,

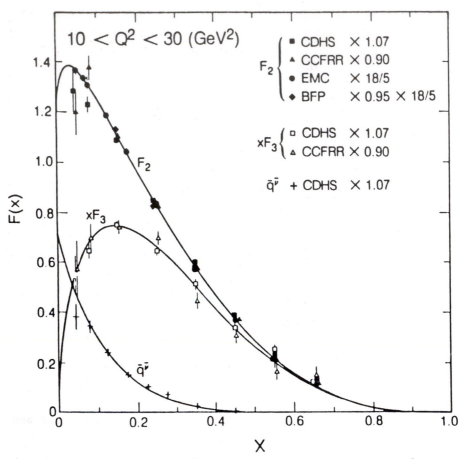

Fig. 3. Structure functions F_2, xF_3, and $\bar{q}^{\bar{\nu}}$, measured in different experiments, for fixed Q^2 versus x, plotted assuming $R=\sigma_L/\sigma_T=0$. The electromagnetic structure function $F_2^{\mu N}$ measured by the EMC (European Muon Collaboration) and the BFP [Berkeley (LBL) – FNAL – Princeton] is compared with the charged-current structure function $F_2^{\nu N}$ using the 18/5 factor from the average charge squared of the quarks. No correction has been applied for the difference between the strange and charm sea quarks, so the interpretation is $F_2 = x[q + \bar{q} - 3/5(s + \bar{s} - c - \bar{c})]$. (In this Q^2 range, $F_2^{\nu N}$ is depleted by a similar amount due to charm threshold effects in the transition s→ c.) The antiquark distribution measured from antineutrino scattering is $\bar{q}^{\bar{\nu}}=x(\bar{u}+\bar{d}+2\bar{s})$. The solid lines have the forms: $F_2=3.9x^{0.55}(1-x)^{3.2}+1.1(1-x)^8$, $xF_3=3.6x^{0.55}(1-x)^{3.2}$, $\bar{q}^{\bar{\nu}}=0.7(1-x)^8$. Relative normalization factors have been fitted to optimize agreement between the different data sets, and absolute changes have been arbitrarily chosen as indicated. [References: CDHS—H. Abramowicz et al., Z. Phys. *C17*, 283 (1983); CCFRR—F. Sciulli, private communication; EMC—J. J. Aubert et al., Phys. Lett. *105B*, 322 (1981); and A. Edwards, private communication; BFP—A. R. Clark et al., Phys. Rev. Lett. *51*, 1826 (1983); and P. Meyers, Ph. D. Thesis, LBL–17108 (1983), Univ. of Calif., Berkeley. Courtesy J. Carr, LBL.]

$$s_{q\bar{q}} = S_{p\bar{p}} \cdot x_p \cdot x_{\bar{p}}.$$

Note that according to Fig. 3, in order to ensure the correct correlation between the quark of the proton (and the antiquark of the antiproton) the energy should be such that $x_p \simeq x_{\bar{p}} \geq 0.25$. Therefore there is one broad optimum

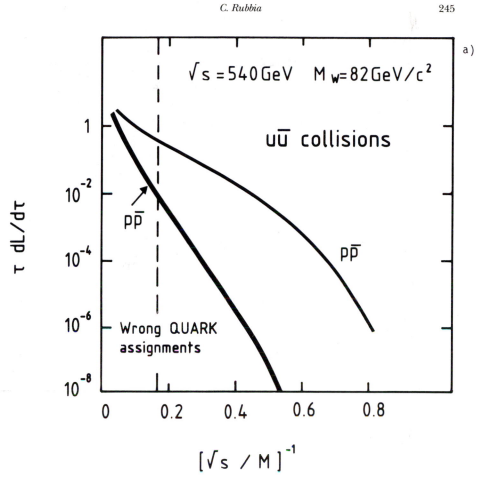

Fig. 4a, b. Production cross-sections of intermediate vector bosons for proton-antiproton collisions. The mass is parametrized with $\tau^{1/2} = \sqrt{s}/M$. Note in Fig. 4a the small probability of wrong quark-antiquark assignments. The prints in Fig. 4b relate to mass predictions for the SU(2) × U(1) model.

energy range for the proton–antiproton collisions for a given W mass. For $m_W=80\,\text{GeV}/c^2$, $\sqrt{S}_{p\bar{p}} \approx 400$–600 GeV. The production cross-section for the process

$$p\bar{p} \to W^{\pm}+X, \quad W^{\pm} \to e^{\pm}+\nu_e$$

(where X denotes the fragmentation of spectator partons) can be easily evaluated by folding the narrow resonance width over the p and $\bar{p}$ momentum distributions (Fig. 4). For $m_W=82$ GeV/c and $\sqrt{S}_{p\bar{p}}=540$ GeV, one finds $\sigma \cdot B=0.54\times10^{-33}$ cm^2.

3. *Proton–antiproton collisions*

The only practical way of achieving centre-of-mass energies of the order of 500 GeV is to collide beams of protons and antiprotons [8]. For a long time such an idea had been considered as unpractical because of the low density of beams when used as targets.

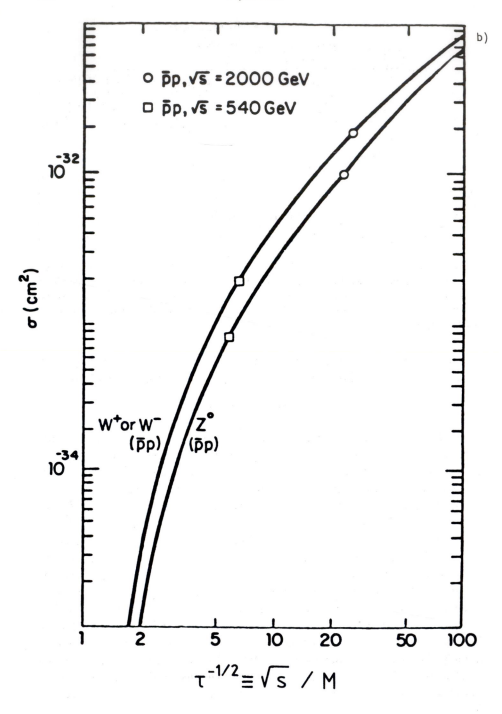

Fig. 4

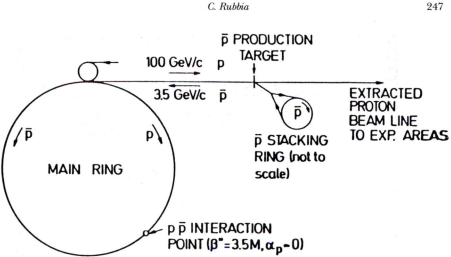

Fig. 5. General layout of the pp̄ colliding scheme, from Ref. [9]. Protons (100 GeV/c) are periodically extracted in short bursts and produce 3.5 GeV/c antiprotons, which are accumulated and cooled in the small stacking ring. Then p̄'s are reinjected in an RF bucket of the main ring and accelerated to top energy. They collide head on against a bunch filled with protons of equal energy and rotating in the opposite direction.

The rate R of events of cross-section σ for two counter-rotating beam bunches colliding head on, with frequency f_0 and n_1 and n_2 particles, is

$$R = (n_1 n_2 f_0/4)(\sigma/\pi\varrho^2),$$

where ϱ is the (common) beam radius, and the numerical factor ¼ takes into account the integration over Gaussian profiles. For our experiment, typically $\varrho = 0.01$ cm and $\sigma = 10^{-34}$ cm². Therefore $(\sigma/\pi\varrho^2) = 3 \times 10^{-31}$, and a very large $n_1 n_2$ product is needed to overcome the 'geometry' effect.

The scheme used in the present experimental programme has been discussed by Rubbia, Cline and McIntyre [9] and is shown in Fig. 5. It makes use of the existing 400 GeV CERN Proton Synchrotron (PS) [10], suitably modified in order to be able to store counter-rotating bunches of protons and antiprotons at an energy of 270 GeV per beam. Antiprotons are produced by collisions of 26 GeV/c protons from the PS onto a solid target. Accumulation in a small 3.5 GeV/c storage ring is followed by stochastic cooling [11] to compress phase space. In Table 1 the parameters of Ref. [9] are given. Taking into account that the original proposal was formulated for another machine, namely the Fermilab synchrotron (Batavia, Ill.) they are quite close to the conditions realised in the SPS conversion. Details of the accumulation of antiprotons are described in the accompanying lecture by Simon van der Meer.

The CERN experiments with proton–antiproton collisions have been the first, and so far the only, example of using a storage ring in which bunched protons and antiprotons collide head on. Although the CERN pp̄ Collider uses *bunched* beams, as do the e^+e^- colliders, the phase-space damping due to synchrotron radiation is now absent. Furthermore, since antiprotons are

Table 1. List of parameters (from Ref. [9])

1. MAIN RING (Fermilab)	
– Beam momentum	250 (400) GeV/c
– Equivalent laboratory energy for (pp̄)	133 (341) TeV
– Accelerating and bunching frequency	53.14 MHz
– Harmonic number	1113
– RF peak voltage/turn	3.3×10^6 V
– Residual gas pressure	$< 0.5 \times 10^{-7}$ Torr
– Beta functions at interaction point	3.5 m
– Momentum compaction at int. point	~ 0 m
– Invariant emittances ($N_p = 10^{12}$)	
– longitudinal	3 eV·s
– transverse	$50 \, \pi \, 10^{-6}$ rad·m
– Bunch length	2.3 m
– Design luminosity	5×10^{29} (8×10^{29}) $cm^{-2} s^{-1}$
2. ANTIPROTON SOURCE (Stochastic Cooling [11])	
– Nominal stored p̄ momentum	3.5 GeV/c
– Circumference of ring	100 m
– Momentum acceptance	0.02
– Betatron acceptances	$100 \, \pi \, 10^{-6}$ rad·m
– Bandwidth of momentum stochastic cooling	400 MHz
– Maximum stochastic accelerating RF voltage	3 000 V
– Bandwidth of betatron stochastic cooling	200 MHz
– Final invariant emittances ($N_{\bar{p}} = 3 \times 10^{10}$)	
– longitudinal	0.5 eV·s
– transverse	$10 \, \pi \, 10^{-6}$ rad·m

scarce, one has to operate the collider in conditions of relatively large beam––beam interactions, which is not the case for the continuous proton beams of the previously operated Intersecting Storage Rings (ISR) at CERN [12]. One of the most remarkable results of the pp̄ Collider has probably been the fact that it has operated at such high luminosity, which in turn means a large beam–beam tune shift. In the early days of construction, very serious concern had been voiced regarding the instability of the beams due to beam–beam interaction. The beam–beam force can be approximated as a periodic succession of extremely non-linear potential kicks. It is expected to excite a continuum of resonances of the storage ring which has, in principle, the density of rational numbers. Reduced to bare essentials, we can consider the case of a weak antiproton beam colliding head on with a strongly bunched proton beam. The increment, due to the angular kick $\Delta x'$, of the action invariant $W = \gamma x^2 + 2\alpha x x' + \beta x'^2$ of an antiproton is $\Delta W = \beta(\Delta x') + 2(\alpha + \beta x')\Delta x'$, and this can be expressed in terms of the 'tune shift', ΔQ as $\Delta x' = 4\pi \Delta Q x/\beta$. If we now assume that the successive kicks are randomized, the second term of ΔW averages to zero, and we get

$$\langle \Delta W/W \rangle = \frac{1}{2}(4\pi \Delta Q)^2.$$

For the design luminosity we need $\Delta Q \sim 0.003$, leading to $(\Delta W/W) = 7.1 \times 10^{-4}$. This is a very large number indeed, giving an e-fold in-

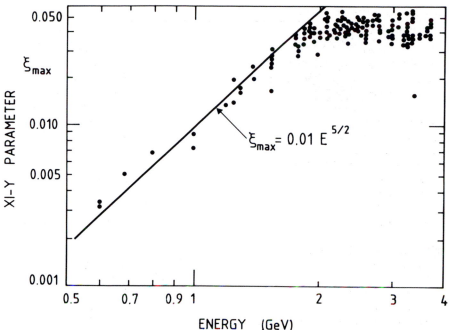

Fig. 6. Maximum allowed beam–beam tune-shift parameter, XI–Y, as a function of energy of the electron–positron collider SPEAR. One can see a dramatic drop in the allowed tune shift at lower energies, as a consequence of the reduced synchrotron damping. Extrapolation to the case of proton–antiproton collisions where the damping is absent and therefore the damping time is constant, is to be identified with the beam lifetime, permitting an infinitesimal tune shift and therefore to an unpractical luminosity.

crease of W in only $1/7.1 \times 10^{-4} = 1.41 \times 10^3$ kicks! Therefore the only reason why the antiproton motion remains stable is because these strong kicks are not random but periodic, and the beam has a long 'memory' which allows them to be added coherently rather than at random. Off-resonance, the effects of these kicks then cancel on the average, giving an overall zero amplitude growth. The beam–beam effects are very difficult, if not impossible, to evaluate theoretically, since this *a priori* purely deterministic problem can exhibit stochastic behaviour and irreversible diffusion-like characteristics.

A measurement at the electron–positron collider SPEAR at Stanford had further aggravated the general concern about the viability of the p$\bar{\text{p}}$ collider scheme. Reducing the energy of the electron collider (Fig. 6) resulted in a smaller value of the maximum allowed tune shift, interpreted as being due to the reduced synchrotron radiation damping. Equating the needed beam lifetime for the p$\bar{\text{p}}$ collider (where damping is absent) with the extrapolated damping time of an e^+e^- collider gives a maximum allowed tune shift $\Delta Q = 10^{-5} \div 10^{-6}$, which is catastrophically low. This bleak prediction was not confirmed by the experience at the collider, where $\Delta Q = 0.003$ per crossing, and six crossings are routinely achieved with a beam luminosity lifetime approaching one day. What, then, is the reason for such a striking contradiction between experiments with protons and those with electrons? The difference is caused by the presence of synchrotron radiation in the latter case. The emission of synchrotron photons is a major source of quick randomization between crossings and leads to a rapid deterioration of the beam emittance. Fortunately, the same phenomenon also provides us with an effective damping mechanism. The p$\bar{\text{p}}$ collider works because *both* the randomizing and the damping mechanisms are absent. This unusually favourable combination of effects has ensured that p$\bar{\text{p}}$ colliders have become viable devices. They have the potential for substantial improvements in the future. The accumulation of more antiprotons would permit us to obtain a substantially larger luminosity, and a project is under way at CERN which is expected to be able to deliver enough antiprotons to accumulate, *in one single day*, the integrated luminosity on which the results presented in this lecture have been based ($\sim$100 nb^{-1}).

4. The detection method

The process we want to observe is the one represented in Fig. 2 b, namely

$$\text{p}+\bar{\text{p}} \to \text{W}^{\pm}+\text{X}, \quad \text{W}^{\pm} \to \text{e}^{\pm}+\nu_{\text{e}}, \tag{3}$$

where X represents the sum of the debris from the interactions of the other protons (spectators). Although the detection of high-energy electrons is relatively straightforward, the observation of neutrino emission is uncommon in colliding-beam experiments. The probability of secondary interactions of the neutrino in any conceivable apparatus is infinitesimal. We must therefore rely on kinematics in order to signal its emission indirectly. This is achieved with an appropriately designed detector [13] which is uniformly sensitive, over the whole solid angle, to all the charged or neutral interacting debris produced by the collision. Since collisions are observed in the centre of mass, a significant momentum imbalance may signal the presence of one or more non-interacting particles, presumably neutrinos.

The method can be conveniently implemented with calorimeters, since their energy response can be made rather uniform for different incident particles. Calorimetry is also ideally suited to the accurate measurement of the energy of the accompanying high-energy electron for process (3). Energy depositions (Fig. 7) in individual cells, E_i, are converted into an energy flow vector $\vec{E}_i = \vec{n}_i E_i$, where $\vec{n}_i$ is the unit vector pointing from the collision point to (the centre of) the cell. Then, for relativistic particles and for an ideal calorimeter response $\Sigma_i \vec{E}_i = 0$, provided no non-interacting particle is emitted. The sum covers the whole solid angle. In reality there are finite residues to the sum: $\Delta \vec{E}_M = \Sigma_i \vec{E}_i$. This quantity is called the 'missing energy' vector. Obviously in the case of a neutrino emission, $\vec{p}_\nu = -\Delta \vec{E}_M$. In the case of process (3) the effect is particularly spectacular, since in the centre of mass of the W the neutrino momentum $p_\nu^* = m_W/2$ is very large.

The practical realization of such a detector [14] is shown in Fig. 8 a. After momentum analysis in a large-image drift chamber in a horizontal magnetic field of 7 000 G oriented normal to the beam directions, six concentric sets of finely segmented calorimeters (Fig. 8 b) surround the collision point, down to

CONSTRUCTION OF ENERGY VECTORS

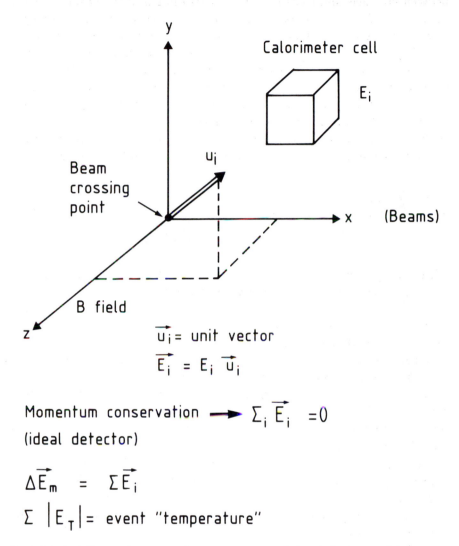

$$\vec{u_i} = \text{unit vector}$$

$$\vec{E_i} = E_i \vec{u_i}$$

Momentum conservation $\longrightarrow$ $\Sigma_i \vec{E_i} = 0$

(ideal detector)

$$\Delta \vec{E_m} = \Sigma \vec{E_i}$$

$$\Sigma |E_T| = \text{event "temperature"}$$

Fig. 7. Principal diagram for constructing energy vectors and the missing energy of the event.

angles of 0.2° with respect to the beam directions. The operation of these calorimeters is shown schematically in Fig. 9 a. The first four segments are sandwiches of lead and scintillator, in which electrons are rapidly absorbed (Fig. 9 b), followed by two sections of iron/scintillator sandwich (which is also the return yoke of the magnetic field). All hadrons are completely absorbed within these calorimeters. Muons are detected by eight planes of large drift chambers which enclose the whole detector volume. If one or more muons are detected, their momenta, measured by magnetic curvature, must be added 'by hand' to the energy flow vector.

The performance of the energy flow measurement has been tested with

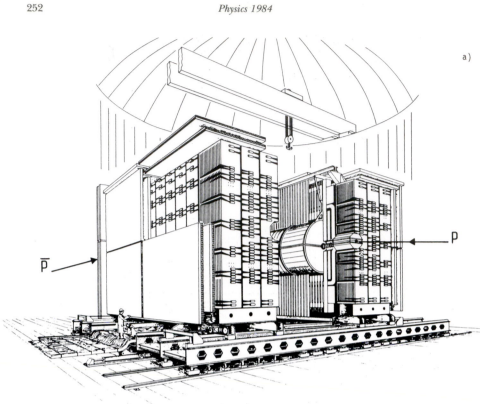

a)

Fig. 8a. The UA1 detector solid-angle is fully covered down to 0.2°.

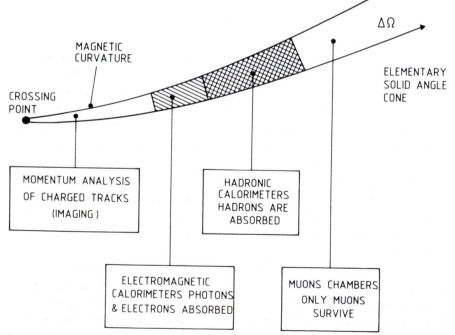

b)

Fig. 8b. The schematic functions of each of the elementary solid-angle elements constituting the detector structure.

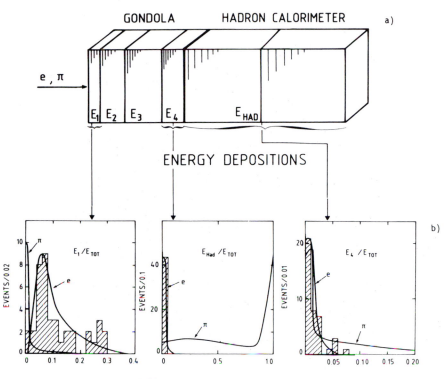

Fig. 9. a) Schema of an elementary solid-angle cell. After four segments of lead/scintillator sandwich, there are two elements of iron/scintillator sandwich, which is also the magnetic field return loop. b) Energy depositions for high-energy pions and electrons. The nature of the particle can be discriminated looking at the transition curve.

standard collisions (minimum bias). Fig. 10 shows how well the vertical component of the missing-energy vector is observed for minimum bias events. The missing energy $\vec{\Delta E}_M$ resolution for each transverse component can be parametrized as $\sigma = 0.43 \sqrt{\Sigma_i E_T^{(i)}}$, where $\Sigma_i E_T^{(i)}$, in units of GeV, is the scalar sum of the transverse components of the energy flow $E_T^{(i)}$. The same parametrization also holds for events which contain high transverse momentum jets, and for which the detector non-uniformities are more critical since energy deposition is highly localized (Fig. 11). The resolution function is shown in Fig. 12, where the missing energy for two-jet events is shown along with a Monte Carlo calculation of the expected distribution based on the expected behaviour of the calorimeters as determined by test-beam data and the measured fragmentation functions of jets.

For a typical event with $\Sigma_i E_T^{(i)} = 80$ GeV, we measure the transverse components of $\vec{\Delta E}_M$ to about 4 GeV. The longitudinal component of the momentum balance will not be used in the present analysis since, in spite of the smallness of the window through which the beam pipes pass ($\leq 0.2°$), energetic particles quite often escape through the aperture.

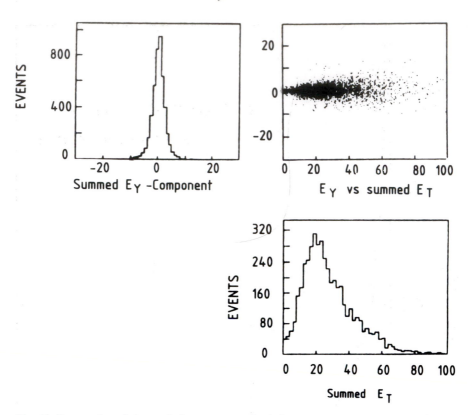

Fig. 10. Scatter-plot of the vertical component of missing transverse energy versus the total transverse energy observed in all calorimeter cells.

5. Observation of the W→e+ν signal

The observation by the UA1 Collaboration [15] of the charged intermediate vector boson was reported in a paper published in February 1983, followed shortly by a parallel paper from the UA2 Collaboration [16]. Mass values were given: $m_W = (80 \pm 5)$ GeV/c^2 (UA1) and $m_W = (80^{+10}_{-6})$ GeV/c^2 (UA2). Since then, the experimental samples have been considerably increased, and one can now proceed much further in understanding the phenomenon. In particular, the assignment of the events to reaction (3) can now be proved rather than postulated. We shall follow here the analysis of the UA1 events [17].

Our results are based on an integrated luminosity of 0.136 pb^{-1}. We first performed an inclusive search for high-energy isolated electrons. The trigger selection required the presence of an energy deposition cluster in the electromagnetic calorimeters at angles larger than 5°, with transverse energy in excess of 10 GeV. In the event reconstruction this threshold was increased to 15 GeV, leading to about 1.5×10^5 beam-beam collision events.

By requiring the presence of an associated, isolated track with $p_T > 7$ GeV/c in the central detector, we reduced the sample by a factor of about 100. Next, a maximum energy deposition (leakage) of 600 MeV was allowed in the hadron calorimeter cells after the electromagnetic counters, leading to a sample of 346

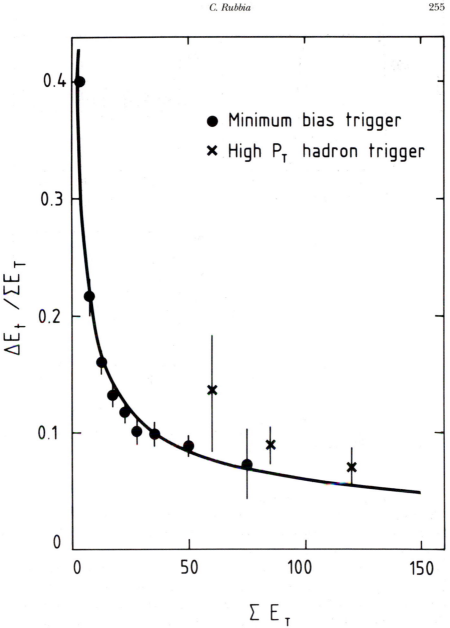

Fig. 11. Missing-energy resolution for minimum-bias and jet events.

events. We then classified events according to whether there was prominent jet activity.

We found that in 291 events there was a clearly visible jet within an azimuthal angle cone $|\Delta\phi|<30°$ opposite to the 'electron' track. These events were strongly contaminated by jet–jet events in which one jet faked the electron signature and had to be rejected. We were left with 55 events without any jet, or with a jet not back-to-back with the 'electron' within 30°. These events had a very clean electron signature (Fig. 13) and a perfect matching between the point of electron incidence and the centroid in the shower detec-

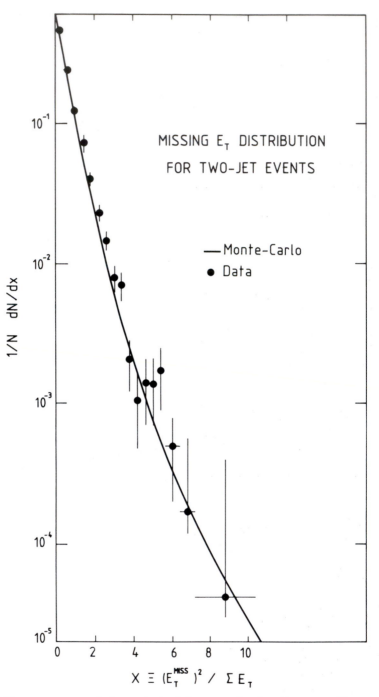

Fig. 12. Transverse energy balance observed for a sample of two-jet events. To convert the horizontal scale to the number of standard deviations (n), use the relationship $n^2 \approx 2x$. Variables have been chosen in such a way as to transform a Gaussian basic response of the calorimeters into a linear plot. The continuous line is the result of a calculation based on the expected calorimeter responses, as measured with test-beam particles.

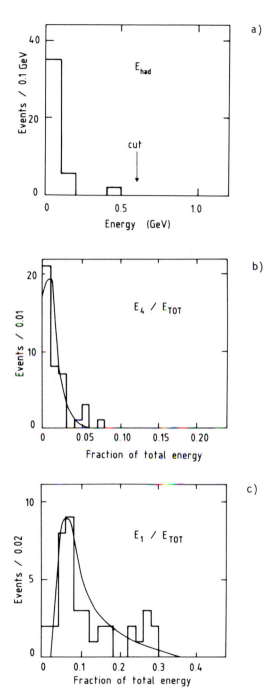

Fig. 13. Distributions showing the quality of the electron signature:

a) The energy deposition in the hadron calorimeter cells behind the 27 radiation lengths (r.l.) of the e.m. shower detector.

b) The fraction of the electron energy deposited in the fourth sampling (6 r.l. deep, after 18 r.l. converter) of the e.m. shower detector. The curve is the expected distribution from test-beam data.

c) As distribution (b) but for the first sampling of the e.m. shower detector (first 6 r.l.).

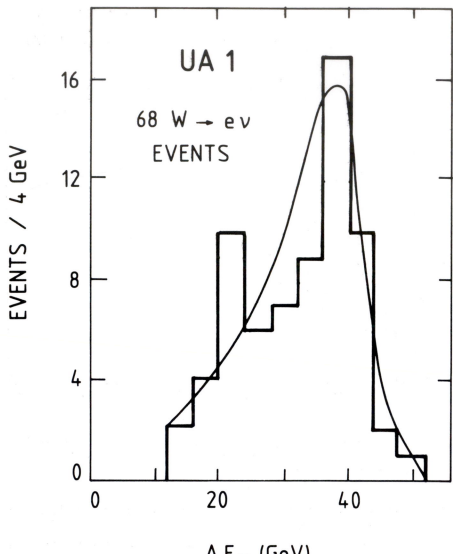

Fig. 14. The distribution of the missing transverse energy for those events in which there is a single electron with $E_T>15$ GeV, and no coplanar jet activity. The curve represents the resolution function for no missing energy normalized to the three lowest missing-energy events.

tors, further supporting the absence of composite overlaps of a charged track and neutral π^0's expected from jets.

The bulk of these events was characterized by the presence of neutrino emission, signalled by a significant missing energy (see Fig. 14). According to the experimental energy resolutions, at most the three lowest missing-energy events were compatible with no neutrino emission. They were excluded by the cut $E_T^{miss}>15$ GeV. We were then left with 52 events.

In order to ensure the best accuracy in the electron energy determination, only those events were retained in which the electron track hit the electromag-

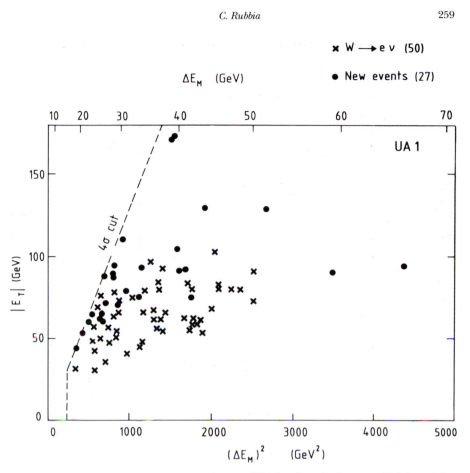

Fig. 15a. Missing transverse energy squared versus ΣE_T for all verified events which have ΔE_m more than 4 st. dev. from zero for all events with $W \rightarrow e+\nu$ decays removed. The events are labelled according to their topology.

netic detectors more than $\pm 15°$ away from their top and bottom edges. The sample was then reduced to 43 events.

An alternative selection was carried out, based on the inclusive presence of a significant missing energy [18]. This is illustrated in Fig. 15a, where all events with missing energy in excess of 4 standard deviations are shown. One can see that previously selected electron events are found as a subset of the sample. However, a significant number of additional events (twenty-seven) were also recorded, in which there was either a jet or an electromagnetic cluster instead of the isolated electrons (Fig. 15b). Evidently the inclusive missing-energy definition implies a broader class of physical phenomena (Fig. 16c) than the simple $W \rightarrow e+\nu$ decay (Figs. 16a, b). As the study of these events [19] is beyond the scope of this lecture, it will not be pursued any further.

We proceeded to a detailed investigation of the events in order to elucidate their physical origin. The large missing energy observed in all of them was interpreted as being due to the emission of one or several non-interacting neutrinos. A very strong correlation in angle and energy was observed (in the plane normal to the colliding beams, where it could be determined accurately),

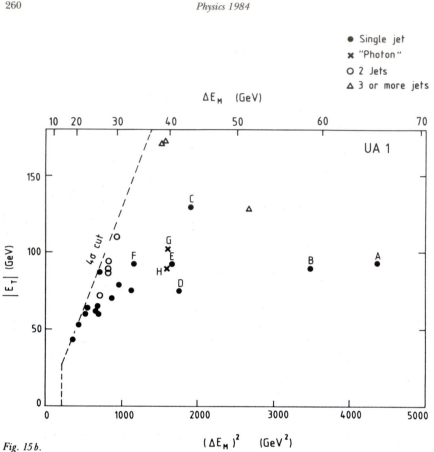

Fig. 15 b.

with the corresponding electron quantities, in a characteristic back-to-back configuration expected from the decay of a massive, slow particle (Figs. 17a, b). This suggested a common physical origin for the electron and for one or several neutrinos.

In order to have a better understanding of the transverse motion of the electron-neutrino(s) system, we studied the experimental distribution of the resultant transverse momentum $p_T^{(W)}$ obtained by adding the neutrino(s) and electron momenta (Fig. 18). The average value was $p_T^{(W)} = 6.3$ GeV/c. Five events which had a visible jet had also the highest values of $p_T^{(W)}$. Transverse momentum balance was almost exactly restored when the vector momentum of the jet was added. The experimental distribution was in good agreement with the many theoretical expectations from quantum chromodynamics (QCD) for the production of a massive state via the Drell–Yan quark-antiquark annihilation [20]. The small fraction (10 %) of events with a jet were then explained as hard gluon bremsstrahlung in the initial state.

Several different hypotheses on the physical origin of the events were tested by looking at kinematical quantities constructed from the transverse variables of the electron and the neutrino(s). We retained two possibilities, namely: i) the two-body decay of a massive particle into the electron and one neutrino,

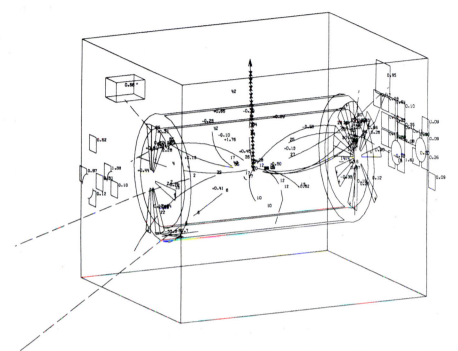

Fig. 16a. Event of the type $W^- \rightarrow e^- + \bar{\nu}_e$. All tracks and calorimeter cells are displayed.

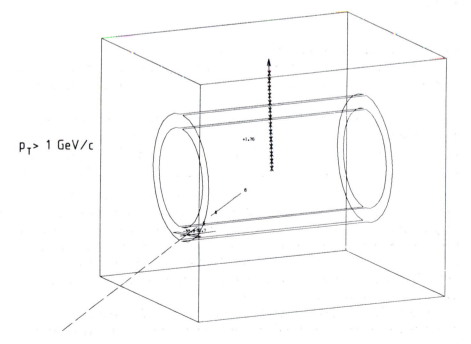

$p_T > 1$ GeV/c

Fig. 16b. The same as picture (a), except that now only particles with $p_T > 1$ GeV/c and calorimeters with $E_T > 1$ GeV are shown.

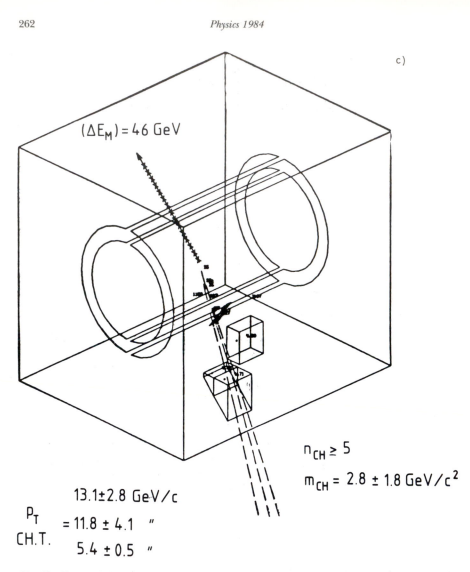

Fig. 16c. Event of the type, jet+missing energy. Only tracks with $p_T>1.5$ GeV/c and cells with $E_T>1.0$ GeV are displayed.

$W \rightarrow e+\nu_e$; and ii) the three-body decay into two, or possibly more, neutrinos and the electron. It can be seen from Figs. 19a and 19b that hypothesis (i) is strongly favoured. At this stage, the experiment could not distinguish between one or several closely spaced massive states.

With the help of a sample of isolated hadrons at large transverse momenta, we estimated in detail the possible sources of background coming from ordinary hadronic interactions, and we concluded that they were negligible (<0.5 events). (For more details on background, we refer the reader to Ref. 20.) However, we expect to get some background events from other decays of the W, namely:

$$W \rightarrow \tau+\nu_\tau[\tau \rightarrow \pi^{\pm}(\pi^0)+\nu_\tau] \qquad (<0.5 \text{ events})$$

or

$$W \rightarrow \tau+\nu_\tau(\tau \rightarrow e+\nu_e+\nu_\tau) \qquad (= 2 \text{ events}).$$

a)

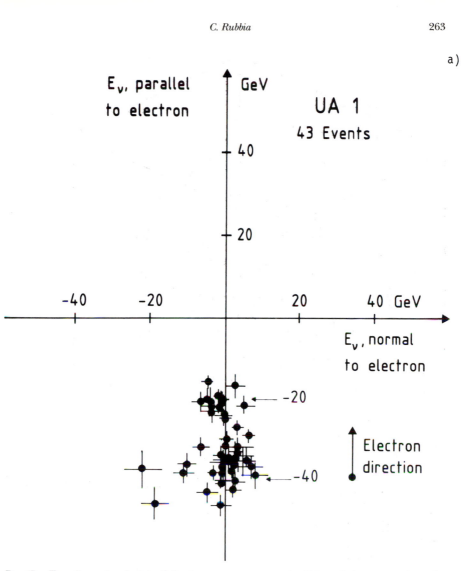

Fig. 17a. Two-dimensional plot of the transverse components of the missing energy (neutrino momentum). Events have been rotated to bring the electron direction to point along the vertical axis. The striking back-to-back configuration of the electron–neutrino system is apparent.

These events were expected to contribute at only the low-p_T part of the electron spectrum, and could even be eliminated in a more restrictive sample.

A value of the W mass can be extracted from the data in a number of ways:

i) It can be obtained from the inclusive transverse momentum distribution of the electrons (Fig. 19a), but the drawback of this technique is that the transverse momentum of the W particle must be known. Taking the QCD predictions [21], in reasonable agreement with experiment, we obtained $m_W = (80.5 \pm 0.5)$ GeV/c^2.

ii) We can define a transverse mass variable, $m_T^2 = 2p_T^{(e)} p_T^{(v)} (1 - \cos \phi)$, with the property $m_T \leqslant m_W$, where the equality would hold for only those events with no longitudinal momentum components. Fitting Fig. 19b to a

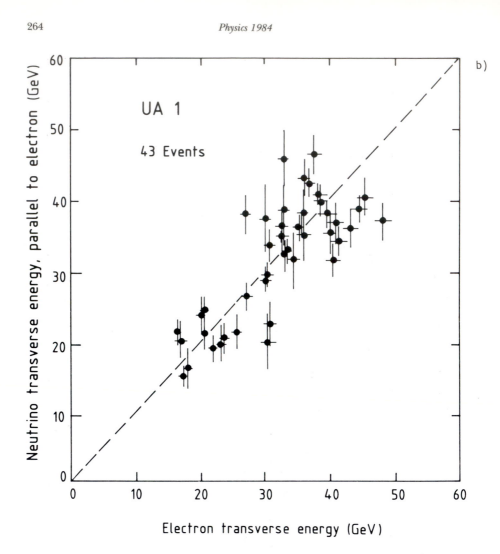

Fig. 17b. Correlation between the electron and neutrino transverse energies. The neutrino component along the electron direction is plotted against the electron transverse energy.

common value of the mass was done almost independently of the transverse motion of the W particles, $m_W = (80.3^{+0.4}_{-1.3})$ GeV/c². It should be noted that the lower part of the distribution in $m_T^{(W)}$ was slightly affected by $W \rightarrow \tau + \nu_\tau$ decays and other backgrounds.

iii) We can define an enhanced transverse mass distribution, selecting only events in which the decay kinematics is largely dominated by the transverse variable with the simple cuts $p_T^{(e)}$, $p_T^{(v)} > 30$ GeV/c. The resultant distribution (Fig. 19 c) then showed a relatively narrow peak at approximately 76 GeV/c². Model-dependent corrections now only contributed to the difference between this average mass value and the fitted m_W value, $m_W = (80.9 \pm 1.5)$ GeV/c². An interesting upper limit to the width of the W was also derived from the distribution, namely $\Gamma_T \lesssim 7$ GeV/c² (90 % confidence level).

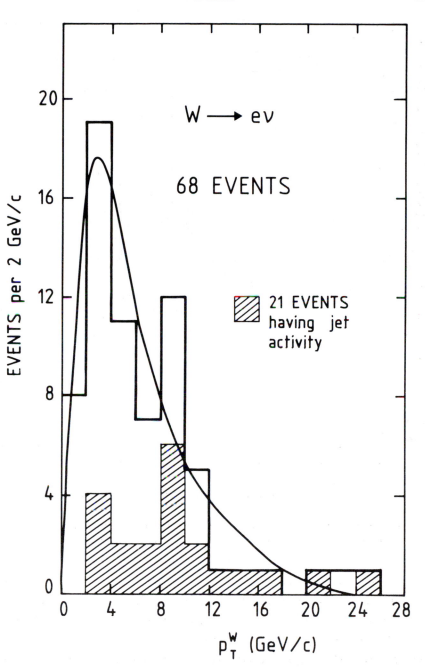

Fig. 18. The transverse momentum distribution of the W derived from our events using the electron and missing transverse energy vectors. The highest $p_T^{(W)}$ events have a visible jet (shown in black in the figure). The data are compared with the theoretical predictions for W production based on QCD (Ref. [21]).

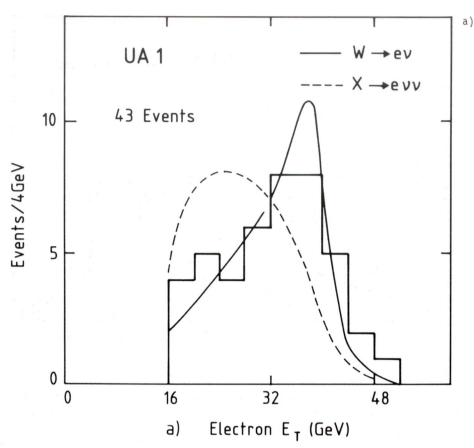

a)

a)   Electron E_T (GeV)

Fig. 19 a. The electron transverse energy distribution. The two curves show the results of a fit of the enhanced transverse mass distribution to the hypotheses W→e+ν and X→e+ν+ν. The first hypothesis is clearly preferred.

The three mass determinations gave very similar results. We preferred to retain the result of method (iii), since we believed it to be the least affected by systematic effects, even if it gave the largest statistical error. Two important contributions had to be added to the statistical errors:

i) *Counter-to-counter calibrations.* They were estimated to be 4 % r.m.s. In the determination of the W mass this effect was greatly attenuated to a negligible level, since many different elements contributed to the event sample.

ii) *Calibration of the absolute energy scale.* This was estimated to be ±3 %, and of course affects both the Z^0 and the W samples by the same multiplicative factor.

Once the decay reaction W→e+$ν_e$ was established, the longitudinal momentum of the electron–neutrino system was determined with a twofold ambiguity for the unmeasured longitudinal component of the neutrino momentum. The overall information of the event was used to establish momentum and energy conservation bounds in order to resolve this ambiguity in 70 % of the cases. Most of the remaining events had solutions which were quite close,

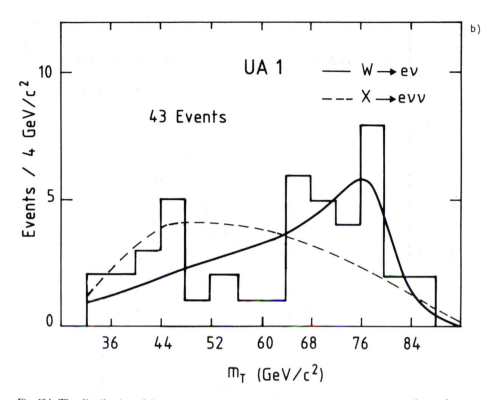

Fig. 19 b. The distribution of the transverse mass derived from the measured electron and neutrino vectors. The two curves show the results of a fit to the hypotheses W → e+ν and X → e+ν+ν.

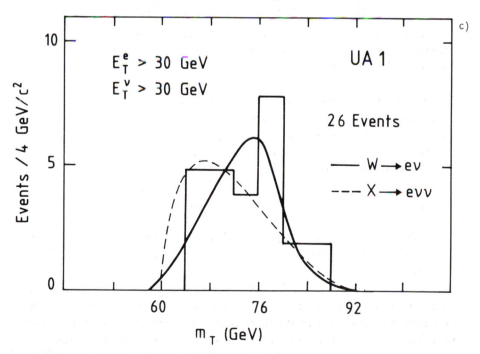

Fig. 19c. The enhanced electron-neutrino transverse mass distribution (see text). The two curves show the results of a fit to the hypotheses W → e + ν and X → e + ν + ν.

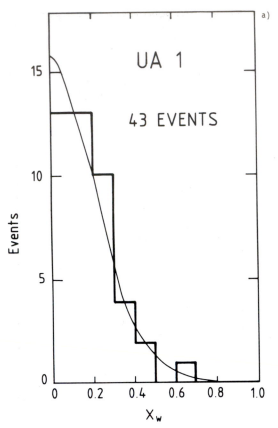

Fig. 20 a. The fractional beam energy x_W carried by the W. The curve is the prediction obtained by assuming that the W has been produced by $q\bar{q}$ fusion. Note that in general there are two kinematic solutions for x_W (see text), which are resolved in 70 % of the events by consideration of the energy flow in the rest of the event. Where this ambiguity has been resolved, the preferred kinematic solution has been the one with the lowest x_W. In the 30 % of the events where the ambiguity is not resolved, the lowest x_W solution has therefore been chosen.

and the physical conclusions were nearly the same for both solutions. The fractional beam energy x_W carried by the W particle is shown in Fig. 20 a, and it appears to be in excellent agreement with the hypothesis of W production in $q\bar{q}$ annihilation [22]. Using the well-known relations $x_W = x_p - x_{\bar{p}}$ and $x_p \cdot x_{\bar{p}} = m_W^2/s$, we determined the relevant parton distributions in the proton and antiproton. It can be seen that the distributions are in excellent agreement with the expected x distributions for quarks and antiquarks in the proton and antiproton, respectively (Figs. 20 b and 20 c). Contributions of the u and d quarks were also neatly separated by looking at the charges of produced W events, since $(u\bar{d}) \rightarrow W^+$ and $(\bar{u}d) \rightarrow W^-$ (Figs. 20 d and 20 e).

6. *Observation of the parity (charge conjugation) violation, and determination of the spin of the W particle*
One of the most relevant properties of weak interactions is the violation of parity and charge conjugation. Evidently the W particle, in order to mediate

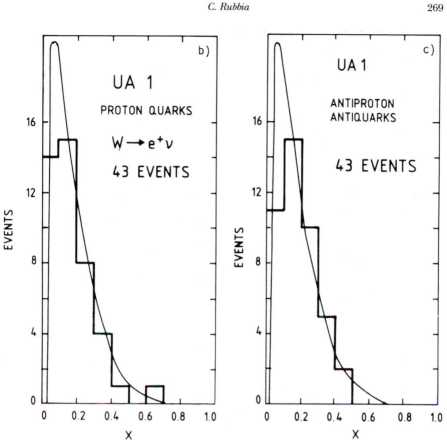

Fig. 20 b. The x-distribution of the proton quarks producing the W by qq fusion. The curve is the prediction assuming qq fusion.
Fig. 20 c. The same as Fig. 20 b for the antiproton quarks.

weak processes, must also exhibit these properties. Furthermore, as already mentioned, the V−A nature of the four-fermion interaction implies the assignment J=1 for its spin. Both of these properties must be verified experimentally. According to the V−A theory, weak interactions should act as a longitudinal polarizer of the W particles, since quarks (antiquarks) are provided by the proton (antiproton) beam. Likewise, decay angular distributions from a polarizer are expected to have a large asymmetry, which acts as a polarization analyser. A strong backward–forward asymmetry is therefore expected, in which electrons (positrons) prefer to be emitted in the direction of the proton (antiproton). In order to study this effect independently of W-production mechanisms, we have looked at the angular distribution of the emission angle θ^* of the electron (positron) with respect to the proton (antiproton) direction in the W centre of mass. Only events with no reconstruction ambiguity can be used. We verified that this does not bias the distribution in the variable $\cos \theta^*$. According to the expectations of V−A theory the distribution should be of the type $(1+\cos \theta^*)^2$, in excellent agreement with the experimental data (Fig. 21). The parity violation parameters and the spin of the W particle can be

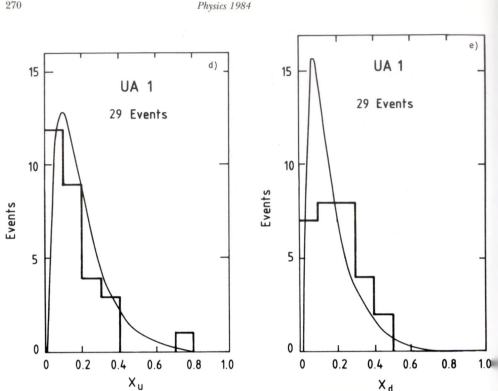

Fig. 20 d. The same as Fig. 20 b but for u(ū) quarks in the proton (antiproton).
Fig. 20 e. The same as Fig. 20 b but for d(ū) quarks in the proton (antiproton).

determined directly. It has been shown by Jacob [23] that for a particle of arbitrary spin J, one expects

$$\langle \cos \theta^* \rangle = \langle \lambda \rangle \langle \mu \rangle / J(J+1),$$

where $\langle \mu \rangle$ and $\langle \lambda \rangle$ are the global helicity of the production system (ud) and of the decay system (eν), respectively.

For V−A, we then have $\langle \lambda \rangle = \langle \mu \rangle = -1$, J=1, leading to the maximal value $\langle \cos \theta^* \rangle = 0.5$. For J=0 it is obvious that $\langle \cos \theta^* \rangle = 0$; and for any other spin value J⩾2, $\langle \cos \theta^* \rangle \leqslant 1/6$. Experimentally we find $\langle \cos \theta^* \rangle = 0.5 \pm 0.1$, which supports *both* the J=1 assignment *and* maximal helicity states at production and decay. Note that the choice of sign $\langle \mu \rangle = \langle \lambda \rangle = \pm 1$ cannot be separated, i.e. right- and left-handed currents, both at production and decay, cannot be resolved without a polarization measurement.

7. *Total cross-section and limits to higher mass W's*
The integrated luminosity of the experiment was 136 nb^{-1}, and it is known to about ±15 % uncertainty. In order to get a clean W→ eν$_e$ sample we selected 47 events with p$_T^{(e)}$>20 GeV/c. The W→ τν$_\tau$ contamination in the sample was estimated to be 2±2 events. The event acceptance was computed to be 0.65, primarily because of i) the p$_T^{(e)}$>20 GeV/c cut (0.80); ii) the jet veto require-

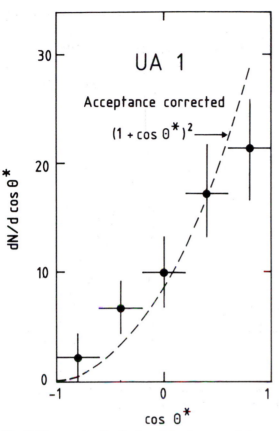

Fig. 21. The angular distribution of the electron emission angle θ* in the rest frame of the W after correction for experimental acceptance. Only those events have been used in which the electron charge is determined and the kinematic ambiguity (see text) has been resolved. The latter requirement has been corrected for in the acceptance calculation.

ment within $\Delta\phi=\pm30°$ (0.96±0.02); iii) the electron-track isolation requirement (0.90±0.07); and iv) the acceptance of events due to geometry (0.94±0.03). The cross-section was then

$$(\sigma\cdot B)_W = 0.53\pm0.08\ (\pm0.09)\ \text{nb},$$

where the last error takes into account systematic errors. This value is in excellent agreement with the expectations for the Standard Model [22]: $(\sigma\cdot B)_W=0.39$ nb.

No event with $p_T^{(e)}$ or $p_T^{(\nu)}$ in excess of the expected distribution for W→eν events was observed. This result can be used to set a limit to the possible existence of very massive W-like objects (W') decaying into electron–neutrino pairs. We found $(\sigma\cdot B)_{W'}\leqslant30$ pb at 90 % confidence level, corresponding to $m_{W'}>170$ GeV/c², when standard couplings and quark distributions were used to evaluate the cross-sections.

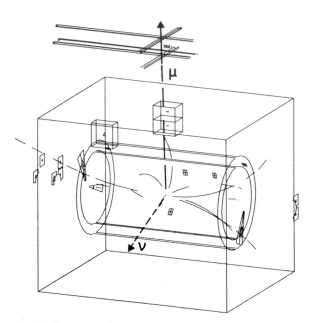

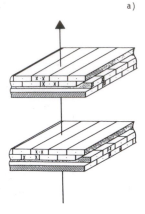

Fig. 22. Examples of decay modes of the W particle:
a) $W \rightarrow \mu + \nu_\mu$; b) $W \rightarrow \tau + \nu_\tau$; c) $W \rightarrow c + \bar{s}$; d) $W \rightarrow t + \bar{b}$ ($t \rightarrow b + e + \nu$). For the events of type (d), one can reconstruct the invariant masses of the W particle and of the decaying t-quark jet (Fig. 22 e).

8. *Universality of the W coupling*

The W field should exhibit a universal coupling strength for all the fundamental lepton doublets and all the quark doublets. This implies—apart from small phase-space corrections—equality of the branching ratios of the decay processes

$$W \rightarrow e \nu_e, \tag{4a}$$

$$W \rightarrow \mu \nu_\mu, \tag{4b}$$

$$W \rightarrow \tau \nu_\tau. \tag{4c}$$

Likewise, in the case of the quark decay channels

$$W \rightarrow u d_C, \tag{4d}$$

$$W \rightarrow c s_C, \tag{4e}$$

$$W \rightarrow t b_C, \tag{4f}$$

where t is the sixth quark (top quark) provided it exists within the kinematic range of reaction (4f). Neglecting phase-space corrections, which are probably

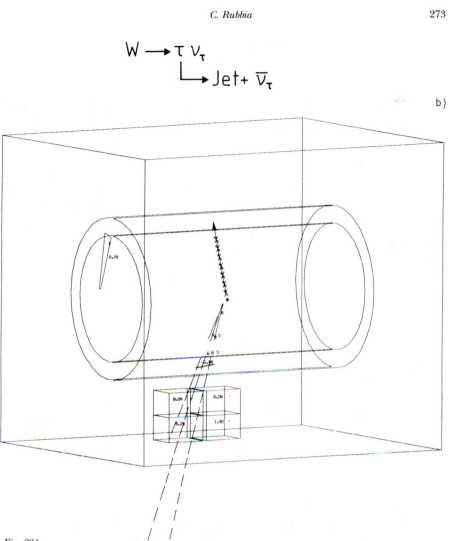

Fig. 22 b.

important for reaction (4 f), we expect equality of the branching ratios, with an overall factor of 3 of enhancement with respect to leptonic channels [(4 a) to (4 c)] due to colour counting. The subscript C in channels (4 d) to (4 f) indicates the presence of the Cabibbo mixing. Reactions (4 a) and (4 d) are implied by the results of Section 5. Reactions (4 b), (4 c), and (4 e) have been observed, and within about ±20 % they appear to have the correct branching ratios. Some events which are believed to be evidence for the process (4 f) have also been reported [24]. They are interpreted for the reaction

$$W \rightarrow t + \bar{b}_C (t \rightarrow b_C + l + \nu) \qquad (l \equiv \text{electron or muon}).$$

The $\bar{b}_C$ and b_C quarks are 'hadronized' into jets. Data are roughly consistent with $m_t \simeq 40$ GeV/c². Examples of reactions (4 b), (4 c), (4 e), and (4 f) are shown in Figs. 22 a–d, respectively.

Therefore, within the limited statistics there is evidence for universality.

c)

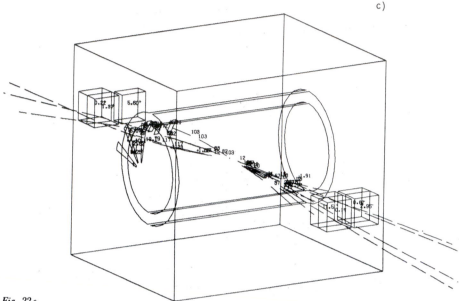

Fig. 22 c.

9. *Can we derive weak interactions from W-particle observations?*

A number of properties of weak interactions as determined by low-energy experiments can now be explained as a consequence of the experimentally observed properties of the W particles. Indeed we know that $W^\pm$ must couple to valence quarks at production and to $(e\nu)$ pairs at decay, which implies the existence of the beta-decay processes $n \rightarrow p + e^- + \nu_e$ and $(p) \rightarrow (n) + e^+ + \nu_e$. The mass value m_W and the cross-section measurement can then be used to calculate G_F, the Fermi coupling constant: $G_F = (1.2 \pm 0.1) \times 10^{-5}$ GeV^{-2}. Thus the W-pole saturates the observed weak interaction rate. The interaction must be vector since $J = 1$, and parity is maximally violated since $\langle \mu \rangle = \langle \lambda \rangle = \pm 1$. The only missing element is the separation between V+A and V−A alternatives. For this purpose a polarization measurement is needed. It may be accomplished in the near future by studying, for instance, the decay $W \rightarrow \tau + \nu_\tau$ and using the τ decay as the polarization analyzer or producing intermediate vector bosons (IVBs) with longitudinally polarized protons.

The universality of couplings and the decay modes of particles of different flavours into different lepton families can also be expected on the basis of the observations of the other decay modes of the W particles.

10. *Observation of the neutral boson Z^0*

We extended our search to the neutral partner Z^0, responsible for neutral currents. As in our previous work, production of IVBs was achieved with proton–antiproton collisions at $\sqrt{s} = 540$ GeV in the UA1 detector, except

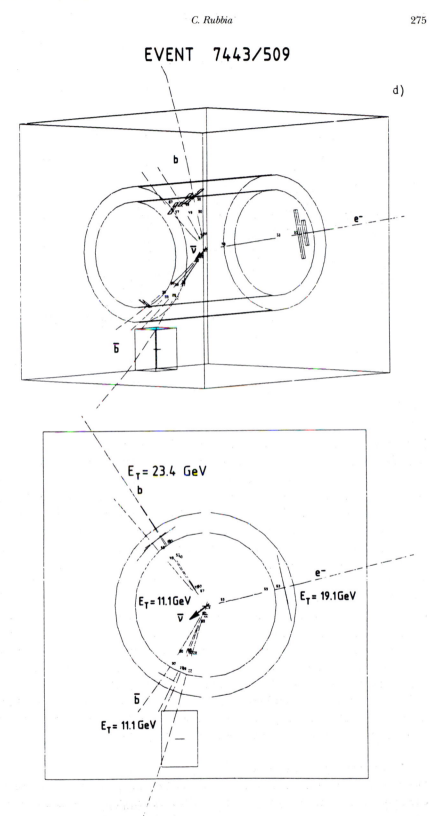

EVENT 7443/509

d)

Fig. 22 d.

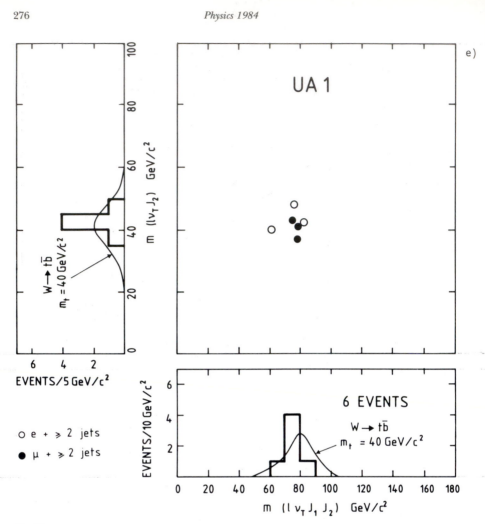

Fig. 22 e.

that we now searched for electron and muon pairs rather than for electron-—neutrino coincidence. The process is then

$$\bar{p} + p \rightarrow Z^0 + X, \quad Z^0 \rightarrow e^+ + e^- \text{ or } \mu^+ + \mu^-.$$

This reaction is approximately a factor of 10 less frequent than the corresponding $W^\pm$ leptonic decay channels. A few events of this type were therefore expected in our muon or electron samples. Evidence for the existence of the Z^0 in the range of masses accessible to the UA1 experiment has also been derived from weak-electromagnetic interference experiments at the highest PETRA energies, where deviations from point-like expectations have been reported (Fig. 23).

We first looked at events of the type $Z^0 \rightarrow e^+ e^-$ [25, 26]. As in the case of the $W^\pm$ search, an electron signature was defined as a localized energy deposition in two contiguous cells of the electromagnetic detectors with $E_T > 25$ GeV, and a small (or no) energy deposition ($\leqslant 800$ MeV) in the

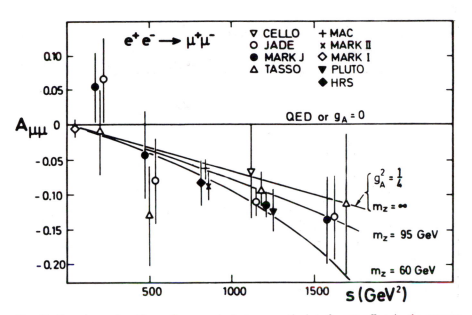

Fig. 23. Experimental evidence for a weak-electromagnetic interference effect in the process $e^+e^- \rightarrow \mu^+\mu^-$ at high-energy colliding beams. It can be seen that data are better fitted if the presence of a finite mass m_Z propagator is assumed.

hadron calorimeters immediately behind them. The isolation requirement was defined as the absence of charged tracks with momenta adding up to more than 3 GeV/c of transverse momentum and pointing towards the electron cluster cells. The effects of the successive cuts on the invariant electron–electron mass are shown in Fig. 24. Four e^+e^- events survived cuts, consistent with a common value of (e^+e^-) invariant mass. One of these events is shown in Figs. 25 and 26. As can be seen from the energy deposition plots (Fig. 27), the dominant feature of the four events is two very prominent electromagnetic energy depositions. All events appear to balance the visible total transverse energy components; namely, there is no evidence for the emission of energetic neutrinos. Except for the one track of event D which travels at less than 15° parallel to the magnetic field, all tracks are shown in Fig. 28, where the momenta measured in the central detector are compared with the energy deposition in the electromagnetic calorimeters. All tracks but one have consistent energy and momentum measurements. The negative track of event C shows a value of (9 ± 1) GeV/c, much smaller than the corresponding deposition of (49 ± 2) GeV. This event can be interpreted as the likely emission of a hard 'photon' accompanying the electron.

The same features are apparent also from the events in which a pair of muons [27] were emitted. A sharp peak (Fig. 29) is visible for high-mass dimuons. Within the statistical accuracy the events are incompatible with additional neutrino emission. They are all compatible with a common mass value:

$$\langle m_{\mu\mu} \rangle = 85.8^{+7.0}_{-5.4} \, \text{GeV}/c^2,$$

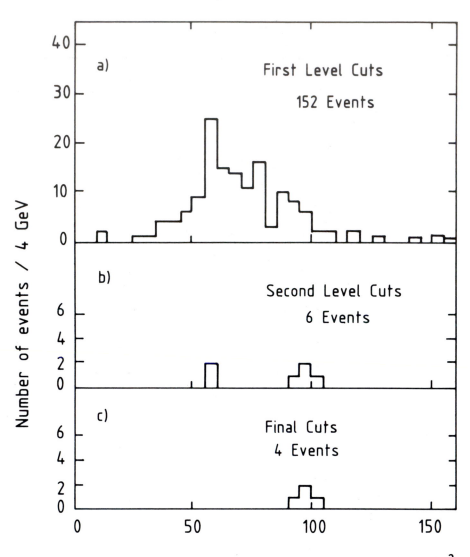

Uncorrected invariant mass cluster pair (GeV/c^2)

Fig. 24. Invariant mass distribution (uncorrected) of two electromagnetic clusters: a) with $E_T > 25$ GeV; b) as above, and a track with $p_T > 7$ GeV/c and projection length of more than 1 cm pointing to the cluster. In addition, a small energy deposition in the hadron calorimeters immediately behind (<0.8 GeV) ensures the electron signature. Isolation is required with $\Sigma\, p_T < 3$ GeV/c for all other tracks pointing to the cluster. c) The second cluster also has an isolated track.

consistent with the value measured for $Z^0 \rightarrow e^+e^-$:

$$\langle m_{ee} \rangle = 95.6 \pm 1.4\ (2.9)\ \text{GeV/c}^2,$$

where the first error accounts for the statistical error and the second for the uncertainty of the overall energy scale of the calorimeters. The average value for the nine Z^0 events found in the UA1 experiment is $m_{z^0} = 93.9 \pm 2.9$ GeV/c^2, where the error includes systematic uncertainties.

$$Z^0 \longrightarrow e^+ e^-$$

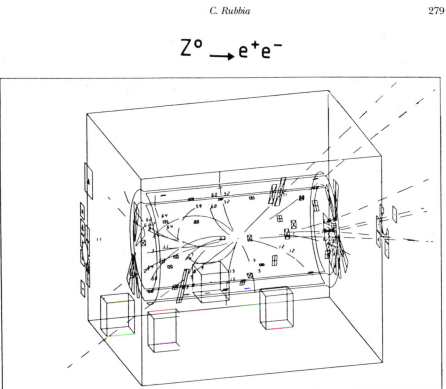

Fig. 25. Event display. All reconstructed vertex-associated tracks and all calorimeter hits are displayed.

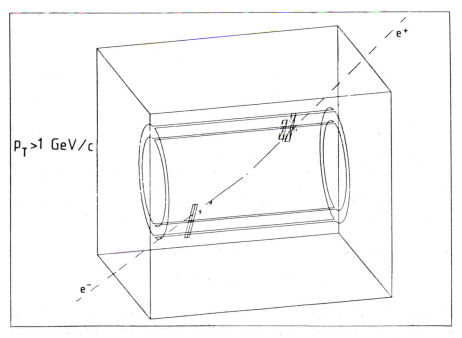

Fig. 26. The same as Fig. 25, but thresholds are raised to $p_T > 2$ GeV/c for charged tracks and $E_T > 2$ GeV for calorimeter hits. We remark that only the electron pair survives these mild cuts.

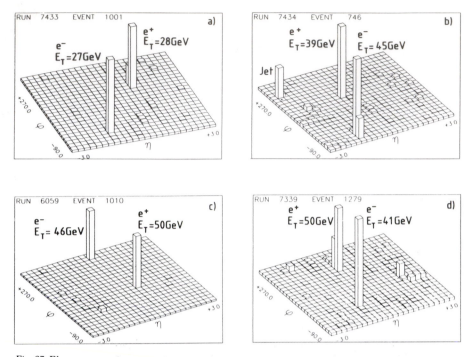

Fig. 27. Electromagnetic energy depositions at angles >5° with respect to the beam direction for the four electron pairs.

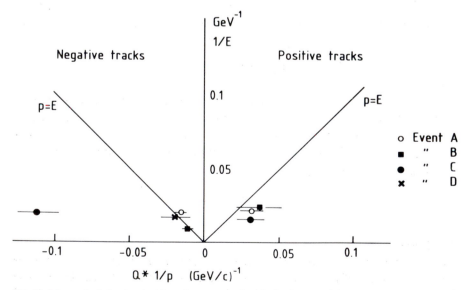

Fig. 28. Magnetic deflection in 1/p units compared with the inverse of the energy deposited in the electromagnetic calorimeters. Ideally, all electrons should lie on the 1/E=1/p line.

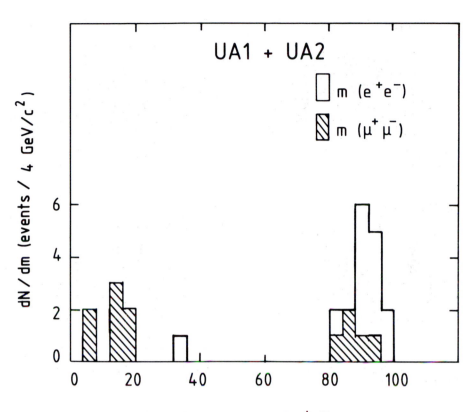

Fig. 29. Invariant mass distribution of dilepton events from UA1 and UA2 experiments. A clear peak is visible at a mass of about 95 GeV/c².

The integrated luminosity for the present data sample is 108 nb^{-1}, with an estimated uncertainty of 15%. With the geometrical acceptance of 0.37, the cross-section, calculated using the four events, is

$$(\sigma \cdot B)_{\mu\mu} = 100 \pm 50 (\pm 15) \, \text{pb},$$

where the last error includes the systematics from the acceptance and from the luminosity. This value is in good agreement both with Standard Model predictions [22] and with our results for $Z^0 \rightarrow e^+e^-$, namely $(\sigma \cdot B)_{ee} = 41 \pm 21 (\pm 7)$ pb. From the electron and the muon channels we obtain the average cross-section of

$$(\sigma \cdot B)_{\ell\ell} = 58 \pm 21 (\pm 9) \, \text{pb}.$$

11. *Comparing theory with experiment*

The experiments discussed in the previous section have shown that the W particle has most of the properties required in order to be the carrier of weak interactions. The presence of a narrow dilepton peak has been seen around 95 GeV/c². Rates and features of the events are consistent with the hypothesis

Table 2. W$^\pm$ and Z^0 parameters from the UA1 and UA2 experiments

	UA1	UA2
N(W$\rightarrow$ eν)	52[a]	37[b]
$\quad$m$_W$ (GeV/c^2)	80.9$\pm$1.5$\pm$2.4	83.1$\pm$1.9$\pm$1.3
$\quad\Gamma_W$ (90 % CL)	$\leqslant$7 GeV	–
$\quad$(σB) (nb)	0.53$\pm$0.08$\pm$0.09	0.53$\pm$0.10$\pm$0.10
N(W$\rightarrow$ $\mu\nu$)	14	
$\quad$m$_W$ (GeV/c^2)	81.0$^{+6}_{-7}$	–
$\quad$(σB) (nb)	0.67$\pm$0.17$\pm$0.15	–
N(Z$^0\rightarrow$ e$^+$e$^-$)	3+1[c]	7+1[c]
$\quad$m$_{Z^0}$ (GeV/c^2)	95.6$\pm$1.4$\pm$2.9	92.7$\pm$1.7$\pm$1.4
$\quad\Gamma_{Z^0}$ (90 % CL)	$\leqslant$8.5 GeV	$\leqslant$6.5 GeV
$\quad$(σB) (nb)	0.05$\pm$0.02$\pm$0.009	0.11$\pm$0.04$\pm$0.02
N(Z$^0\rightarrow$ $\mu^+\mu^-$)	4+1[c]	
$\quad$m$_{Z^0}$ (GeV/c^2)	85.6$\pm$6.3	–
$\quad$(σB) (nb)	0.105$\pm$0.05$\pm$0.15	–
sin$^2\theta_w$=38.5/m$_W$	0.226$\pm$0.015	0.216$\pm$0.010$\pm$0.007
ϱ=[m$_w$/m$_z$ cos θ_w]^2	0.968$\pm$0.045	1.02$\pm$0.06

[a] $\quad$p$_T^\nu$>15 GeV/c
[b] $\quad$p$_T^c$>25 GeV/c
[c] $\quad$Z$^0\rightarrow\ell^+\ell^-\gamma$ (E$_\gamma$>20 GeV)

that the neutral partner of the W$^\pm$ has indeed been observed. At present the statistics are not sufficient to test the form of the interaction experimentally; neither has parity violation been detected. However, the precise values of the masses of Z^0 and W$^\pm$ now available constitute a critical test of the idea of unification between weak and electromagnetic forces, and in particular of the predictions of the SU(2)$\times$U(1) theory of Glashow, Weinberg and Salam [6]. A careful account of systematic errors is needed in order to evaluate an average between the mass determination for the two collider experiments, UA1 and UA2 [28]. Table 2 summarizes all experimental information related to W$^\pm$ and Z^0.

The charged vector boson mass is

$$m_{W^\pm} = (80.9\pm1.5)\ \text{GeV/c}^2 \quad \text{(statistical errors only)},$$

to which a 3 % energy scale uncertainty must be added. In this report a value for the Z^0 mass, m$_{z^0}$=(95.1$\pm$2.5) GeV/c^2, has been given. Neglecting systematic errors, a mass value is found with somewhat smaller errors:

$$m_{z^0} = (95.6\pm1.4)\ \text{GeV/c}^2 \quad \text{(statistical errors only)},$$

to which the same scale uncertainty as that for the W$^\pm$ applies. The quoted errors include: i) the neutral width of the Z^0 peak, which is found to be Γ<8.5 GeV/c^2 (90 % confidence level); ii) the experimental resolution of counters; and iii) the r.m.s. spread between calibration constants of individual elements.

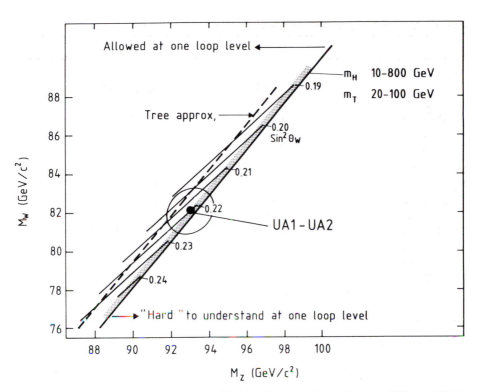

Fig. 30. Comparison between the Standard Model and the experimental results (UA1 and UA2 combined). Theory is from Ref. [29].

It should be remarked that the masses of the IVBs have the following prediction:

$$m_W = \left[\pi\alpha/\sqrt{2}\, G_F \sin^2\theta_W (1-\Delta r)\right]^{1/2},$$

$$m_Z = m_W/\cos\theta_W,$$

where the value Δr represents the effect of the higher-order radiative corrections, and the second equation can be used as a definition of the Weinberg angle θ_W. Since G_F and α are known, θ_W can be eliminated between equations:

$$m_Z = m_W/(1-A^2/m_W^2)^{1/2},$$

$$\Delta r = A^2 m_Z^2/\left[m_W^2(m_W+m_Z)(m_Z-m_W)\right],$$

$$A = (37.2810\pm0.0003)\,\text{GeV}.$$

Radiative corrections are quite large [29] and detectable at the present level of accuracy. Calculations of order $O(\alpha)+O(\alpha^2\ln m)$ give the following result:

$$\Delta r = 0.0696\pm0.0020,$$

which is insensitive to the parameters

$$\sin^2 \theta_W = 0.217,$$

$$m_t = 40 \text{ GeV}/c^2, \quad m_b = 5 \text{ GeV}/c^2.$$

The main effect can be understood as α being a running coupling constant, namely:

$$\alpha = 1/137.035962, \quad \text{at } q^2 = 0,$$
$$\alpha = 1/137.5, \quad \quad \text{at } q^2 = m_W^2.$$

In Fig. 30 we have plotted m_Z against m_W. The elliptical shape of the errors reflects the uncertainty in the energy scale. It can be seen that there is excellent agreement with the expectations of the SU(2)×U(1) Standard Model [29].

We can then extract the renormalized value of $\sin^2 \theta_W$ at mass scale m_W. Inserting the value of m_W one finds

$$\sin^2 \theta_W = 0.220 \pm 0.009,$$

In excellent agreement with the renormalized value of $\sin^2 \theta_W = 0.215 \pm 0.014$ deduced from neutral-current experiments. Using the information of the Z^0 mass, one can determine the parameter ϱ, related immediately to the isospin of the Higgs particle:

$$\varrho = m_W^2 / m_{Z^0}^2 \cos^2 \theta_W.$$

Using the experimental values, one finds

$$\varrho = 1.000 \pm 0.036,$$

in perfect agreement with the prediction of $\varrho = 1$ for a Higgs doublet. Let us point out that ϱ deviates from 1 at most by 3 %, owing to radiative corrections involving possible new fermion generations. The present value seems to indicate no such new fermion families.

We conclude that, within errors, the observed experimental values are completely compatible with the SU(2)×U(1) model, thus supporting the hypothesis of a unified electroweak interaction.

ACKNOWLEDGEMENTS

This lecture is based on the work of the UA1 Collaboration team, and I would like to express my appreciation of their remarkable achievements which have led to so many exciting results. At present the following persons are members of the collaboration: G. Arnison, A. Astbury, B. Aubert, C. Bacci, A. Bezaguet, R. K. Bock, T. J. V. Bowcock, M. Calvetti, P. Catz, P. Cennini, S. Centro, F. Ceradini, S. Cittolin, D. Cline, C. Cochet, J. Colas, M. Corden, D. Dallman,

D. Dau, M. DeBeer, M. Della Negra, M. Demoulin, D. Denegri, A. Diciaccio, D. DiBitonto, L. Dobrzynski, J. D. Dowell, K. Eggert, E. Eisenhandler, N. Ellis, P. Erhard, H. Faissner, M. Fincke, G. Fontaine, R. Frey, R. Frühwirth, J. Garvey, S. Geer, C. Ghesquiere, P. Ghez, K. L. Giboni, W. R. Gibson, Y. Giraud-Heraud, A. Givernaud, A. Gonidec, G. Grayer, T. Hansl-Kozanecka, W. J. Haynes, L. O. Hertzberger, C. Hodges, D. Hoffmann, H. Hoffmann, D. J. Holthuizen, R. J. Homer, A. Honma, W. Jank, G. Jorat, P. I. P. Kalmus, V. Karimaki, R. Keeler, I. Kenyon, A. Kernan, R. Kinnunen, W. Kozanecki, D. Kryn, F. Lacava, J. P. Laugier, J. P. Lees, H. Lehmann, R. Leuchs, A. Leveque, D. Linglin, E. Locci, J. J. Malosse, T. Markiewicz, G. Maurin, T. McMahon, J. P. Mendiburu, M. N. Minard, M. Mohammadi, M. Moricca, K. Morgan, H. Muirhead, F. Muller, A. K. Nandi, L. Naumann, A. Norton, A. Orkin-Lecourtois, L. Paoluzi, F. Pauss, G. Piano Mortari, E. Pietarinen, M. Pimiä, J. P. Porte, E. Radermacher, J. Ransdell, H. Reithler, J. P. Revol, J. Rich, M. Rijssenbeek, C. Roberts, J. Rohlf, P. Rossi, C. Rubbia, B. Sadoulet, G. Sajot, G. Salvi, G. Salvini, J. Sass, J. Saudraix, A. Savoy-Navarro, D. Schinzel, W. Scott, T. P. Shah, D. Smith, M. Spiro, J. Strauss, J. Streets, K. Sumorok, F. Szoncso, C. Tao, G. Thompson, J. Timmer, E. Tscheslog, J. Tuominiemi, B. Van Eijk, J. P. Vialle, J. Vrana, V. Vuillemin, H. D. Wahl, P. Watkins, J. Wilson, R. Wilson, C. E. Wulz, Y. G. Xie, M. Yvert and E. Zurfluh.

REFERENCES

1. E. Fermi, Ric. Sci. *4* (2), 491 (1933), reprinted *in* E. Fermi, Collected Papers (eds. E. Segré et al.) (University of Chicago Press, Chicago, Ill., 1962), Vol. 1, p. 538; Z. Phys. *88*, 161 (1934); English translation: F. L. Wilson, Am. J. Phys. *36* 1150 (1968).
2. M. Gell-Mann and M. Levy, Nuovo Cimento *16*, 705 (1960).
3. N. Cabibbo, Phys. Rev. Lett. *10*, 531 (1963).
   M. Kobayashi and K. Maskawa, Progr. Theor. Phys. *49*, 652 (1973).
4. S. S. Gerschtein and J. R. Zel'dovich, Z. Eksp. Teor. Fiz. *29*, 698 (1955).
   R. P. Feynman and M. Gell-Mann, Phys. Rev. *109*, 193 (1958).
   E. C. G. Sudarshan and R. E. Marshak, Phys. Rev. *109*, 1860 (1958).
   J. J. Sakurai, Nuovo Cimento *7*, 649 (1958).
5. O. Klein, *in* Proc. Symp. on Les Nouvelles Théories de la Physique, Warsaw, 1938 (Institut International de Coopération Intellectuelle, Paris, 1939), p. 6.
   O. Klein, Nature *161*, 897 (1948).
6. S. L. Glashow, Nucl. Phys. *22*, 579 (1961).
   S. Weinberg, Phys. Rev. Lett. *19*, 1264 (1967).
   A. Salam, Proc. 8th Nobel Symposium (ed. N. Svartholm) (Almqvist and Wiksell, Stockholm, 1968), p. 367.
7. T. D. Lee and C. N. Yang, Phys. Rev. *119*, 1410 (1960).
8. The first realistic scheme for colliding beams was discussed by D. W. Kerst et al., Phys. Rev. *102*, 590 (1956). The first suggestion for proton–antiproton intersecting beams was given by G. I. Budker, Proc. Int. Symp. on Electron and Positron Storage Rings, Saclay, 1966 (eds. H. Zyngier and E. Crémieu-Alcan) (PUF, Paris, 1966), p. II-1-1. G. I. Budker, At. Energ. *22*, 346 (1967).
9. C. Rubbia, P. McIntyre and D. Cline, Proc. Int. Neutrino Conf., Aachen, 1976 (eds. H. Faissner, H. Reithler and P. Zerwas) (Vieweg, Braunschweig, 1977), p. 683.
10. The 300 GeV programme, CERN/1050 (14 January 1972).

11. S. van der Meer, Internal Report CERN ISR–PO/72–31 (1972).

D. Möhl, G. Petrucci, L. Thorndahl and S. van der Meer, Phys. Rep. *58*, 73 (1980).

12. Design study of intersecting storage rings (ISR) for the CERN Proton Synchrotron, CERN AR/Int. SG/64–9 (1964).

13. A. Astbury et al., A 4π solid-angle detector for the SPS used as a proton–antiproton collider at a centre-of-mass energy of 540 GeV, Proposal, CERN–SPSC 78–6/P92 (1978).

14. The UA1 Collaboration is preparing a large and comprehensive report on the detector (1984). For details see:

M. Barranco Luque et al., Nucl. Instrum. Methods *176*, 175 (1980).

M. Calvetti et al., Nucl. Instrum. Methods *176*, 255 (1980).

M. Calvetti et al., Proc. Int. Conf. on Instrumentation for Colliding Beam Physics, Stanford, 1982 (SLAC–250, Stanford, 1982), p. 16.

M. Calvetti et al., IEEE Trans. Nucl. Sci. *NS–30*, 71 (1983).

J. Timmer, The UA1 detector, Antiproton proton physics and the W discovery, presented at the 3rd Moriond Workshop, La Plagne, France, March 1983 (Ed. Tran Thanh Van) (Editions Frontières, Gif-sur-Yvette, France), p. 593.

E. Locci, Thèse de doctorat ès sciences, Paris, 1984, unpublished.

M. J. Corden et al., Phys. Scr. *25*, 5 and 11 (1982).

M. J. Corden et al., Rutherford preprint RL–83–116 (1983).

K. Eggert et al., Nucl. Instrum. Methods *176*, 217 (1980).

J. C. Santiard, CERN EP Internal Report 80–04 (1980).

K. Eggert et al., Nucl. Instrum. Methods *176*, 223 (1980) and *188*, 463 (1981).

G. Arnison et al., Phys. Lett. *121B* 77 (1983).

G. Arnison et al., Phys. Lett. *128B* 336 (1983).

The UA2 Collaboration have described their detector, which however makes a more limited use of the missing-energy concept, *in* Proc. 2nd Int. Conf. on Physics in Collisions, Stockholm, 1982 (Plenum Press, New York, 1983), p. 67.

A. G. Clark, Proc. Int. Conf. on Instrumentation for Colliding Beam Physics, Stanford, 1982 (SLAC–250, Stanford, 1982), p. 169.

B. Mansoulié, The UA2 apparatus at the CERN p̄p Collider, Antiproton proton physics and the W discovery, presented at the 3rd Moriond Workshop, La Plagne, France, March 1983 (Ed. Tran Thanh Van) (Editions Frontières, Gif-sur-Yvette, France), p. 609.

15. G. Arnison et al., Phys. Lett. *122B*, 103 (1983).

16. M. Banner et al., Phys. Lett. *122B*, 476 (1983).

17. G. Arnison et al., Phys. Lett. *129B*, 273 (1983).

C. Rubbia, Proc. Int. Europhysics Conf. on High-Energy Physics, Brighton, 1983 (eds. J. Guy and C. Costain) (Rutherford Appleton Lab., Didcot, 1983), p. 860.

18. G. Arnison et al., Phys. Lett. *139B*, 115 (1984).

19. C. Rubbia, Physics results of the UA1 Collaboration at the CERN proton–antiproton collider, Proc. Int. Conf. on Neutrino Physics and Astrophysics, Dortmund, 1984 (ed. K. Kleinknecht), p. 1.

20. J. F. Owens and E. Reya, Phys. Rev. *D17*, 3003 (1978).

F. Halzen and W. Scott, Phys. Lett. *78B*, 318 (1978).

F. Halzen, A. D. Martin and D. M. Scott, Phys. Rev. *D25*, 754 (1982).

P. Aurenche and R. Kinnunen, Annecy preprint LAPP–TH–78 (1983).

A. Nakamura, G. Pancheri and Y. Srivastava, Frascati preprint LNF–83–44 (1983).

21. G. Altarelli, R. K. Ellis, M. Greco and G. Martinelli, Nucl. Phys. *B246*, 12 (1984).

G. Altarelli, G. Parisi and R. Petronzio, Phys. Lett. *76B*, 351 and 356 (1978).

22. F. E. Paige, *in* Proc. Topical Workshop on the Production of New Particles in Super High Energy Collisions, Madison, 1979 (eds. V. Barger and F. Halzen) (AIP, New York, 1979).

F. E. Paige and S. D. Protopopescu, ISAJET program, BNL 29777 (1981).

G. Altarelli, R. K. Ellis and G. Martinelli, Nucl. Phys. *B143*, 521 (1978); (E) *B196*, 544 (1978); *B157*, 461 (1979).

J. Kubar-Andre and F. E. Paige, Phys. Rev. *D19*, 221 (1979).

23. M. Jacob, Nuovo Cimento *9*, 826 (1958).

   M. Jacob, unpublished (see, for instance, in C. Rubbia, Proc. 9th Topical Conf. on Particle Physics, Honolulu, 1983).

24. G. Arnison et al., Phys. Lett. *147B*, 493 (1984).

25. G. Arnison et al., Phys. Lett. *126B*, 398 (1983).

26. P. Bagnaia et al., Phys. Lett. *129B*, 150 (1983).

27. G. Arnison et al., Phys. Lett. *147B*, 241 (1984).

28. C. Rubbia, Proc. Int. Europhysics Conf. on High-Energy Physics, Brighton, 1983 (eds. J. Guy and C. Costain) (Rutherford Appleton Lab., Didcot, 1983), p. 860.

29. M. Consoli, Proc. Third Topical Workshop on pp̄ Collider Physics, Rome, 1983 (eds. C. Bacci and G. Salvini) (CERN 83–04, Geneva, 1983), p. 478.

   M. Consoli, S. L. Presti and L. Maiani, Univ. Catania preprint PP/738–14/1/1983.

   M. Veltman, Proc. Int. Europhysics Conf. on High-Energy Physics, Brighton, 1983 (eds. J. Guy and C. Costain) (Rutherford Appleton Lab., Didcot, 1983), p. 880.

   W. Marciano, BNL preprint 33819 (1983).

Simon van der Meer

SIMON VAN DER MEER

I was born in 1925, in The Hague, the Netherlands, as the third child of Pieter van der Meer and Jetske Groeneveld, both of Frisian origin. I had three sisters.

My father was a schoolteacher and my mother came from a teacher's family. Under these conditions it is not astonishing that learning was highly prized; in fact, my parents made sacrifices to be able to give their children a good education.

I visited the Gymnasium in The Hague and passed my final examination (in the sciences section) in 1943. Because the Dutch universities had just been closed at that time under the German occupation, I spent the next two years attending the humanities section of the Gymnasium. Meanwhile, my interest in physics and technology had been growing; I dabbled in electronics, equipped the parental home with various gadgets and assisted my brilliant and inspiring physics teacher (U. Ph. Lely) with the preparation of numerous demonstrations.

From 1945 onwards, I studied "Technical Physics" at the University of Technology, Delft, where I specialized in measurement and regulation technology under C. J. D. M. Verhagen. The physics taught in this newly created subsection of an old and established engineering school, although of excellent quality, was of necessity somewhat restricted and I have often felt regrets at not having had the intensive physics training that many of my colleagues enjoyed. Nevertheless, if I have at times been able to make original contributions in the accelerator field, I cannot help feeling that to a certain extent my slightly amateur approach in physics, combined with much practical experience, was an asset.

After obtaining my engineering degree in 1952, I worked in the Philips Research Laboratory, Eindhoven, mainly on high-voltage equipment and electronics for electron microscopes. In 1956 I moved to Geneva to join the recently founded European Organization for Nuclear Research (CERN), where I have been working ever since on many different projects, in an agreeable and stimulating international atmosphere.

To start with, my work (under the leadership of J. B. Adams and C. A. Ramm) was concerned mainly with technical design: poleface windings, multipole correction lenses for the 28 GeV synchrotron and their power supplies. My interest in matters more directly concerned with the handling of particles was growing, in the meantime, stimulated by many contacts with people understanding accelerators. After working for a year on a separated antiproton beam (1960), I proposed a high-current, pulsed focusing device ("horn") aimed at increasing the intensity of a beam of neutrinos, then at the centre of interest at

CERN and elsewhere. The design of this monster, together with the associated neutrino flux calculations kept me busy until 1965, when I joined a small group, led by F. J. M. Farley, preparing the second "g-2" experiment for measuring the anomalous magnetic moment of the muon. I designed the small storage ring used and participated at all stages of the experiment proper, including part of the data treatment. This was an invaluable experience; not only did I learn the principles of accelerator design, but I also got acquainted with the lifestyle and way of thinking of experimental high-energy physicists.

From 1967 to 1976 I returned to more technical work when I was responsible for the magnet power supplies, first of the Intersecting Storage Rings (ISR) and then of the 400 GeV synchrotron (SPS). I kept up with accelerator ideas, however, and worked (during my ISR period) on a method for the luminosity calibration of storage rings and on stochastic cooling. The latter was, of course, aimed at increasing the ISR luminosity, but practical application seemed difficult at the time, mainly because the high beam intensity in the ISR would have made the cooling very slow. After developing a primitive theory (1968) I therefore did not pursue this subject. However, the work was taken up by others and in 1974 the first experiments were done in the ISR.

In 1976, Cline, McIntyre, Mills, and Rubbia proposed to use the SPS or the Fermilab ring as a pp̄ collider. Accumulation of the needed antiprotons would clearly require cooling. At this time, my work on the SPS power supplies had just come to an end; I joined a study group on the pp̄ project and an experimental team studying cooling in a small ring (ICE). The successful experiments in this ring and the work by Sacherer on theory and by Thorndahl on filter cooling showed that p̄ accumulation by stochastic stacking was feasible. The collider project was approved and I became joint project leader with R. Billinge for the accumulator construction. Since then, I have worked with the group that commissioned and improved the ring and that is now preparing the construction of a second ring to increase the p̄ stacking rate by an order of magnitude. As a spin-off from this work, I proposed the stochastic extraction method that is now used (in a much improved form) in the Low-Energy Antiproton Ring (LEAR).

In the meantime, in 1966, while skiing with friends in the Swiss mountains, I met my wife-to-be Catharina M. Koopman and after a very brief interval we decided to marry. This was certainly one of the best decisions I ever made; my life has since been far more interesting and colourful. We have two children: Esther (1968) and Mathijs (1970).

(*added in 1991*): In 1990 I retired from CERN.

Honours
Loeb Lecturer, Harvard University, 1981.
Duddell Metal, Institute of Physics, 1982.
Honorary Degree, Geneva University, 1983.
Honorary Degree, Amsterdam University, 1984.
Foreign Honorary Member, American Academy of Arts and Sciences, 1984.
Correspondent, Royal Netherlands Academy of Sciences, 1984.

STOCHASTIC COOLING AND THE ACCUMULATION OF ANTIPROTONS

Nobel lecture, 8 December, 1984

by

SIMON VAN DER MEER
CERN, CH-1211 Geneva 23, Switzerland

1. *A general outline of the pp̄ project*

The large project mentioned in the motivation of this year's Nobel award in physics includes in addition to the experiments proper described by C. Rubbia, the complex machinery for colliding high-energy protons and antiprotons (Fig. 1). Protons are accelerated to 26 GeV/c in the PS machine and are used to produce p̄'s in a copper target. An accumulator ring (AA) accepts a batch of these with momenta around 3.5 GeV/c every 2.4 s. After typically a day of accumulation, a large number of the accumulated p̄'s ($\sim 10^{11}$) are extracted from the AA, reinjected into the PS, accelerated to 26 Gev/c and transferred to the large (2.2 km diameter) SPS ring. Just before, 26 Gev/c protons, also from the PS, have been injected in the opposite direction. Protons and antiprotons are then accelerated to high energy (270 or 310 Gev) and remain stored for many hours. They are bunched (in 3 bunches of about 4 ns duration each) so that collisions take place in six well-defined points around the SPS ring, in two of which experiments are located. The process is of a complexity that could only be mastered by the effort and devotion of several hundreds of people. Only a small part of it can be covered in this lecture, and I have chosen to speak about stochastic cooling, a method that is used to accumulate the antiprotons, and with which I have been closely associated.

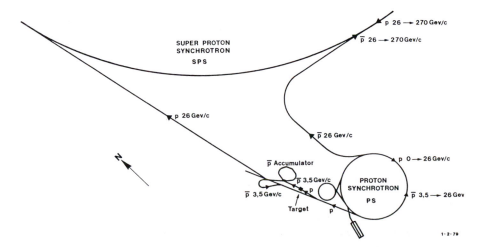

Fig. 1. Overall layout of the pp̄ project.

2. *Cooling, why and how?*

A central notion in accelerator physics is phase space, well-known from other areas of physics. An accelerator or storage ring has an acceptance that is defined in terms of phase volume. The antiproton accumulator must catch many antiprotons coming from the target and therefore has a large acceptance; much larger than the SPS ring where the p̄'s are finally stored. The phase volume must therefore be reduced and the particle density in phase space increased. On top of this, a large density increase is needed because of the requirement to accumulate many p̄ batches. In fact, the density in 6-dimensional phase space is boosted by a factor 10^9 in the AA machine.

This seems to violate Liouville's theorem that forbids any compression of phase volume by conservative forces such as the electromagnetic fields that are used by accelerator builders. In fact, all that can be done in treating particle beams is to distort the phase volume without changing the density anywhere.

Fortunately, there is a trick — and it consists of using the fact that particles are points in phase space with empty space in between. We may push each particle towards the centre of the distribution, squeezing the empty space outwards. The small-scale density is strictly conserved, but in a macroscopic sense the particle density increases. This process is called cooling because it reduces the movements of the particles with respect to each other.

Of course, we can only do this if we have information about the individual particle's position in phase space and if we can direct the pushing action against the individual particles. Without these two prerequisites, there would be no reason why particles rather than empty space would be pushed inwards. A stochastic cooling system therefore consists of a sensor (pick-up) that acquires electrical signals from the particles, and a so-called kicker that pushes the particles and that is excited by the amplified pick-up signals.

Such a system resembles Maxwell's demon, which is supposed to reduce the entropy of a gas by going through a very similar routine, violating the second law of thermodynamics in the process. It has been shown by Szilard[1] that the measurement performed by the demon implies an entropy increase that compensates any reduction of entropy in the gas. Moreover, in practical stochastic cooling systems, the kicker action is far from reversible; such systems are therefore even less devilish than the demon itself.

3. *Qualitative description of betatron cooling*

The cooling of a single particle circulating in a ring is particularly simple. Fig. 2 shows how it is done in the horizontal plane. (Horizontal, vertical and longitudinal cooling are usually decoupled.)

Under the influence of the focusing fields the particle executes betatron oscillations around its central orbit. At each passage of the particle a so-called differential pick-up provides a short pulse signal that is proportional to the distance of the particle from the central orbit. This is amplified and applied to the kicker, which will deflect the particle. If the distance between pick-up and kicker contains an odd number of quarter betatron wavelengths and if the gain is chosen correctly, any oscillation will be cancelled. The signal should arrive at

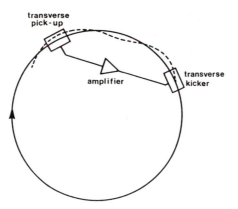

Fig. 2. Cooling of the horizontal betatron oscillation of a single particle.

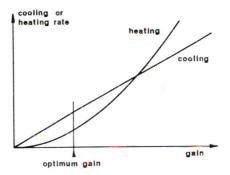

Fig. 3. Variation with system gain of the coherent cooling and incoherent heating effect,

the kicker at the same time as the particle; because of delays in the cabling and amplifiers, the signal path must cut off a bend in the particle's trajectory.

In practice, there will not just be one particle, but a very large number (e.g. 10^6 or 10^{12}). It is clear that even with the fastest electronics their signals will overlap. Nevertheless, each particle's individual signal will still be there and take care of the cooling. However, we must now reduce the gain of the system because all the other particles whose signals overlap within one system response time will have a perturbing (heating) effect, as they will in general have a random phase with respect to each other. Fortunately, the perturbing effect is on average zero and it is only its second-order term that heats (i.e. increases the mean square of the amplitude). This is proportional to the square of the gain, whereas the cooling effect—each particle acting on itself—varies linearly with gain. As illustrated in Fig. 3, we may always choose the gain so that the cooling effect predominates.

4. *Simplified analysis of transverse cooling*

We shall now analyse the process sketched above in a somewhat approximative way, neglecting several effects that will be outlined later. The purpose is to get some feeling about the possibilities without obscuring the picture by too much detail.

In the first place, we shall assume a system with constant gain over a bandwidth W and zero gain outside this band. A signal passed by such a system may be described completely in terms of 2W samples per unit time. If we have N particles in the ring and their revolution time is T, each sample will on average contain

$$N_s = N/2WT \qquad (1)$$

particles. We may now consider the system from two viewpoints:

a) we may look at each individual particle and combine the cooling by its own signal with the heating by the other particles,

b) we may look at the samples as defined above and treat each sample as the single particle of Fig. 1; this is justified because the samples are just resolved by the system.

The two descriptions are equivalent and yield the same result. For the moment, we shall adopt b). Incidentally, the name "stochastic cooling" originated[2] because from this viewpoint we treat a stochastic signal from random samples. However, viewpoint a) is more fundamental; cooling is not a stochastic process.

The pick-up detects the average position of each sample $\bar{x}$ and the gain will be adjusted so that this is reduced to zero, so that for each particle x is changed into $x - \bar{x}$. Averaging over many random samples, we see that the mean square $\overline{x^2}$ is changed into

$$\overline{(x-\bar{x})^2} = \overline{x^2} - \overline{\bar{x}^2}.$$

Therefore, the decrement of $\overline{x^2}$ per turn is $\overline{\bar{x}^2}/\overline{x^2} = 1/N_s$ and the cooling rate (expressed as the inverse of cooling time) is $1/\tau = 1/N_s T$. In fact, we have to divide this by four. One factor 2 occurs because the betatron oscillation is not always maximum at the pick-up as shown in Fig. 2. Both at the pick-up and at the kicker we therefore lose by a factor equal to the sine of a random phase angle; the average of $\sin^2$ is $1/2$. Another factor 2 is needed because it is usual to define cooling rate in terms of amplitude rather than its square. So we have, using (1)

$$\frac{1}{\tau} = \frac{1}{4N_s T} = \frac{W}{2N} . \qquad (2)$$

This result, although approximative, shows that stochastic cooling is not a practical technique for proton accelerators; for a typical accelerator $N \approx 10^{13}$, so that even with a bandwidth of several GHz the cooling would be much too slow compared to the repetition rate. In storage rings, however, the available time is longer and sometimes the intensity is lower, so that the technique may become useful.

5. *Mixing and thermal noise*

In deriving the cooling rate, we assumed that all samples have a random population, without correlation between successive turns. The main reason why the sample populations change is the spread in energy between the particles, which results in a revolution frequency spread. The particles overtake each other, and if the spread of revolution time is large compared to the sample duration, we speak of "good mixing"; in this case the derivation above is valid. In practice, it is rarely possible to achieve this ideal situation. In particular with strongly relativistic particles a large spread of revolution frequency can only be obtained by a large spread in orbit diameter; for a given aperture this reduces the momentum spread that is accepted by the machine.

We may see how bad mixing influences the cooling by replacing the correction $\bar{x}$ in the derivation of the cooling rate by a smaller amount $g\bar{x}$. As a result we find in the same way

$$\frac{1}{\tau} = \frac{W}{2N} \, (2g - g^2).$$

$$(3)$$

Clearly, this is largest for $g = 1$.

It can be shown that the two terms correspond to the coherent, cooling effect (each particle cooled by its own signal) and the incoherent, heating effect from the other particles[3]. It is the second one that increases by bad mixing, because of the correlation between samples at successive turns. It may also increase if thermal noise is added to the signal (usually originating in the low-level amplifier attached to the pick-up). Thus, we may define a mixing factor M ($= 1$ for perfect mixing) and a thermal noise factor U (equal to the noise/ signal power ratio) and obtain

$$\frac{1}{\tau} = \frac{W}{2N} \, (2g - g^2(M+U)).$$

By optimizing g (now < 1) we find

$$\frac{1}{\tau} = \frac{W}{2N(M+U)} \, .$$

$$(4)$$

6. *Frequency domain analysis*

This qualitative analysis may be made much more precise by considering the process from the frequency (instead of time) domain standpoint[4-5].

Each particle produces in the pick-up (considered to be ideal) a delta-function signal at each passage. For a sum pick-up, where the signal is independent of the transverse position, the Fourier transformation into the frequency domain results in a contribution at each harmonic of the revolution frequency (Fig. 4) while for a difference pick-up the modulation by the betatron oscillation splits up each line into two components[5]. For a collection of many particles with slightly different revolution frequencies, these lines spread out into bands, called Schottky bands because they represent the noise due to the finite number of charge carriers as described by Schottky[6].

The width of these bands increases towards higher frequency. The total power is the same for each band. The power density is therefore lower for the wider bands at high frequency up to the point where they start to overlap; beyond this point the bands merge and their combined density is constant with frequency. This is illustrated in Fig. 5 for so-called longitudinal lines (from a sum pick-up).

The cooling process may now be seen as follows. Firstly, each particle will cool itself with its own (coherent) signal. This means that at the frequency of each of its Schottky lines the phase of the corresponding sine-wave signal must be correct at the kicker, so that the latter exerts its influence in the right direction. Secondly, the other particles produce an incoherent heating effect at

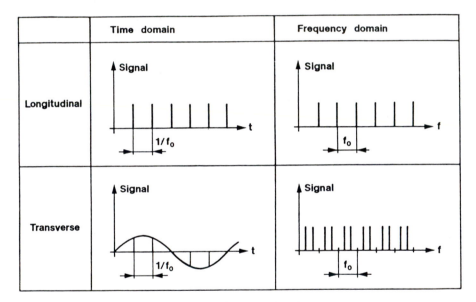

Fig. 4. Schottky signals in time domain and frequency domain.

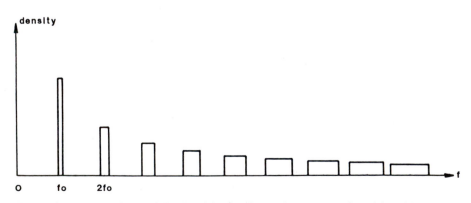

Fig. 5. Longitudinal Schottky bands originating from a large group of particles with slightly different revolution frequencies. At high frequencies the bands overlap.

each Schottky line proportional to the noise power density around that line[7]. Thus, only particles with frequencies very near to those of the perturbed particle will contribute. Any power density from thermal noise must of course be added to the Schottky power density.

For obtaining optimum cooling, the gain at each Schottky band should be adjusted so as to achieve an optimum balance between these two effects. If the bands are separated, the low-frequency ones have a higher density. This requires a lower gain and leads to less cooling for these bands. This is exactly the same effect that we called "bad mixing" in the time domain. At higher frequencies where the bands overlap we have good mixing and the gain should be independent of frequency.

Note that the picture given here (i.e. heating only caused by signals near the particle's Schottky frequencies) is completely different from the time-domain picture, where it seemed that particles in the same sample all contribute, independent of their exact revolution frequency. In fact, the latter is only true if the mixing is perfect and the samples are statistically independent. In the more general case, it turns out that both the optimum gain and the optimum cooling rate per line are inversely proportional to the density dN/df around that line, rather than to the total number of particles N. In the time domain treatment this was expressed by the mixing factor M, but the dependence of the parameters on frequency was lost.

There is yet another mixing effect that we have neglected so far. While moving from the pick-up to the kicker, each sample will already mix to a certain extent with its neighbours. This harmful effect may be described in the frequency domain as a phase lag increasing with frequency (particles with higher revolution frequency arrive too early at the kicker, so that their signal is too late). It appears quite difficult to correct this by means of filters at each Schottky band; on the other hand, in practical cases the effect is usually not very serious[8].

7. *Beam feedback*
Another aspect that we have not yet considered is essential for the correct analysis of a cooling system. This is the feedback loop formed by the cooling chain together with the beam response (Fig. 6). Any signal on the kicker will

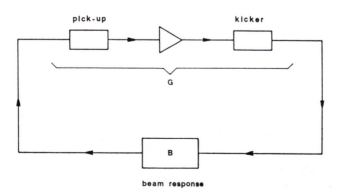

Fig. 6. Beam feedback effect. The loop is closed by the coherent beam response B.

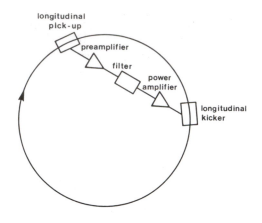

Fig. 7. Filter cooling.

modulate the beam coherently (in position for a transverse kicker, in energy and density for a longitudinal one). The modulation is smoothed by mixing, but some of it will always remain at the pick-up, closing the feedback loop.

The beam response is a well-known effect from the theory of instabilities in accelerator rings. For cooling purposes, because the exciting and detecting points are separated in space[5, 9], the treatment is slightly different. This is not the place to discuss the details; it may, however, be said that the response as a function of frequency can be calculated if the particle distribution versus revolution frequency is given, as well as some of the ring parameters.

It is found that for separated Schottky bands and with negligible thermal noise the optimum gain for cooling corresponds to an open-loop gain with an absolute value of unity and that the phase angle of the amplifier chain response must be opposite to the phase of the beam response[8]. As a result of this, it turns out that in the centre of the distribution the optimum loop gain becomes -1 for transverse cooling. The coherent feedback will then halve the amplitude of the Schottky signals as soon as the system is switched on. This is a convenient way of adjusting the gain; the correct phase may be checked by interrupting the loop somewhere and measuring its complete response with a network analyser[10].

8. *Longitudinal cooling*

So far, I have mainly discussed transverse cooling, i.e. reducing the betatron oscillations. Longitudinal cooling reduces the energy spread and increases the longitudinal density. This process, as it turns out, is most important for accumulating antiprotons.

One method of longitudinal cooling (sometimes called "Palmer cooling"[11]) is very similar to the one of Fig. 2. Again, we use a differential pick-up, now placed at a point where the dispersion is high, so that the particle position depends strongly on its momentum. The kicker must now give longitudinal kicks.

a)

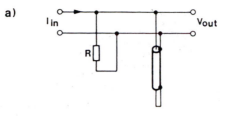

Fig. 8a. Simple transmission-line filter.

b)

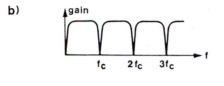

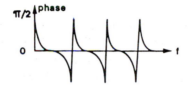

Fig. 8b. Amplitude and phase response vs. frequency.

A different method is to use a sum pick-up (Fig. 7) and to discriminate between particles of different energy by inserting a filter into the system ("Thorndahl method"[12]). This works because the Schottky frequencies of particles with different energy are different; the filter must cause a phase change of 180° in the middle of each band, so that particles from both sides will be pushed towards the centre. Such a filter may be made by using transmission lines whose properties vary periodically with frequency. The simple filter of Fig. 8a may serve as an example. The line, shorted at the far end, behaves as a short-circuit at all resonant frequencies, which may be made to coincide with the centres of the Schottky bands. Just above these frequencies the line behaves as an inductance, just below as a capacitance; thus, the phase jump of 180° is achieved (Fig. 8b). For relativistic particles, the length of the line must be equal to half the ring's circumference. More complicated filters, using several lines and/or active feedback circuits may sometimes be useful[10].

The advantage of the filter method, especially for low-intensity beams, is that the attenuation at the central frequencies is now obtained after the preamplifier, instead of before it as with a difference pick-up. The signal-to-noise ratio is therefore much better. Also, at frequencies below about 500 MHz where ferrites may be used, sum pick-ups may be made much shorter than differential ones, so that more may fit into the same space. This again gives a better signal-to-noise ratio. Of course, for filter cooling to be practical, the Schottky bands must be separated (bad mixing).

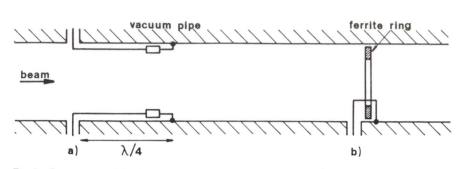

Fig. 9. Loop-type and ferrite ring-type pick-ups (or kickers). Note that for loop-type kickers the beam direction should be inverted.

9. *Pick-ups and kickers*

Cooling systems often have an octave bandwidth, with the highest frequency equal to twice the lowest one. Pick-ups with a reasonably flat response may consist of coupling loops that are a quarter wavelength long in the middle of the band (Fig. 9a). At the far end, a matching resistor equal to the characteristic impedance prevents reflections (or, seen in the frequency domain, ensures a correct phase relationship between beam and signal). Two loops at either side of the beam may be connected in common or differential mode for use as a sum or differential pick-up. The same structure may function as pick-up or kicker. Sum pick-ups or kickers may also consist of a ferrite frame with one or more coupling loops around it (Fig. 9b).

At high frequencies (typically > 1 GHz), slot-type pick-ups or kickers[13] become interesting (Fig. 10). The field from the particles couples to the transmission line behind the slots. If the latter are shorter than $\lambda/2$, the coupling is weak and the contributions from each slot may all be added together, provided the velocity along the line is equal to the particle velocity.

The signal-to-noise ratio at the pick-ups may be improved by using many of these elements and adding their output power in matched combiner circuits. A

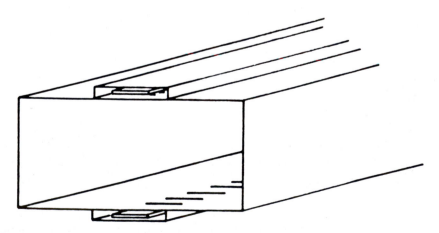

Fig. 10. Slot-type pick-up or kicker. One end of the transmission line is terminated with its own characteristic impedance.

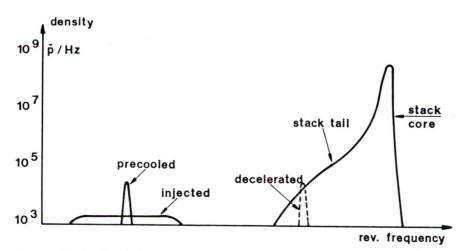

Fig. 11. Density distribution vs. revolution frequency in the Antiproton Accumulator. On the right, the stack; on the left, the newly injected batch, before and after precooling.

further improvement may be obtained by cryogenic cooling of the matching resistors and/or the preamplifiers.

Using many kickers reduces the total power required. The available power is sometimes a limitation to the cooling rate that may be obtained.

Fig. 12. Inside of a vacuum tank with precooling kickers at the left and space for the stack at the right. The ferrite frames of the kickers are open in the centre of the picture; they can be closed by the ferrite slabs mounted on the shutter that rotates around a pivot at the far right. Water tubes for cooling the ferrite may be seen.

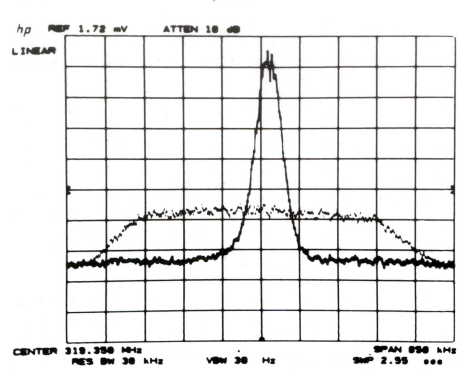

Fig. 13. Precooling 6×10^6 p̄'s in 2 seconds. Longitudinal Schottky band at the 170th harmonic (314 MHz) before and after cooling.

10. *Accumulation of antiprotons; stochastic stacking*

It is now possible to explain how the antiproton accumulator works. It should, however, be made clear first that stochastic cooling is not the only method available for this purpose. In fact, already in 1966, Budker[14] proposed a pp̄ collider scheme where the cooling was to be done by his so-called electron cooling method. A cold electron beam superimposed on the p̄ beam cools it by electromagnetic interaction (scattering). We originally also planned to use this idea; it turns out, however, that it needs particles with low energy to work well with large-emittance beams. An additional ring to decelerate the antiprotons would then have been needed. The simpler stochastic method, using a single ring at fixed field was preferred.

In Fig. 11 we see how the particle density depends on revolution frequency (or energy, or position of the central orbit; the horizontal axis could represent any of these). On the right, the so-called stack, i.e. the particles that have already been accumulated. On the left, the low-density beam that is injected every 2.4 seconds. The latter is separated in position from the stack in those regions of the circumference where the dispersion of the lattice is large. In such a place the injection kicker can therefore inject these particles without kicking the stack. Also, the pick-ups and kickers used for the first cooling operation (longitudinal precooling) are placed here so that they do not see the stack. They consist, in fact, of ferrite frames surrounding the injected beam (Fig. 12). The pick-ups are therefore sum pick-ups (200 in total, each 25 mm long in beam direction) and the Thorndahl type of cooling, with a filter, is used[15]. Figure 13 shows how the distribution is reduced in width by an order of magnitude within 2 seconds. The number of antiprotons involved is about 6×10^6, the band used is $150-500$ MHz.

After this precooling, one leg of the ferrite frames is moved downwards by a fast actuator mechanism[16] so that the precooled beam can be bunched by RF and decelerated towards the low-frequency tail of the stack (Fig. 11). The whole process, including the upward movement of the "shutter" to restore the pick-ups and kickers, takes 400 ms. The RF is then slowly reduced[17] so that the particles are debunched and deposited in the stack tail.

They must be removed from this place within the next 2.4 seconds because Liouville's theorem prevents the RF system from depositing the next batch at the same place without simultaneously removing what was there before. A further longitudinal cooling system, using the $250-500$ MHz band, therefore pushes these particles towards higher revolution frequencies, up against the density gradient[18].

This so-called stack tail system should have a gain that depends on energy (or revolution frequency). In fact, the density gradient increases strongly towards the stack core (note the logarithmic scale), and the gain for optimum cooling should vary inversely with this. We achieve this by using as pick-ups small quarter-wave coupling loops, positioned underneath and above the tail region, in such a place that they are sensitive to the extreme tail, but much less to the far-away dense core. This results in a bad signal-to-noise ratio for the region nearer to the core. Therefore, two sets of pick-ups are used, each at a different radial position and each with its own preamplifier and gain adjustment. With this set-up we obtain fast cooling at the stack edge where the particles are deposited, and slow cooling at the dense core, where we can afford it because the particles remain there for hours.

A problem is that the tail systems must be quite powerful to remove the particles fast enough. As a result, their kickers will also disturb the slowly-cooled stack core (the Schottky signals do not overlap with the core frequencies, but the thermal noise does). The problem exists because the kickers must be at a point where the dispersion is zero to prevent them from exciting horizontal betatron oscillations. They therefore kick all particles (tail or core) equally.

A solution is found by using transmission-line filters as described above to suppress the core frequencies in the tail cooling systems. These filters also rotate the phase near the core region in an undesirable way; this does not matter, however, because the cooling of the core is done by a third system of larger bandwidth ($1-2$ GHz).

While the particles move towards the core, they are also cooled horizontally and vertically, first by tail cooling systems, then by $1-2$ GHz core systems. The layout of the various cooling circuits is shown in Fig. 14. In the general view of Fig. 15, some of the transmission lines transporting the signals for the pick-ups to the kickers may be seen.

When the stack contains a sufficient number of antiprotons (typically 2×10^{11}), a fraction of these ($\sim 30\%$) is transferred to the PS and from there to

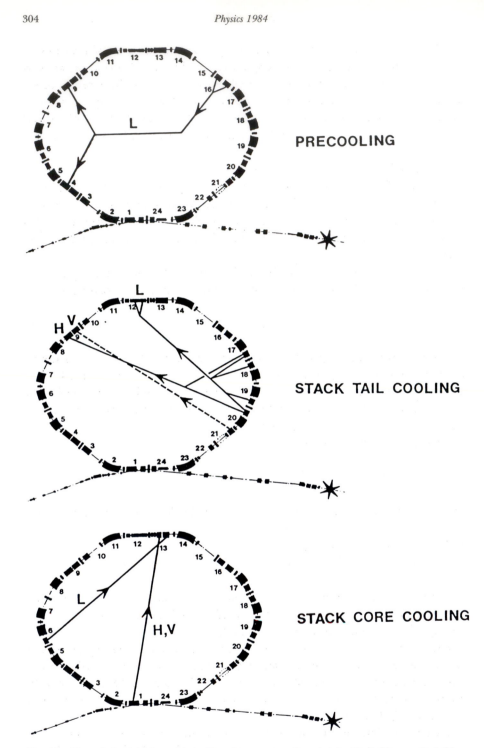

PRECOOLING

STACK TAIL COOLING

STACK CORE COOLING

Fig. 14. Plan of the AA ring with its 7 cooling systems. L = longitudinal, V = vertical, H = horizontal.

the SPS machine. This is done by bunching a part of the stack, of a width that may be adjusted by properly choosing the RF bucket area[19]. These are accelerated until they are on the same orbit where normally particles are injected. They can then be extracted without disturbing the remaining stack. This process is repeated (at present three times); each time one RF bucket of the SPS is filled. The remaining p̄'s form the beginning of the next stack.

11. *Design of longitudinal cooling systems; Fokker-Planck equation*

The main difference between transverse and longitudinal cooling systems is that the latter will change the longitudinal distribution on which the incoherent (heating) term depends, as well as effects such as the beam feedback. This complicates the theory; still, everything can be calculated if all parameters are given.

It is convenient to define the flux ϕ, i.e. the number of particles passing a certain energy (or frequency) value per unit time. It may be shown[5] that

$$\phi = F\Psi - D\delta\Psi/\delta f_0, \tag{5}$$

where Ψ is the density dN/df_0 while F and D are slowly varying constants, depending on various system parameters as well as on the particle distribution. The first term represents the coherent cooling, the second one the incoherent (diffusion) effect that has the effect of pushing the particles down the gradient under the influence of perturbing noise.

By using the continuity equation

$$\delta\Psi/\delta t + \delta\phi/\delta f_0 = 0,$$

expressing that no particles are lost, we find the Fokker-Planck-type equation

$$\frac{\delta\Psi}{\delta t} = -\frac{\delta}{\delta f_0}(F\Psi) + \frac{\delta}{\delta f_0}(D\frac{\delta\Psi}{\delta f_0}) \tag{6}$$

that allows us to compute the evolution of the density versus revolution frequency f_0 and time given the initial distribution. The particles deposited at the edge are introduced as a given flux at that point.

The constants F and D depend on many system parameters (pick-up and kicker characteristics, amplifier gain, filter response, beam distribution, etc.). Their value is found through summing the contributions of all Schottky bands. Analytic solutions of (6) do not exist in practice and a complicated numerical treatment is indicated.

Such calculations resulted in the design of the antiproton stacking system. At the time this was done, tests in a small experimental ring (ICE) had confirmed the cooling in all planes at time scales of the order of 10 seconds. However, it was not possible to check the stacking system (increasing the density by four orders of magnitude) in any way, and it may be argued that we took a certain risk by starting the project without being able to verify this aspect. Fortunately,

Fig. 15. View of the Antiproton Accumulator before it was covered by concrete slabs. The silvered material around the vacuum tanks is insulation, needed because everything may be heated to 300° to obtain ultra-high vacuum. The transmission lines crossing the ring and carrying the cooling signals may be seen.

everything behaved according to theory and although the number of p̄'s injected is smaller than was hoped for by a factor 3.5, the cooling works largely as expected.

12. *Other applications of stochastic cooling; future developments*

At present, stochastic cooling is used at CERN in the p̄ accumulator and in the low energy ring (LEAR) where the p̄'s may be stored after deceleration in the PS. Before the intersecting storage rings (ISR) were closed down last year, they also used the antiprotons and contained cooling equipment.

In the SPS where the high-energy collisions take place, cooling would be attractive because it would improve the beam lifetime and might decrease its cross-section. However, a difficulty is formed by the fact that the beam is bunched in this machine; the bunches are narrow (3×4 ns). In fact, owing to the bunching each Schottky band is split up into narrow, dense satellite bands and the signals from different bands are correlated[20]. Nevertheless, a scheme is being considered that might improve the lifetime to a certain extent[21].

In the United States, a p̄ accumulator complex similar to the CERN one and also using stochastic cooling is being constructed[22]. This machine is expected to have a stacking rate an order of magnitude higher than the CERN one because it uses a higher primary energy to produce the antiprotons and higher frequencies to cool them. In the meantime, we are building a second ring at CERN, surrounding the present accumulator (Fig. 16), with a similar performance. It will have stronger focusing, so increasing both transverse acceptances by at least a factor 2, and the longitudinal one by a factor 4. The increased focusing strengths will diminish the mixing; consequently, higher frequencies (up to 4 GHz) will be used for cooling. The present AA will be used to contain the stack and its cooling systems will also be upgraded.

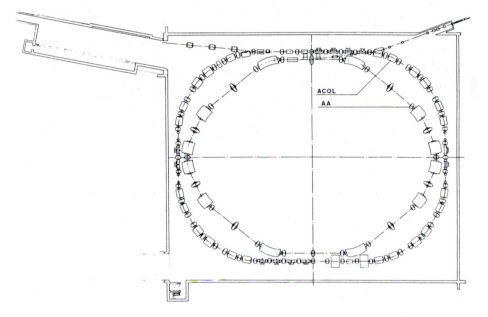

Fig. 16. The new ACOL ring (under construction) around the AA. This ring will increase the stacking rate by an order of magnitude. The stack will still be kept in the AA ring.

ACKNOWLEDGEMENTS

The development of the stochastic cooling theory owes much to H. G. Hereward, D. Möhl, F. Sacherer, and L. Thorndahl. The latter also made important contributions to the construction of most cooling systems at CERN and it is doubtful if the accumulator would have been feasible without his invention of the filter method. It is also a pleasure to acknowledge the invaluable contributions of G. Carron (hardware), L. Faltin (slot pick-ups) and C. Taylor (1−2 GHz systems).

During the construction of the Antiproton Accumulator, R. Billinge was joint project leader with myself and it is mostly because of his contributions to the design and his able management that the machine was ready in a record time (2 years) and worked so well.

REFERENCES

1. L. Szilard, Über die Entropieverminderung in einem thermodynamischen System bei Eingriffen intelligenter Wesen, Zeitsch. f. Physik *53* (1929) 840.
2. S. van der Meer, Stochastic damping of betatron oscillations in the ISR, CERN/ISR-PO/ 72–31 (1972).
3. D. Möhl, Stochastic cooling for beginners, Proceedings of the CERN Accelerator School on Antiprotons for Colliding Beam Facilities, CERN 84–15 (1984).
4. F. Sacherer, Stochastic cooling theory, CERN/ISR-TH/78–11 (1978).
5. D. Möhl, G. Petrucci, L. Thorndahl, S. van der Meer, Physics and technique of stochastic cooling, Phys. Reports *58* (1980) 73.
6. W. Schottky, Ann. Physik *57* (1918) 541.
7. H. G. Hereward, The elementary theory of Landau damping, CERN 65–20 (1965).
8. S. van der Meer, Optimum gain and phase for stochastic cooling systems, CERN/PS-AA/83–48 (1983).
9. S. van der Meer, A different formulation of the longitudinal and transverse beam response, CERN/PS-AA/80-4 (1980).
10. S. van der Meer, Stochastic cooling in the Antiproton Accumulator, IEEE Trans. Nucl. Sci. *NS28* (1981) 1994.
11. R. B. Palmer, BNL, Private communication (1975).
12. G. Carron, L. Thorndahl, Stochastic cooling of momentum spread by filter techniques, CERN/ISR-RF/78–12 (1978).
13. L. Faltin, Slot-type pick-up and kicker for stochastic beam cooling, Nucl. Instrum. Methods *148* (1978) 449.
14. G. I. Budker, Proc. Int. Symp. on Electron and Positron Storage Rings, Saclay, 1966, p. 11–1-1; Atomn. Energ. *22* (1967) 346.
    G. I. Budker et al., Experimental studies of electron cooling, Part. Acc. *7* (1976) 197.
15. S. van der Meer, Precooling in the Antiproton Accumulator, CERN/PS-AA/78–26 (1978).
16. D. C. Fiander, S. Milner, P. Pearce, A. Poncet, The Antiproton Accumulator shutters: design, technology and performance, CERN/PS/84–23 (1984).
17. R. Johnson, S. van der Meer, F. Pedersen, G. Shering, Computer control of RF manipulations in the CERN Antiproton Accumulator, IEEE Trans. Nucl. Sci. NS-30 (1983) 2290.
18. S. van der Meer, Stochastic stacking in the Antiproton Accumulator, CERN/PS-AA/78–22 (1978).
19. R. Johnson, S. van der Meer, F. Pedersen, Measuring and manipulating an accumulated stack of antiprotons in the CERN Antiproton Accumulator, IEEE Trans. Nucl. Sci. NS-30 (1983) 2123.
20. H. Herr, D. Möhl, Bunched beam stochastic cooling, CERN/EP/Note 79–34 (1979).
21. D. Boussard, S. Chattopadhyay, G. Dôme, T. Linnecar, Feasibility study of stochastic cooling of bunches in the SPS, CERN/SPS/84–4 (1984).
22. Design Report Tevatron I project, Fermi National Accelerator Laboratory, Batavia, Ill. (1983).

Physics 1985

KLAUS VON KLITZING

for the discovery of the quantized Hall effect

THE NOBEL PRIZE FOR PHYSICS

Speech by Professor STIG LUNDQVIST of the Royal Academy of Sciences.
Translation from the Swedish text

Your Majesties, Your Royal Highnesses, Ladies and Gentlemen,

This year's Nobel Prize for Physics has been awarded to Professor Klaus von Klitzing for the discovery of the quantized Hall effect.

This discovery is an example of these unexpected and surprising discoveries that now and then take place and which make research in the sciences so exciting. The Nobel Prize is sometimes an award given to large projects, where one has shown great leadership and where one with ingenuity combined with large facilities and material resources has experimentally verified the correctness of theoretical models and their predictions. Or, one has succeeded through creation of new theoretical concepts and methods to develop theories for fundamental problems in physics that resisted all theoretical attempts over a long period of time. However, now and then things happen in physics that no one can anticipate. Someone discovers a new phenomon or a new fundamental relation in areas of physics where no one expects anything exciting to happen.

This was exactly what happened when Klaus von Klitzing in February 1980 was working on the Hall effect at the Hochfelt-Magnet-Labor in Grenoble. He discovered from his experimental data that a relation which had been assumed to hold only approximately seemed to hold with an exceptionally high accuracy and in this way the discovery of the quantized Hall effect was made.

The discovery by von Klitzing has to do with the relation between electric and magnetic forces in nature and has a long history. Let us go back to 1820, when the Danish physicist H. C. Ørsted found that an electric current in a wire influenced a compass needle and made it change its direction. He discovered this phenomenon in a class with his students. No one had seen a relation between electric and magnetic forces before. More than 50 years later a young American physicist, E. H. Hall, speculated that the magnetic force might influence the charge carriers in a metallic wire placed in a magnetic field and give rise to an electric voltage across the wire. He was able to show that when sending an electric current through a strip of gold there was a small voltage across the wire in a direction perpendicular both to the current and the magnetic field. That was the discovery of the Hall effect.

The Hall effect is now a standard method frequently used to study semiconductor materials of technical importance, and the effect is described in all textbooks in solid state physics. The experiment is in principle very simple and requires only a magnetic field plus instruments to measure current and voltage. If one varies the magnetic field, the current and voltage will change in a completely regular way and no surprising effects are expected to happen.

von Klitzing studied the Hall effect under quite extreme conditions. He used an extremely high magnetic field and cooled his samples to just a couple of degrees above the absolute zero point of temperature. Instead of the regular change one would expect, he found some very characteristic steps with plateaus in the conductivity. The values at these plateaus can with extremely high accuracy be expressed as an integer times a simple expression that just depends of two fundamental constants: the electric elementary charge and Planck's constant which appear everywhere in quantum physics.

The result represents a quantization of the Hall effect—a completely unexpected effect. The accuracy in his results was about one part in ten million, which would correspond to measuring the distance between Stockholm and von Klitzing's home station Stuttgart with an accuracy of a few centimeters.

The discovery of the quantized Hall effect is a beautiful example of the close interrelation between the highly advanced technology in the semiconductor industry and fundamental research in physics. The samples used by von Klitzing were refined versions of a kind of transistor we have in our radios. His samples, however, had to satisfy extremely high standards of perfection and could only be made by using a highly advanced technique and refined technology.

The quantized Hall effect can only be observed in a two-dimensional electron system. Two-dimensional electron systems do not occur in nature. However, the development in semiconductor technology has made possible the realization of a two-dimensional electron system. In the kind of transistor that von Klitzing used, some of the electrons are bound to the interface between two parts of the transistor. At sufficiently low temperature the electrons can move only along the interface and one has effectively a two-dimensional electron system.

von Klitzing's discovery of the quantized Hall effect attracted immediately an enormous interest. Because of the extremely high accuracy the effect can be used to define an international standard for electric resistance. The metrological possibilities are of great importance and have been subject to detailed studies at many laboratories all over the world.

The quantized Hall effect is one of the few examples, where quantum effects can be studied in ordinary macroscopic measurements. The underlying detailed physical mechanisms are not yet fully understood. Later experiments have revealed completely new and unexpected properties and the study of two-dimensional systems is now one of the most challenging areas of research in physics.

Professor von Klitzing,

On behalf of the Royal Swedish Academy of Sciences I wish to convey our warmest congratulations and ask you to receive your prize from the hands of His Majesty the King.

KLAUS v. KLITZING

Born 28th June 1943 in Schroda (Posen), German nationality.

February 1962:

  Abitur in Quakenbrück

April 1962 to March 1969:

  Technical University Braunschweig Diploma in Physics.

  Title of diploma work: "Lifetime Measurements on InSb" (Prof. F. R. Keßler)

May 1969 to Nov. 1980:

  University Würzburg (Prof. Dr. G. Landwehr)

  Thesis work about: "Galvanomagnetic Properties of Tellurium in Strong Magnetic Fields" (Ph.D. in 1972).

  Habilitation in 1978.

  The most important publication related to the Nobel Prize appeared in: Phys. Rev. Letters *45*, 494 (1980).

  Research work at the Clarendon Laboratory, Oxford (1975 to 1976) and High Magnetic Field Laboratory, Grenoble (1979 to 1980)

Nov. 1980 to Dec. 1984:

  Professor at the Technical University, München

Since January 1985:

  Director at the Max-Planck-Institut für Festkörperforschung, Stuttgart.

THE QUANTIZED HALL EFFECT

Nobel lecture, December 9, 1985

by

KLAUS von KLITZING,

Max-Planck-Institut für Festkörperforschung, D-7000 Stuttgart 80

1. *Introduction*

Semiconductor research and the Nobel Prize in physics seem to be contradictory since one may come to the conclusion that such a complicated system like a semiconuctor is not useful for very fundamental discoveries. Indeed, most of the experimental data in solid state physics are analyzed on the basis of simplified theories, and very often the properties of a semiconductor device is described by empirical formulas since the microscopic details are too complicated. Up to 1980 nobody expected that there exists an effect like the Quantized Hall Effect, which depends exclusively on fundamental constants and is not affected by irregularities in the semiconductor like impurities or interface effects.

The discovery of the Quantized Hall Effect (QHE) was the result of systematic measurements on silicon field effect transistors—the most important device in microelectronics. Such devices are not only important for applications but also for basic research. The pioneering work by Fowler, Fang, Howard and Stiles [1] has shown that new quantum phenomena become visible if the electrons of a conductor are confined within a typical length of 10 nm. Their discoveries opened the field of two-dimensional electron systems which since 1975 is the subject of a conference series [2]. It has been demonstrated that this field is important for the description of nearly all optical and electrical properties of microelectronic devices. A two-dimensional electron gas is absolutely necessary for the observation of the Quantized Hall Effect, and the realization and properties of such a system will be discussed in section 2. In addition to the quantum phenomena connected with the confinement of electrons within a two-dimensional layer, another quantization—the Landau quantization of the electron motion in a strong magnetic field—is essential for the interpretation of the Quantized Hall Effect (section 3). Some experimental results will be summarized in section 4 and the application of the QHE in metrology is the subject of section 5.

2 *Two-Dimensional Electron Gas*

The fundamental properties of the QHE are a consequence of the fact that the energy spectrum of the electronic system used for the experiments is a *discrete* energy spectrum. Normally, the energy E of mobile electrons in a semiconductor is quasicontinuous and can be compared with the kinetic energy of free electrons with wave vector k but with an effective mass m*

$$E = \frac{\hbar^2}{2m^*} (k_x^2 + k_y^2 + k_z^2) \tag{1}$$

If the energy for the motion in one direction (usually z-direction) is fixed, one obtains a quasi-two-dimensional electron gas (2DEG), and a strong magnetic field perpendicular to the two-dimensional plane will lead—as discussed later—to a fully quantized energy spectrum which is necessary for the observation of the QHE.

A two-dimensional electron gas can be realized at the surface of a semiconductor like silicon or gallium arsenide where the surface is usually in contact with a material which acts as an insulator (SiO_2 for silicon field effect transistors and, e.g. $Al_xGa_{1-x}As$ for heterostructures). Typical cross sections of such devices are shown in Fig 1. Electrons are confined close to the surface of the semiconductor by an electrostatic field F_z normal to the interface, originating from positive charges (see Fig. 1), which causes a drop in the electron potential towards the surface.

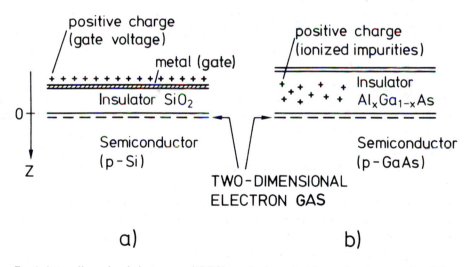

Fig. 1. A two-dimensional electron gas (2DEG) can be formed at the semiconductor surface if the electrons are fixed close to the surface by an external electric field. Silicon MOSFETs (a) and GaAs-$Al_xGa_{1-x}As$ heterostructures (b) are typical structures used for the realization of a 2DEG.

If the width of this potential well is small compared to the de Broglie wavelength of the electrons, the energy of the carriers are grouped in so-called electric subbands E_i corresponding to quantized levels for the motion in z-direction, the direction normal to the surface. In lowest approximation, the electronic subbands can be estimated by calculating the energy eigenvalues of an electron in a triangular potential with an infinite barrier at the surface ($z=0$) and a constant electric field F_s for $z \geq 0$, which keeps the electrons close to the surface. The result of such calculations can be approximated by the equation

$$E_j = \left(\frac{\hbar^2}{2m^*}\right)^{1/3} \cdot \left(\frac{3}{2} \pi e F_s\right)^{2/3} \cdot \left(j + \frac{3}{4}\right)^{2/3} \qquad (2)$$

$$j = 0, 1, 2\ldots$$

In some materials, like silicon, different effective masses m^* and $m^{*'}$ may be present which leads to different series E_j and E_j'.

Equation (2) must be incorrect if the energy levels E_j are occupied with electrons, since the electric field F_s will be screened by the electronic charge.

For a more quantitative calculation of the energies of the electric subbands it is necessary to solve the Schrödinger equation for the actual potential V_z' which changes with the distribution of the electrons in the inversion layer. Typical results of such calculation for both silicon MOSFETs and GaAs-heterostructures are shown in Fig. 2 [3,4]. Usually, the electron concentration of the two-dimensional system is fixed for a heterostructure (Fig. 1b) but can be varied in a MOSFET by changing the gate voltage.

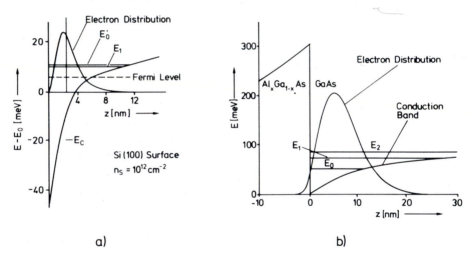

a)   b)

Fig. 2. Calculations of the electric subbands and the electron distribution within the surface channel of a silicon MOSFET (a) and a GaAs-Al$_x$Ga$_{1-x}$As heterostructure [3, 4].

Experimentally, the separation between electric subbands, which is of the order of 10 meV, can be measured by analyzing the resonance absorption of electromagnetic waves with a polarization of the electric field perpendicular to the interface [5].

At low temperatures (T<4 K) and small carrier densities for the 2DEG (Fermi energy E_F relative to the lowest electric subbands E_0 small compared with the subband separation E_1-E_0) only the lowest electric subband is occupied with electrons (electric quantum limit), which leads to a strictly two-dimensional electron gas with an energy spectrum

$$E = E_0 + \frac{\hbar^2 k_\parallel^2}{2m^*} \qquad (3)$$

where $k_\parallel$ is a wavevector within the two-dimensional plane.

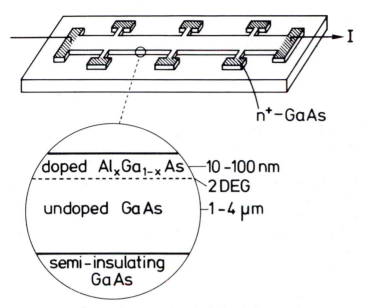

Fig. 3. Typical shape and cross-section of a GaAs-Al$_x$Ga$_{1-x}$As heterostructure used for Hall effect measurements.

For electrical measurements on a 2DEG, heavily doped n$^+$-contacts at the semiconductor surface are used as current contacts and potential probes. The shape of a typical sample used for QHE-experiments (GaAs-heterostructure) is shown in Fig. 3. The electrical current is flowing through the surface channel, since the fully depleted Al$_x$Ga$_{1-x}$As acts as an insulator (the same is true for the SiO$_2$ of a MOSFET) and the p-type semiconductor is electrically separated from the 2DEG by a p-n junction. It should be noted that the sample shown in Fig. 3 is basically identical with new devices which may be important for the next computer generation [6]. Measurements related to the Quantized Hall Effect which include an analysis and characterization of the 2DEG are therefore important for the development of devices, too.

3. Quantum Transport of a 2DEG in Strong Magnetic Fields

A strong magnetic field B with a component B$_z$ normal to the interface causes the electrons in the two-dimensional layer to move in cyclotron orbits parallel to the surface. As a consequence of the orbital quantization the energy levels of the 2DEG can be written schematically in the form

$$E_n = E_o + (n + 1/2)\hbar\omega_c + s \cdot g \cdot \mu_B \cdot B \tag{4}$$
$$n = 0,1,2,...$$

with the cyclotron energy $\hbar\omega_c = \hbar eB/m^*$, the spin quantum numbers $s = \pm 1/2$ the Landé factor g and the Bohr magneton μ_B.

The wave function of a 2DEG in a strong magnetic field may be written in a form where the y-coordinate y_0 of the center of the cyclotron orbit is a good quantum number [7].

$$\psi = e^{ikx}\Phi_n(y-y_o) \tag{5}$$

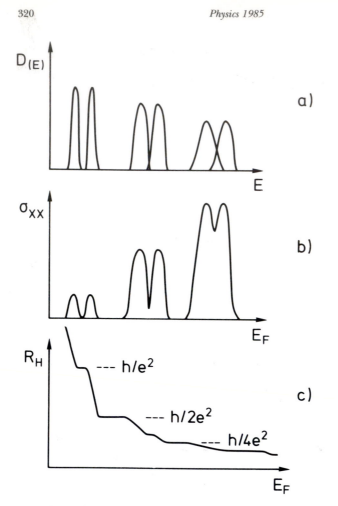

Fig. 4. Sketch for the energy dependence of the density of states (a), conductivity σ_{xx} (b), and Hall resistance R_H (c) at a fixed magnetic field.

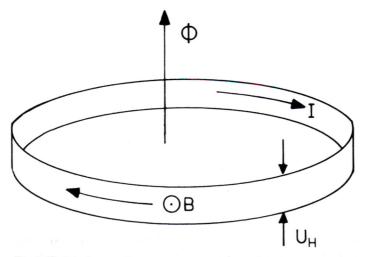

Fig. 5. Model of a two-dimensional metallic loop used for the derivation of the quantized Hall resistance.

where Φ_n is the solution of the harmonic-oscillator equation

$$\frac{1}{2m^*}\left[p_y^2 + (eB)^2y^2\right]\phi_n = (n + \tfrac{1}{2})\,\hbar\omega_c\,\phi_n \tag{6}$$

and y_0 is related to k by

$$y_0 = \hbar k/eB \tag{7}$$

The degeneracy factor for each Landau level is given by the number of center coordinates y_0 within the sample. For a given device with the dimension $L_x \cdot L_y$, the center coordinates y_0 are separated by the amount

$$\Delta y_0 = \frac{\hbar}{eB}\,\Delta k = \frac{\hbar}{eB}\frac{2\pi}{L_x} = \frac{h}{eBL_x} \tag{8}$$

so that the degeneracy factor $N_0 = L_y/\Delta y_0$ is identical with $N_0 = L_xL_yeB/h$ the number of flux quanta within the sample. The degeneracy factor per unit area is therefore:

$$N = \frac{N_0}{L_xL_y} = \frac{eB}{h} \tag{9}$$

It should be noted that this degeneracy factor for each Landau level is independent of semiconductor parameters like effective mass.

In a more general way one can show [8] that the commutator for the center coordinates of the cyclotron orbit $[x_o,y_o] = i\hbar/eB$ is finite, which is equivalent to the result that each state occupies in real space the area $F_0 = h/eB$ corresponding to the area of a flux quantum.

The classical expression for the Hall voltage U_H of a 2DEG with a surface carrier density n_s is

$$U_H = \frac{B}{n_s \cdot e} \cdot I \tag{10}$$

where I is the current through the sample. A calculation of the Hall resistance $R_H = U_H/I$ under the condition that i energy levels are fully occupied $(n_s = iN)$, leads to the expression for the quantized Hall resistance

$$R_H = \frac{B}{iN \cdot e} = \frac{h}{ie^2} \tag{11}$$

$$i = 1, 2, 3\ldots$$

A quantized Hall resistance is always expected if the carrier density n_s and the magnetic field B are adjusted in such a way that the filling factor i of the energy levels (Eq. 4)

$$i = \frac{n_s}{eB/h} \tag{12}$$

is an integer.

Under this condition the conductivity σ_{xx} (current flow in the direction of the electric field) becomes zero since the electrons are moving like free particles exclusively perpendicular to the electric field and no diffusion (originating from scattering) in the direction of the electric field is possible. Within the self-consistent Born approximation [9] the discrete energy spectrum broadens as shown in Fig. 4a. This theory predicts that the conductivity σ_{xx} is mainly proportional to the square of the density of states at the Fermi energy E_F which leads to a vanishing conductivity σ_{xx} in the quantum Hall regime and quantized plateaus in the Hall resistance R_H (Fig. 4c).

The simple one-electron picture for the Hall effect of an ideal two-dimensional system in a strong magnetic field leads already to the correct value for the quantized Hall resistance (Eq. 11) at integer filling factors of the Landau levels. However, a microscopic interpretation of the QHE has to include the influences of the finite size of the sample, the finite temperature, the electron-electron interaction, impurities and the finite current density (including the inhomogenious current distribution within the sample) on the experimental result. Up to now, no corrections to the value h/ie^2 of the quantized Hall resistance are predicted if the conductivity σ_{xx} is zero. Experimentally, σ_{xx} is never exactly zero in the quantum Hall regime (see section 4) but becomes unmeasurably small at high magnetic fields and low temperatures. A quantitative theory of the QHE has to include an analysis of the longitudinal conductivity σ_{xx} under real experimental conditions, and a large number of publications are discussing the dependence of the conductivity on the temperature, magnetic field, current density, sample size etc. The fact that the value of the quantized Hall resistance seems to be exactly correct for $\sigma_{xx} = 0$ has led to the conclusion that the knowledge of microscopic details of the device is not necessary for a calculation of the quantized value. Consequently Laughlin [10] tried to deduce the result in a more general way from gauge invariances. He considered the situation shown in Fig. 5. A ribbon of a two-dimensional system is bent into a loop and pierced everywhere by a magnetic field B normal to its surface. A voltage drop U_H is applied between the two edges of the ring. Under the condition of vanishing conductivity σ_{xx} (no energy dissipation), energy is conserved and one can write Faraday's law of induction in a form which relates the current I in the loop to the adiabatic derivative of the total energy of the system E with respect to the magnetic flux Φ threading the loop

$$I = \frac{\partial E}{\partial \phi} \tag{13}$$

If the flux is varied by a flux quantum $\Phi_0 = h/e$, the wavefunction enclosing the flux must change by a phase factor 2π corresponding to a transition of a state with wavevector k into its neighbour state $k + (2\pi)/(L_x)$, where L_x is the circumference of the ring. The total change in energy corresponds to a transport of states from one edge to the other with

$$\Delta E = i \cdot e \cdot U_H \tag{14}$$

The integer i corresponds to the number of filled Landau levels if the free electron model is used, but can be in principle any positive or negative integer number.

From Eq. (13) the relation between the dissipationless Hall current and the Hall voltage can be deduced

$$I = i \cdot e \cdot U_H / \phi_0 = i \frac{e^2}{h} \cdot U_H \tag{15}$$

which leads to the quantized Hall resistance $R_H = \dfrac{h}{ie^2}$.

In this picture the main reason for the Hall quantization is the flux quantization h/e and the quantization of charge into elementary charges e. In analogy, the fractional quantum Hall effect, which will not be discussed in this paper, is interpreted on the basis of elementary excitations of quasiparticles with a charge e* = $\frac{e}{3}, \frac{e}{5}, \frac{e}{7}$ etc.

The simple theory predicts that the ratio between the carrier density and the magnetic field has to be adjusted with very high precision in order to get exactly integer filling factors (Eq. 12) and therefore quantized values for the Hall resistance. Fortunately, the Hall quantization is observed not only at special magnetic field values but in a wide magnetic field range, so that an accurate fixing of the magnetic field or the carrier density for high precision measurements of the quantized resistance value is not necessary. Experimental data of such Hall plateaus are shown in the next section and it is believed that localized states are responsible for the observed stabilization of the Hall resistance at certain quantized values.

After the discovery of the QHE a large number of theoretical paper were published discussing the influence of localized states on the Hall effect [11−14] and these calculations demonstrate that the Hall plateaus can be explained if localized states in the tails of the Landau levels are assumed. Theoretical investigations have shown that a mobility edge exists in the tails of Landau levels separating extended states from localized states [15−18]. The mobility edges are located close to the center of a Landau level for long-range potential fluctuations. Contrary to the conclusion reached by Abrahams, et al [19] that all states of a two-dimensional system are localized, one has to assume that in a strong magnetic field at least one state of each Landau level is extended in order to observe a quantized Hall resistance. Some calculations indicate that the extended states are connected with edge states [17].

In principle, an explanation of the Hall plateaus without including localized states in the tails of the Landau levels is possible if a reservoir of states is present outside the two-dimensional system [20, 21]. Such a reservoir for electrons, which should be in equilibrium with the 2DEG, fixes the Fermi energy within the energy gap between the Landau levels if the magnetic field or the number of electrons is changed. However, this mechanism seems to be more unlikely than localization in the the tails of the Landau levels due to disorder. The following discussion assumes therefore a model with extended and localized states within one Landau level and a density of states as sketched in Fig. 6.

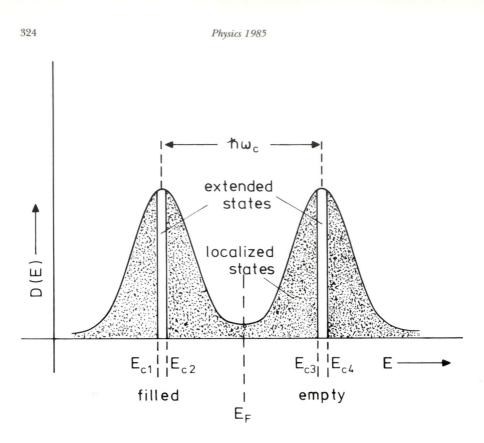

Fig. 6. Model for the broadened density of states of a 2 DEG in a strong magnetic field. Mobility edges close to the center of the Landau levels separate extended states from localized states.

4. *Experimental Data*

Magneto-quantum transport measurements on two-dimensional systems are known and published for more than 20 years. The first data were obtained with silicon MOSFETs and at the beginning mainly results for the conductivity σ_{xx} as a function of the carrier density (gate voltage) were analyzed. A typical curve is shown in Fig. 7. The conductivity oscillates as a function of the filling of the Landau levels and becomes zero at certain gate voltages V_g. In strong magnetic fields σ_{xx} vanishes not only at a fixed value V_g but in a range ΔV_g, and Kawaji was the first one who pointed out that some kind of immobile electrons must be introduced [22], since the conductivity σ_{xx} remains zero even if the carrier density is changed. However, no reliable theory was available for a discussion of localized electrons, whereas the peak value of σ_{xx} was well explained by calculations based on the self-consistent Born approximation and short-range scatterers which predict $\sigma_{xx} \sim (n + 1/2)$ independent of the magnetic field.

The theory for the Hall conductivity is much more complicated, and in the lowest approximation one expects that the Hall conductivity σ_{xy} deviates from the classical curve $\sigma_{xy}^{0} = - \dfrac{n_s e}{B}$ (where n_s is the total number of electrons in the two-dimensional system per unit area) by an amount $\Delta\sigma_{xy}$ which depends mainly on the third power of the density of states at the Fermi energy [23].

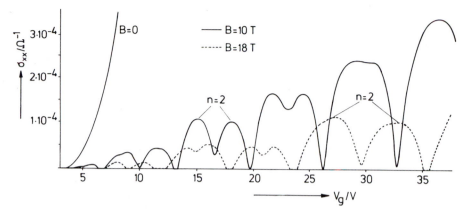

Fig. 7. Conductivity σ_{xx} of a silicon MOSFET at different magnetic fields B as a function of the gate voltage V_g.

However, no agreement between theory and experiment was obtained. Today, it is believed, that $\Delta\sigma_{xy}$ is mainly influenced by localized states, which can explain the fact that not only a positive but also a negative sign for $\Delta\sigma_{xy}$ is observed. Up to 1980 all experimental Hall effect data were analyzed on the basis of an incorrect model so that the quantized Hall resistance, which is already visible in the data published in 1978 [24] remained unexplained.

Whereas the conductivity σ_{xx} can be measured directly by using a Corbino disk geometry for the sample, the Hall conductivity is not directly accessible in an experiment but can be calculated from the longitudinal resistivity ϱ_{xx} and the Hall resistivity ϱ_{xy} measured on samples with Hall geometry (see Fig. 3):

$$\sigma_{xy} = -\frac{\rho_{xy}}{\rho_{xx}^2 + \rho_{xy}^2}, \ \sigma_{xx} = \frac{\rho_{xx}}{\rho_{xx}^2 + \rho_{xy}^2} \tag{16}$$

Fig. 8 shows measurements for ϱ_{xx} and ϱ_{xy} of a silicon MOSFET as a function of the gate voltage at a fixed magnetic field. The corresponding σ_{xx}- and σ_{xy}-data are calculated on the basis of Eq. (16).

The classical curve $\sigma_{xy}^0 = -\dfrac{n_s e}{B}$ in Fig. 8 is drawn on the basis of the incorrect model, that the experimental data should lie always below the classical curve ($=$ fixed sign for $\Delta\sigma_{xy}$) so that the plateau value $\sigma_{xy} = $ const. (observable in the gate voltage region where σ_{xx} becomes zero) should change with the width of the plateau. Wider plateaus should give smaller values for $|\sigma_{xy}|$. The main discovery in 1980 was [23] that the value of the Hall resistance in the plateau region is not influenced by the plateau width as shown in Fig. 9. Even the aspect ratio L/W (L = length, W = width of the sample), which influences normally the accuracy in Hall effect measurements, becomes unimportant as shown in Fig. 10. Usually, the measured Hall resistance R_H^{exp} is always smaller than the theoretical value $R_H^{theor} = \rho_{xy}$ [26, 27]

$$R_H^{exp} = G \cdot R_H^{theor} \quad G < 1 \tag{17}$$

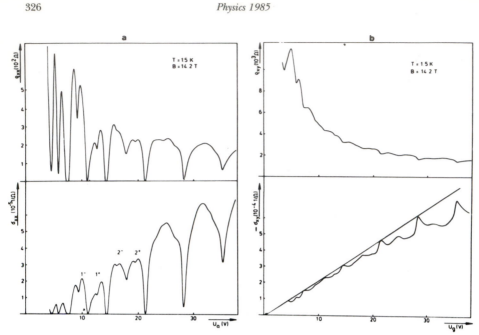

Fig. 8. Measured ϱ_{xx}- and ϱ_{xy}-data of a silicon MOSFET as a function of the gate voltage at $B = 14.2$ T together with the calculated σ_{xx}- and σ_{xy}-curves.

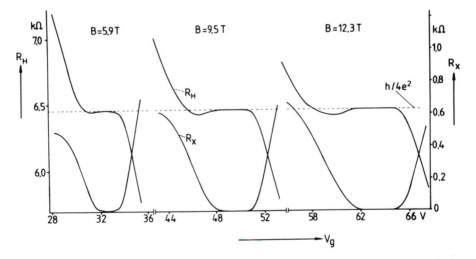

Fig. 9. Measurements of the Hall resistance R_H and the resistivity R_x as a function of the gate voltage at different magnetic field values. The plateau values $R_H = h/4e^2$ are independent of the width of the plateaus.

However, as shown in Fig. 11, the correction 1-G becomes zero (independent of the aspect ratio) if $\sigma_{xx} \to 0$ or the Hall angle θ approaches 90° ($\tan \theta = \dfrac{\sigma_{xy}}{\sigma_{xx}}$).

This means that any shape of the sample can be used in QHE-experiments as long as the Hall angle is 90° (or $\sigma_{xx} = 0$). However, outside the plateau region ($\sigma_{xx} \sim \rho_{xx} \neq 0$) the measured Hall resistance $R_H^{exp} = \dfrac{U_H}{I}$ is indeed always

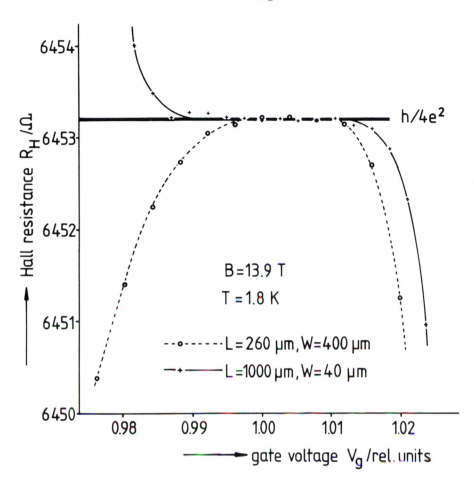

Fig. 10. Hall resistance R_H for two different samples with different aspect ratios L/W as a function of the gate voltage (B = 13.9 T).

smaller than the theoretical ρ_{xy}-value [28]. This leads to the experimental result that an additional minimum in R_H^{exp} becomes visible outside the plateau region as shown in Fig. 9, which disappears if the correction due to the finite length of the sample is included (See Fig. 12). The first high-precision measurements in 1980 of the plateau value in $R_H(V_g)$ showed already that these resistance values are quantized in integer parts of $h/e^2 = 25812.8 \ \Omega$ within the experimental uncertainty of 3 ppm.

The Hall plateaus are much more pronounced in measurements on GaAs-$Al_xGa_{1-x}As$ heterostructures, since the small effective mass m* of the electrons in GaAs (m*(Si)/m*(GaAs) > 3) leads to a relatively large energy splitting between Landau levels (Eq. 4), and the high quality of the GaAs-$Al_xGa_{1-x}As$ interface (nearly no surface roughness) leads to a high mobility μ of the electrons, so that the condition $\mu B > 1$ for Landau quantizations is fulfilled already at relatively low magnetic fields. Fig. 13 shows that well-developed Hall plateaus are visible for this material already at a magnetic field strength of 4 tesla. Since a finite carrier density is usually present in heterostructures, even

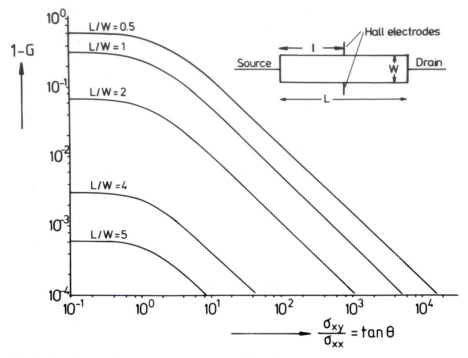

Fig. 11. Calculations of the correction term G in Hall resistance measurements due to the finite length to width ratio L/W of the device (l/L = 0,5).

at a gate voltage $V_g = 0V$, most of the published transport data are based on measurements without applied gate voltage as a function of the magnetic field. A typical result is shown in Fig. 14. The Hall resistance $R_H = \varrho_{xy}$ increases steplike with plateaus in the magnetic field region where the longitudinal resistance ϱ_{xx} vanishes. The width of the ϱ_{xx}-peaks in the limit of zero temperature can be used for a determination of the amount of extended states and the analysis [29] shows, that only few percent of the states of a Landau level are not localized. The fraction of extended states within one Landau level decreases with increasing magnetic field (Fig. 15) but the number of extended states within each level remains approximately constant since the degeneracy of each Landau level increases proportionally to the magnetic field.

At finite temperatures ϱ_{xx} is never exactly zero and the same is true for the slope of the ϱ_{xy}-curve in the plateau region. But in reality, the slope $d\varrho_{xy}/dB$ at $T<2K$ and magnetic fields above 8 Tesla is so small that the ϱ_{xy}-value stays constant within the experimental uncertainty of $6 \cdot 10^{-8}$ even if the magnetic field is changed by 5%. Simultaneously the resistivity ϱ_{xx} is usually smaller than $1m\Omega$. However, at higher temperatures or lower magnetic fields a finite resistivity ϱ_{xx} and a finite slope $d\varrho_{xy}/dn_s$ (or $d\varrho_{xy}/dB$) can be measured. The data are well described within the model of extended states at the energy position of the undisturbed Landau level E_n and a finite density of localized states between the Landau levels (mobility gap). Like in amorphous systems, the temperature dependence of the conductivity σ_{xx} (or resistivity ϱ_{xx}) is thermally activated

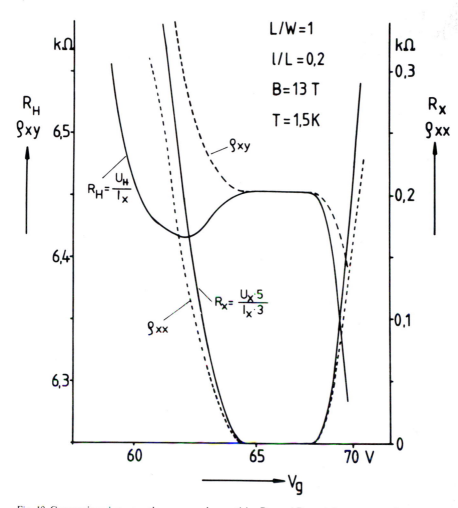

Fig. 12. Comparison between the measured quantities R_H and R_x and the corresponding resistivity components ϱ_{xy} and ϱ_{xx}, respectively.

with an activation energy E_a corresponding to the energy difference between the Fermi energy E_F and the mobility edge. The largest activation energy with a value $E_a = 1/2\hbar\omega_c$ (if the spin splitting is negligibly small and the mobility edge is located at the center E_n of a Landau level) is expected if the Fermi energy is located exactly at the midpoint between two Landau levels.

Experimentally, an activated resistivity

$$\varrho_{xx} \sim \exp\left[-(E_a/kT)\right] \tag{18}$$

is observed in a wide temperature range for different two-dimensional systems (deviations from this behaviour, which appear mainly at temperatures below 1K, will be discussed separately) and a result is shown in Fig. 16. The activation energies (deduced from these data) are plotted in Fig. 17 for both, silicon MOSFETs and GaAs-Al$_x$Ga$_{1-x}$As heterostructures as a function of the

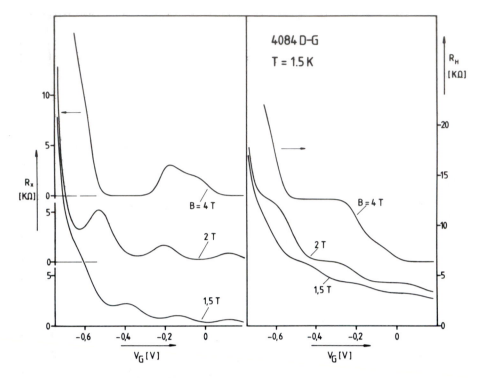

Fig. 13. Measured curves for the Hall resistance R_H and the longitudinal resistance R_x of a GaAs-$Al_xGa_{1-x}As$ heterostructure as a function of the gate voltage at different magnetic fields.

magnetic field and the data agree fairly well with the expected curve $E_a = 1/2\hbar\omega_c$. Up to now, it is not clear whether the small systematic shift of the measured activation energies to higher values originates from a temperature dependent prefactor in Eq. (18) or is a result of the enhancement of the energy gap due to many body effects.

The assumption, that the mobility edge is located close to the center of a Landau level E_n is supported by the fact that for the samples used in the experiments only few percent of the states of a Landau level are extended [29]. From a systematic analysis of the activation energy as a function of the filling factor of a Landau level it is possible to determine the density of states $D(E)$ [30]. The surprising result is, that the density of states (DOS) is finite and approximately constant within 60 % of the mobility gap as shown in Fig. (18). This background DOS depends on the electron mobility as summarized in Fig. (19).

An accurate determination of the DOS close to the center of the Landau level is not possible by this method since the Fermi energy becomes temperature dependent if the DOS changes drastically within the energy range of 3kT. However, from an analysis of the capacitance C as a function of the Fermi energy the peak value of the DOS and its shape close to E_n can be deduced [31, 32].

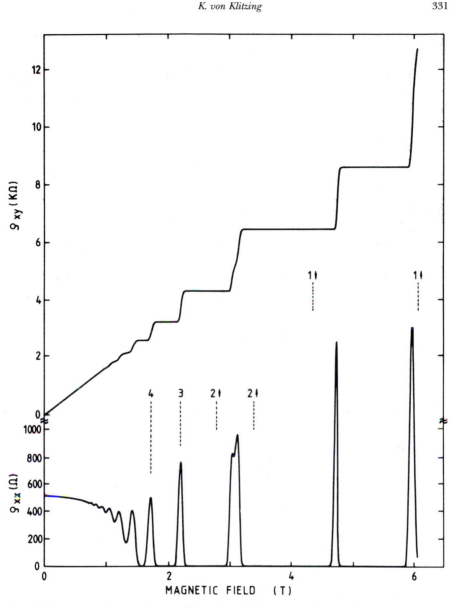

Fig. 14. Experimental curves for the Hall resistance $R_H = \varrho_{xy}$ and the resistivity $\varrho_{xx} \sim R_x$ of a heterostructure as a function of the magnetic field at a fixed carrier density corresponding to a gate voltage $V_g = 0V$. The temperature is about 8mK.

This analysis is based on the equation

$$\frac{1}{C} = \frac{1}{e^2 \cdot D\,(E_F)} + \text{const.} \tag{19}$$

The combination of the different methods for the determination of the DOS leads to a result as shown in Fig. (20). Similar results are obtained from other experiments, too [33, 34] but no theoretical explanation is available.

If one assumes that only the occupation of extended states influences the Hall effect, than the slope $d\varrho_{xy}/dn_s$ in the plateau region should be dominated

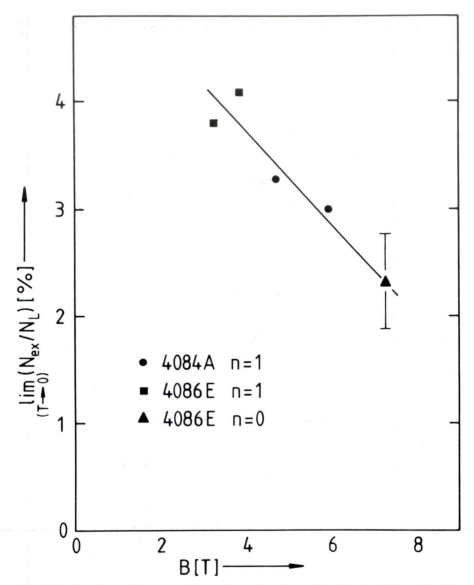

Fig. 15. Fraction of extended states relative to the number of states of one Landau level as a function of the magnetic field.

by the same activation energy as found for $\varrho_{xx}(T)$. Experimentally [35], a one to one relation between the minimal resistivity ϱ_{xx}^{min} at integer filling factors and the slope of the Hall plateau has been found (Fig. 21) so that the flatness of the plateau increases with decreasing resistivity, which means lower temperature or higher magnetic fields.

The temperature dependence of the resistivity for Fermi energies within the mobility gap deviates from an activated behaviour at low temperatures, typically at T<1K. Such deviations are found in measurements on disordered

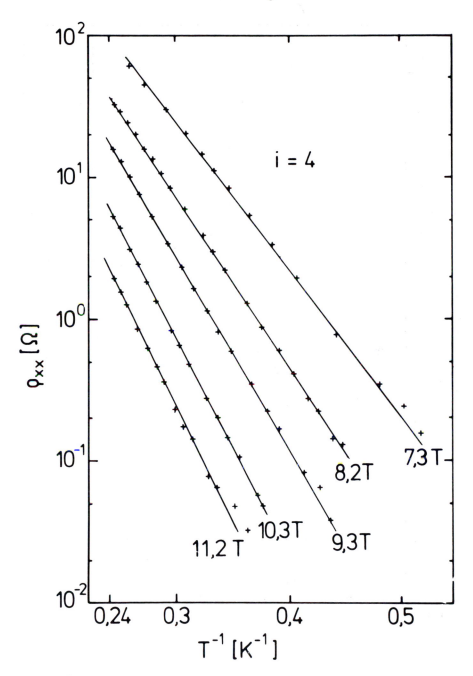

Fig 16. Thermally activated resistivity ϱ_{xx} at a filling factor i = 4 for a silicon MOSFET at different magnetic field values.

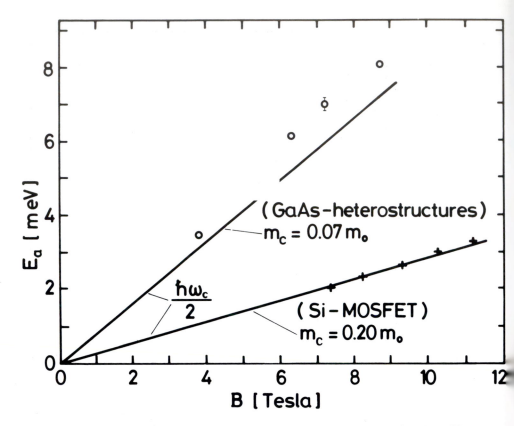

Fig. 17. Measured activation energies at filling factors i = 2 (GaAs heterostructure) or i = 4 (Si-MOSFET) as a function of the magnetic field. The data are compared with the energy $0,5\hbar\omega_c$.

systems, too, and are interpreted as variable range hopping. For a two-dimensional system with exponentially localized states a behaviour

$$\varrho_{xx} \sim \exp\left[-(T_0/T)^{1/3}\right] \tag{20}$$

is expected. For a Gaussian localization the following dependence is predicted [36, 37]

$$\rho_{xx} \sim \frac{1}{T} \exp\left\{-(T_0/T)^{1/2}\right\} \tag{21}$$

The analysis of the experimental data demonstrates (Fig. 22) that the measurements are best described on the basis of Eq. (21). The same behaviour has been found in measurements on another two-dimensional system, on InP-InGaAs heterostructures [38].

The contribution of the variable range hopping (VRH) process to the Hall effect is negligibly small [39] so that experimentally the temperature dependence of $d\varrho_{xy}/dn_s$ remains thermally activated even if the resistivity ϱ_{xx} is dominated by VRH.

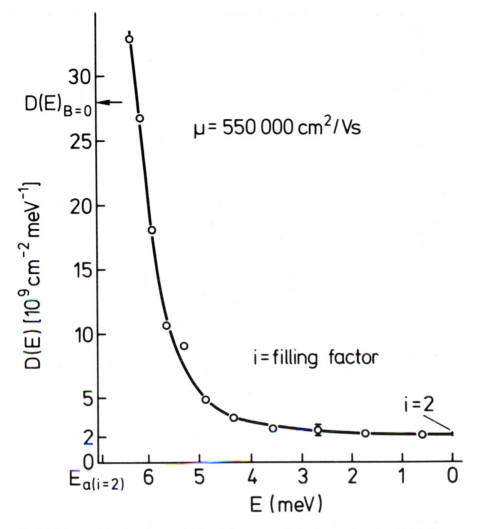

Fig. 18. Measured density of states (deduced from an analysis of the activated resisitivity) as a function of the energy relative to the center between two Landau levels (GaAs-heterostructure).

The QHE breaks down if the Hall field becomes larger than about $E_H = 60V/cm$ at magnetic fields of 5 Tesla.

This corresponds to a classical drift velocity $v_D = \dfrac{E_H}{B} \approx 1200m/s$. At the critical Hall field E_H (or current density j) the resistivity increases abruptly by orders of magnitude and the Hall plateau disappears. This phenomenon has been observed by different authors for different materials [40−47]. A typical result is shown in Fig. 23. At a current density of $j_c = 0,5$ A/m the resistivity ϱ_{xx} at the center of the plateau (filling factor i = 2) increases drastically. This instability, which develops within a time scale of less than 100 ns seems to originate from a runaway in the electron temperature but also other mechanism like electric field dependent delocalization, Zener tunneling or emission of

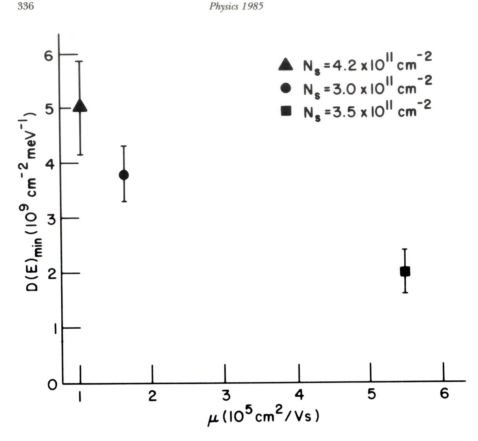

Fig. 19. Background density of states as a function of the mobility of the device.

acoustic phonons, if the drift velocity exceeds the sound velocity, can be used for an explanation [48–50].

Fig. 23 shows that ϱ_{xx} increases already at current densities well below the critical value j_c which may be explained by a broadening of the extended state region and therefore a reduction in the mobility gap ΔE. If the resistivity ϱ_{xx} is thermally activated and the mobility gap changes linearly with the Hall field (which is proportional to the current density j) then a variation

$$\ln \varrho_{xx} \sim j$$

is expected. Such a dependence is seen in Fig. 24 but a quantitative analysis is difficult since the current distribution within the sample is usually inhomogenious and the Hall field, calculated from the Hall voltage and the width of the sample, represents only a mean value. Even for an ideal two-dimensional system an inhomogenious Hall potential distribution across the width of the sample is expected [51–53] with an enhancement of the current density close to the boundaries of the sample.

The experimental situation is still more complicated as shown in Fig. 25. The potential distribution depends strongly on the magnetic field. Within the plateau region the current path moves with increasing magnetic field across the

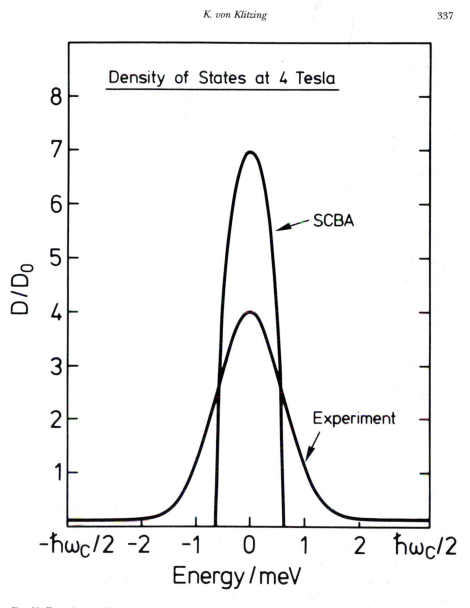

Fig. 20. Experimentally deduced density of states of a GaAs heterostructure at B = 4T compared with the calculated result based on the self-consistent Born approximation (SCBA).

width of the sample from one edge to the other one. A gradient in the carrier density within the two-dimensional system seems to be the most plausible explanation but in addition an inhomogeneity produced by the current itself may play a role. Up to now, not enough microscopic details about the two-dimensional system are known so that at present a microscopic theory, which describes the QHE under real experimental conditions, is not available. However, all experiments and theories indicate that in the limit of vanishing resistivity ϱ_{xx} the value of the quantized Hall resistance depends exclusively on fundamental constants. This leads to a direct application of the QHE in metrology.

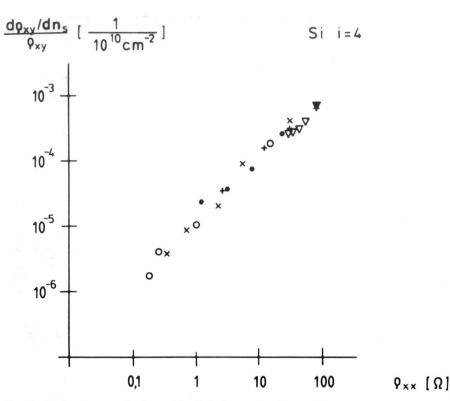

$$\frac{d\varrho_{xy}/dn_s}{\varrho_{xy}} \quad [\ \frac{1}{10^{10}cm^{-2}} \ ]$$

Si i = 4

ϱ_{xx} [Ω]

Fig. 21. Relation between the slope of the Hall plateaus $d\varrho_{xy}/dn_s$ and the corresponding ϱ_{xx}-value at integer filling factors.

5. *Application of the Quantum Hall Effect in Metrology*

The applications of the Quantum Hall Effect are very similar to the applications of the Josephson-Effect which can be used for the determination of the fundamental constant h/e or for the realization of a voltage standard. In analogy, the QHE can be used for a determination of h/e^2 or as a resistance standard. [54].

Since the inverse fine structure constant α^{-1} is more or less identical with h/e^2 (the proportional constant is a fixed number which includes the velocity of light), high precision measurements of the quantized Hall resistance are important for all areas in physics which are connected with the finestructure constant.

Experimentally, the precision measurement of α is reduced to the problem of measuring an electrical resistance with high accuracy and the different methods and results are summarized in the Proceedings of the 1984 Conference on Precision Electromagnetic Measurements (CPEM 84) [55]. The mean value of measurements at laboratories in three different countries is

$$\alpha^{-1} = 137,035988 \pm 0.00002$$

The internationally recommended value (1973) is

$$\alpha^{-1} = 137,03604 \pm 0.00011$$

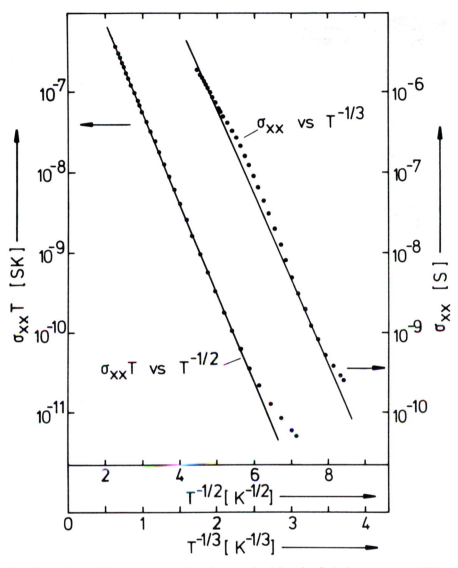

Fig. 22. Analysis of the temperature dependent conductivity of a GaAs heterostructure (filling factor i = 3) at T < 0,2K.

and the preliminary value for the finestructure constant based on a new least square adjustment of fundamental constants (1985) is

$$\alpha^{-1} = 137,035991 \pm 0.000008$$

Different groups have demonstrated that the experimental result is within the experimental uncertainty of less than $3.7 \cdot 10^{-8}$ independent of the material (Si, GaAs, $In_{0.53}Ga_{0.47}As$) and of the growing technique of the devices (MBE or MOCVD) [56]. The main problem in high precision measurements of α is — at present — the calibration and stability of the reference resistor. Fig. 26 shows the drift of the maintained 1Ω-resistor at different national laboratories. The

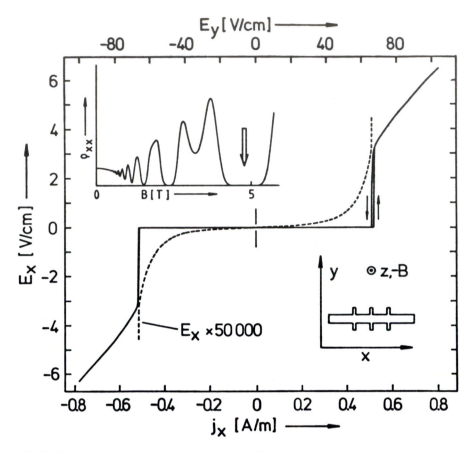

Fig. 23. Current-voltage characteristic of a GaAs-Al$_x$Ga$_{1-x}$As heterostructure at a filling factor i = 2 (T = 1,4K). The device geometry and the ϱ_{xx}(B)-curve are shown in the inserts.

very first application of the QHE is the determination of the drift coefficient of the standard resistors since the quantized Hall resistance is more stable and more reproducible than any wire resistor. A nice demonstration of such an application is shown in Fig. 27. In this experiment the quantized Hall resistance R_H has been measured at the "Physikalisch Technische Bundesanstalt" relative to a reference resistor R_R as a function of time. The ratio R_H/R_R changes approximately linearly with time but the result is independent of the QHE-sample. This demonstrates that the reference resistor changes its value with time. The one standard deviation of the experimental data from the mean value is only $2.4 \cdot 10^{-8}$ so that the QHE can be used already today as a relative standard to maintain a laboratory unit of resistance based on wire-wound resistors. There exists an agreement that the QHE should be used as an absolute resistance standard if three independent laboratories measure the same value for the quantized Hall resistance (in SI-units) with an uncertainty of less than $2 \cdot 10^{-7}$. It is expected that these measurements will be finished until the end of 1986.

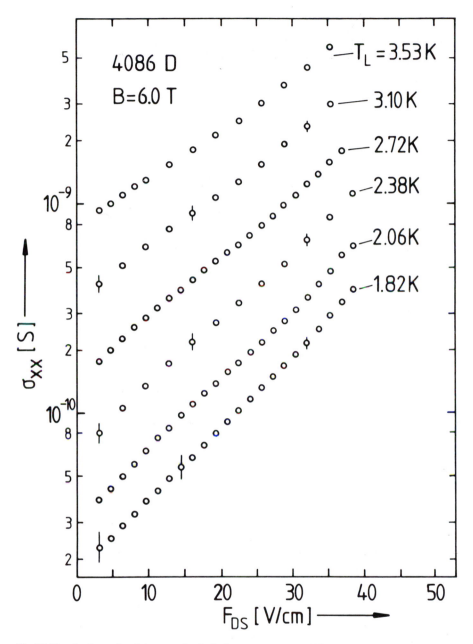

Fig. 24. Nonohmic conductivity σ_{xx} of a GaAs heterostructure at different temperatures T_L (filling factor $i = 2$). An instability is observed at source-drain fields larger than 40 V/cm.

Acknowledgements

The publicity of the Nobel Prize has made clear that the research work connected with the Quantum Hall Effect was so successful because a tremendous large number of institutions and individuals supported this activity. I would like to thank all of them and I will mention by name only those scientists who supported my research work at the time of the discovery of the QHE in

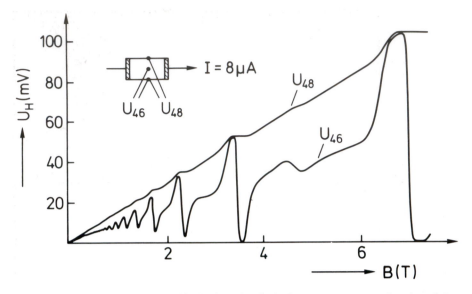

Fig. 25. Measured Hall potential distribution of a GaAs heterostructure as a function of the magnetic field.

1980. Primarily, I would like to thank G. Dorda (Siemens Forschungslaboratorien) and M. Pepper (Cavendish Laboratory, Cambridge) for providing me with high quality MOS-devices. The continuous support of my research work by G. Landwehr and the fruitful discussions with my coworker, Th. Englert, were essential for the discovery of the Quantum Hall Effect and are greatfully acknowledged.

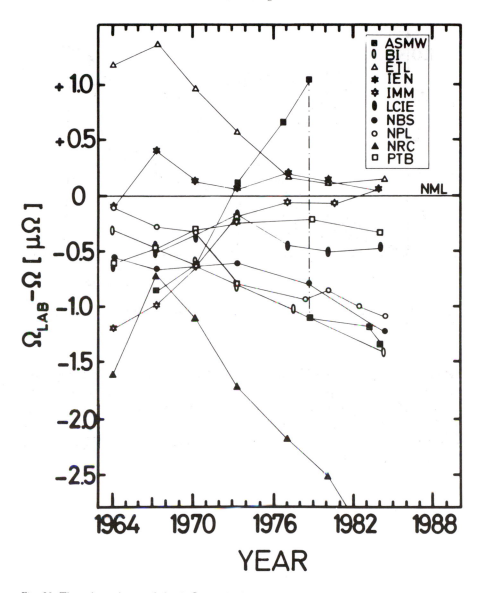

Fig. 26. Time dependence of the 1 Ω standard resistors maintained at the different national laboratories.

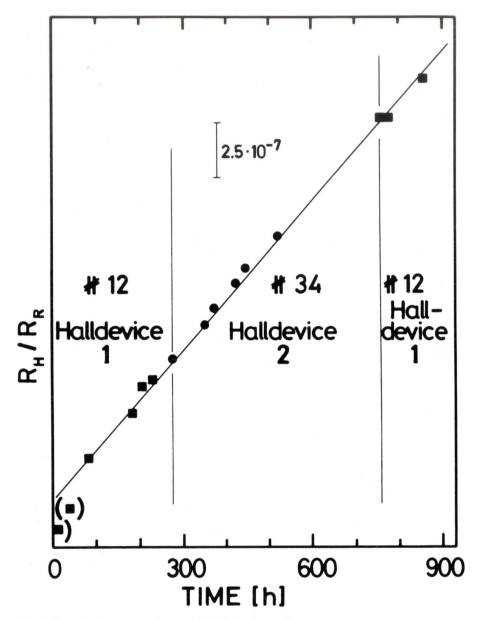

Fig. 27. Ratio R_H/R_R between the quantized Hall resistance R_H and a wire resistor R_R as a function of time. The result is time dependent but independent of the Hall device used in the experiment.

REFERENCES

[1] A.B. Fowler, F.F. Fang, W.E. Howard and P.J. Stiles Phys. Rev. Letter *16*, 901 (1966)

[2] For a review see: Proceedings of the Int. Conf. on Electronic Properties of Two-Dimensional Systems, Surf. Sci. *58*, (1976), *73* (1978), *98* (1980, *113* (1982), *142* (1984)

[3] F. Stern and W.E. Howard, Phys. Rev. *163*, 816 (1967)

[4] T. Ando, J. Phys. Soc., Jpn. *51*, 3893 (1982)

[5] J.F. Koch, Festkörperprobleme (Advances in Solid State Physics), H.J. Queisser, Ed. (Pergamon-Vieweg, Braunschweig, 1975) Vol. XV, p. 79

[6] T. Mimura, Surf. Science *113*, 454 (1982)

[7] R.B. Laughlin, Surface Science *113*, 22 (1982)

[8] R. Kubo, S.J. Miyake and N. Hashitsume, in Solid State Physics, Vol. 17, 269 (1965), F. Seitz and D. Turnball, Eds., (Academic Press, New York, 1965)

[9] T. Ando, J. Phys. Soc. Jpn. *37*, 1233 (1974)

[10] R. B. Laughlin in Springer Series in Solid State Sciences *53*, p. 272, G. Bauer, F. Kuchar and H. Heinrich, Eds. (Springer Verlag, 1984)

[11] R.E. Prange, Phys. Rev. B *23*, 4802 (1981)

[12] H. Aoki and T. Ando, Solid State Commun. *38*, 1079 (1981)

[13] J. T. Chalker, J. Phys. C *16*, 4297 (1983)

[14] W. Brenig, Z. Phys. *50B*, 305 (1983)

[15] A. Mac Kinnon, L. Schweitzer and B. Kramer, Surf. Sci. *142*, 189 (1984)

[16] T. Ando, J. Phys. Soc. Jpn. *52*, 1740 (1983)

[17] L. Schweitzer, B. Kramer and A. Mac Kinnon, J. Phys. C *17*, 4111 (1984)

[18] H. Aoki and T. Ando, Phys. Rev. Letters *54*, 831 (1985)

[19] E. Abrahams, P.W. Anderson, D.C. Licciardello and T.V. Ramakrishnan, Phys. Rev. Letters *42*, 673 (1979)

[20] G.A. Baraff and D.C. Tsui, Phys. Rev. B *24*, 2274 (1981)

[21] T. Toyoda, V. Gudmundsson and Y. Takahashi, Phys. Letters *102A*, 130 (1984)

[22] S. Kawaji and J. Wakabayashi, Surf. Schi. *58*, 238 (1976)

[23] S. Kawaji, T. Igarashi and J. Wakabayashi, Progr. in Theoretical Physics *57*, 176 (1975)

[24] T. Englert and K. v. Klitzing, Surf. Sci. *73*, 70 (1978)

[25] K. v. Klitzing, G. Dorda and M. Pepper, Phys. Rev. Letters *45*, 494 (1980)

[26] K. v. Klitzing, H. Obloh, G. Ebert, J. Knecht and K. Ploog, Prec. Measurement and Fundamental Constants II, B.N. Taylor and W.D. Phillips, Eds., Natl. Burl. Stand. (U.S.), Spec. Publ. *617*, (1984) p. 526

[27] R.W. Rendell and S.M. Girvin, Prec. Measurement and Fundamental Constants II, B.N. Taylor and W.D. Phillips. Eds. Natl. Bur. Stand. (U.S.), Spec. Publ. *617*, (1984) p. 557

[28] K. v. Klitzing, Festkörperprobleme (Advances in Solid State Physics), XXI, 1 (1981), J. Treusch, Ed., (Vieweg, Braunschweig)

[29] G. Ebert, K. v. Klitzing, C. Probst and K. Ploog, Solid State Commun. *44*, 95 (1982)

[30] E. Stahl, D. Weiss, G. Weimann, K. v. Klitzing and K. Ploog, J. Phys. C *18*, L 783 (1985)

[31] T.P. Smith, B.B. Goldberg, P.J. Stiles and M. Heiblum, Phys. Rev. B *32*, 2696 (1985)

[32] V. Mosser, D. Weiss, K. v. Klitzing, K. Ploog and G. Weimann to be published in Solid State Commun.

[33] E. Gornik, R. Lassnig, G. Strasser, H.L. Störmer, A.C. Gossard and W. Wiegmann, Phys. Rev. Lett. *54*, 1820 (1985)

[34] J.P. Eisenstein, H.L. Störmer, V. Navayanamurti, A.Y. Cho and A.C. Gossard, Yamada Conf. XIII on Electronic Properties of Two-Dimensional Systems, p. 292 (1985)

[35] B. Tausendfreund and K. v. Klitzing, Surf. Science *142*, 220 (1984)

[36] M. Pepper, Philos, Mag. *37*, 83 (1978)

[37] Y. Ono, J. Phys. Soc. Jpn. *51*, 237 (1982)

[38] Y. Guldner, J.P. Hirtz, A. Briggs, J.P. Vieren, M. Voos and M. Razeghi, Surf. Science *142*, 179 (1984)

[39] K.I. Wysokinski and W. Brenig, Z. Phys. B − Condensed Matter *54*, 11 (1983)

[40] G. Ebert and K. v. Klitzing, J. Phys. C 5441 (1983)

[41] M.E. Cage, R.F. Dziuba, B.F. Field, E.R. Williams, S.M. Girvin, A.C. Gossard, D.C. Tsui and R.J. Wagner, Phys. Rev. Letters *51*, 1374 (1983)

[42] F. Kuchar, G. Bauer, G. Weimann and H. Burkhard, Surf. Science *142*, 196 (1984)

[43] H.L. Störmer, A.M. Chang, D.C. Tsui and J.C.M. Hwang, Proc. 17th ICPS, San Francisco 1984

[44] H. Sakaki, K. Hirakawa, J. Yoshino, S.P. Svensson, Y. Sekiguchi, T. Hotta and S. Nishii, Surf. Science *142*, 306 (1984)

[45] K. v. Klitzing, G. Ebert, N. Kleinmichel, H. Obloh, G. Dorda and G. Weimann, Proc. 17th ICPS, San Francisco 1984

[46] F. Kuchar, R. Meisels, G. Weimann and H. Burkhard, Proc. 17th ICPS, San Francisco 1984

[47] V.M. Pudalov and S.G. Semenchinsky, Solid State Commun. *51*, 19 (1984)

[48] P. Streda and K. v. Klitzing, J. Phys. C *17*, L 483 (1984)

[49] O. Heinonen, P.L. Taylor and S.M. Girvin, Phys. Rev. B *30*, 3016 (1984)

[50] S.A. Trugman, Phys. Rev. B *27*, 7539 (1983)

[51] A.H. Mac Donold, T.M. Rice and W.F. Brinkman, Phys. Rev. B *28*, 3648 (1983)

[52] O. Heinonen and P.L. Taylor, Phys. Rev. B *32*, 633 (1985)

[53] J. Riess, J. Phys. C *17*, L 849 (1984)

[54] K. v. Klitzing and G. Ebert, Metrologia *21*, 11 (1985)

[55] High-precision measurements of the quantized Hall resistance are summarized in: IEEE Trans. Instrum. Meas. IM *34*, pp. 301–327

[56] F. Delahaye, D. Dominguez, F. Alexandre, J. P. Andre, J. P. Hirtz and M. Razeghi, Metrologia *22*, 103 (1986)

Physics 1986

ERNST RUSKA

*for his fundamental work in electron optics, and for the design of the first
electron microscope*

GERD BINNIG and HEINRICH ROHRER

for their design of the scanning tunneling microscope

THE NOBEL PRIZE FOR PHYSICS

Speech by Professor SVEN JOHANSSON of the Royal Academy of Sciences.
Translation from the Swedish text

Your Majesties, Your Royal Highnesses, Ladies and Gentlemen,

The problem of the basic structure of matter has long interested man but it was not until the time of the Greek philosophers that the problem took on a scientific character. These ideas reached their culmination in Democritos' theory which postulated that atoms were the building blocks of matter. All this was, however, mere speculation, and it was first the early science and technology of Western Europe which made it possible to tackle the problem experimentally.

The first major breakthrough came with the invention of the microscope. The significance of the microscope in the fields of, for example, biology and medicine is well known, but it did not provide a means of studying the basic nature of matter. The reason is that there is a limit to the amount of detail one can see in a microscope. This is connected with the wave nature of light. In the same way as ocean waves are not affected, to any great degree, by small objects, but only by larger ones, for example a breakwater, light will not produce a picture of an object that is too small. The limit is set by the wavelength of light which is about 0.0005 mm. We know that an atom is 1000 times smaller. It is clear, therefore, that something radically new was needed in order to be able to see an atom.

This new development was the electron microscope. The electron microscope is based on the principle that a short coil of a suitable construction, carrying an electric current, can deflect electrons in the same way that a lens deflects light. A coil can therefore give an enlarged image of an object that is irradiated with electrons. The image can be registered on a fluorescent screen or a photographic film. In the same way that lenses can be combined to form a microscope, it was found that an electron microscope could be constructed of coils. As the electrons used in an electron microscope have a much shorter wavelength than light, it is thus possible to reach down to much finer details. Several scientists, among them Hans Busch, Max Knoll, and Bodo von Borries, contributed to the development of the instrument, but Ernst Ruska deserves to be placed foremost. He built in 1933 the first electron microscope with a performance significantly better than that of an ordinary light microscope. Developments since then have led to better and better instruments. The importance, in many areas of research, of the invention of the electron microscope should, by now, be well known.

The microscope can be regarded as an extension of the human eye. But sight is not the only sense we use to orientate us in our surroundings, another is feeling. With modern technology it is possible to construct equipment that is based on the principle of feeling, using, for example, a sort of mechanical finger. The "finger" may be a very fine needle which is moved across the surface of the structure to be investigated. By registering the needle's movements in the vertical direction as it traverses the surface, a sort of topographical map is obtained, which, in principle, is equivalent to the image obtained in a electron microscope. It is clear that this is a rather coarse method of microscopical investigation and no one had expected any revolutionary developments in this field. However, two basic improvements led to a breakthrough. The most important of these was that a method for keeping the tip of the needle at a very small and exact constant distance from the surface was developed, thus eliminating the mechanical contact between the needle and the surface, which was a limiting factor. This was achieved using the so-called tunnelling effect. This involves applying a potential between the needle tip and the surface so that an electric current flows between the needle and the surface without actually touching them, provided that the tip of the needle and the surface are close enough together. The magnitude of the current is strongly dependent on the distance, and can therefore be used to keep the needle a certain distance above the surface with the aid of a servo mechanism, typically 2−3 atomic diameters. It was also decisive that it turned out to be possible to produce extremely fine needles so that the tip consists of only a few atoms. It is clear that if such a fine tip is moved across a surface at a height of a few atomic diameters the finest atomic details in the surface structure can be registered. It is as if one were feeling the surface with an infinitely fine finger. A crystal surface which appears completely flat in a microscope is seen with this instrument to be a plain on which atoms rise like hills in a regular pattern.

Attempts by Russell Young and co-workers to realize these ideas revealed enormous experimental difficulties. The scientists who finally mastered these difficulties were Gerd Binnig and Heinrich Rohrer. Here it was a question of moving the needle over the surface of the sample and registering its vertical position, with great precision and without disturbing vibrations. The data obtained are then printed out, in the form of a topographic map of the surface, by a computer. The investigation may be concerned with a crystal surface, whose structure is of interest in microelectronic applications. Another example is the investigation of the adsorption of atoms on a surface. It has also been found to be possible to study organic structures, for example, DNA molecules and viruses. This is just the beginning of an extremely promising and fascinating development. The old dream from antiquity of a visible image of the atomic structure of matter is beginning to look like a realistic possibility, thanks to progress in modern microscopy.

Professor Ruska, Dr Binnig, Dr Rohrer!

In Ihrer bahnbrechenden Arbeit haben Sie den Grund für die entscheidenden Entwicklungen moderner Mikroskopie gelegt. Es ist jetzt möglich, die kleinsten Einzelheiten der Struktur von Materie zu erkennen. Dies ist von grösster Bedeutung − nicht nur in der Physik, sondern auch in vielen anderen Bereichen der Wissenschaft.

Es gereicht mir zur Ehre und Freude, Ihnen die herzlichsten Glückwünsche der Königlich Schwedischen Akademie der Wissenschaften zu übermitteln. Darf ich Sie nun bitten vorzutreten, um Ihren Preis aus der Hand Seiner Majestät des Königs entgegenzunehmen.

Dr. Ernst Auerke

ERNST RUSKA

I was born on 25 December 1906 in Heidelberg as the fifth of seven children of Professor Julius Ruska and his wife Elisbeth (*née* Merx). After graduating from grammar school in Heidelberg I studied electronics at the Technical College in Munich, studies which I began in the autumn of 1925 and continued two years later in Berlin. I received my practical training from Brown-Boveri & Co in Mannheim and Siemens & Halske Ltd in Berlin. Whilst still a student at the Technical College in Berlin I began my involvement with high voltage and vacuum technology at the Institute of High Voltage, whose director was Professor Adolf Matthias. Under the direct tutelage of Dr Max Knoll and together with other doctoral students I worked on the development of a high performance cathode ray oscilloscope. On the one hand my interest lay principally in the development of materials for the building of vacuum instruments according to the principles of construction; on the other it lay in continuing theoretical lectures and practical experiments in the optical behaviour of electron rays.

My first completed scientific work (1928–9) was concerned with the mathematical and experimental proof of Busch's theory of the effect of the magnetic field of a coil of wire through which an electric current is passed and which is then used as an electron lens. During the course of this work I recognised that the focal length of the waves could be shortened by use of an iron cap. From this discovery the polschuh lens was developed, a lens which has been used since then in all magnetic high-resolution electron microscopes. Further work, conducted together with Dr Knoll, led to the first construction of an electron microscope in 1931. With this instrument two of the most important processes for image reproduction were introduced — the principles of emission and radiation. In 1933 I was able to put into use an electron microscope, built by myself, that for the first time gave better definition than a light microscope. In my Doctoral thesis of 1934 and for my university teaching thesis (1944), both at the Technical College in Berlin, I investigated the properties of electron lenses with short focal lengths.

Since the further technical development of electron microscopes could not be the task of a college institute — whose resources would have been far overstretched — I went to work in industry in the field of electron optics. From 1933 to 1937 I was with Fernseh Ltd in Berlin-Zehlendorf and was responsible for the development of television receivers and transmitters, as well as photoelectric cells with secondary amplification. Convinced of the great practical importance of electron microscopy for pure and applied research I attempted during this time to continue the development of high-resolution electron microscopes with larger materials, this time working with Dr Bodo von Borries. This work was made possible in 1936–7 by Siemens & Halske. In Berlin–Spandau in 1937 we set up the Laboratory for Electron Optics and developed there until 1939 the first customised electron microscopes (the 'Siemens Super Microscope'). Parallel to the development of this instrument my brother, Dr Med. Helmut Ruska, and his colleagues worked on its application, particularly in the medical and biological fields. In order to promote its usage in different scientific areas as quickly as possible we suggested to Siemens that they set up a visiting institute for research work to be carried out using electron microscopy. This institute was founded in 1940. From this institute, in which we worked together with both German and foreign scientists, around 200 scientific papers were published before the end of 1944. My task consisted in the development and production of the electron microscope, such that by the beginning of 1945 around 35 institutions were equipped with one.

In the years following 1945 I, together with a majority of new colleagues, reconstituted the Institute of Electron Optics in Berlin-Siemensstadt, which had been disbanded due to bombing, so that by 1949 electron microscopes were again being built. This new period of development led in 1954 to 'Elmiskop 1', which since then has been used in over 1200 institutions the world over. At the same time I sought the further physical development of the electron microscope by working at other scientific institutions. Thus from August 1947 to December 1948 I worked at the German Academy of Sciences in Berlin-Buch in the Faculty of Medicine and Biology, then from January 1949 as Head of Department at what is today the Fritz Haber Institute of the Max Planck Society in Berlin-Dahlem. Here on 27 June 1957 I was made Director of the Institute for Electron Microscopy, after I had given up my position with Siemens in 1955. I retired on 31 December 1974.

From 1949 until 1971 I held lectures on the basic principles of electron optics and electron microscopy at both the Free University and the Technical University of Berlin. My publications in the area of electron optics and electron microscopy include several contributions to books and over 100 original scientific papers.

(*added by the editor*): Ernst Ruska died on May 25, 1988.

THE DEVELOPMENT OF THE ELECTRON MICROSCOPE AND OF ELECTRON MICROSCOPY

Nobel lecture, December 8, 1986

by

ERNST RUSKA

Max-Eyth-Strasse 20, D-1000 BERLIN 33

A. Parents' house, family

A month ago, the Nobel Foundation sent me its yearbook of 1985. From it I learnt that many Nobel lectures are downright scientific lectures, interspersed with curves, synoptic tables and quotations. I am somewhat reluctant to give here such a lecture on something that can be looked up in any modern schoolbook on physics. I will therefore not so much report here on physical and technical details and their connections but rather on the human experiences — some joyful events and many disappointments which had not been spared me and my colleagues on our way to the final breakthrough. This is not meant to be a complaint though; I rather feel that such experiences of scientists in quest of new approaches are absolutely understandable, or even normal.

In such a representation I must, of course, consider the influence of my environment, in particular of my family. There have already been some scientists in my family: My father, Julius Ruska, was a historian of sciences in Heidelberg and Berlin; my uncle, Max Wolf, astronomer in Heidelberg; his assistant, a former pupil of my father and my godfather, August Kopff, Director of the Institute for astronomical calculation of the former Friedrich-Wilhelm University in Berlin. A cousin of my mother, Alfred Hoche, was Professor for Psychiatry in Freiburg/Breisgau; my grandfather from my mother's side, Adalbert Merx, theologian in Gießen and Heidelberg.

My parents lived in Heidelberg and had seven children. I was the fifth, my brother Helmut the sixth. To him I had particularly close and friendly relations as long as I can remember. Early, optical instruments made a strong impression on us. Several times Uncle Max had shown us the telescopes at the observatory on the Königstuhl near Heidelberg headed by him. With the light microscope as well we soon had impressive, yet contradictory, relations. In the second floor of our house, my father had two study rooms connected by a broad sliding door which usually was open. One room he used for his scientific historical studies relating to classical philology, the other for his scientific interests, in particular mineralogy, botany and zoology. When our games with neighbours' kids in front of the house became too noisy, he would knock at the window panes. This usually only having a brief effect, he soon knocked a second time, this time considerably louder. At the third knock, Helmut and I had to come to his room and sit still on a low wooden stool, dos à dos, up to one hour at 2 m distance from his desk. While doing so we would see on a table in the other room the pretty yellowish wooden box that housed my father's big Zeiss microscope, which we were strictly forbidden to touch. He sometimes demonstrated to us interesting objects under the microscope, it is true; for good reasons, however, he feared that childrens' hands would damage the objective or the specimen by clumsy manipulation of the coarse and fine drive. Thus, our first relation to the value of microscopy was not solely positive.

B. School, vocational choice

Much more positive was, several years later, the excellent biology instruction my brother had through his teacher Adolf Leiber and the very thorough teaching I received through my teacher Karl Reinig. To my great pleasure I recently read an impressive report on Reinig's personality in the Memoirs of a two-years-older student at my school, the later theoretical physicist Walter Elsasser. Even today I remember the profound impression Reinig's comments made upon me when he explained that the movement of electrons in an electrostatic field followed the same laws as the movement of inert mass in gravitational fields. He even tried to explain to us the limitation of microscopical resolution due to the wavelength of light. I certainly did not clearly understand all this then, because soon after that on one of our many walks through the woods around Heidelberg I had a long discussion on that subject with my brother Helmut, who already showed an inclination to medicine, and my classmate Karl Deißler, who later studied medicine as well.

In our College (Humanistisches Gymnasium), we had up to 17 hours of Latin, Greek and French per week. In contrast to my father, who was extremely gifted for languages, I produced only very poor results in this field. My father, at that time teacher at the same school, daily learnt about my minus efforts from his colleagues and blamed me for being too lazy, so that I had some sorrowful school years. My Greek teacher, a fellow student of my father, had a more realistic view of things: He gave me for my confirmation the book "Hinter Pflug und Schraubstock" (Behind plow and vise) by the Swabian "poet" engineer Max Eyth (1836–1906). I had always been fascinated by technical progress; in particular I was later interested in the development of aeronautics, the construction of airships and air planes. The impressive book of Max Eyth definitely prompted me to study engineering. My father, having studied sciences at the universities of Straßburg, Berlin and Heidelberg, obviously regarded study at a Technical High School as not being adequate and offered me one physics semester at a university. I had, however, the strong feeling that engineering was more to my liking and refused.

C. The cathode-ray oscillograph and the short coil

After I had studied two years electrotechnical engineering in Munich, my father received a call to become head of a newly founded Institute for the

History of Sciences in Berlin in 1927. Thus, after my pre-examination in Munich I came to Berlin for the second half of my studies. Here I specialized in high-voltage techniques and electrical plants and heard, among others, the lectures of Professor Adolf Matthias. At the end of the summer term in 1928 he told us about his plan of setting up a small group of people to develop from the Braun tube an efficient cathode-ray oscillograph for the measurement of very fast electrical processes in power stations and on open-air high-voltage transmission lines. Perhaps with the memory of my physics school lesson in the back of my head, I immediately volunteered for this task and became the youngest collaborator of the group, which was headed by Dr. Ing. Max Knoll. My first attempts with experimental work had been made in the practical physics course at the Technical High School in Munich under Professor Jonathan Zenneck, and now in the group of Max Knoll. As a newcomer I was first entrusted with some vacuum-technical problems which were important to all of us. Through the personality of Max Knoll, there was a companionable relationship in the group, and at our communal afternoon coffee with him the scientific day-to-day-problems of each member of the group were openly discussed. As I did not dislike calculations, and our common aim was the development of cathode-ray oscillographs for a desired measuring capability, I wanted to devise a suitable method of dimensioning such cathode-ray oscillographs in my "Studienarbeit"—a prerequisite for being allowed to proceed to the Diploma examination.

The most important parameters for accuracy of measurement and writing speed af cathode-ray oscillographs are the diameter of the writing spot and its energy density. To produce small and bright writing spots, the electron beams emerging divergently from the cathode had to be concentrated in a small writing spot on the fluorescent screen of the cathode-ray oscillograph. For this, already Rankin in 1905 [1] used a short dc-fed coil, as had been used by earlier experimentalists with electron beams (formerly called "glow" or "cathode rays"). Even before that, Hittorf (1869) and Birkeland (1896) used the rotationally symmetric field lying in front of a cylindrical magnet pole for focussing cathode rays. A more precise idea of the effect of the axially symmetric, i. e. inhomogeneous magnet field of such poles or coils on the electron bundle alongside of their axes had long been unclear.

Therefore, Hans Busch [3] at Jena calculated the electron trajectories in such an electron ray bundle and found that the magnetic field of the short coil has the same effect on the electron bundle as has the convex glass lens with a defined focal length on a light bundle. The focal length of this "magnetic electron lens" can be changed continuously by means of the coil current. Busch wanted to check experimentally his theory but for reasons of time he could not carry out new experiments. He made use of the experimental results he had already obtained sixteen years previously in Göttingen. These were, however, in extremely unsatisfactory agreement with the theory. Perhaps this was the reason that Busch did not draw at least the practical conclusion from his lens theory to image some object with such a coil.

In order to account more precisely for the properties of the writing spot of a cathode-ray oscillograph produced by the short coil, I checked Busch's lens theory with a simple experimental arrangement under better, yet still inadequate, experimental conditions (Fig. 1) and thereby found a better but still not entirely satisfactory agreement of the imaging scale with Busch's theoretical

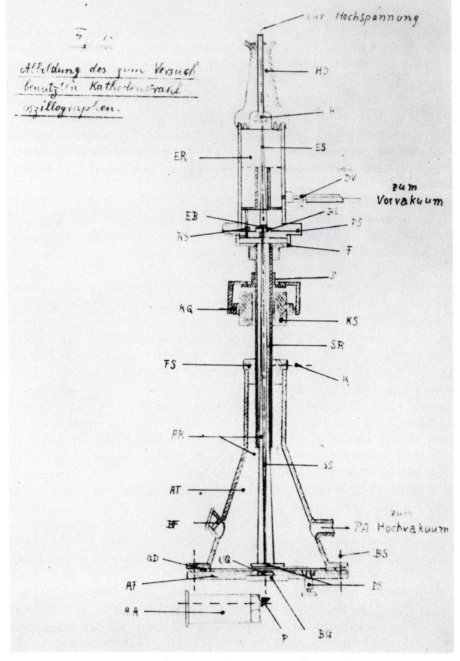

Fig. 1: Sketch by the author (1929) of the cathode ray tube for testing the imaging properties of the non-uniform magnetic field of a short coil[4, 5].

expectation. The main reason was that I had used a coil of the dimensions of Busch's coil whose field distribution along the axis was much too wide. My Studienarbeit [4], submitted to the Faculty for Electrotechnical Engineering in 1929, contained numerous sharp images with different magnifications of an electron-irradiated anode aperture of 0.3 mm diameter which had been taken by means of the short coil ("magnetic electron lens")—i. e. the first recorded electron-optical images.

Busch's equation for the focal length of the magnetic field of a short coil implied that a desired focal length could be produced by the fewer ampere turns the more the coil field was limited to a short region alongside the axis, because in that case the field maximum is increased. It was therefore logical for me as a prospective electrotechnical engineer to suitably envelop the coil with an iron coating, with a ring-shaped gap in the inner tube. Measurements at such a coil immediately showed that the same focal length had been reached with markedly fewer ampere turns [4, 5]. Vice versa, in this manner a shorter focal length can, of course, also be obtained by an equal number of ampere turns.

D. Why I pursued the magnetic electron lens for the electron microscope

In my Diploma Thesis (1930) I was to search for an electrostatic replacement for the magnetic concentration of the divergent electron ray bundle, which would probably be easier and cheaper. To this end, Knoll suggested experimental investigation of an arrangement of hole electrodes with different electrical potential for which he had taken out a patent a year before [6]. We discussed the shape of the electric field between these electrodes, and I suggested that because of the mirror-like symmetry of the electrostatic field of the electrodes on either side of the lens centre, a concentrating effect of the curved equipotential planes in the hole area could not take place. I only had the field geometry in mind then. But this conclusion was wrong. I overlooked that as a consequence of the considerably varying electron velocity on passage through such a field arrangement, a concentration of the divergent electron bundle must, in fact, occur. Knoll did not notice this error either. Therefore I pursued another approach in my Diploma Thesis [7]. I made the electron bundle pass a bored-out spherical condenser with fine-meshed spherically shaped grids fixed over each end of the bore. With this arrangement I obtained laterally inverted images in the correct imaging scale. Somewhat later I found a solution which was unfortunately only theoretically correct. In analogy to the refraction of the light rays on their passage through the optical lens at their surfaces ("Grenzflä-chen), I wanted to use, for the electrical lens, the potential steps at corresponding surfaces, which are shaped like glasses lenses [8]. Thus, the energy of the electron beams is temporarily changed—just like light beams on passage through optical lenses. For the realization of this idea, on each side of the lens two closely neighboured fine-meshed grids of the shape of optical lenses are required which must be kept on electrical potentials different from each other. First attempts confirmed the rightness of this idea, but at the same time also the practical inaptness of such grid lenses because of the too-strong absorption of the electron beam at the four grids and due to the field distribution by the wires.

As a consequence of my false reasoning and the experimental disappointment I decided to continue with the magnetic lens. I only report this in so much detail to show that occasionally it can be more a matter of luck than of superior intellectual vigor to find a better—or perhaps the only acceptable way. The approach of the transmission electron microscope with electron lenses of electrostatic hole electrodes was later pursued by outstanding experimentalists in other places and led to considerable initial success. It had, however, to be abandoned because the electrostatic lens was for physical reasons inferior to the magnetic electron lens.

E. The invention of the electron microscope

After obtaining my Degree (early 1931), the economic situation had become very difficult in Germany and it seemed not possible to find a satisfactory position at a University or in industry. Therefore I was glad that I could at least continue my unpaid position as doctorand in the high-voltage institute. After having shown in my Studienarbeit of 1929 that sharp and magnified images of electron-irradiated hole apertures could be obtained with the short coil, I was now interested in finding out if such images—as in light optics— could be further magnified by arranging a second imaging stage behind the first stage. Such an apparatus with two short coils was easily put together (Fig. 2) and in April 1931 I obtained the definite proof that it was possible (Fig. 3). This apparatus is justifiably regarded today as the first electron microscope even though its total magnification of $3.6 \times 4.8 = 14.4$ was extremely modest.

The first proof had thus been given that—apart from light and glass lenses— images of irradiated specimens could be obtained also by electron beams and magnetic fields, and this in even more than one imaging stage. But what was the use of such images if even grids of platinum or molybdenum were burnt to cinders at the irradiation level needed for a magnification of only $17.4 \times$. Not wishing to be accused of showmanship, Max Knoll and I agreed to avoid the term *electron microscope* in the lecture Knoll gave in June 1931 on the progress in the construction of cathode ray oscillographs where he also, for the first time, described in detail my electron-optical investigations [9, 10]. But, of course, our thoughts were circling around a more efficient microscope. The resolution limit of the light microscope due to the length of the light wave which had been recognized 50 years before by Ernst Abbe and others could, because of lack of light, not be important at such magnifications. Knoll and I simply hoped for extremely low dimensions of the electrons. As engineers we did not know yet the thesis of the "material wave" of the French physicist de Broglie [11] that had been put forward several years earlier (1925). Even physicists only reluctantly accepted this new thesis. When I first heard of it in summer 1931, I was very much disappointed that now even at the electron microscope the resolution should be limited again by a wavelength (of the "Materiestrahlung"). I was immediately heartened, though, when with the aid of the de Broglie equation I became satisfied that these waves must be around five orders of

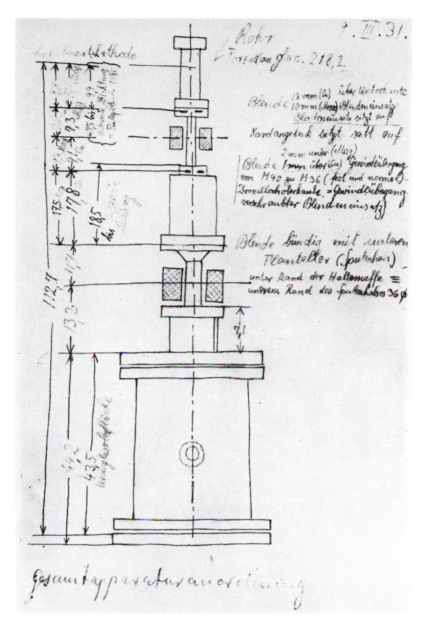

Fig. 2: Sketch by the author (9 March 1931) of the cathode ray tube for testing one-stage and two-stage electron-optical imaging by means of two magnetic electron lenses (electron microscope) [8].

magnitude shorter in length than light waves. Thus, there was no reason to abandon the aim of electron microscopy surpassing the resolution of light microscopy.

Physics 1986

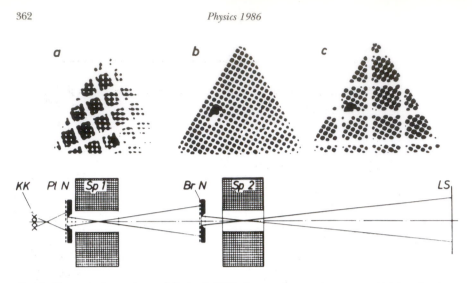

Fig. 3: First experimental proof (7 April 1931) that speciemens (aperture grids) irradiated by electrons can be imaged in magnified form not only in one but also in more than one stage by means of (magnetic) electron lenses.
(U = 50 kV). [8].
a) one-stage image of the platinum grid in front of coil 1 by coil 1; M = 13 ×
b) one-stage image of the bronze grid in front of coil 2 by coil 2; M = 4.8 ×
c) two-stage image of the platinum grid in front of coil 1 by coil 2; M = 17.4 × together with the one-stage image of the bronze grid in front of coil 2 by coil 2; M = 4.8 ×
   kk Cold cathode; Pt N Platinum grid; Sp 1 coil 1;
   Br N Bronze grid; Sp 2 coil 2; LS Fluorescent screen

In 1932 Knoll and I dared to make a prognosis of the resolution limit of the electron microscope [12]. Assuming that the equation for the resolution limit of the light microscope is valid also for the material wave of the electrons, we replaced the wave length of the light by the wave length of electrons at an accelerating voltage of 75 kV and inserted into the Abbe relation the imaging aperture of 2×10^{-2} rad which is what we had used previously. This imaging aperture is still used today. Thereby, that early we came up with a resolution limit of $2.2 \, \text{Å} = 2.2 \times 10^{-10}$ m, a value that was in fact obtained 40 years later.

Of course, at that time our approach was not taken seriously by most of the experts. They rather regarded it as a pipe-dream. I myself felt that it would be very hard to overcome the efforts still needed—mainly the problem of specimen heating. In April 1932, M. Knoll had taken up a position with Telefunken (Berlin) involving developmental work in the field of television.

In contrast to many biologists and medical scientists, my brother Helmut, who had almost completed his medical studies, believed in considerable progress for these disciplines should we be successful. With his confidence in a successful outcome he encouraged me to overcome the expected difficulties. In a next step I had to show that it was possible to obtain sufficiently high magnifications to prove a better-than-light-microscope resolution. To this effect a coil shape had to be developed whose magnetic field was compressed to a length that small of the coil axis to allow short focal lengths as are needed for

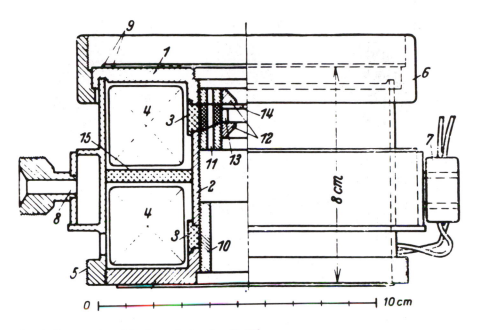

Fig. 4: Cross-section of the first polepiece lens [4, 15].

highly magnified images in not too great a distance behind the coil. The technical solution for this I had already given in my Studienarbeit of 1929 with the iron-clad coil. In 1932 I applied—together with my friend and co-doctorand Bodo v. Borries—for a patent on the optimization of this solution[13], the "Polschuhlinse", which is used in all magnetic electron microscopes today. Its realization and the measuring of the focal lengths which could be verified with it were subject of my thesis [14]. It was completed in August 1933, and in my measurements I obtained focal lengths of 3 mm for electron rays of 75 kV acceleration (Fig. 4). Of course, now with these lenses I immediately wanted to design a second electron microscope with much higher resolving power. To carry out this task I obtained by the good offices of Max v. Laue for the second half year of 1933 a stipend of Reichsmark 100 per month from the Notgemeinschaft der Deutschen Wissenschaft to defray running costs and personal expenses. Since I had completed the new instrument by the end of November (Fig. 5), I felt I ought to return my payment for December. To my great joy, however, I was allowed to keep the money "as an exception". Nevertheless, this certainly was the cheapest electron microscope ever paid for by a German organization for the promotion of science.

For reasons explained in the beginning of the next chapter, I accepted a position in industry on 1 December 1933. Therefore I could only make a few images with this instrument which magnified 12 000 × [15], but I noticed a decisive fact which gave me hope for the future: Even very thin specimens yielded sufficient contrast, yet no longer by absorption but solely by diffraction of the electrons, whereby—as is known—the specimens are heated up considerably less.

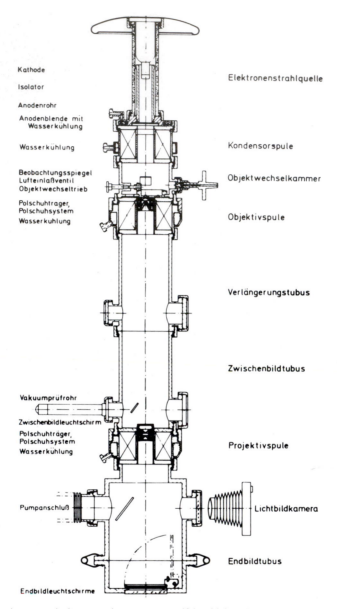

Kathode

Isolator

Anodenrohr

Anodenblende mit
   Wasserkühlung

Wasserkühlung

Beobachtungsspiegel
Lufteinlaßventil
Objektwechseltrieb

Polschuhträger,
Polschuhsystem
Wasserkühlung

Vakuumprüfrohr

Zwischenbildleuchtschirm
Polschuhträger,
Polschuhsystem
Wasserkühlung

Pumpanschluß

Endbildleuchtschirme

Elektronenstrahlquelle

Kondensorspule

Objektwechselkammer

Objektivspule

Verlängerungstubus

Zwischenbildtubus

Projektivspule

Lichtbildkamera

Endbildtubus

Fig. 5: First (two-stage) electron microscope magnifying higher than the light microscope. Cross-section of the microscope column (Re-drawn 1976) [15].

F: How the industrial production of electron microscopes came to be

I also realized, however, that the further development of a practically useful instrument with better resolution would require a longer period of time and enormous costs. In view of the results achieved there was little hope of obtaining financial support from any side for the time being. I was prepared for a longer dry spell and decided to approach the goal of a commercial instrument later, together with Bodo v. Borries and my brother Helmut. Therefore, I

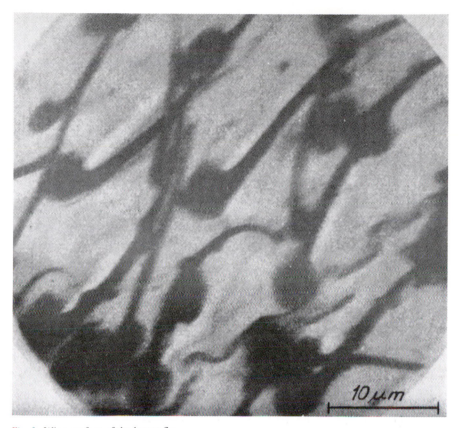

Fig. 6: Wing surface of the house fly.
(First internal photography, U = 60 kV, M_{el} = 2200)
(Driest, E., and Müller, H.O.: Z. Wiss. Mikroskopie 52, 53−57 (1935).

accepted a position with the Fernseh AG in Berlin-Zehlendorf where I was engaged in the development of Braun tubes for image pick-up and display tubes. In order to better coordinate our efforts to obtain financial support for the production of commercial electron microscopes, I convinced Bodo v. Borries to give up his position at the Rheinisch-Westfälische Elektrizitätswerke at Essen and return to Berlin. Here, he found a position at Siemens-Schuckert in 1934. We approached many governmental and industrial research facilities for financial help.

During this period, first electron micrographs appeared of biological specimens. Heinz Otto Müller (student in electrotechnical engineering) and Friedrich Krause (medical student) worked at the instrument I had built in 1933, and they published increasingly better results (Figs. 6 to 9). Unfortunately these two very gifted young scientists did not survive the II. World War.

At Brussels Ladislaus Marton had built his first horizontal microscope and obtained relatively low magnifications of biological specimens [17]. In 1936 he built a second instrument, this time with a vertical column [18].

Fig. 7: Diatoms *Amphipleura pellucida*.
(U = 53 kV, M_{el} = 3500, δ'' = 130 nm)
(F. Krause in: Busch, H., and Brüche, E.: Beiträge zur Elektronenoptik, 55−61, Verl. Joh. Ambrosius Barth, Leipzig 1937)

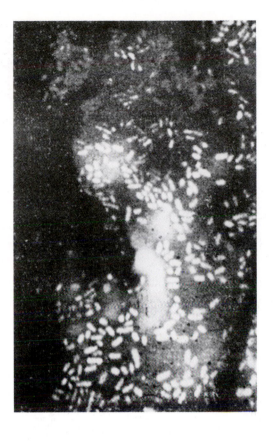

├─────── *10 μm* ───────┤

Fig. 8: Bacteria (culture infusion), fixed with formalin and embedded in a supporting film stained with a heavy metal salt
(U = 73.5 kV, M_{el} = 2000)
(Krause, F.: Naturwissenschaften 25, 817–825 (1937)).

In spite of these more recent publications, it took us three years to be successful in our quest for financial support through the professional assessment of Helmut Ruska's former clinical teacher, Professor Dr. Richard Siebeck, Director of the I. Medical Clinic of the Berlin Charité. I quote two paragraphs of his assessment of 2 October 1936 [19]:

"If these things were to be realised it hardly needs to be emphasised that the advances in the field of research into the causes of disease would be of immediate practical interest to the doctor. It would deeply affect real problems concerned to a large extent with diseases of growing clinical significance and thus of great importance for public health.

Should the possibilities of microscopical resolution exceed the assumed values by a factor of a hundred, the scientific consequences would be incalculable. What seems attainable now, I consider to be so important, and success seems to me so close, that I am ready and willing to advise on medical research work and to collaborate by making available the resources of my Institute".

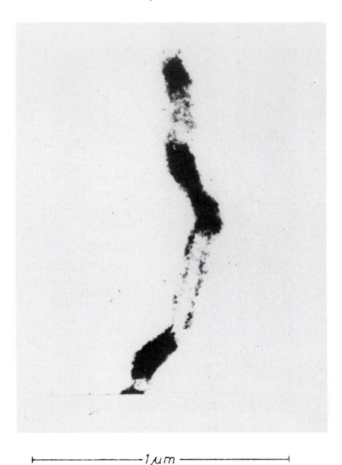

$\vdash\!\!-\!\!-\!\!-\!\!-\!\!-\!\!-\!\!-\!\!-\!\!-\!\!1\mu m\!-\!\!-\!\!-\!\!-\!\!-\!\!-\!\!-\!\!\dashv$

Fig. 9.: Iron Whisker
(U = 79 kW, M_{el} = 3100)
(Beischer, D., and Krause, F.: Naturwissenschaften 25,
825−829 (1937)).

This expertise impressed Siemens in Berlin and Carl Zeiss in Jena, and they were both ready to further the development of industrial electron microscopes. We suggested the setting up of a common development facility in order to make use of the electrotechnical expertise of Siemens and the know-how in precision engineering of Zeiß, but unfortunately the suggestion was refused and so we decided in favour of Siemens. As first collaborators we secured Heinz Otto Müller for the practical development and Walter Glaser from Prag as theorist. We started in 1937, and in 1938 we had completed two prototypes with condenser and polepieces for objective and projective as well as airlocks for specimens and photoplates. The maximum magnification was 30 000× [20]. One of these instruments was immediately used for first biological investigations by Helmut Ruska and several medical collaborators. (H. Ruska was released from Professor Siebeck for our work at Siemens.) Unfortunately, for reasons of time I cannot give here a survey of this fruitful publication period.

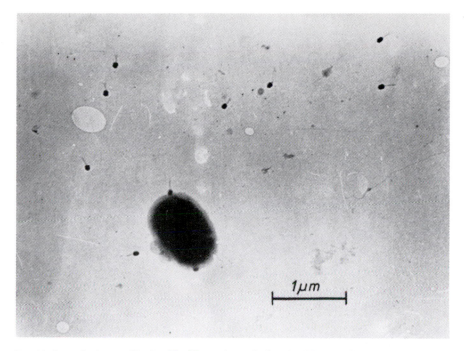

Fig. 10: Bacteriophages. (Ruska, H.: Naturwissenschaften 29, 367–368 (1941) and Arch. Ges. Virusforsch 2, 345–387 (1942).

In 1940, upon our proposal Siemens set up a guest laboratory, headed by Helmut Ruska, with four electron microscopes for visiting scientists. Helmut Ruska could show first images of bacteriophages in 1940. An image taken somewhat later (Fig. 10) clearly shows the shape of these tiny hostile bacteria. This laboratory was destroyed during an air raid in the autumn of 1944.

Very gradually now interest in electron microscopy was growing. A first sales success for Siemens has been achieved in 1938 when the chemical industry which was represented largely by IG Farbenindustrie placed orders for an instrument in each of their works in Hoechst, Leverkusen, Bitterfield and Wolfen. The instrument was only planned at the time, however not yet built or even tested. By the end of 1939 the first serially produced Siemens instrument [21] had been delivered to Hoechst (Fig. 11). The instrument No 26 was, by the way, delivered to Professor Arne Tiselius in Uppsala in autumn 1943. By Februrary 1945 more than 30 electron microscopes had been built in Berlin and delivered. Thus, now also independent representatives of various medical and biological disciplines could form their own opinions about the future prospects of electron microscopy. The choice of specimens was still limited though, since sufficiently thin sections were not yet available. The end of the war terminated the close cooperation with my brother and B. v. Borries.

Fig. 11: The first serially produced electron microscope, by Siemens. General view [21].

G. Development of electron microscopy after 1945

Our laboratory had to be reconstructed completely. I could start working with mainly new coworkers as early as June 1945. In spite of difficult conditions in Berlin and Germany, newly developed electron microscopes [22] could be delivered by the end of 1949. In 1954 Siemens had regained its former leading position with the "Elmiskop" [23] (Fig. 12 and Fig. 13). This instrument had, for the first time, two condenser lenses allowing thermal protection of the specimen by irradiating only the small region that is required for the desired final magnification. Since now, for a final magnification of $100\,000 \times$ a specimen field of only 1 μm must be irradiated for an image of 10 cm diameter in contrast to earlier irradiation areas of about 1 mm diameter, the power of the electron beam converted into heat in the object can be reduced down to the millionth part. The specimens are heated up just to the extent that the heat power produced can be radiated into the entire region around the object. If the heat power is low, a lower temperature rise with respect to the environment results.

The new instrument was, however, a big disappointment at first when we realized that at this "small region radiation" the image of the specimen fields, which was now no longer hot, became so dark within seconds that all initially visible details disappeared. Investigations then showed that minor residual gases in the evacuated instrument, particularly hydrocarbons, condensed on the cold inner planes of the instrument, i.e. they now even condensed on the specimen itself. The image of the resulting C layer in the irradiated specimen field becomes darker with increasing thickness of the layer. Happily, also this hurdle could, after some time, be surmounted by relatively simple means: The entire environment of the specimen was cooled by liquid air so that the specimen was still markedly warmer than its environment, even without being heated up by the beam. Thus, the residual gases of hydrocarbons condensed on the low-cooled planes and no longer on the specimen.

Along with the successful solution of this problem, another difficulty, that of specimen thickness, had also surprisingly been overcome by newly developed "ultramicrotomes". Instead of the ground steel knives whose blades were not sufficiently smooth due to crystallization, glass fracture edges were used which had no crystalline unevenness. The usual mechanical translation of the material perpendicular to the knife is—because of mechanical backlash or even oil layers—not sufficiently precise for the desired very small displacements of $\sim 10^{-5}$ mm. Smallest displacements free of flaws were obtained by thermal extension of a rod at whose ends the specimen to be cut was fastened. In order to keep the extremely thin sections smooth, they were dropped into an alcoholic solution immediately after being cut so that they remained entirely flat. Moreover, more suitable fixing agents had been found for the new cutting techniques. The development of these new ultramicrotomes considerably reduced the limitation in the choise of specimens for electron microscopy. For 25 years now, almost all disciplines furthered by light microscopy have also been able to benefit from electron microscopy.

During the last decades, electron microscopy has been advanced in many

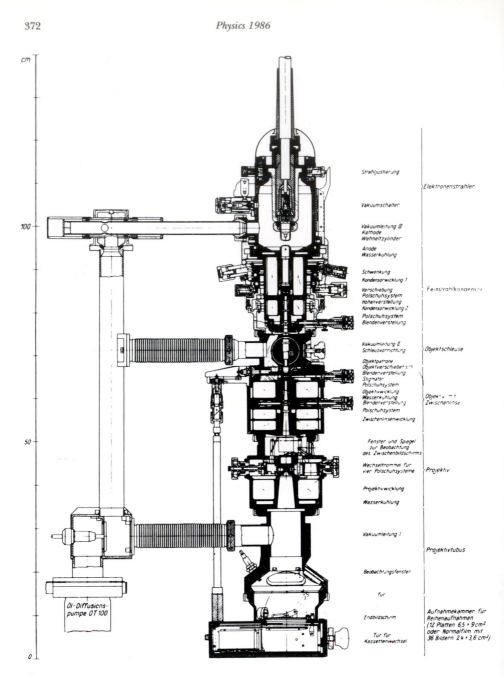

cm

100

50

Fig. 12: The first serially produced 100 kV-Electron microscope with two condenser lenses for "small region radiation" by Siemens. (cross-section) [23].

countries by numerous leading scientists and engineers through new ideas and procedures. I can here only give a few examples: Fig. 14 shows a cross-section through an electron microscope with single-field condenser objective, the specimen being in the field maximum of a magnetic polepiece lens [24]. Thereby, the region of increasing magnetic field in front of the specimen behaves like a condenser of short focal length and the decreasing field region behind the

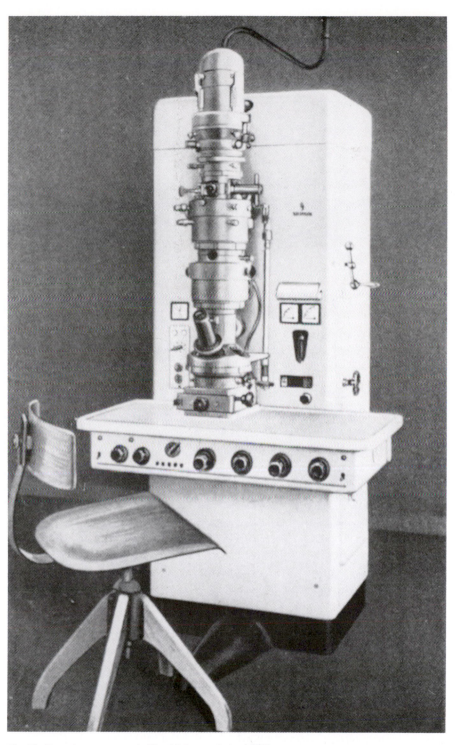

Fig. 13: Same instrument as in Fig. 12 (general view) [23].

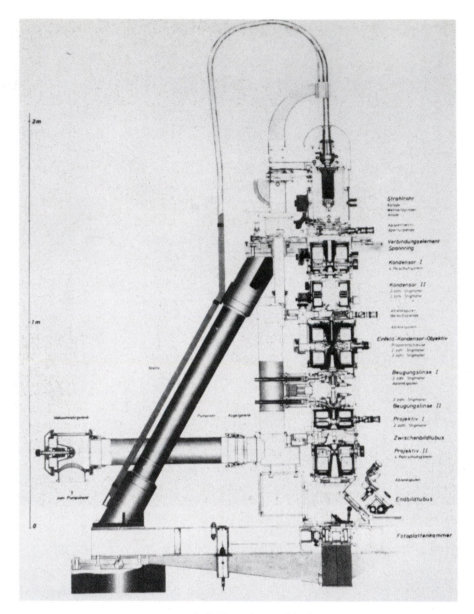

Fig. 14: Electron microscope with single-field condenser objective.

specimen as an objective of equal focal length. With this arrangement both lenses have a particularly small spherical aberration. Fig. 15 gives a view of the same instrument. Fig. 16 shows an image obtained with this instrument of a platelet of a gold crystal. One can clearly see lattice planes separated by a distance of 1.4 Å. Two such instruments have been further developed in the Institute for Electron Microscopy, which had been set up for me in 1957 by the Max-Planck-Gesellschaft after I had left Siemens. Fig. 17 shows a 3 MV high-voltage instrument developed by Japan Electron Optics Laboratory Co. Ltd. With such instruments whose development was mainly promoted by Gaston

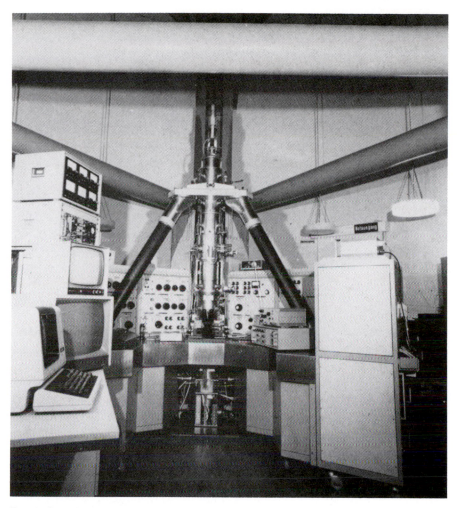

Fig. 15: Same instrument as in Fig. 14 (general view) [24].

Dupouy (1900–1985), apart from extremely high costs, special problems occur in the stabilization of the acceleration voltage and with the protection of the operators against X rays. The aim of the development of these instruments was the investigation of thicker specimens, but now that the problem of stabilizing the high voltages has been overcome, also the resolution has been improved by the shorter material wave length of particularly highly accelerated electrons, so that thinner specimens can also be investigated.

For quite some time now, the cryotechnique—put forward mainly by Fernandez-Moran in the USA—has been of increasing importance. With this technique specimens cooled down to very low temperatures can be studied, because they are more resistant to higher electron doses, i.e. the mobility inside the specimen is very much reduced compared to room temperature. Thus, even after unavoidable ionization, the molecules keep their structure for a long time. In the last years it has been possible to image very beam-sensitive crystals in a cryomicroscope with a resolution of 3.5 Å [25, 26] (Fig. 18) [27].

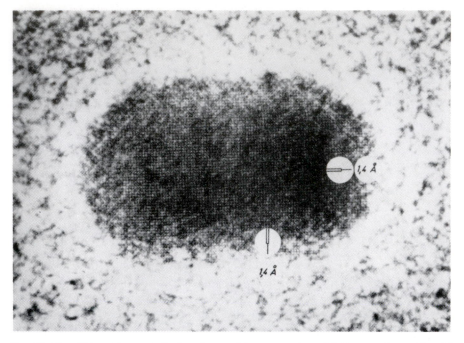

Fig. 16: Plate-like gold crystal, lattice planes with a separation of 0.14 nm, taken with axial illumination.
(U = 100 kV, M_{el} = 800 000); taken (1976) by K. Weiss and F. Zemlin with the 100 kV transmission electron microscope with single-field condenser objective at the Fritz-Haber-Institut of the Max-Planck-Gesellschaft.

The specimens were cooled down to $-269°$ C. Direct imaging with sufficient contrast is not possible because the specimen is destroyed at the beam dose needed for normal exposure. Therefore, many very low-dose images are recorded and averaged. Such a single image is very noisy but still contains sufficient periodical information. The evaluation procedure is the following: First, the microgram is digitized using the densitometer so that each image point is given a number which describes the optical density. The underexposed image of the whole crystal is divided like a checkerboard by the computer and then a large number—in our case 400—of these image sub-regions is cross-correlated and summed up by the computer. The resulting image corresponds to a sufficiently exposed micrograph. On the left part in Fig. 18, the initial noisy image of a paraffin crystal is seen; the right side shows the averaged image. Each white point is the image of a paraffin molecule. The long paraffin molecules $C_{44}H_{90}$ are vertical to the image plane. With this procedure electron micrographical images can be processed by the computer. It is even possible to image three-dimensional protein crystals with very high resolution [27]. The computer is a powerful tool in modern electron microscopy.

I cannot go into detail concering the transmission electron microscopes with electrostatic lenses, the scanning electron microscopes which are widely used mainly for the study of surfaces as well as transparent specimens, the great

Fig. 17: 1 MV Electron Microscope (Japan Electron Optics Laboratory Co. Ltd.)

importance of various image processing methods carried out partly by the computer, the field-electron microscope and the ion microscope.

The development of the electron microscopy of today was mainly a battle against the undesired consequences of the same properties of electron rays which paved the way for sub-light-microscopical resolution. Thus, for instance, the short material wavelength—prerequisite for good resolution—is coupled with the undesired high electron energy which causes specimen damage. The

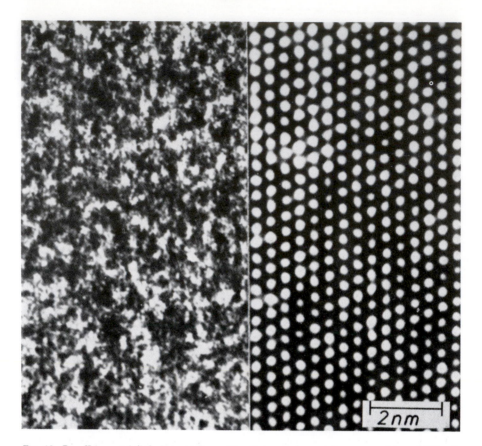

Fig. 18: Paraffin crystal (left: image taken with minimum dose, right: superposition of 400 sub-regions of the left image by means of the computer. [25].

deflectability in the magnetic field, a precondition for lens imaging, can also limit the resolution if the alternating magnetic fields in the environment of the microscope are not sufficiently shielded by the electron microscopy. We should not, therefore, blame those scientists today who did not believe in electron microscopy at its beginning. It is a miracle that by now the difficulties have been solved to an extent that so many scientific disciplines today can reap its benefits.

REFERENCES

1. Rankin, R.: The cathode ray oscillograph. The Electric Club J, II, 620−631 (1905).
2. Hittorf, W.: Über die Elektrizitätsleitung der Gase, I. und II. (On the electrical conductivity of gases.) Ann Physik. Chemie, 16, 1−31 und 197−234 (1869), Münster, 9 Oct. 1868.
3. Busch, H.: Über die Wirkungsweise der Konzentrierungsspule bei der Braunschen Röhre. (On the mode of action of the concentrating coil in the Braun tube.) Arch. Elektrotechnik 18, 583−594 (1927), Jena, Physikalisches Institut, März 1927.
4. Ruska, E.: Über eine Berechnungsmethode des Kathodenstrahloszillographen auf Grund der experimentell gefundenen Abhängigkeit des Schreibfleckdurchmessers von der Stellung der Konzentrierspule. (On a method of designing a cathode ray oscillograph on the basis of the experimentally found dependence of the writing spot diameter on the position of the concentrating coil.) Carried out from 1 November 1928 in the High Voltage Laboratory of the Technische Hochschule Berlin (Director Prof. A. Matthias). Student Project thesis (117 pp.) submitted 10 May 1929.
5. Ruska, E., and Knoll, M.: Die magnetische Sammelspule für schnelle Elektronenstrahlen. (The magnetic concentrating coil for fast electron beams.) Z. techn. Physik 12, 389−400 and 448 (1931), submitted 18 April 1931.
6. Knoll, M.: Vorrichtung zur Konzentrierung des Elektronenstrahls eines Kathodenstrahloszillographen. (Devise for concentrating the electron beam of a cathode ray oscillograph.) German Patent NO. 690809, patented on 10 November 1929, granted on 11 April 1940.
7. Ruska, E.: Untersuchung elektrostatischer Sammelvorrichtungen als Ersatz der magnetischen Konzentrierspulen bei Kathodenstrahl-Oszillographen. (Investigation of electrostatic concentrating devices as a substitute for the magnetic concentrating coils in cathode ray oscillographs.) Begun on 18 July 1930 in the High Voltage Laboratory of the Technological University of Berlin (Direktor Prof. Dr. A. Matthias) and submitted on 23 December 1930 as a Diploma Project (pp. 1−90).
8. Knoll, M. and Ruska, E.: Beitrag zur geometrischen Elektronenoptik I und II. (Contribution to geometrical electron optics.) Ann. Physik 12, 607−640 and 641−661 (1932), submitted 10 September 1931.
9. Knoll, M.: Berechnungsgrundlagen und neuere Ausführungsformen des Kathodenstrahloszillographen. (The basis of design and new forms of construction of the cathode ray oscillograph.) Manuscript of a lecture in the Cranz-Colloquium at the Technological University of Berlin on 4 June 1931, pp. 1−26.
10. Ernst Ruska: "The Early Development of Electron Lenses and Electron Microscopy", S. Hirzel Verlag Stuttgart (1980), see pp. 113−116.
11. De Broglie, L.: Recherches sur la théorie des quanta. (Researches on the theory of quanta.) Thèse, Paris: Masson & Cie. 1924. Ann. de Physiques 3, 22−128 (1925).
12. Knoll, M., and Ruska, E.: Das Elektronenmikroskop. (The electron microscope.) Z. Physik 78, 318−339 (1932), submitted 16 June 1932.
13. v. Borries, B., and Ruska, E.: Magnetische Sammellinse kurzer Feldlänge. (Magnetic converging lens of short field length.) German Patent No. 680284, patented on 17 March 1932; Patent granted on 3 August 1939.
14. Ruska, E.: Über ein magnetisches Objektiv für das Elektronenmikroskop. (On a magnetic objective lens for the electron microscope.) Dissertation of the Technological University of Berlin, submitted 31 August 1933. Z. Physik 89, 90−128 (1934), submitted 5 March 1934.
15. Ruska, E.: Über Fortschritte im Bau und in der Leistung des magnetischen Elektronenmikroskops. (On progress in the construction and performance of the magnetic electron microscope.) Z. Physik 87, 580−602 (1934), submitted 12 Dec. 1933.
16. see 10), pp. 120−122.
17. Marton, L.: La microscope électronique des objects biologiques. (Electron microscopy of biological objects.) Acad. roy. de Belg. Bull de la Cl. des Sci., Ser. 5, 20, 439−446 (1934), Université libre de Bruxells, Mai 1934.

18. Marton, L.: Le microscope électronique. (The electron microscope.) Rev. de Micro-
    biol. appl. 2, 117–124 (1936).
19. see 10), pp. 123–124.
20. Borries, B.v., and Ruska, E.: Vorläufige Mitteilung über Fortschritte im Bau und in
    der Leistung des Übermikroskops. (Preliminary communication on advances in the
    construction and performance of the Ultramicroscope.) Wiss. Veröff. a.d. Siemens-
    Werken 17, 99–106 (1938), submitted 29 Feb. 1938.
21. Borries, B.v., and Ruska, E.: Ein Übermikroskop für Forschungsindustrie. (An
    Ultramicroscope for industrial research.) Naturwissenschaften 27, 577–582 (1939),
    submitted 24 June 1939.
22. Ruska, E.: Über neue magnetische Durchstrahlungs-Elektronenmikroskope im
    Strahlspannungsbereich von 40 . . . 220 kV, Teil I. (On new magnetic transmission-
    electron microscopes for beam voltages between 40 and 200 kV.) Kolloid-Zeitschrift
    116, 103–120 (1950), submitted 15 Dec. 1949.
23. Ruska, E., and O. Wolff: Ein hochauflösendes 100-kV-Elektronenmikroskop mit
    Kleinfelddurchstrahlung (A high-resolution transmission electron microscope (100
    kV) with small-area illumination.) Zeitschrift für wissenschaftl. Mikroskopie und
    mikroskopische Technik 62, 466–509 (1956), submitted 19 July 1955.
24. Riecke, W.D., and Ruska, E.: A 100 kV transmission electron microscope with
    single-field condenser objective. VI. Int. Congress for Electron Microscopy, Kyoto,
    Japan, I, 19–20 (1966).
25. Zemlin, F., Reuber, E., Beckmann, E., Zeitler, E. and Dorset, D.L.: Molecular
    Resolution Electron Micrographs of Monolamellar Paraffin Crystals. Science 229,
    461–462 (1985).
26. Dietrich, I., Fox, F., Knapek, E., Lefranc, G., Nachtrieb, K., Weyl, R. and Zerbst,
    H.: Improvements in electron microscopy by application of superconductivity,
    Ultramicroscopy 2, 241–249 (1971).
27. Henderson, R., and Unwin, P.N.T.: Three-dimensional model of purple membrane
    obtained by electron microscopy, Nature 257, 28–32 (1975).

GERD BINNIG

I was born in Frankfurt, W. Germany, on 7.20., '47 as the first of two sons. My childhood was very much influenced by the Second World War, which had only just ended. We children had great fun playing among the ruins of the demolished buildings, but naturally were too young to realize that much more than just buildings had been destroyed.

Until the age of 31, I lived partly in Frankfurt and partly in Offenbach, a nearby city. I attended school in both cities, and it was in Frankfurt that I started to study physics. Already as a child about 10 years of age, I had decided to become a physicist without actually knowing what it involved. While studying physics, I started to wonder whether I had really made the right choice. Especially theoretical physics seemed so technical, so relatively unphilosophical and unimaginative. In those years, I concentrated more on playing music with friends in a beat-band rather than on physics. My mother had introduced me to classical music very early in life, and I believe this played an important role in my subsequent development. Unfortunately, I started playing the violin rather late, at the age of 15 only, but thoroughly enjoyed being a member of our school orchestra. My brother was responsible for my transition from classics to beat by his perpetually immersing me with the sounds of the Beatles and the Rolling Stones, until I finally really liked that kind of music, and even started composing songs and playing in various beat-bands. In this way, I first learned how difficult teamwork can be, how much fun it is to be creative, and how unpredictable the reaction of an audience can be.

My education in physics gained some significance when I began my diploma work in Prof. Dr. W. Martienssen's group, under Dr. E. Hoenig's guidance. I realized that actually *doing* physics is much more enjoyable than just learning it. Maybe 'doing it' is the right way of learning, at least as far as I am concerned.

I have always been a great admirer of Prof. Martienssen, especially of his ability to grasp and state the essence of the scientific context of a problem. Dr. Hoenig introduced me to experimenting, and exhibited great patience when I asked him very stupid questions in trying to catch up on what I had missed over all the previous years.

In 1969, Lore Wagler became my wife. We had both been studying for quite a long time — Lore is now a psychologist — so only recently did we decide to have children: a daughter born in Switzerland in 1984, and a son born in California in 1986. This was the absolute highlight and most wonderful experience of my whole life. However, fatherhood is not without its sacrifice. For the time being, nearly all my hobbies, like music (singing, playing the guitar and the violin), and sports (soccer, tennis, skiing, sailing and playing golf) have had to take a back seat.

It was in 1978 that Lore—my private psychotherapist—convinced me to accept an offer from the IBM Zurich Research Laboratory to join a physics group. This turned out to be an extremely important decision, as it was here I met Heinrich Rohrer. His way of viewing physics, combined with his humanity and sense of humor, fully restored my somewhat lost curiosity in physics. My years at Rüschlikon, and in IBM Research in general, have been very exciting, not only because of the development of the STM, but also because of the stimulating and pleasant atmosphere created by the people working there, and by those responsible. Working together in a team with Heini Rohrer, Christoph Gerber and Edmund Weibel was an extraordinarily delightful experience, and one for which I shall be eternally grateful. It is also extremely gratifying that our work was recognized far afield. We were first awarded the German Physics Prize, the Otto Klung Prize, the Hewlett Packard Prize, the King Faisal Prize, and now the ultimate crown, the Nobel Prize for Physics. Life certainly does not become easier for a scientist once his work has exceeded a certain significance. But while prizes do add some complications, I must admit they also have their compensations!

(*added in 1991*): In 1990 I joined the Supervisory Board of the Daimler Benz Holding and presently I am involved in a few political activities.

Heini Rob-ro

HEINRICH ROHRER

I was born in Buchs, St. Gallen, Switzerland on 6.6., '33 as the third child, half an hour after my twin sister. We were fortunate to enjoy a carefree childhood with a sound mixture of freedom, school and farm work. In 1949, the family moved to Zurich and our way of life changed from country to town. My finding to physics was rather accidental. My natural bent was towards classical languages and natural sciences, and only when I had to register at the ETH (Swiss Federal Institute of Technology) in autumn 1951, did I decide in favor of physics. In the next four years, Professors G. Busch, W. Pauli, and P. Scherrer taught me the rudiments. In autumn 1955, I started work on my Ph.D. Thesis and it was fortuitous that Jörgen Lykke Olsen trusted me to measure the length changes of superconductors at the magnetic-field-induced superconducting transition. He had already pioneered the field with measurements on the discontinuity of Young's modulus. Following in his footsteps, I lost all respect for angstroms. The mechanical transducers were very vibration sensitive, and I learned to work after midnight, when the town was asleep. My four graduate years were a most memorable time, in a group of distinguished graduate students always receptive for fun, and including the interruptions by my basic training courses in the Swiss mountain infantry.

In summer 1961, Rose-Marie Egger became my wife, and her stabilizing influence has kept me on an even keel ever since. Our honeymoon trip led us to the United States where I spent two post-doc years working on thermal conductivity of type-II superconductors and metals in the group of Professor Bernie Serin at Rutgers University in New Jersey. Then in the summer of 1963, Professor Ambros Speiser, Director of the newly founded IBM Research Laboratory in Rüschlikon, Switzerland, made me an offer to join the physics effort there. Encouraged by Bruno Lüthi, who later became a Professor at the University of Frankfurt, and, at the time, strongly recommended the hiring of Gerd Binnig, I accepted to start in December 1963, after having responded to the call of the wild in the form of a four-month camping trip through the USA.

My first couple of years in Rüschlikon were spent studying mainly Kondo systems with magnetoresistance in pulsed magnetic fields. End of the sixties, Keith Blazey interested me to work on $GdAlO_3$, an antiferromagnet on which he had done optic experiments. This started a fruitful cooperation on magnetic phase diagrams, which eventually brought me into the field of critical phenomena. Encouraged by K. Alex Müller, who had pioneered the critical-phenomena effort in our Laboratory, I focused on the bicritical and tetracritical behavior and finally on the random-field problem. These were most enjoyable years, during which so many patient colleagues taught me physics. I left them with some regret, when I ventured with Gerd to discover new shores. We found them. Thank you, Gerd.

In 1974/75, I spent a sabbatical year with Professor Vince Jaccarino and Dr. Alan King at the University of California in Santa Barbara, to get a taste of nuclear magnetic resonance. We solved a specific problem on the bicritical point of MnF_2, their home-base material. We traded experience, NMR and critical phenomena. Rose-Marie and I also took the opportunity at the beginning and end of my sabbatical to show the USA to our two daughters, Doris and Ellen, on two extended camping trips from coast to coast.

In all the years with IBM Research, I have especially appreciated the freedom to pursue the activities I found interesting, and greatly enjoyed the stimulus, collegial cooperation, frankness, and intellectual generosity of two scientific communities, namely, in superconductivity and critical phenomena. I should also like to take this opportunity to thank the many, many friends, teachers, and seniors who have contributed towards my scientific career in any way whatsoever, and most particularly my mother for her unstinting aid and assistance, especially when times were difficult.

SCANNING TUNNELING MICROSCOPY – FROM BIRTH TO ADOLESCENCE

Nobel lecture, December 8, 1986

by

GERD BINNIG AND HEINRICH ROHRER

IBM Research Division, Zurich Research Laboratory, 8803 Rüschlikon, Switzerland

We present here the historic development of Scanning Tunneling Microscopy; the physical and technical aspects have already been covered in a few recent reviews and two conference proceedings [1] and many others are expected to follow in the near future. A technical summary is given by the sequence of figures which stands alone. Our narrative is by no means a recommendation of how research should be done, it simply reflects what we thought, how we acted and what we felt. However, it would certainly be gratifying if it encouraged a more relaxed attitude towards doing science.

Perhaps we were fortunate in having common training in superconductivity, a field which radiates beauty and elegance. For scanning tunneling microscopy, we brought along some experience in tunneling [2] and angstroms [3], but none in microscopy or surface science. This probably gave us the courage and light-heartedness to start something which should "not have worked in principle" as we were so often told.

"After having worked a couple of years in the area of phase transitions and critical phenomena, and many, many years with magnetic fields, I was ready for a change. Tunneling, in one form or another had intrigued me for quite some time. Years back, I had become interested in an idea of John Slonczewski to read magnetic bubbles with tunneling; on another occasion, I had beeen involved for a short time with tunneling between very small metallic grains in bistable resistors, and later I watched my colleagues struggle with tolerance problems in the fabrication of Josephson junctions. So the local study of growth and electrical properties of thin insulating layers appeared to me an interesting problem, and I was given the opportunity to hire a new research staff member, Gerd Binnig, who found it interesting, too, and accepted the offer. Incidentally, Gerd and I would have missed each other, had it not been for K. Alex Müller, then head of Physics, who made the first contacts [1]."

The original idea then was not to build a microscope but rather to perform spectroscopy locally on an area less than 100 Å in diameter.

"On a house-hunting expedition, three months before my actual start at IBM, Heini Rohrer discussed with me in more detail his thoughts on inhomogeneities on surfaces, especially those of thin oxide layers grown on metal

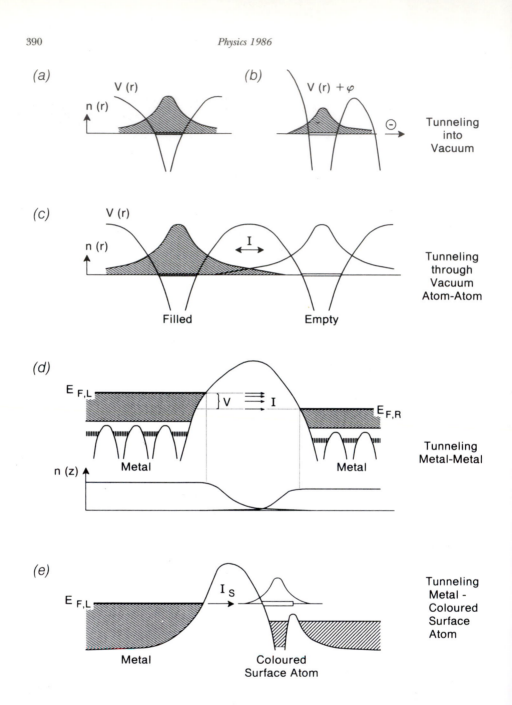

Fig. 1. Tunneling. (a) The wave function of a valence electron in the Coulomb potential well of the atom core plus other valence electrons extends into the vacuum; it "tunnels" into the vacuum. (b) Exposed to an electric field, φ, the electron can tunnel through the potential barrier and leaves the atom. (c) If two atoms come sufficiently close, then an electron can tunnel back and forth through the vacuum or potential barrier between them. (d) In a metal, the potential barriers between the atoms in the interior are quenched and electrons move freely in energy bands, the conduction bands. At the surface, however, the potential rises on the vacuum side forming the tunnel barrier through which an electron can tunnel to the surface atom of another metal close by. The voltage V applied between the two metals produces a difference between the Fermi levels $E_{F,L}$ and $E_{F,R}$, thus providing empty states on the right for the electrons tunneling from the left side. The

Oxide Junction

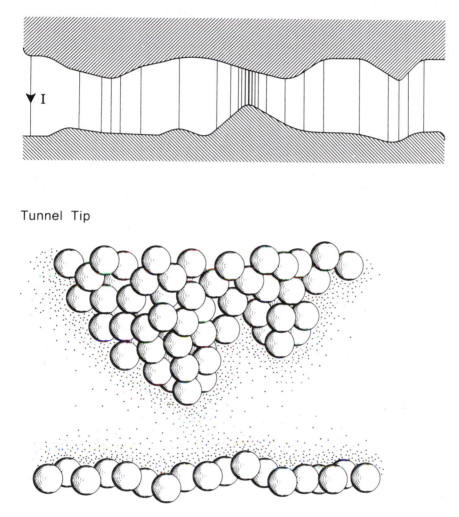

Tunnel Tip

Fig. 2. The principle. The tunneling transmittivity decreases exponentially with the tunneling distance, in vacuum about a factor 10 for every Å. In an oxide tunnel junction, most of the current flows through narrow channels of small electrode separation. With one electrode shaped into a tip, the current flows practically only from the front atoms of the tip, in the best case from a specific orbital of the apex atom. This gives a tunnel-current filament width and thus a lateral resolution of atomic dimensions. The second tip shown is recessed by about two atoms and carries about a million times less current.

resulting tunnel current is roughly of the form $I = f(V) \cdot \exp(- \sqrt{\varnothing} \cdot s)$. The f(V) contains a weighted joint local density of states of tip and object, the exponential gives the transmittivity with $\varnothing$ the averaged tunnel barrier height in eV, and s the separation of the two metals in Å. Here f(V) and $\sqrt{\varnothing}$ are material properties obtained by measuring dlnI/dV and dlnI/ds. (e) A simple case of local spectroscopy. A characteristic state, the "color", of a surface species is observed by the onset of the tunnel-current contribution IΣ, [see Lang, N. D. (1987) Phys. Rev. Lett. *58*, 45, and references therein].

surfaces. Our discussion revolved around the idea of how to study these films locally, but we realized that an appropriate tool was lacking. We were also puzzling over whether arranging tunneling contacts in a specific manner would give more insight on the subject. As a result of that discussion, and quite out of the blue at the LT15 Conference in Grenoble—still some weeks before I actually started at IBM—an old dream of mine stirred at the back of my mind, namely, that of vacuum tunneling. I did not learn until several years later that I had shared this dream with many other scientists, who like myself, were working on tunneling spectroscopy. Strangely enough, none of us had ever talked about it, although the idea was old in principle." Actually, it was 20 years old, dating back to the very beginning of tunneling spectroscopy [4]. Apparently, it had mostly remained an idea and only shortly after we had started, did Seymour Keller, then a member of the IBM Research Division's Technical Review Board and an early advocate of tunneling as a new research area in our Laboratory, draw our attention to W.A. Thompson's attempting vacuum tunneling with a positionable tip [5].

We became very excited about this experimental challenge and the opening up of new possibilities. Astonishingly, it took us a couple of weeks to realize that not only would we have a local spectroscopic probe, but that scanning would deliver spectroscopic and even topographic images, i.e., a new type of microscope. The operating mode mostly resembled that of stylus profilometry [6], but instead of scanning a tip in mechanical contact over a surface, a small gap of a few angstroms between tip and sample is maintained and controlled by the tunnel current flowing between them. Roughly two years later and shortly before getting our first images, we learned about a paper by R. Young *et al.* [7] where they described a type of field-emission microscope they called "topografiner". It had much in common with our basic principle of operating the STM, except that the tip had to be rather far away from the surface, thus on high voltage producing a field-emission current rather than a tunneling current and resulting in a lateral resolution roughly that of an optical microscope. They suggested to improve the resolution by using sharper field-emission tips, even attempted vacuum tunneling, and discussed some of its exciting prospects in spectroscopy. Had they, even if only in their minds, combined vacuum tunneling with scanning, and estimated *that* resolution they would probably have ended up with the new concept, Scanning Tunneling Microscopy. They came closer than anyone else.

Mid-January 1979, we submitted our first patent disclosure on STM. Eric Courtens, then deputy manager of physics at the IBM Rüschlikon Laboratory, pushed the disclosure to a patent application with "thousands of future STM's". He was the first believer in our cause. Shortly afterwards, following an in-house seminar on our STM ideas, Hans-Jörg Scheel became the third.

For the technical realization of our project, we were fortunate in securing the craftsmanship of Christoph Gerber. "Since his joining IBM in 1966, Christoph had worked with me (HR) on pulsed high-magnetic fields, on phase diagrams, and on critical phenomena. By the end of 1978, we were quite excited about our first experimental results on the random-field problem, but when asked to participate in the new venture, Christoph did not hesitate an instant. He always liked things which were out of the ordinary, and, incidentally, was the second believer. This left me and the random-field problem without his diligent technical support. About a year later, Edi Weibel was the next one to join in, which left another project without technical support. Finally, I completed the team, leaving the random-field problem to others."

During the first few months of our work on the STM, we concentrated on the main instrumental problems and their solutions [8]. How to avoid mechanical vibrations that move tip and sample against each other? Protection against vibrations and acoustical noise by soft suspension of the microscope within a vacuum chamber. How strong are the forces between tip and sample? This seemed to be no problem in most cases. How to move a tip on such a fine scale? With piezoelectric material, the link between electronics and mechanics, avoiding friction. The continuous deformation of piezomaterial in the angstrom and subangstrom range was established only later by the tunneling experiments themselves. How to move the sample on a fine scale over long distances from the position of surface treatment to within reach of the tip? The 'louse'. How to avoid strong thermally excited length fluctuations of the sample and especially the tip? Avoid whiskers with small spring constants. This led to a more general question, and the most important one: What should be the shape of the tip and how to achieve it? At the very beginning, we viewed the tip as a kind of continuous matter with some radius of curvature. However, we very soon realized that a tip is never smooth because of the finite size of atoms, and because tips are quite rough unless treated in a special way. This roughness implies the existence of minitips as we called them, and the extreme sensitivity of the tunnel current on tip-sample separation then selects the minitip reaching closest to the sample.

Immediately after having obtained the first stable STM images showing remarkably sharp monoatomic steps, we focused our attention onto atomic resolution. Our hopes of achieving this goal were raised by the fact that vacuum tunneling itself provides a new tool for fabricating extremely sharp tips: The very local, high fields obtainable with vacuum tunneling at a few volts only can be used to shape the tip by field migration or by field evaporation. Gently touching the surface is another possibility. All this is not such a controlled procedure as tip sharpening in field-ion microscopy, but it appeared to us to be too complicated to combine STM with field-ion microscopy at this stage. We hardly knew what field-ion microscopy was, to say nothing of working with it. We had no means of controlling exactly the detailed shape of the tip. We repeated our trial-and-error procedures until the structures we observed became sharper and sharper. Sometimes it worked, other times it did not.

But first we had to demonstrate vacuum tunneling. In this endeavor, apart from the occurrence of whiskers, the most severe problem was building vibrations. To protect the STM unit also against acoustical noise, we installed the vibration-isolation system within the vacuum chamber. Our first set-up was designed to work at low temperatures and in ultra-high vacuum (UHV). Low

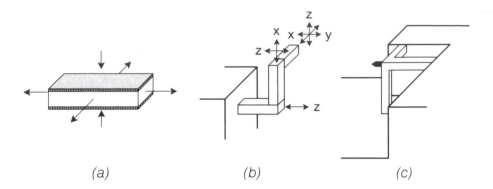

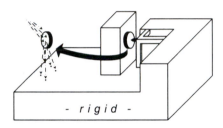 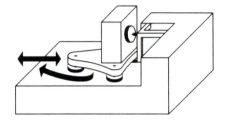

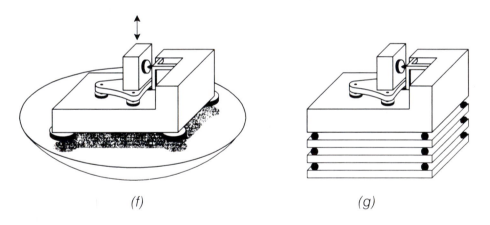

Fig. 3. The instrument. (a) A voltage applied to two electrodes contracts or expands the piezo-electric material in between. The practical total excursion of a piezo is usually in the region of micrometers. (b) A frictionless x-y-z piezodrive, which is quite vibration sensitive. (c) A rigid tripod is at present the piezodrive most used apart from the single-tube scanner. (d) Tripod and sample holder are installed on a rigid frame. The sample has to be cleared from the tip for preparation and sample transfer. (e) Positioning of the sample to within reach of the piezodrive was originally achieved with a piezoelectric 'louse' with electrostatically clampable feet. Magnetic-driven positioners and differential screws are also now in use. (f) In the first vibration-isolation system, the tunnel unit with permanent magnets levitated on a superconducting lead bowl. (g) The simple and presently widely used vibration protection with a stack of metal plates separated by viton — a UHV-compatible rubber spacer.

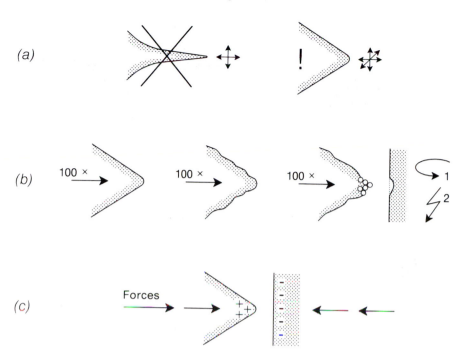

Fig. 4. Tips. (a) Long and narrow tips, or whiskers, are vibration sensitive and thermally excited. (b) A mechanically ground or etched tip shows sharp minitips, only one of which usually carries the tunnel current. Further sharpening was initially achieved with gentle contact (1), later with field evaporation (2). (c) Electrostatic and interatomic forces between tip and sample do not deform a blunt tip, or a rigid sample, but they make the tunnel gap mechanically unstable when the tip carries a whisker. The response of soft materials like graphite or organic matter to such forces, however, can be appreciable and has to be taken into account.

temperatures guaranteed low thermal drifts and low thermal length fluctuations, but we had opted for them mainly because our thoughts were fixed on spectroscopy. And tunneling spectroscopy was a low-temperature domain for both of us with a Ph.D. education in superconductivity. The UHV would allow preparation and retention of well-defined surfaces. The instrument was beautifully designed with sample and tip accessible for surface treatments and superconducting levitation of the tunneling unit for vibration isolation. Construction and first low-temperature and UHV tests took a year, but the instrument was so complicated, we never used it. We had been too ambitious, and it was only seven years later that the principal problems of a low-temperature and UHV instrument were solved [9]. Instead, we used an exsicator as vacuum chamber, lots of Scotch tape, and a primitive version of superconducting levitation wasting about 20 l of liquid helium per hour. Emil Haupt, our expert glassblower, helped with lots of glassware and, in his enthusiasm, even made the lead bowl for the levitation. Measuring at night and hardly daring to breathe from excitement, but mainly to avoid vibrations, we obtained our first clear-cut **exponential dependence of the tunnel current I on tip-sample separation s characteristic for tunneling. It was the portentous night of March 16, 1981.**

Constant Current Mode

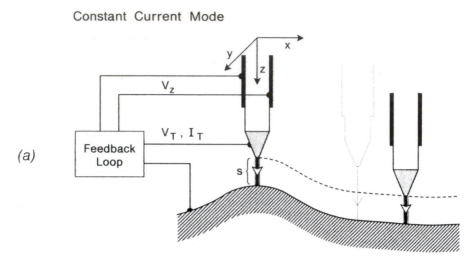

(a)

$$V_z \, (\, V_x \, , V_y \,) \; \longrightarrow \; z \, (\, x \, , y \,)$$

(b)

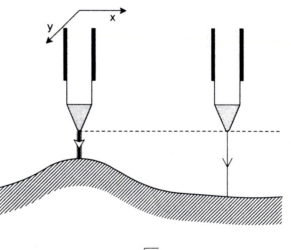

$$\ln I \, (\, V_x \, , V_y \,) \; \longrightarrow \; \sqrt{\overline{\Phi}} \; \cdot z \, (\, x \, , y \,)$$

Fig. 5. Imaging. (a) In the constant current mode, the tip is scanned across the surface at constant tunnel current, maintained at a pre-set value by continuously adjusting the vertical tip position with the feedback voltage V_z. In the case of an electronically homogeneous surface, constant current essentially means constant s. (b) On surface portions with denivellations less than a few Å—corresponding to the dynamic range of the current measurement—the tip can be rapidly scanned at constant average z-position. Such "current images" allow much faster scanning than in (a) but require a separate determination of $\sqrt{\varnothing}$ to calibrate z. In both cases, the tunnel voltage and/or the z-position can be modulated to obtain in addition, dlnI/dV and/or dlnI/ds, respectively.

So, 27 months after its conception the Scanning Tunneling Microscope was born. During this development period, we created and were granted the necessary elbow-room to dream, to explore, and to make and correct mistakes. We did not require extra manpower or funding, and our side activities produced acceptable and publishable results. The first document on STM was the March/April 1981 in-house Activity Report.

A logarithmic dependence of the tunnel current I on tip-sample separation s alone was not yet proof of vacuum tunneling. The slope of ln I versus s should correspond to a tunnel-barrier height of $\emptyset \approx 5$ eV, characteristic of the average workfunctions of tip and sample. We hardly arrived at 1 eV, indicating tunneling through some insulating material rather than through vacuum. Fortunately, the calibration of the piezosensitivity for small and fast voltage changes gave values only half of those quoted by the manufacturers. This yielded a tunnel-barrier height of more than 4 eV and thus established vacuum tunneling. This reduced piezosensitivity was later confirmed by careful calibration with H.R. Ott from the ETH, Zurich, and of S.Vieira of the Universidad Autónoma, Madrid [10].

U. Poppe had reported vacuum tunneling some months earlier [11], but his interest was tunneling spectroscopy on exotic superconductors. He was quite successful at that but did not measure I(s). Eighteen months later, we were informed that E.C. Teague, in his Thesis, had already observed similar I(s) curves which at that time were not commonly available in the open-literature [12].

Our excitement after that March night was quite considerable. Hirsh Cohen, then Deputy Director of our Laboratory, spontaneously asked us "What do you need?", a simple and obvious question people only rarely dare to ask. "Gerd immediately wanted to submit a post-deadline contribution [13] to the LT16 Conference to be held in Los Angeles in September. He was going there anyway with his superconducting strontium titanate, and I was sure we would have some topographic STM images by then. And indeed we had. I arranged an extended colloquium tour through the USA for Gerd, but about three weeks before his departure, a friend warned him, that once the news became public, hundreds of scientists would immediately jump onto the STM bandwagon. They did—a couple of years later. After two extended discussions on a weekend hike, he nevertheless became convinced that it was time for the STM to make its public appearance." Our first attempt to publish a letter failed. "That's a good sign", Nico Garcia, a Visiting Professor from the Universidad Autónoma de Madrid, Spain consoled us.

After this first important step with a complete STM set-up, it took us only three months, partly spent waiting for the high-voltage power supplies for the piezos, to obtain the first images of monosteps [14] on a $CaIrSn_4$ single crystal grown by R. Gambino. Here, the main problem was getting rid of the whiskers we continually created by bumping the tip into the surface. Now we were ready to turn to surface science, first to resolve surface reconstructions. We built a UHV-compatible STM (no longer with Scotch tape!) and as a quick trial, operated it in vacuum suspended from a rubber band. The results indicated that superconducting levitation might be unnecessary.

That was the state of the art for the publicity tour through the USA in September '81. Most reactions were benevolent, some enthusiastic, and two even anticipated the Nobel prize, but the STM was apparently still too exotic for any active outside engagement.

Next, we protected the STM from vibrations by a double-stage spring system with eddy-current damping [8], and incorporated it in a UHV chamber not in use at that moment. We added sputtering and annealing for sample treatment, but no other surface tool to characterize and monitor the state of the sample or tip could yet be combined with that STM. Although the superconducting levitation served for three months only, it was cited for years. It would appear that something complicated is much easier to remember!

A most intriguing and challenging surface-science problem existed, namely, the 7×7 reconstruction of the Si(111) surface. A class of fashionable models contained rather rough features which should be resolvable by the STM. So we started to chase after the 7×7 structure, and succumbed to its magic. At first, with no success. The STM would function well, sometimes with resolutions clearly around 5 Å, but not our surface preparation. We occasionally found quite nice patterns with monolayer step lines [8] but usually the surface always looked rough and disordered on an atomic scale. One image even foreshadowed the 7×7 by a regular pattern of depressions, the precursors of the characteristic corner holes. However, a single event is too risky to make a case for a new structure obtained with a new method. But it boosted our confidence.

By spring '82, STM was already a subject talked about. Supposedly, an image of a vicinal surface expertly prepared with a regular step sequence would have eased the somewhat reserved attitude of the surface-science community. We, however, thought that the mono-, double-, and triple-steps of the CaIrSn$_4$ with atomically flat terraces [14] and the step lines of Si(111) [8] were convincing and promising enough. And instead of wasting further time on uninteresting step lines, we preferred to attack surface reconstructions with known periodicities and with a reasonable chance of learning and contributing something new.

For easier sample preparation and because the demand on resolution was only 8 Å, we changed to a gold single crystal, namely, the (110) surface known to produce a 1×2 reconstruction. This seemed to be well within reach of the STM resolution from what we had learned from the silicon step lines. Although some time earlier, we had returned to Karl-Heinz Rieder, the Laboratory's surface-science expert, his Si single crystal in a kind of droplet form, it did not deter him from proposing this gold experiment which meant lending us his Au crystal, and some weeks later we added another droplet to his collection! But in between, with his advice on surface preparation, we succeeded in resolving the 1×2 structure [15]. Contrary to expectations, we also had to struggle with resolution, because Au transferred from the surface even if we only touched it gently with our tip. The mobility of Au at room temperature is so high that rough surfaces smooth out after a while, i. e., really sharp Au-coated tips cease to exist. We should like to mention here that later, for measurements on Au(100), we formed sharp Au tips by field evaporation of Au atoms from sample to tip, and could stabilize them by a relatively high field resulting from a 0.8 V tunnel voltage.

In the case of the Au(110) surface, the atomic resolution was rather a matter of good luck and perseverance. It jumped from high to low in an unpredictable manner, which was probably caused by migrating adatoms on the tip finding a stable position at the apex for a while. We also observed an appreciable disorder leading to long but narrow ribbons of the 1 × 2 reconstruction mixed with ribbons of 1 × 3 and 1 × 4 reconstructions and step lines. Nevertheless, these experiments were the first STM images showing atomic rows with atomic resolution perpendicular to the rows. The disorder, intrinsic on this surface, but in its extent criticized from the surface-science point of view, demonstrated very nicely the power of STM as a local method, and about a year later played an important role in testing the first microscopic theories of scanning tunneling microscopy.

With gold, we also performed the first spectroscopy experiment with an STM. We wanted to test a prediction regarding the rectifying I-V characteristic of a sample-tip tunnel junction induced by the geometric asymmetry [16]. Unfortunately, the sample surface became unstable at around 5 V, sample positive, and the small asymmetry observed in this voltage range could also have been due to other reasons. But with reversed polarity, the voltage could be swept up to 20 V producing a whole series of marked resonant surface states [8]. We consider the gold exercise during spring and early summer of '82 a most important step in the development of the method, and the STM had already exceeded our initial expectations. We had also won our first believers outside the Laboratory, Cal Quate from Stanford University [17] and Paul Hansma from the University of California at Santa Barbara [18]. We gave numerous talks on the Au work, and it attracted some attention but all in all, there was little action. We did not even take the time to write a paper—the 7 × 7 was waiting!

Meanwhile, we had also made the first attempts at chemical imaging: Small Au islands on silicon. The islands were visible as smooth, flat hills on a rough surface in the topography, but they were also clearly recognizable as regions with enhanced tunnel-barrier height [8]. Thus, the Au islands were imaged thanks to their different surface electronic properties. It would certainly have been interesting to pursue this line, but we knew that, in principle, it worked,— and the 7 × 7 was still waiting!

We started the second 7 × 7 attempt in autumn 1982 taking into consideration the advice of Franz Himpsel not to sputter the surface. This immediately worked and we observed the 7 × 7 wherever the surface was flat. We were absolutely enchanted by the beauty of the pattern.

"I could not stop looking at the images. It was like entering a new world. This appeared to me as the unsurpassable highlight of my scientific career and therefore in a way its end. Heini realized my mood and whisked me away for

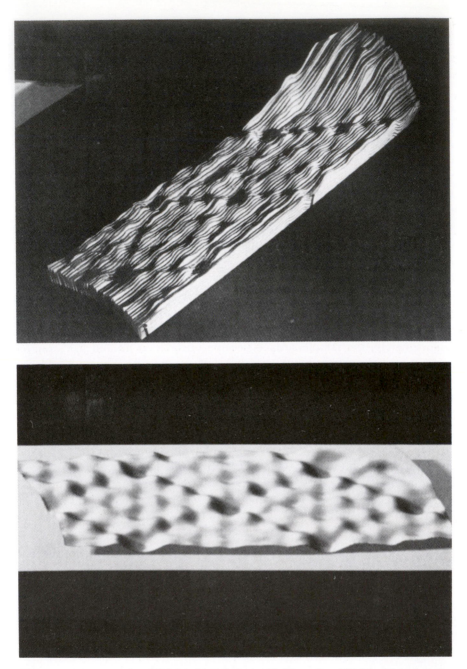

Fig. 6. 7 × 7 reconstruction of Si(111). (a) Relief assembled from the original recorder traces, from Ref. [19], © 1983 The American Physical Society, and (b) processed image of the 7 × 7 reconstruction of Si(111). Characteristic of the rhombohedral surface unit cell are the corner hole and the 12 maxima, the adatoms. In the processed image, the six adatoms in the right half of the rhombi appear higher. This is an electronic inequivalence on the surface owing to a structural left–right inequivalence in the underlying layers. The reconstruction extends undisturbed to the immediate vicinity of the large "atom hill" on the right.

some days to St. Antönien, a charming village high up in the Swiss mountains, where we wrote the paper on the 7 × 7."

We returned convinced that this would attract the attention of our colleagues, even of those not involved with surface science. We helped by presenting both an unprocessed relief model assembled from the original recorder traces with scissors, Plexiglass and nails, and a processed top view; the former for credibility, the latter for analysis and discussion [19]. It certainly did help, with the result that we practically stopped doing research for a while. We were inundated with requests for talks, and innumerable visitors to our Laboratory were curious to know how to build an STM. However, the number of groups that seriously got started remained small. It seemed there was still a conflict between the very appealing, conceptual easiness of displaying individual atoms in three-dimensional real space direct by recorder traces, and the intuitive reservation that, after all, it just could not be that simple.

Our result excluded all the numerous models that existed, and strangely enough also some that followed. Only one came very close: The adatom model by W. Harrison [20] with just the number of adatoms not quite right. Nowadays, a variation of the adatom model where deeper layers are also reconstructed besides the characteristic 7 × 7 adatom pattern [21], is generally accepted and compatible with most results obtained by various experimental methods like ion channeling [22], transmission electron diffraction [23], and more detailed STM results from other groups [24].

The 7 × 7 experiments also accelerated the first theoretical efforts of STM on a microscopic level. Tersoff and Hamman, and Baratoff [25] applied Bardeen's transfer Hamiltonian formalism to the small geometries of tip and an atomically corrugated surface. Garcia, Ocal, and Flores, and Stoll, Baratoff, Selloni, and Carnevali worked out a scattering approach [26]. The two approaches converged; they consoled us by roughly confirming our intuitive view on tunneling in small geometries by simply scaling down planar tunneling, and they certainly improved the acceptance of STM in physics circles. The theoretical treatments concentrated on the nonplanar aspect of tunneling of free electrons, and the STM results on Au(110), still unpublished, served as a testing ground. They remained unpublished for quite some time, since the flashy images of the 7 × 7 silicon surface somehow overshadowed the earlier Au(110) experiments. One reaction to the first attempt to publish them was: "... The paper is virtually devoid of conceptual discussion let alone conceptual novelty ... I am interested in the behavior of the surface structure of gold and the other metals in the paper. Why should I be excited about the results in this paper? ..." It was certainly bad publication management on our part, but we were not sufficiently familiar with a type of refereeing which searches for weak points, innocently ignoring the essence.

The gold and silicon experiments showed that STM in surface science would benefit greatly from additional, in-situ surface characterization, in particular low-energy electron diffraction (LEED). We had already learned that surfaces, even elaborately prepared, were frequently not as uniform and flat as generally assumed. The in-situ combination of LEED with STM proved extremely helpful, avoiding searching when there was nothing to be searched, and it gave us the opportunity to learn about and work with LEED and Auger electron spectroscopy (AES). The combination of STM with other established surface-science techniques also settled a concern frequently mentioned: How much did our STM images really have in common with surfaces characterized otherwise? We did not share this concern to such a degree, as we had also learned that reconstructions extended unchanged to the immediate vicinity of defect areas, and because we could detect most contaminants or defects individually. Thus, for us, the combined instrumentation was more a practical than a scientific issue.

After a short but interesting excursion with the new STM/LEED/AES combination into resolving and understanding the (100) surface of Au [27], we proceeded into the realms of chemistry. Together with A. Baró, a Visiting Professor from Universidad Autónoma de Madrid, Spain, who also wanted to familiarize himself with the technique, we observed the oxygen-induced 2 × 1 reconstruction of Ni(110) [28], interpreting the pronounced and regularly arranged protrusions we saw as individual oxygen atoms. We had seen atomic-scale features before, which could be interpreted as adsorbates or adsorbate clusters but they were more a nuisance than a matter of interest. The oxygen on Ni experiments demonstrated that the oxygen overlayer was not irreversibly changed by the imaging tunnel tip. This was a most significant result in regard to observing, studying and performing surface chemistry with an STM tip. About a year later, when studying the oxygen-induced 2 × 2 reconstructed Ni(100) surface, we observed characteristic current spikes which we could attribute to oxygen diffusing along the surface underneath the tip [29]. We noted that the same type of spikes had already been present in our earlier images of oxygen-covered Ni(110), but had been discarded at that time. Not only could diffusing atoms be observed individually, but their migration could be correlated to specific surface features like step lines or bound oxygen atoms, imaged simultaneously. Towards the end of 1983, we also started to probe the possibilities of STM in biology together with H. Gross from the ETH, Zurich. We could follow DNA chains lying on a carbon film deposited on a Ag-coated Si wafer [30].

That year ended with a most pleasant surprise: On Friday December 9, we received a telegram from the secretary of the King Faisal Foundation, followed on Monday by a phone call from the secretary of the European Physical Society announcing the King Faisal Prize of Science and the Hewlett Packard Euro-physics Prize, respectively. "The day the telegram arrived, Gerd was in Berlin delivering the Otto Klung Prize lecture. It was also my twentieth anniversary with IBM." This was an encouraging sign that Scanning Tunneling Micro-scopy was going to make it. It also brought a new flood of requests.

In the summer of 1984, we were finally ready to assume what we had set out to do in autumn 1978, before the notion of microscopy had ever evolved, namely, performing local spectroscopy. Together with H. Fuchs and F. Salvan, we investigated the clean 7 × 7 [1, 31] and the $\sqrt{3} \times \sqrt{3}$ Au reconstructions on Si(111) [31], and—right back to the heart of the matter—a thin oxide film on Ni

[1,32]. We could see that surfaces are electronically structured as known, for example, from photoemission experiments, and that we could resolve these electronic structures in space on an atomic scale. We called this (and still do) the color of the atoms. Indeed, the oxide layers were inhomogeneous and most clearly visible in scanning tunneling spectroscopy (STS) images. On the 7 × 7, we could see by STS down to the second layer, and observe individual dangling bonds between the adatoms [1]. At that time, C. Quate and his group already had an STM running, and they had performed local spectroscopy; not yet with atomic resolution but a low temperature [33]. They had measured the energy gap of a superconductor, and later even plotted its spatial dependence. Spectroscopic imaging was not really surprising, yet it was an important development. We now had the tools to fully characterize a surface in terms of topographic and electronic structure. Although it is usually quite an involved problem to separate the property of interest from a set of STM and STS measurements, our vision of the scanning tunneling microscope had become true. But nevertheless, we heard that this view was not generally shared. Rumors reached us that scientists would bet cases of champagne that our results were mere computer simulations! The bets were probably based on the fact the STM was already three years old, and atomic resolution was still our exclusive property. This was also our concern, but in another way. In late summer '83, Herb Budd, promoter of the IBM Europe Institute and an enthusiastic STM supporter, had asked us to run an STM Seminar in summer 1984 within the framework of the Institute. This meant one week with 23 lectures in front of a selected audience of the European academia. At that time, there was no way whatsoever of filling 23 hours, let alone of committing 23 speakers. A year later, we agreed, full of optimism for summer '85. In December '84, on Cal Quate's initiative, nine representatives of the most advanced STM groups came together for a miniworkshop in a hotel room in Cancun. It was a most refreshing exchange of ideas, but there was still no other atomic resolution, and thus not a sufficient number of lectures in sight for the Seminar.

In the following few months, the situation changed drastically. R. Feenstra and coworkers came up first with cleaved GaAs [34], C.F. Quate's group with the 1 × 1 structure on Pt(100) [35], and J. Behm, W. Hoesler, and E. Ritter with the hexagonal phase on Pt(100) [36]. At the American Physical Society March Meeting in 1985, P. Hansma presented STM images of graphite structures of atomic dimensions [37], and when J. Golovchenko unveiled the beautiful results on the various reconstructions of Ge films deposited on Si(111) [38], one could have heard a pin drop in the audience. The atomic resolution was official and scanning tunneling microscopy accepted. The IBM Europe Institute Seminar in July turned into an exclusive workshop for STM'ers, and comprised some 35 original contributions, not all of them on atomic resolution, but already more than in March [39]. "A watershed of ideas" as Cal Quate expressed it.

Our story so far has dealt mainly with the striving for structural and electronic imaging in a surface-science environment with atomic resolution.

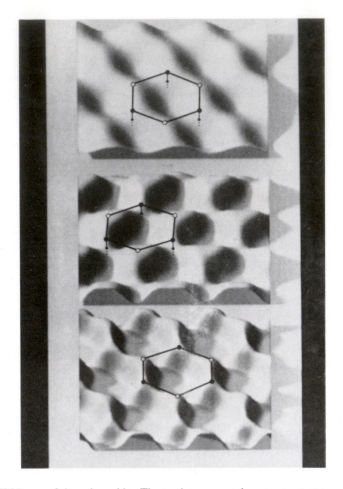

Fig. 7. STM image of cleaved graphite. The top image was taken at a constant tunnel current of 1 nA and at 50 mV. The corrugation traced by the tip reflects the local density of states (LDOS) at the Fermi level and *not* the positions of atoms, which form a flat honeycomb lattice as indicated. The LDOS at the atoms bound to the neighbors in the second layer (open circles) is lower than at the "free" atoms. The image is thus rather a spectroscopic than a topographic one. The middle image is a "current image" showing essentially the same pattern. In the bottom current image, taken closer to the surface, the two inequivalent atoms appear practically identical. This peculiar behavior is compatible with a different local elastic response of the two types of carbon atoms to the interatomic force exerted by the tip compensating for their different LDOS. A local perturbation of the electronic structure might also be important.

Individual atoms had been seen before with field-ion microscopy, and dealt with individually by the atom probe technique [40]. The beauty of these techniques is relativized by the restriction to distinct atom sites on fine tips made from a rather limited selection of materials. Similarly, electron microscopy, the main source of present-day knowledge on submicron structures in practically all areas of science, technology, and industry, has advanced to the atomic level. Imaging of individual atoms or atomic structures, however, is still reserved for specific problems, expertise, and extraordinary equipment. The appeal and the impact of STM lie not only in the observation of surfaces atom by atom, but also in its widespread applicability, its conceptual and instrumental simplicity and its affordability, all of which have resulted in a relaxed and almost casual perception of atoms and atomic structures.

But there are many other aspects, maybe less spectacular but nonetheless significant, which have made STM an accepted and viable method now pursued in many areas of science and technology.

The instruments themselves have become simpler and smaller. Their greatly reduced size allows easy incorporation into other systems, for instance, into a scanning electron microscope [41]. One type of instrument retains accurate sample positioning but is sufficiently rigid for in-situ sample and tip exchange. Other instruments are so rigid they are even insensitive to vibrations when immersed in liquid nitrogen [42], and even small enough to fit through the neck of a liquid-helium storage vessel [43]. These humming-birds of STM, some concepts of which reach back to the squeezable tunnel junctions [18], can also operate at television speed on relatively flat surfaces using single-tube scanners [43, 44]. Also tip preparation has advanced to a level where well-defined pyramidal tips ending with one [45] or more [46] atoms can be fabricated in a UHV environment. Such tips are particularly important for investigations of nonperiodic structures, disordered systems and rough surfaces. They are also interesting in their own right, for example, as low-energy electron and ion point sources.

Outside the physics and surface-science communities, the various imaging environments and imaging capabilities seem as appealing as atomic resolution. Images obtained at ambient-air pressure were first reported in 1984 [47], followed by imaging in cryogenic liquids [42], under distilled water [48], in saline solutions [48], and in electrolytes [49]. Scanning tunneling potentiometry appears to have become an interesting technique to study the potential distribution on an atomic scale of current-carrying microstructures [50]. More recent advances include interatomic-force imaging with the atomic-force microscope [51], with which the structure and elastic properties of conductors and insulators are obtained, and combined imaging of electronic and elastic properties of soft materials [52]. Also the use of spin polarized electron tunneling to resolve magnetic surface structures is being explored.

Finally, we revert to the point where the STM originated: The performance of a local experiment, at a preselected position and on a very small spatial scale down to atomic dimensions. Besides imaging, it opens, quite generally, new possibilities for experimenting, whether to study nondestructively or to modify

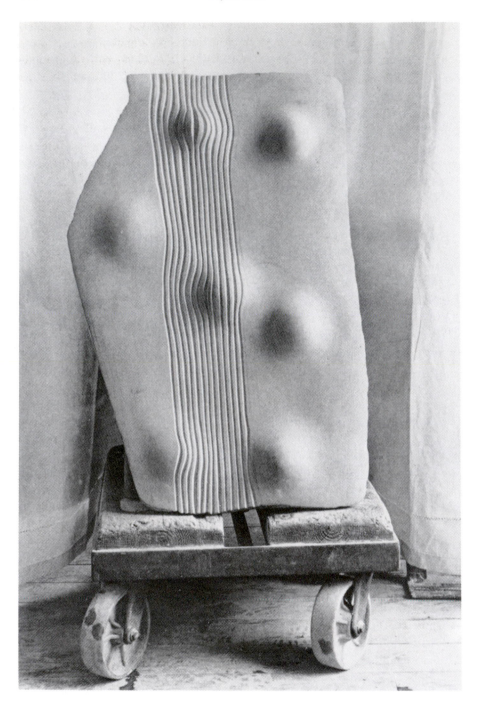

Fig. 8. Artist's conception of spheres. Art and Science are both products of the creativity of Man, and the beauty of nature is reflected in both. Ruedi Rempfler, the sculptor, found his interpretation in the deformation of a surface. It was the tension of the sphere in its environment which fascinated him, more than the mere portrayal of its shape. An independent creation, its visual and conceptual similarity with Fig. 6 is astounding. Original sculpture by Ruedi Rempfler, photograph courtesy of Thomas P. Frey.

locally: Local high electric fields, extreme current densities, local deformations, measurements of small forces down to those between individual atoms, just to name a few, ultimately to handle atoms [53] and to modify individual molecules, in short, to use the STM as a Feynman Machine [54]. This area has not yet reached adolescence.

The STM's "Years of Apprenticeship" have come to an end, the fundamentals have been laid, and the "Years of Travel" begin. We should not like to speculate where it will finally lead, but we sincerely trust that the beauty of atomic structures might be an inducement to apply the technique to those problems where it will be of greatest service solely to the benefit of mankind. Alfred Nobel's hope, our hope, everybody's hope.

ACKNOWLEDGEMENT

We should like to thank all those who have supported us in one way or another, and those who have contributed to the development of Scanning Tunneling Microscopy, and express our appreciation of the pleasant and collegial atmosphere existing in the STM community. Thanks are also due to Dilys Brüllmann for her diligent handling of our manuscripts from the start and for her careful reading of this manuscript, and to Erich Stoll for processing Figs. 6 and 7 using ideas of R. Voss.

REFERENCES

[1] For reviews, see Binnig, G., and Rohrer, H. (1986) IBM J. Res. Develop. *30*, 355; Golovchenko, J. A. (1986) Science *232*, 48; Behm, R. J., and Hoesler, W. (1986) *Physics and Chemistry of Solid Surfaces,* Vol VI, (Springer Verlag, Berlin), p. 361; Hansma, P. K., and Tersoff, J. (1987) J. Appl. Phys. *61*, RI; Proceedings of the STM Workshops in Oberlech, Austria, July 1–5, 1985, IBM J. Res. Develop. *30*, Nos. 4 and 5 (1986); Proceedings of STM '86, Santiago de Compostela, Spain, July 14–18, 1986, Surface Sci., *181*, Nos. 1 and 2 (1987). An article combining technical and biographical details is presented in Dordick, Rowan L. (1986) IBM Research Magazine *24*, 2.

[2] Binnig, G. K., and Hoenig, H. E. (1978) Z. Phys. B *32*, 23.

[3] Rohrer, H. (1960) Helv. Phys. Acta *33*, 675.

[4] Giaever, I. (1974) Rev. Mod. Phys. *46*, 245.

[5] Thompson W. A., and Hanrahan, S. F. (1976) Rev. Sci. Instrum. *47*, 1303.

[6] Williamson, B. P. (1967) Proc. Inst. Mech. Eng. *182*, 21; Guenther, K. H., Wierer, P. G., and Bennett, J. M. (1984) Appl. Optics *23*, 3820.

[7] Young, R., Ward, J., and Scire, F. (1972) Rev. Sci. Instrum. *43*, 999.

[8] For technical details, see Binnig, G., and Rohrer, H. (1982) Helv. Phys. Acta *55*, 726; *idem* (1983) Surface Sci. *126*, 236; *idem* (1985) Sci. Amer. *253*, 50.

[9] Marti, O. (1986) Ph.D. Thesis No. 8095, ETH Zurich, Switzerland; Marti, O., Binnig, G., Rohrer, H., Salemink, H., (1987) Surface Sci., *181*, 230.

[10] Ott, H. R., and Rohrer, H. (1981) unpublished; Vierira, S. (1986) IBM J. Res. Develop. *30*, 553.

[11] Poppe, U. (1981) Verhandl. DPG (VI) *16*, 476.

[12] Teague, E. C. (1978) Dissertation, North Texas State Univ., Univ. Microfilms International, Ann Arbor, Mich., p. 141; *idem* (1978) Bull. Amer. Phys. Soc. *23*, 290; *idem* (1986) J. Res. Natl. Bur. Stand. *91*, 171.

[13] Binnig, G., Rohrer, H., Gerber, Ch., and Weibel, E. (1982) Physica B *109 & 110*, 2075.

[14] Binnig, G., Rohrer, H., Gerber, Ch., and Weibel, E. (1982) Phys. Rev. Lett. *49*, 57.

[15] Binnig, G., Rohrer, H., Gerber, Ch., and Weibel, E. (1983) Surface Sci. *131*, L379.

[16] Miskowsky, N. M., Cutler, P. H., Feuchtwang, T. E. Shepherd, S. J., Lucas, A. A., and Sullivan, T. E. (1980) Appl. Phys. Lett. *37*, 189.

[17] Quate, C. F. (1986) Physics Today *39*, 26.

[18] Moreland, J., Alexander, S., Cox, M., Sonnenfeld, R., and Hansma, P. K. (1983) Appl. Phys. Lett. *43*, 387. Actually, Paul Hansma was indisposed and could not attend the first seminar given on STM in the USA. However, his students attended, and with them Paul built the squeezable tunnel junction.

[19] Binnig, G., Rohrer, H., Gerber, Ch., and Weibel, E. (1983) Phys. Rev. Lett. *50*, 120.

[20] Harrison, W. A. (1976) Surface Sci. *55*, 1.

[21] Takayanagi, K., Tanishiro, Y., Takahashi, M., and Takahashi, S. (1985) J. Vac. Sci. Tech. A *3*, 1502.

[22] Tromp, R. M., and van Loenen, E. J. (1985) Surface Sci. *155*, 441.

[23] Tromp, R. M. (1985) Surface Sci. *155*, 432, and references therein.

[24] Becker, R. S., Golovchenko, J. A., McRae, E. G., and Swartzentruber, B. S. (1985) Phys. Rev. Lett. *55*, 2028; Hamers, R. J., Tromp, R. M., and Demuth, J. E. (1986) Phys. Rev. Lett. *56*, 1972.

[25] Tersoff, J., and Hamann, D. R. (1983) Phys. Rev. Lett. *50*, 1998; Baratoff, A. (1984) Physica *127B*, 143.

[26] Garcia, N., Ocal, C., and Flores, F. (1983) Phys. Rev. Lett. *50*, 2002; Stoll, E., Baratoff, A., Selloni, A., and Carnevali, P. (1984) J. Phys. C *17*, 3073.

[27] Binnig; G., Rohrer, H., Gerber, Ch., and Stoll, E. (1984) Surface Sci. *144*, 321.

[28] Baró, A. M., Binnig, G., Rohrer, H., Gerber, Ch., Stoll, E., Baratoff, A., and Salvan, F. (1984) Phys. Rev. Lett. *52*, 1304.

[29] Binnig, G., Fuchs, H., and Stoll, E. (1986) Surface Sci., *169*, L295.

[30] Binnig, G., and Rohrer, H. (1984) *Trends in Physics,* editors, J. Janta and J. Pantoflicek (European Physical Society) p. 38.

[31] Baratoff, A., Binnig, G., Fuchs, H., Salvan, F., and Stoll, E. (1986) Surface Sci., *168*, 734.

[32] Binnig, G., Frank, K. H., Fuchs, H., Garcia, N., Reihl, B., Rohrer, H., Salvan, F., and Williams, A. R. (1985) Phys. Rev. Lett. *55*, 991; Garcia, R., Saenz, J. J., and Garcia, N. (1986) Phys. Rev. B *33*, 4439.

[33] de Lozanne, A. L., Elrod, S. A., and Quate, C. F. (1985) Phys. Rev. Lett. *54*, 2433.

[34] Feenstra, R. M., and Fein, A. P. (1985) Phys. Rev. B *32*, 1394.

[35] Elrod, S. A., Bryant, A., de Lozanne, A. L., Park, S., Smith, D., and Quate, C. F. (1986) IBM J. Res. Develop. *30*, 387.

[36] Behm, R. J., Hoesler, W., Ritter, E., and Binnig, G. (1986) Phys. Rev. Lett. *56*, 228.

[37] Hansma, P. K. (1985) Bull. Amer. Phys. Soc. *30*, 251.

[38] Becker, R. S., Golovchenko, J. A., and Swartzentruber, B. S. (1985) Phys. Rev. Lett. *54*, 2678.

[39] Proceedings published (1986) IBM J. Res. Develop. *30*, 4/5.

[40] For a review, see Ernst, N., and Ehrlich, G. (1986) *Topics in Current Physics,* Vol. 40, editor U. Gonser (Springer Verlag, Berlin) p. 75.

[41] Gerber, Ch., Binnig, G., Fuchs, H., Marti, O., and Rohrer, H. (1986) Rev. Sci. Instrum. *57*, 221.

[42] Coleman, R. V., Drake, B., Hansma, P. K., and Slough, G. (1985) Phys. Rev. Lett. *55*, 394.

[43] Smith, D. P. E., and Binnig, G. (1986) Rev. Sci. Instrum. *57*, 2630.

[44] Bryant, A., Smith, D. P. E., and Quate, C. F. (1986) Appl. Phys. Lett. *48*, 832.

[45] Fink, H.-W., (1986) IBM J. Res. Develop. *30*, 460.

[46] Kuk, Y., and Silverman, P. J. (1986) Appl. Phys. Lett. *48*, 1597.

[47] Baró, A. M., Miranda, R., Alaman, J., Garcia, N., Binnig, G., Rohrer, H., Gerber, Ch,. and Carrascosa, J. L. (1985) Nature *315*, 253.

[48] Sonnenfeld, R., and Hansma, P. K. (1986) Science *232*, 211.

[49] Sonnenfeld, R., and Schardt, B. C. (1986) Appl. Phys. Lett. *49*, 1172.

[50] Muralt, P., and Pohl, D. W. (1986) Appl. Phys. Lett. *48*, 514.

[51] Binnig, G., Quate, C. F., and Gerber, Ch. (1986) Phys. Rev. Lett. *56*, 930.

[52] Soler, J. M., Baró, A. M., Garcia, N., and Rohrer, H. (1986) Phys. Rev. Lett. *57*, 444; Dürig, U., Gimzewski, J. K., and Pohl, D. W. (1986) Phys. Rev. Lett. *57*, 2403.

[53] Becker, R. S., Golovchenko, J. A., and Swartzentruber, B. S. (1987) Nature *325*, 419.

[54] Feynman, R. P. (1960) Engr. and Sci., *22*, February; Hameroff, S., Schneiker, C., Scott, A., Jablonka, P., Hensen, T., Sarid, D., and Bell, S. (1987).

Physics 1987

J GEORG BEDNORZ and K ALEXANDER MÜLLER

for their important breakthrough in the discovery of superconductivity in ceramic materials

THE NOBEL PRIZE FOR PHYSICS

Speech by Professor GÖSTA EKSPONG of the Royal Academy of Sciences.
Translation from the Swedish text

Your Majesties, Your Royal Highnesses, Ladies and Gentlemen.

The Nobel Prize for Physics has been awarded to Dr. Georg Bednorz and Professor Dr. Alex Müller by the Royal Swedish Academy of Sciences "for their important breakthrough in the discovery of superconductivity in ceramic materials". This discovery is quite recent— less than two years old—but it has already stimulated research and development throughout the world to an unprecedented extent. The discovery made by this year's laureates concerns the transport of electricity without any resistance whatsoever and also the expulsion of magnetic flux from superconductors.

Common experience tells us that bodies in motion meet resistance in the form of friction. Sometimes this is useful, occasionally unwanted. One could save energy, that is to say fuel, by switching off the engine of a car when it had attained the desired speed, were it not for the breaking effect of friction. An electric current amounts to a traffic of a large number of electrons in a conductor. The electrons are compelled to elbow and jostle among the atoms which usually do not make room without resistance. As a consequence some energy is converted into heat. Sometimes the heat is desirable as in a hot plate or a toaster, occasionally it is undesirable as when electric power is produced and distributed and when it is used in electromagnets, in computers and in many other devices.

The Dutch scientist Heike Kamerlingh-Onnes was awarded the Nobel Prize for Physics in 1913. Two years earlier he had discovered a new remarkable phenomenon, namely that the electric resistance of solid mercury could completely disappear. Superconductivity, as the phenomenon is called, has been shown to occur in some other metals and alloys.

Why hasn't such an energy saving property already been extensively applied? The answer is, that this phenomenon appears only at very low temperatures; in the case of mercury at −269 degrees Celsius, which means 4 degrees above the absolute zero. Superconductivity at somewhat higher temperatures has been found in certain alloys. However, in the 1970's progress seemed to halt at about 23 degrees above the absolute zero. It is not possible to reach this kind of temperatures whithout effort and expense. The dream of achieving the transport of electricity without energy losses has been realized only in special cases.

Another remarkable phenomenon appears when a material during cooling crosses the temperature boundary for superconductivity. The field of a nearby magnet is expelled from the superconductor with such force that the magnet can become levitated and remain floating in the air. However, the dream of frictionless trains based on levitated magnets has not been realisable on a large scale because of the difficulties with the necessarily low temperatures.

Dr. Bednorz and Professor Müller started some years ago a search for superconductivity in materials other than the usual alloys. Their new approach met with success early last year, when they found a sudden drop towards zero resistance in a ceramic material consisting of lanthanum-barium-copper oxide. Sensationally, the boundary temperature was 50 % higher than ever before, as measured from absolute zero. The expulsion of magnetic flux, which is a sure mark of superconductivity, was shown to occur in a following publication.

When other experts had overcome their scientifically trained scepticism and had carried out their own control experiments, a large number of scientists decided to enter the new line of research. New ceramic materials were synthesized with superconductivity at temperatures such that the cooling suddenly became a simple operation. New results from all over the world flooded the international scientific journals, which found difficulties in coping with the situation. Research councils, industries and politicians are busily considering means to best promote the not so easy development work in order to benefit from the promising possibilities now in sight.

Scientists strive to describe in detail how the absence of resistance to the traffic of electrons is possible and to find the traffic rules, i. e. the laws of nature, which apply. The trio of John Bardeen, Leon Cooper and Robert Schrieffer found the solution 30 years ago in the case of the older types of superconductors and were awarded the Nobel Prize for Physics in 1972. Superconductivity in the new materials has reopened and revitalized the scientific debate in this field.

Herr Dr Bednorz und Herr Professor Müller:

In Ihren bahnbrechenden Arbeiten haben Sie einen neuen, sehr erfolgreichen Weg für die Erforschung und die Entwicklung der Supraleitung angegeben. Sehr viele Wissenschaftler hohen Ranges sind zurzeit auf dem Gebiet tätig, das Sie eröffnet haben.

Mir ist die Aufgabe zugefallen, Ihnen die herzlichsten Glückwünsche der Königlich Schwedischen Akademie der Wissenschaften zu übermitteln. Darf ich Sie nun bitten vorzutreten um Ihren Preis aus der Hand Seiner Majestät des Königs entgegenzunehmen.

J. Georg Bednorz

J. GEORG BEDNORZ

I was born in Neuenkirchen, North-Rhine Westphalia, in the Federal Republic of Germany on May 16, 1950, as the fourth child of Anton and Elisabeth Bednorz. My parents, originating from Silesia, had lost sight of each other during the turbulences of World War II, when my sister and two brothers had to leave home and were moved westwards. I was a latecomer completing our family after its joyous reunion in 1949.

During my childhood, my father, a primary school teacher and my mother, a piano teacher, had a hard time to direct my interest to classical music. I was more practical-minded and preferred to assist my brothers in fixing their motorcycles and cars, rather than performing solo piano exercises. At school it was our teacher of arts who cultivated that practical sense and helped to develop creativity and team spirit within the class community, inspiring us to theater and artistic performances even outside school hours. I even discovered my interest in classical music at the age of 13 and started playing the violin and later the trumpet in the school orchestra.

My fascination in the natural sciences was roused while learning about chemistry rather than physics. The latter was taught in a more theoretical way, whereas in chemistry, the opportunity to conduct experiments on our own, sometimes even with unexpected results, was addressing my practical sense.

In 1968, I started my studies in chemistry at the University of Münster, but somehow felt lost due to the impersonal atmosphere created by the large number of students. Thus I soon changed my major to cristallography, that field of mineralogy which is located between chemistry and physics.

In 1972, Prof. Wolfgang Hoffmann and Dr. Horst Böhm, my teachers, arranged for me to join the IBM Zurich Research Laboratory for three months as a summer student. It was a challenge for me to experience how my scientific education could be applied in reality. The decision to go to Switzerland set the course for my future. The physics department of which I became a member was headed by K. Alex Müller, whom I met with deep respect. I was working under the guidance of Hans Jörg Scheel, learning about different methods of crystal growth, materials characterization and solid state chemistry. I soon was impressed by the freedom even I as a student was given to work on my own, learning from mistakes and thus losing the fear of approaching new problems in my own way.

After my second visit in 1973, I came to Rüschlikon for six months in 1974 to do the experimental part of my diploma work on crystal growth and characterization of $SrTiO_3$, again under the guidance of Hans Jörg Scheel. The perovskites were Alex Müller's field of interest and, having followed my work, he encouraged me to continue my research on this class of materials.

In 1977, after an additional year in Münster, I joined the Laboratory of Solid State Physics at the Swiss Federal Institute of Technology (ETH) in Zurich and started my Ph.D. thesis under the supervision of Prof. Heini Gränicher and K. Alex Müller. I gratefully remember the time at the ETH and the family-like atmosphere in the group, where Hanns Arend provided a continuous supply of ideas. It was also the period during which I began to interact more closely with Alex and learned about his intuitive way of thinking and his capability of combining ideas to form a new concept.

In 1978, Mechthild Wennemer followed me to Zurich to start her Ph.D. at the ETH, but more importantly to be my partner in life. I had met her in 1974 during our time together at the University of Münster. Since then she has acted as a stabilizing element in my life and is the best adviser for all decisions I make, sharing the up's and down's in an unselfish way.

I completed my work on the crystal growth of perovskite-type solid solutions and investigating them with respect to structural, dielectric and ferroelectric properties, and joined IBM in 1982. This was the end of a ten-year approach which had begun in 1972.

The intense collaboration with Alex started in 1983 with the search for a high-T_c superconducting oxide; in my view, a long and thorny but ultimately successful path. We both realized the importance of our discovery in 1986, but were surprised by the dramatic development and changes in both the field of science and in our personal lives.

(added in 1991):

Honours

Thirteenth Fritz London Memorial Award (1987), Dannie Heineman Prize (1987), Robert Wichard Pohl Prize (1987), Hewlett-Packard Europhysics Prize (1988), The Marcel Benoist Prize (1986), Nobel Prize for Physics (1987), APS International Prize for Materials Research (1988), Minnie Rosen Award, the Viktor Mortiz Goldschmidt Prize and the Otto Klung Prize

K. Alex Müller

K. ALEX MÜLLER

I was born in Basle, Switzerland, on 20th April 1927. The first years of my life were spent with my parents in Salzburg, Austria, where my father was studying music. Hereafter, my mother and I moved to Dornach near Basle to the home of my grandparents, and from there to Lugano in the italian-speaking part of Switzerland. Here, I attended school and thus became fluent in the Italian language.

My mother died when I was eleven years old, and I attended the Evangelical College in Schiers, situated in a mountain valley in eastern Switzerland. I remained there until I obtained my baccalaureate (Matura) seven years later. This means I arrived in Schiers just before the Second World War started, and left just after it terminated. This was indeed quite a unique situation for us youngsters. Here, in a neutral country, we followed the events of the war worldwide, even in discussion groups in the classes. These college years in Schiers were of significance for my career.

The school was liberal in the spirit of the nineteenth century, and intellectually quite demanding. We were also very active in sports, I especially so in alpine skiing. In my spare time, I became quite involved in building radios and was so fascinated that I really wanted to become an electrical engineer. However, in view of my abilities, my chemistry tutor, Dr. Saurer, eventually convinced me to study physics.

At the age of 19, I did my basic military training in the Swiss army. Upon its completion, I enrolled in the famous Physics and Mathematics Department of the Swiss Federal Institute of Technology (ETH) in Zurich. Our freshman group was more than three times the normal size. We were called the "atom-bomb semester", as just prior to our enrollment nuclear weapons had been used for the first time, and many students had become interested in nuclear physics. The basic course was taught by Paul Scherrer and his vivid demonstrations had a lasting effect on my approach to physics. Other courses were in part not as illuminating, so that, despite good grades, I once seriously considered switching to electrical engineering. However, Dr. W. Känzig, responsible for the advanced physics practicum, convinced me to continue. In the later semesters, Wolfgang Pauli, whose courses and examinations I took, formed and impressed me. He was truly a wise man with a deep understanding of nature and the human being. I did my diploma work under Prof. G. Busch on the Hall effect of gray tin, now known as a semimetal, and, prompted by his fine lectures, also became acquainted with modern solid-state physics.

After obtaining my diploma, following my interest in applications, I worked for one year in the Department of Industrial Research (AFIF) of the ETH on the Eidophor large-scale display system. Then I returned to Prof. Busch's group as an assistant and started my thesis on paramagnetic resonance (EPR).

At one point, Dr. H. Gränicher suggested I look into the, at that time, newly synthesized double-oxide $SrTiO_3$. I found and identified the EPR lines of impurity present in Fe^{3+}.

In spring of 1956, just before starting the latter work, Ingeborg Marie Louise Winkler became my wife. She has always had a substantial influence in giving me confidence in all my undertakings, and over the past 30 years has been my mentor and good companion, always showing interest in my work. Our son Eric, now a dentist, was born in the summer of 1957, six months before I submitted my thesis.

After my graduation in 1958, I accepted the offer of the Battelle Memorial Institute in Geneva to join the staff. I soon became the manager of a magnetic resonance group. Some of the more interesting investigations were conducted on layered compounds, especially on radiation damage in graphite and alkali-metal graphites. The general manager in Geneva, Dr. H. Thiemann, had a strong personality, and his ever-repeated words "one should look for the extraordinary" made a lasting impression on me. Our stay in Geneva was most enjoyable for the family, especially for two reasons: the charm of the city and the birth of our daughter Silvia, now a kindergarten teacher.

While in Geneva, I became a Lecturer (with the title of Professor in 1970) at the University of Zurich on the recommendation of Prof. E. Brun, who was forming a strong NMR group. Owing to this lectureship, Prof. A. P. Speiser, on the suggestion of Dr. B. Lüthi, offered me a position as a research staff member at the IBM Zurich Research Laboratory, Rüschlikon, in 1963. With the exception of an almost two-year assignment, which Dr. J. Armstrong invited me to spend at IBM's Thomas J. Watson Research Center in Yorktown Heights, N.Y., I have been here ever since. For almost 15 years, research on $SrTiO_3$ and related perovskite compounds absorbed my interest: this work, performed with Walter Berlinger, concerned the photochromic properties of various doped transition-metal ions and their chemical binding, ferroelectric and soft-mode properties, and later especially critical and multicritical phenomena of structural phase transitions. In parallel, Dr. Heinrich Rohrer was studying such effects in the antiferromagnetic system of $GdAlO_3$. It was an intense and also, from a personal point of view, happy and satisfying time. While I was on sabbatical leave at the Research Center, he and Dr. Gerd Binnig started the Scanning Tunneling Microscope (STM) project. Just before leaving for the USA, I had been involved in the hiring of Dr. Binnig. Upon my return to Rüschlikon, I closely followed the great progress of the STM project, especially as from 1972 onwards, I was in charge of the physics groups.

The desire to devote more time to my own work prompted me to step down as manager in 1985. This was possible because in 1982 the company had honored me with the status of IBM Fellow. The ensuing work is summarized in Georg Bednorz's part of the Lecture. As he describes there, he joined our Laboratory to pursue his diploma work, on $SrTiO_3$ of course! Ever since making his acquaintance, I have deeply respected his fundamental insight into materials, his human kindness, his working capacity and his tenacity of purpose!

(*added in 1991*):

Honorary degrees

Doctor of Science, University of Geneva, Switzerland (1987), Faculty of Physics, the Technical University of Munich, Germany (1987), Università degli Studi di Pavia, Italy (1987), University of Leuven, Belgium (1988), Boston University, USA (1988), Tel Aviv University, Israel (1988), the Technical University of Darmstadt, Germany (1988), University of Nice, France (1989), Universidad Politecnica, Madrid, Spain (1989), University of Bochum, Germany (1990), and Università degli Studi di Roma, Italy (1990)

Honours

Foreign Associate Member, the Academy of Sciences, USA (1989), Special Tsukuba Award (1989), Thirteenth Fritz London Memorial Award (1987), Dannie Heineman Prize (1987), Robert Wichard Pohl Prize (1987), Hewlet-Packard Europhysics Prize (1988), Marcel-Benoist Prize (1986), Nobel Prize in Physics (1987), APS International Prize for New Materials Research (1988), and the Minnie Rosen Award (1988)

PEROVSKITE-TYPE OXIDES –
THE NEW APPROACH TO HIGH-T$_c$
SUPERCONDUCTIVITY

Nobel lecture, December 8, 1987

by

J. GEORG BEDNORZ and K. ALEX MÜLLER

IBM Research Division, Zurich Research Laboratory, 8803 Rüschlikon, Switzerland

PART 1: THE EARLY WORK IN RÜSCHLIKON

In our Lecture, we take the opportunity to describe the guiding ideas and our effort in the search for high-T$_c$ superconductivity. They directed the way from the cubic niobium-containing alloys to layered copper-containing oxides of perovskite-type structure. We shall also throw some light onto the circumstances and the environment which made this breakthrough possible. In the second part, properties of the new superconductors are described.

The Background

At IBM's Zurich Research Laboratory, there had been a tradition of more than two decades of research efforts in insulating oxides. The key materials under investigation were perovskites like $SrTiO_3$ and $LaAlO_3$, used as model crystals to study structural and ferroelectric phase transitions. The pioneering ESR experiments by Alex Müller (KAM) [1.1] and W. Berlinger on transition-metal impurities in the perovskite host lattice brought substantial insight into the local symmetry of these crystals, i.e., the rotations of the TiO_6 octahedra, the characteristic building units of the lattice.

One of us (KAM) first became aware of the possibility of high-temperature superconductivity in the 100 K range by the calculations of T. Schneider and E. Stoll on metallic hydrogen [1.2]. Such a hydrogen state was estimated to be in the 2−3 Megabar range. Subsequent discussions with T. Schneider on the possibility of incorporating sufficient hydrogen into a high-dielectric-constant material like $SrTiO_3$ to induce a metallic state led, however, to the conclusion that the density required could not be reached.

While working on my Ph.D. thesis at the Solid State Physics Laboratory of the ETH Zurich, I (JGB) gained my first experience in low-temperature experiments by studying the structural and ferroelectric properties of perovskite solid-solution crystals. It was fascinating to learn about the large variety of properties of these materials and how one could change them by varying their compositions. The key material, pure $SrTiO_3$, could even be turned into a superconductor if it were reduced, i.e., if oxygen were partially removed from its lattice [1.3]. The transition temperature of 0.3 K, however, was too low to create large excitement in the world of superconductivity research. Nevertheless, it was interesting that superconductivity occurred at all, because the carrier densities were so low compared to superconducting NbO, which has carrier densities like a normal metal.

My personal interest in the fascinating phenomenon of superconductivity was triggered in 1978 by a telephone call from Heinrich Rohrer, the manager of a new hire at IBM Rüschlikon, Gerd Binnig. With his background in superconductivity and tunneling, Gerd was interested in studying the superconductive properties of $SrTiO_3$, especially in the case when the carrier density in the system was increased. For me, this was the start of a short but stimulating collaboration, as within a few days I was able to provide the IBM group with Nb-doped single crystals which had an enhanced carrier density compared to the simply reduced material. The increase in T_c was exciting for us. In the Nb-doped samples $n = 2 \times 10^{20}$ cm^{-3}, the plasma edge lies below the highest optical phonon, which is therefore unshielded [1.4]. The enhanced electron-phonon coupling led to a T_c of 0.7 K [1.5]. By further increasing the dopant concentration, the T_c even rose to 1.2 K, but this transpired to be the limit, because the plasma edge passes the highest phonon. Gerd then lost his interest in this project, and with deep disappointment I realized that he had started to develop what was called a scanning tunneling microscope (STM). However, for Gerd and Heinrich Rohrer, it turned out to be a good decision, as everyone realized by 1986 at the latest, when they were awarded the Nobel Prize in Physics. For my part, I concentrated on my thesis.

It was in 1978 that Alex (KAM), my second supervisor, took an 18-month sabbatical at IBM's T. J. Watson Research Center in Yorktown Heights, NY, where he started working in the field of superconductivity. After his return in 1980, he also taught an introductory course at the University of Zurich. His special interest was the field of granular superconductivity, an example being aluminum [1.6], where small metallic grains are surrounded by oxide layers acting as Josephson junctions. In granular systems, the T_c's were higher, up to 2.8 K, as compared to pure Al with $T_c = 1.1$ K.

Involvement with the Problem

It was in fall of 1983, that Alex, heading his IBM Fellow group, approached me and asked whether I would be interested in collaborating in the search for superconductivity in oxides. Without hesitation, I immediately agreed. Alex later told me he had been surprised that he hardly had to use any arguments to convince me; of course, it was the result of the short episode of my activities in connection with the superconducting $SrTiO_3$—he was knocking on a door already open. And indeed, for somebody not directly involved in pushing T_c's to the limit and having a background in the physics of oxides, casual observation of the development of the increase of superconducting transition temperatures, shown in Figure 1.1, would naturally lead to the conviction that intermetallic compounds should not be pursued any further. This because since 1973 the highest T_c of 23.3 K [1.7] could not be raised. But nevertheless, the fact that superconductivity had been observed in several complex oxides evoked our special interest.

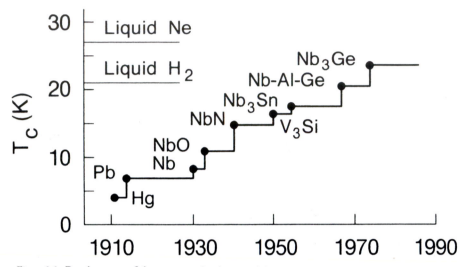

Figure 1.1. Development of the superconducting transition temperatures after the discovery of the phenomenon in 1911. The materials listed are metals or intermetallic compounds and reflect the respective highest T_c's.

The second oxide after $SrTiO_3$ to exhibit surprisingly high T_c's of 13 K was discovered in the Li-Ti-O system by Johnston *et al.* [1.8] in 1973. Their multiphase samples contained a $Li_{1+x}Ti_{2-x}O_4$ spinel responsible for the high T_c. Owing to the presence of different phases and difficulties in preparation, the general interest remained low, especially as in 1975 Sleight *et al.* [1.9] discovered the $BaPb_{1-x}Bi_xO_3$ perovskite also exhibiting a T_c of 13 K. This compound could easily be prepared as a single phase and even thin films for device applications could be grown, a fact that triggered increased activities in the United States and Japan. According to the BCS theory [1.10]

$$k_x T_c = 1.13 \hbar \omega_D \, e^{-1/(N(E_F)) \, xV^*},$$

both mixed-valent oxides, having a low carrier density $n = 4 \times 10^{21}/cm^3$ and a comparatively low density of states per unit cell $N(E_F)$ at the Fermi level, should have a large electron-phonon coupling constant V^*, leading to the high T_c's. Subsequently, attempts were made to raise the T_c in the perovskite by increasing $N(E_F)$ via changing the Pb:Bi ratio, but the compound underwent a metal-insulator transition with a different structure, thus these attempts failed.

We in Rüschlikon felt and accepted the challenge as we expected other metallic oxides to exist where even higher T_c's could be reached by increasing $N(E_F)$ and/or the electron-phonon coupling. Possibly we could enhance the latter by polaron formation as proposed theoretically by Chakraverty [1.11] or

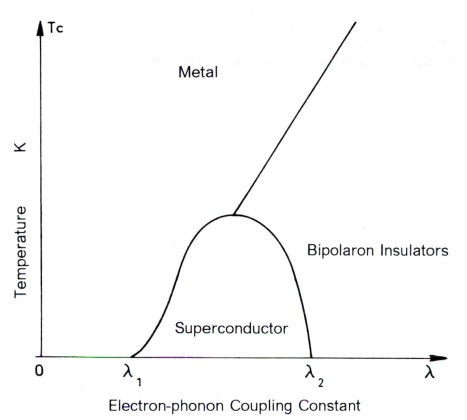

Figure 1.2. Phase diagram as a function of electron-phonon coupling strength. From [1.11], © Les Editions de Physique 1979.

by the introduction of mixed valencies. The intuitive phase diagram of the coupling constant $\lambda = N(E_F)^x V^*$ versus T proposed by Chakraverty for polaronic contributions is shown in Figure 1.2. There are three phases, a metallic one for small λ and an insulating bipolaronic one for large λ, with a superconductive phase between them, i.e., a metal-insulator transition occurs for large λ. For intermediate λ, a high-T_c superconductor might be expected. The question was, in which systems to look for superconductive transitions.

The Concept

The guiding idea in developing the concept was influenced by the Jahn-Teller (JT) polaron model, as studied in a linear chain model for narrow-band intermetallic compounds by Höck *et al.* [1.12].

The Jahn-Teller (JT) theorem is well-known in the chemistry of complex units. A nonlinear molecule or a molecular complex exhibiting an electronic degeneracy will spontaneously distort to remove or reduce this degeneracy. Complexes containing specific transition-metal (TM) central ions with special valency show this effect. In the linear chain model [1.12], for small JT distortions with a stabilization energy E_{JT} smaller than the bandwidth of the metal, only a slight perturbation of the traveling electrons is present. With increasing

E_{JT}, the tendency to localization is enhanced, and for E_{JT} being of the magnitude of the bandwidth, the formation of JT polarons was proposed.

These composites of an electron and a surrounding lattice distortion with a high effective mass can travel through the lattice as a whole, and a strong electron-phonon coupling exists. In our opinion, this model could realize the Chakraverty phase diagram. Based on the experience from studies of isolated JT ions in the perovskite insulators, our assumption was that the model would

Copper Ions in the Oxide Octahedron

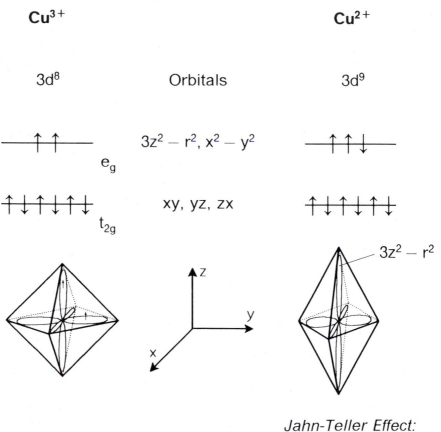

Jahn-Teller Effect:
Elongation of
the Octahedron

Figure 1.3. Schematic representation of electron orbitals for octahedrally coordinated copper ions in oxides. For Cu^{3+} with $3d^8$ configuration, the orbitals transforming as base functions of the cubic e_g group are half-filled, thus a singlet ground state is formed. In the presence of Cu^{2+} with $3d^9$ configuration, the ground state is degenerate, and a spontaneous distortion of the octahedron occurs to remove this degeneracy. This is known as the Jahn-Teller effect.

also apply to the oxides, our field of expertise, if they could be turned into conductors. We knew there were many of them. Oxides containing TM ions with partially filled e_g orbitals, like Ni^{3+}, Fe^{4+} or Cu^{2+} exhibit a strong JT effect, Figure 1.3, and we considered these as possible candidates for new superconductors.

The Search and Breakthrough

We started the search for high-T_c superconductivity in late summer 1983 with the La-Ni-O system. $LaNiO_3$ is a metallic conductor with the transfer energy of the JT-e_g electrons larger than the JT stabilization energy, and thus the JT distortion of the oxygen octahedra surrounding the Ni^{3+} is suppressed [1.13]. However, already the preparation of the pure compound brought some surprises, as the material obtained by our standard coprecipitation method [1.14] and subsequent solid-state reaction turned out to be sensitive not only to the chemicals involved [1.15] but also to the reaction temperatures. Having overcome all difficulties with the pure compound, we started to partially substitute the trivalent Ni by trivalent Al to reduce the metallic bandwidth of the Ni ions and make it comparable to the Ni^{3+} Jahn-Teller stabilization energy. With increasing Al concentration, the metallic characteristics (see Figure 1.4) of the pure $LaNiO_3$ gradually changed, first giving a general increase in the resistivity and finally with high substitution leading to a semiconducting behavior with a transition to localization at low temperatures. The idea did not seem to work out the way we had thought, so we considered the introduction of some internal strain within the $LaNiO_3$ lattice to reduce the bandwidth. This we realized by replacing the La^{3+} ion by the smaller Y^{3+} ion, keeping the Ni site unaffected. The resistance behavior changed in a way we had already recorded in the previous case, and at that point we started wondering whether the target at which we were aiming really did exist. Would the path we decided to embark upon finally lead into a blind alley?

It was in 1985 that the project entered this critical phase, and it probably only survived because the experimental situation, which had generally hampered our efforts, was improved. The period of sharing another group's equipment for resistivity measurements came to an end as our colleague, Pierre Guéret, agreed to my established right to use a newly set-up automatic system. Thus, the measuring time was transferred from late evening to normal working hours. Toni Schneider, at that time acting manager of the Physics department, supported the plans to improve the obsolete x-ray analytical equipment to simplify systematic phase analysis, and in addition, we had some hopes in our new idea, involving another TM element encountered in our search, namely, copper. In a new series of compounds, partial replacement of the JT Ni^{3+} by the non-JT Cu^{3+} increased the absolute value of the resistance, however the metallic character of the solid solutions was preserved down to 4 K [1.13]. But again, we observed no indication of superconductivity. The time to study the literature and reflect on the past had arrived.

It was in late 1985 that the turning point was reached. I became aware of an article by the French scientists C. Michel, L. Er-Rakho and B. Raveau, who

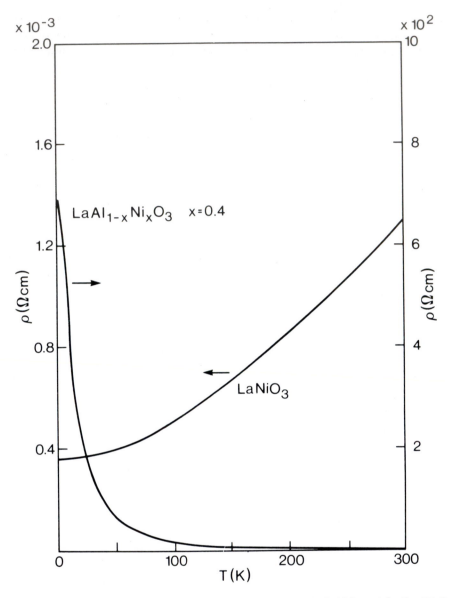

Figure 1.4. Temperature dependence of the resistivity for metallic $LaNiO_3$ and $LaAl_{1-x}Ni_xO_3$, where substitution of Ni^{3+} by Al^{3+} leads to insulating behavior for $x = 0.4$.

had investigated a Ba-La-Cu oxide with perovskite structure exhibiting metallic conductivity in the temperature range between 300 and $-100°$ C [1.16]. The special interest of that group was the catalytic properties of oxygen-deficient compounds at elevated temperatures [1.17]. In the Ba-La-Cu oxide with a perovskite-type structure containing Cu in two different valencies, all our concept requirements seemed to be fulfilled.

I immediately decided to proceed to the ground-floor laboratory and start preparations for a series of solid solutions, as by varying the Ba/La ratio one would have a sensitive tool to continuously tune the mixed valency of copper.

Within one day, the synthesis had been performed, but the measurement had to be postponed, owing to the announcement of the visit of Dr. Ralph Gomory, our Director of Research. These visits always kept people occupied for a while, preparing their presentations.

Having lived through this important visit and returning from an extended vacation in mid-January 1986, I recalled that when reading about the Ba-La-Cu oxide, it had intuitively attracted my attention. I decided to restart my activities in measuring the new compound. When performing the four-point

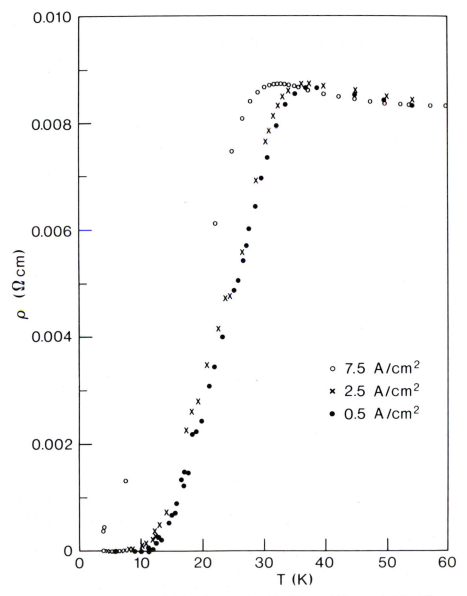

Figure 1.5. Low-temperature resistivity of a sample with x(Ba) = 0.75, recorded for different current densities. From [1.19], © Springer-Verlag 1986.

resistivity measurement, the temperature dependence did not seem to be anything special when compared with the dozens of samples measured earlier. During cooling, however, a metallic-like decrease was first observed, followed by an increase at low temperatures, indicating a transition to localization. My inner tension, always increasing as the temperature approached the 30 K range, started to be released when a sudden resistivity drop of 50% occurred at 11 K. Was this the first indication of superconductivity?

Alex and I were really excited, as repeated measurements showed perfect reproducibility and an error could be excluded. Compositions as well as the thermal treatment were varied and within two weeks we were able to shift the onset of the resistivity drop to 35 K, Figure 1.5. This was an incredibly high value compared to the highest T_c in the Nb_3Ge superconductor.

We knew that in the past there had been numerous reports on high-T_c superconductivity which had turned out to be irreproducible [1.7], therefore prior to the publication of our results, we asked ourselves critical questions about its origin. A metal-to-metal transition, for example, was unlikely, owing to the fact that with increasing measuring current the onset of the resistivity drop was shifted to lower temperatures. On the contrary, this behavior supported our interpretation that the drop in $\varrho(T)$ was related to the onset of

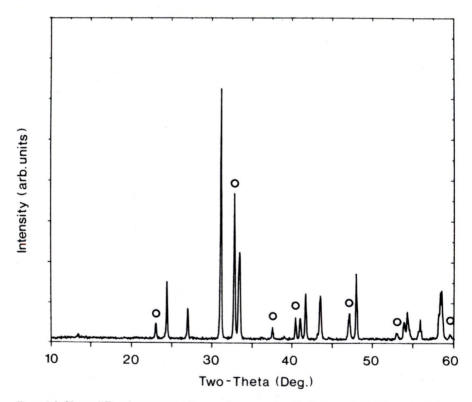

Figure 1.6. X-ray diffraction pattern of a two-phase sample with Ba:La = 0.08. The second phase occurring together with the K_2NiF_4-type phase is indicated by open circles. From [1.20], © 1987 Pergamon Journals Ltd.

superconductivity in granular materials. These are, for example, polycrystal-line films of $BaPb_{1-x}Bi_xO_3$ [1.18] exhibiting grain boundaries or different crystallographic phases with interpenetrating grains as in the Li-Ti oxide [1.7]. Indeed, x-ray diffraction patterns of our samples revealed the presence of at least two different phases (see Figure 1.6). Although we started the preparation process of the material with the same cation ratios as the French group, the wet-chemical process did not lead to the same result. This later turned out to be a stroke of luck, in the sense that the compound we wanted to form was not superconducting. The dominating phase could be identified as having a layered perovskite-like structure of K_2NiF_4-type as seen from Figure 1.7. The diffraction lines of the second phase resembled that of an oxygen-deficient perovskite with a three-dimensionally connected octahedra network. In both structures, La was partially replaced by Ba, as we learned from an electron microprobe analysis which Dr. Jürg Sommerauer at the ETH Zurich performed for us as a favor. However, the question was "which is the compound where the mixed valency of the copper leads to the superconductive transition?"

We had difficulties in finding a conclusive answer at the time; however, we rated the importance of our discovery so high that we decided to publish our findings, despite the fact that we had not yet been able to perform magnetic measurements to show the presence of the Meissner-Ochsenfeld effect. Thus, our report was cautiously entitled "Possible High T_c Superconductivity in the Ba-La-Cu-O System" [1.19]. We approached Eric Courtens, my manager at the time, who in late 1985 had already strongly supported our request to purchase a DC Squid Magnetometer, and who is on the editorial board of *Zeitschrift für Physik*. In this capacity, we solicited his help to receive and submit the paper, although, admittedly, it did involve some gentle persuasion on our part!

Alex and I then decided to ask Dr. Masaaki Takashige whether he would be interested in our project. Dr. Takashige, a visiting scientist from Japan, had joined our Laboratory in February 1986 for one year. He was attached to Alex's Fellowship group, and I had given him some support in pursuing his activities in the field of amorphous oxides. As he was sharing my office, I was able to judge his reaction, and realized how his careful comments of skepticism changed to supporting conviction while we were discussing the results. We had found our first companion.

Following this, while awaiting delivery of the magnetometer, we tried hard to identify the superconducting phase by systematically changing the composition and measuring the lattice parameters and electrical properties. We found strong indications that the Ba-containing La_2CuO_4 was the phase responsible for the superconducting transition in our samples. Starting from the ortho-rhombically distorted host lattice, increasing the Ba substitution led to a continuous variation of the lattice towards a tetragonal unit cell [1.20], see Figure 1.8. The highest T_c's were obtained with a Ba concentration close to this transition (Figure 1.9), whereas when the perovskite phase became dominant, the transition was suppressed and the samples showed only metallic character-istics.

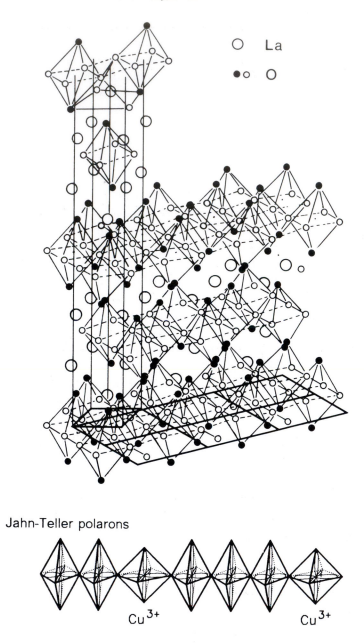

Figure 1.7. Structure of the orthorhombic La_2CuO_4. Large open circles represent the lanthanum atoms, small open and filled circles the oxygen atoms. The copper atoms (not shown) are centered on the oxygen octahedra. From [1.29], © 1987 by the American Association for the Advancement of Science.

The lower part shows schematically how in a linear chain substitution of trivalent La by a divalent alkaline-earth element would lead to a symmetric change of the oxygen polyhedra in the presence of Cu^{3+}.

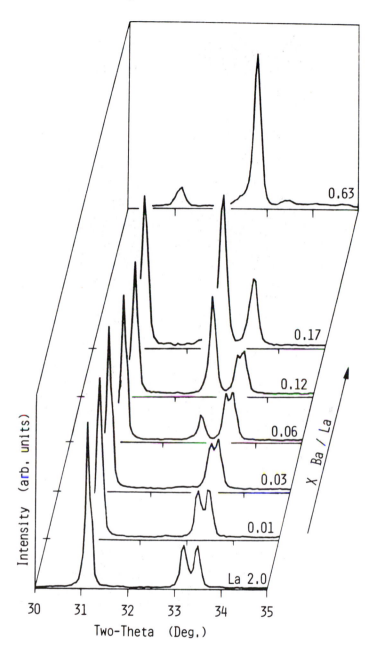

Figure 1.8. Characteristic part of the x-ray diffraction pattern, showing the orthorhombic-to-tetragonal structural phase transition with increasing Ba:La ratio. Concentration axis not to scale. From [1.20], © 1987 Pergamon Journals Ltd.

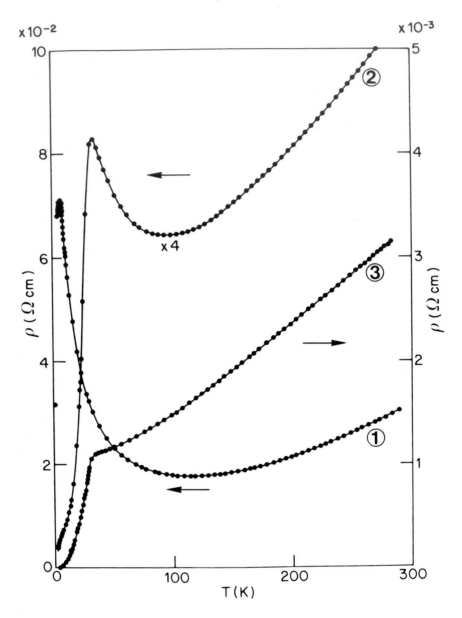

Figure 1.9. Resistivity as a function of temperature for La_2CuO_{4-y}: Ba samples with three different Ba:La ratios. Curves 1, 2, and 3 correspond to ratios of 0.03, 0.06, and 0.07, respectively. From [1.20], © 1987 Pergamon Journals Ltd.

Finally, in September 1986, the susceptometer had been set up and we were all ready to run the magnetic measurements. To ensure that with the new magnetometer we did not measure any false results, Masaaki and I decided to gain experience on a known superconductor like lead rather than starting on our samples. The Ba-La-Cu oxide we measured first had a low Ba content, where metallic behavior had been measured down to 100 K and a transition to localization occurred at lower temperatures. Accordingly, the magnetic susceptibility exhibited Pauli-like positive, temperature-independent and Curie-Weiss behavior at low temperatures, as illustrated by Figure 1.10. Most importantly, within samples showing a resistivity drop, a transition from para- to dia-magnetism occurred at slightly lower temperatures, see Figure 1.11, indicating that superconductivity-related shielding currents existed. The diamagnetic transition started below what is presumably the highest T_c in the samples as indicated by theories [1.21, 1.22] describing the behavior of percolative super-conductors. In all our samples, the transition to the diamagnetic state was systematically related to the results of our resistivity measurements. The final proof of superconductivity, the presence of the Meissner-Ochsenfeld effect, had

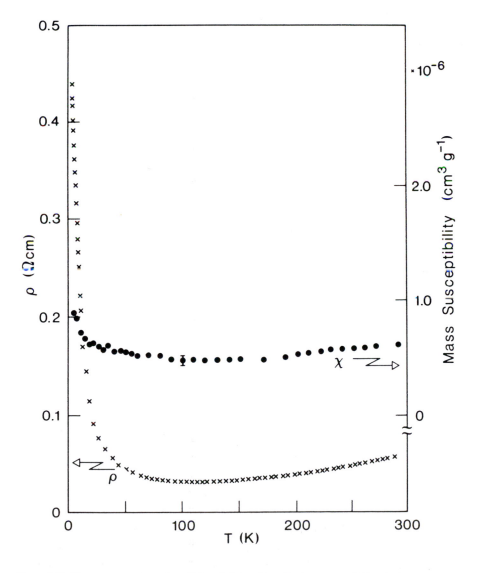

Figure 1.10. Temperature dependence of resistivity (x) and mass susceptibility (●) of sample 1. From [1.23], © Les Editions Editions de Physique 1987.

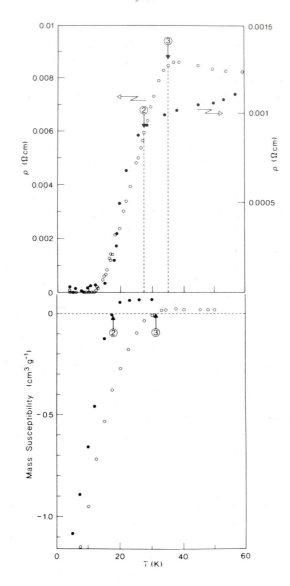

Figure 1.11. Low-temperature restivity and susceptibility of (La-Ba)-Cu-O samples 2 (●) end 3 (○) from [1.23]. Arrows indicate the onset of the resistivity and the paramagnetic-to-diamagnetic transition, respectively. From [1.23], © Les Editions de Physiqe 1987.

been demonstrated. Combining the x-ray analysis, resistivity and susceptibility measurements, it was now possible to clearly identify the Ba-doped La_2CuO_4 as the superconducting compound.

First Responses and Confirmations

The number of our troops was indeed growing. Richard Greene at our Research Center in Yorktown Heights had learned about our results and became excited. He had made substantial contributions in the field of organic superconductors and wanted to collaborate in measuring specific-heat data on our samples. We initiated an exchange of information, telefaxing the latest results of our research and sending samples. Realizing that our first paper had appeared in the open literature, we rushed to get the results of our susceptibility data written up for publication.

The day we made the final corrections to our report turned out to be one of the most remarkable days in the history of our Laboratory. Alex, Masaaki and I were sitting together, when the announcement was made over our P.A. system that the 1986 Nobel Prize for Physics had been awarded to our colleagues Gerd Binnig and Heinrich Rohrer. With everything prepared for the submission of our paper [1.23], for one more day we could forget about our work, and together with the whole Laboratory celebrate the new laureates. The next day we were back to reality, and I started to prepare a set of samples for Richard Greene. Praveen Chaudhari, our director of Physical Science in Yorktown Heights, took them with him the same evening.

Later in November, we received the first response to our latest work from Professor W. Buckel, to whom Alex had send a preprint with the results of the magnetic measurements. His congratulations on our work were an encouragement, as we began to realize that we would probably have a difficult time getting our results accepted. Indeed, Alex and I had started giving talks about our discovery and, although the presence of the Meissner effect should have convinced people, at first we were met by a skeptical audience. However, this period turned out be very short indeed.

We continued with the magnetic characterization of the superconducting samples and found interesting properties related to the behavior of a spin glass [1.24]. We then intensively studied the magnetic field and time dependences of the magnetization, before finally starting to realize an obvious idea, namely, to replace La also by other alkaline-earth elements like Sr and Ca. Especially Sr^{2+} had the same ionic radius as La^{3+}. We began experiments on the new materials which indicated that for the Sr-substituted samples T_c was approaching 40 K and the diamagnetism was even higher, see Figure 1.12(a) and (b), [1.25]. It was just at that time that we learned from the Asahi-Shinbun International Satellite Edition [November 28, 1986] that the group of Professor Tanaka at the University of Tokyo had repeated our experiments and could confirm our result [1.26]. We were relieved, and even more so when we received a letter from Professor C. W. Chu at the University of Houston, who was also convinced that within the Ba-La-Cu-O system superconductivity occurred at 35 K [1.27]. Colleagues who had not paid any attention to our work at all suddenly became alert. By applying hydrostatic pressure to the samples, Professor Chu was able to shift the superconductive transition from 35 to almost 50 K [1.27]. Modification of the original oxides by introducing the smaller Y^{3+} for the larger La^{3+} resulted in a giant jump of T_c to 92 K in multiphase samples [1.28], Figure 1.13. At a breathtaking pace, dozens of groups now repeated these experiments, and after an effort of only a few days the new superconducting compound could be isolated and identified. The resistive transition in the new $YBa_2Cu_3O_7$ compound was complete at 92 K, Figure 1.14, and even more impressive was the fact that the Meissner effect

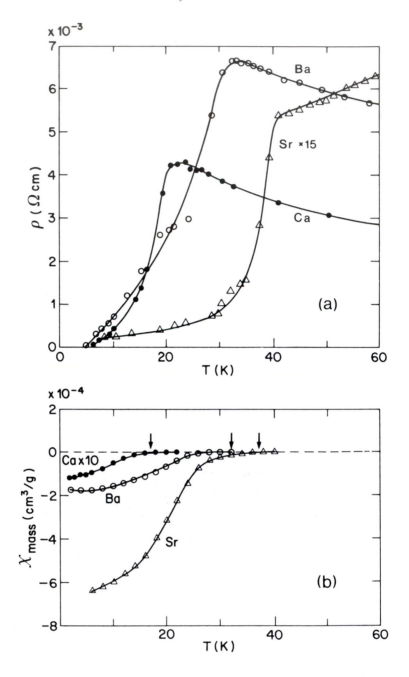

Figure 1.12. (a) Resistivity as a function of temperature for Ca (●), Sr (△), and Ba (○) substitution with substituent−to−La ratios of 0.2/1.8, ＃ 0.2/1.8, 0.15/1.85, respectively. The Sr curve has been vertically expanded by a factor of 15. (b) Magnetic susceptibility of these samples. The substituents are Ca (●), Sr (△), and Ba (○), with total sample masses of 0.14, 0.21, and 0.13 g, respectively. The Ca curve has been expanded by a factor of 10. Arrows indicate onset temperatures. From [1.25], ©️ 1987 by the American Association for the Advancement of Science.

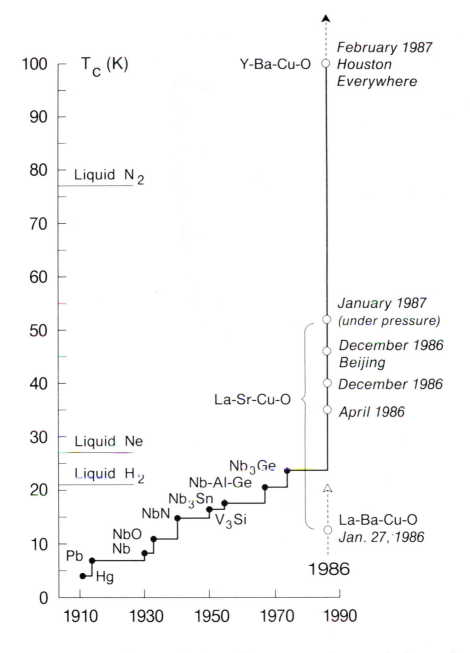

Figure 1.13. Evolution of the superconductive transition temperature subsequent to the discovery of the phenomenon. From [1.29], © 1987 by the American Association for the Advancement of Science.

could now be demonstrated without any experimental difficulties with liquid nitrogen as the coolant. Within a few months, the field of superconductivity had experienced a tremendous revival, with an explosive development of T_c's which nobody can predict where it will end.

An early account of the discovery appeared in the September 4, 1987, issue of *Science*, which was dedicated to science in Europe [1.29].

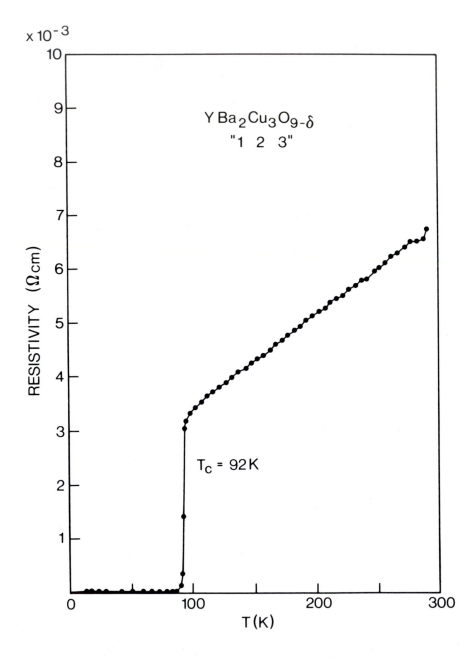

Figure 1.14. Resistivity of a single-phase $YBa_2Cu_3O_7$ sample as a function of temperature.

References Part 1

[1.1] Müller, K. A. (1971) in: *Structural Phase Transitions and Soft Modes*, editors E. J. Samuelsen, E. Andersen, and J. Feder (Universitetsforlaget, Oslo) p. 85.

[1.2] Schneider, T. and Stoll, E. (1971) Physica *55*, 702.

[1.3] Schooley, J. F., Frederikse, H. P. R., Hosler, W. R., and Pfeiffer, E. R. (1967) Phys. Rev. *159*, 301.

[1.4] Baratoff, A. and Binnig, G. (1981) Physica *108B*, 1335; Baratoff, A., Binnig, G., Bednorz, J. G., Gervais, F., and Servoin, J. L. (1982) in: *Superconductivity in d- and f-Band Metals, Proceedings IV Conference on Superconductivity in d- and f-Band Metals*, editors W. Buckel and W. Weber (Kernforschungszentrum Karlsruhe) p. 419.

[1.5] Binnig, G., Baratoff, A., Hoenig, H. E., and Bednorz, J. G. (1980) Phys. Rev. Lett. *45*, 1352.

[1.6] Müller, K. A., Pomerantz, M., Knoedler, C. M., and Abraham, D. (1980) Phys. Rev. Lett. *45*, 832, and references therein.

[1.7] Beasley, M. R. and Geballe, T. H. (1984) Phys. Today *36*(10), 60; Muller, J. (1980) Rep. Prog. Phys. *43*, 663.

[1.8] Johnston, D. C., Prakash, H., Zachariasen, W. H., and Viswanathan, R. (1973) Mat. Res. Bull. *8*, 777.

[1.9] Sleight, A. W., Gillson, J. L., and Bierstedt, P. E. (1975) Solid State Commun. *17*, 27.

[1.10] Bardeen, J., Cooper, L. N., and Schrieffer, J. R. (1957) Phys. Rev. *108*, 1175.

[1.11] Chakraverty, B. K. (1979) J. Physique Lett. *40*, L99; *idem* (1981) J. Physique *42*, 1351.

[1.12] Höck, K.-H., Nickisch, H., and Thomas, H. (1983) Helv. Phys. Acta *56*, 237.

[1.13] Goodenough, J. B. and Longo, M. (1970, "Magnetic and other properties of oxide and related compounds" In: *Landoldt-Boernstein New Series. Vol. III/4a: Crystal and Solid State Physics*, editors K. H. Hellwege and A. M. Hellwege (Springer-Verlag, Berlin, Heidelberg, New York), p. 262, Fig. 73.

[1.14] Bednorz, J. G., Müller, K. A., Arend, H., and Gränicher, H. (1983) Mat. Res. Bull. *18*, 181.

[1.15] Vasanthacharya, N. Y., Ganguly, P., Goodenough, J. B., and Rao, C. N. R. (1984) J. Phys. C: Solid State Phys. *17*, 2745.

[1.16] Michel, C., Er-Rakho, L., and Raveau, B. (1985) Mat. Res. Bull. *20*, 667.

[1.17] Michel, C. and Raveau, B. (1984) Rev. Chim. Min. *21*, 407. It is of interest that back in 1973, Goodenough, J. B., Demazeaux, G., Pouchard, M., and Hagenmüller, P. (1973) J. Solid State Chem. *8*, 325, in Bordeaux, and later Shaplygin, I. S., Kakhan, B. G., and Lazarev, V. B. (1979) Russ. J. Phys. Chem. *24*(6), 1478, pursued research on layered cuprates with catalysis applications in view.

[1.18] Suzuki, M., Murakami, T., and Inamura, T. (1981) Shinku *24*, 67; Enomoto, Y., Suzuki, M., Murakami, T., Inukai, T., and Inamura, T. (1981) Jpn. J. Appl. Phys. *20*, L661.

[1.19] Bednorz, J. G. and Müller, K. A. (1986) Z. Phys. B *64*, 189.

[1.20] Bednorz, J. G., Takashige, M., and Müller K. A. (1987) Mat. Res. Bull. *22*, 819.

[1.21] Bowman, D. R. and Stroud, D. (1984) Phys. Rev. Lett. *52*, 299.

[1.22] Ebner, C. and Stroud, D. (1985) Phys. Rev. B *31*, 165, and references therein.

[1.23] Bednorz, J. G., Takashige, M., and Müller, K. A. (1987) Europhys. Lett. *3*, 379.

[1.24] Müller, K. A., Takashige, M., and Bednorz, J. G. (1987) Phys. Rev. Lett. *58*, 1143.

[1.25] Bednorz, J. G., Müller, K. A., and Takashige, M. (1987) Science *236*, 73.

[1.26] Takagi, H., Uchida, S., Kitazawa, K., and Tanaka, S. (1987) Jpn. J. Appl. Phys. *26*, L123; Uchida, S., Takagi, H., Kitazawa, K., and Tanaka, S. (1987) *ibid.*, L151.

[1.27]   Chu, C. W., Hor, P. H., Meng, R. L., Gao, L., Huang, Z. J., and Wang, Y. Q. (1987) Phys. Rev. Lett. *58*, 405.

[1.28]   Wu, M. K., Ashburn, J. R., Torng, C. J., Hor, P. H., Meng, R. L., Goa, L., Huang, Z. J., Wang, Y. Q., and Chu, C. W. (1987) Phys. Rev. Lett. *58*, 908; Hor, P. H., Gao, L., Meng, R. L., Huang, Z. J., Wang, Y. Q., Forster, K., Vassilious, J., Chu, C. W., Wu, M. K., Ashburn, J. R., and Torng, C. J. (1987) *ibid.*, 911.

[1.29]   Müller, K. A. and Bednorz, J. G. (1987) Science *237*, 1133.

PART 2: PROPERTIES OF THE NEW SUPERCONDUCTORS

In the second part, properties of the new layered oxygen superconductors were described. Since their discovery, summarized in the first part, a real avalanche of papers has been encountered; thus it would be beyond the scope of this Lecture to review all of them here. A forthcoming international conference in Interlaken, Switzerland, in February 1988, is intended to fulfill this task and will be chaired by one of us. Therefore, only a selected number of experiments were presented in Stockholm; those judged of importance at this time for the understanding of superconductivity in the layered copper oxides. In some of them, the laureates themselves were involved, in others not. Owing to the frantic activity in the field, it may be possible that equivalent work with priority existed unbeknown to us. Should this indeed be the case, we apologize and propose that the following be read for what it is, namely, a write-up of the lecture given, including the transparencies shown.

After the existence of the new high-T_c superconductors had been confirmed, one of the first questions was "What type of superconductivity is it?" Does one again have Cooper pairing [2.1] or not? This question could be answered in the affirmative. The earliest experiment to come to our knowledge was that of the Saclay-Orsay collaboration. Estève *et al.* [2.2] measured the I-V characteristics of sintered La $_{1.85}$Sr$_{0.15}$CuO$_4$ ceramics using nonsuperconducting Pt-Rh, Cu or Ag contacts. In doing so, they observed weak-link characteristics internal to the superconductor, to which we shall subsequently revert. Then they applied microwaves at $\varkappa = 9.4$ GHz and observed Shapiro steps [2.3] at $V_s = 19$ μV intervals. From the well-known Josephson formula [2.4]

$$V_s = h\varkappa/q, \qquad (2.1)$$

they obtained q = 2e, i.e., Cooper pairs were present. Figure 2.1 illustrates these steps. From the fundamental London equations, the flux ϕ through a ring is quantized [2.5]

$$\phi = n\phi_0$$
$$\phi_0 = hc/q. \qquad (2.2)$$

The clearest experiment, essentially following the classical experiments in 1961, was carried out in Birmingham, England, by C. E. Gough *el al.* [2.6]. They detected the output of an r.f.-SQUID magnetometer showing small integral numbers of flux quantum ϕ_0 jumping into and out of the ring of Y$_{1.2}$Ba$_{0.8}$CuO$_4$, see Figure 2.2. The outcome clearly confirmed that q = 2e.

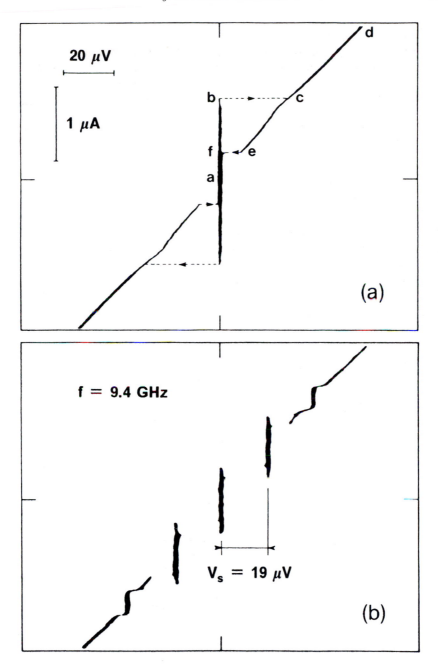

Figure 2.1. (a) Oscilloscope trace of a current-voltage characteristic obtained at 4.2 K with an aluminum tip on a $La_{1.85}Sr_{0.15}CuO_4$ sample. Letters a through f indicate sense of trace. Dashed lines have been added to indicate the switching between the two branches. (b) Steps induced by a microwave irradiation at frequency $f = 9.4$ GHz. All other experimental conditions identical with those of (a). From [2.2] © Les Editions de Physique 1987.

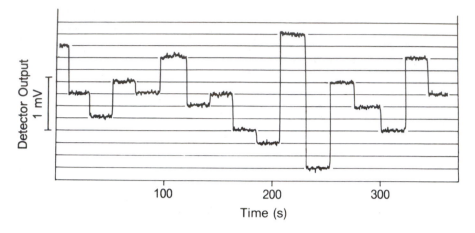

Figure 2.2. Output of the r.f.-SQUID magnetometer showing small integral numbers of flux quanta jumping in to and out of a ring. Reprinted by permission from [2.6], copyright © 1987 Macmillan Magazines Ltd.

To understand the mechanism, it was of relevance to know the nature of the carrier charge present. In La_2CuO_4 doped very little, the early measurements [1.23] showed localization upon doping with divalent Ba^{2+} or Sr^{2+} and Ca $^{2+}$; it was most likely that these ions substituted for the trivalent La^{3+} ions. Thus, from charge-neutrality requirements, the compounds had to contain holes. Subsequent thermopower and Hall-effect measurements confirmed this assumption [2.7]. The holes were thought to be localized on the Cu ions. Because the copper valence is two in the stoichiometric insulator La_2CuO_4, doping would create Cu^{3+} ions. Thus a mixed Cu^{2+}/Cu^{3+} state had to be present. By the same argument, this mixed-valence state ought also to occur in $YBa_2CuO_{7-\delta}(\delta \sim 0.1)$. Early photo-electron core-level spectra (XPS and UPS) by Fujimori *el al.* [2.8] and Bianconi *et al.* [2.9] in $(La_{1-x}Sr_x)_2CuO_{4-y}$ and $YBa_2La_3Cu_{6.7}$ did not reveal a $\underline{2p}$ $3d^8$ final state owing to a Cu^{3+} $3d^8$ state (the underlining indicates a hole). However, the excitation was consistent with the formation of holes L in the oxygen-derived band, i.e., a predominant $3d^9$ $\underline{L}$ configuration for the formal Cu^{3+} state. Photo-x-ray absorption near the edge structure was also interpreted in the same manner by comparison to other known Cu compounds. Emission spectra by Petroff's group [2.10] pointed in the same direction since the excitation thresholds were compatible with the presence of holes in Cu-O hybrid bands. From their data, both groups concluded that strong correlation effects were present for the valence carriers. However, these results were challenged by other groups working in the field, partially because the spectra involved the interpretation of Cu-atom satellites. A beautifully direct confirmation of the presence of holes on the oxygen p-levels, like $\underline{L}$, was carried out by Nücker *et al.* [2.11]. These authors investigated the core-level excitation of oxygen 1s electrons into empty 2p states of oxygen at 528 eV. This is an oxygen-specific experiment. If no holes are present on the p-level, no absorption will occur. Figure 2.3 summarizes their data on $La_{2-x}Sr_xCuO_4$ and $YBa_2Cu_3O_{7-\delta}$. It is shown **that for x ≈ 0 and δ = 0.5**, no

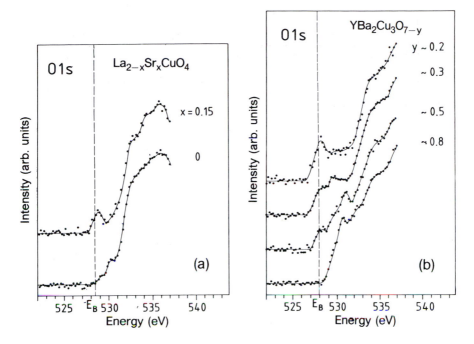

Figure 2.3. Oxygen 1s absorption edges of (a) $La_{2-x}Sr_xCuO_4$ and (b) $YBa_2Cu_3O_{7-y}$ measured by energy-loss spectroscopy. The binding energy of the O 1s level, as determined by x-ray induced photoemission, is shown by the dashed line. In the framework of an interpretation of the spectra by the density of unoccupied states, this line would correspond to the Fermi energy. From [2.11], © 1988 The American Physical Society.

oxygen p-holes are present and thus no absorption is observed, whereas upon increasing x or reducing δ, a 2p hole density at the Fermi level is detected in both compounds.

Of substantial interest is the dependence of the transition temperature on the hole concentration. The electron deficiency is hereafter written in the form $[Cu-O]^+$ as a peroxide complex in which the probability of the hole is about 70 % $3d^9$ $\underline{2p}$, as discussed above, and 30 % $3d^8$ as recently inferred from an XPS study [2.12]. Hall-effect data are difficult to analyze in the presence of two-band conductivity, which is possible in these copper-oxide compounds, owing to the well-known compensation effects. Therefore, M. W. Shafer, T. Penney and B. L. Olson [2.13] determined the concentration by wet chemistry according to the reaction $[Cu-O]^+ + Fe^{2+} \leftrightarrows Cu^{2+} + Fe^{3+} + O^{2-}$ in the $La_{2-x}Sr_xCuO_{4-\delta}$ compound. Figure 2.4 shows a plot of T_c vs $[Cu-O]^+$ concentration with a maximum of 35 K of 15 % total copper present. There is also a clear threshold at about 5 %. From the study, it is apparent that 15–16 % $[Cu-O]^+$ is the maximum number of holes the La_2CuO_4 structure accepts. Beyond this concentration, oxygen vacancies are formed. The relationship between T_c and $[Cu-O]^+$ in $La_{2-x}Sr_xCuO_4$ was extended to $YBa_2Cu_3O_{7-\delta}$. The inset of Figure 2.4 illustrates the results under the assumption that two layers in the 123 compound are active for $\delta \approx 0.1$ and $\delta \approx 0.3$, i.e., T_c's of 92 K and 55 K [2.14]. The latter transition, first reported by Tarascon

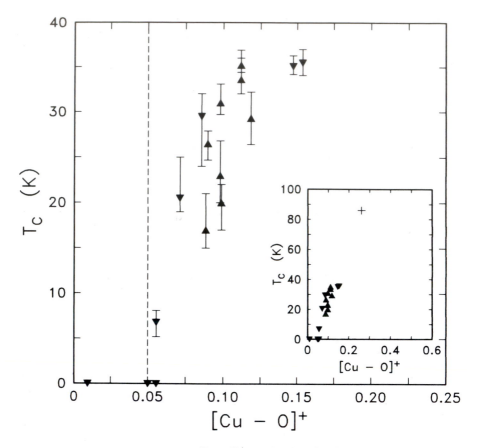

Figure 2.4. T_c vs the hole concentration $[Cu-O]^+$, as a fraction of total copper. Down triangles are
for compositions with x<0.15, up triangles for x>0.15. Inset shows same data plus points for single
$YBa_2Cu_3O_{6.6}$ sample as discussed in the text. From [2.13], © 1987 The American Physical Society.

and coworkers, could be well evidenced by near-room-temperature plasma
oxidation of the oxygen-deficient Y-compound [2.14].

The $La_{2-x}Sr_xCuO_{4-y}$ with its less complicated structure allows easier testing of models. Its magnetic properties below the hole threshold concentration,
$x = 0.05$, are of special interest. For $x = 0$, the susceptibility $\chi(T) = M(T)/H$
exhibits a maximum at low fields of $H = 0.05$ Tesla below 300 K. This maximum increases in height and shifts to lower temperatures for higher magnetic
fields up to 4.5 T [2.15] as seen in Figure 2.5(a). Such behavior is indicative of
spin density waves or antiferromagnetic fluctuations. Indeed, neutron-diffrac-
tion experiments by Vaknin *et al.* [2.16] proved three-dimensional (3-D) antiferromagnetic ordering up to 240 K depending on oxygen stoichiometry (i.e.,
hole concentration). The structure is shown in Figure 2.5(b). Subsequent
neutron-scattering experiments on a single crystal revealed a novel two-dimensional (2-D) antiferromagnetic correlation well above and also *below* the 3-D
Néel temperature of T_N as shown in Figure 2.6. This instantaneous (not time-
averaged) ordering was seen even above room temperature [2.17]. The existence of antiferromagnetism (A.F.) supports models in which holes lead either

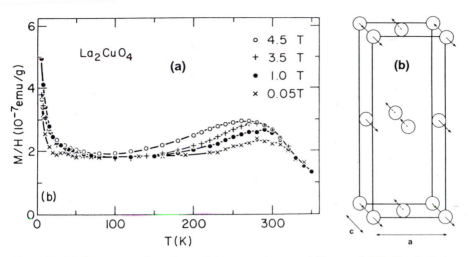

Figure 2.5. (a) Temperature dependence of the magnetic susceptibility χ = M/H of La_2CuO_4 in different fields H. From [2.15], © 1987 Pergamon Journals Ltd. (b) Spin structure of antiferromagnetic La_2CuO_{4-y}. Only copper sites in the orthorhombic unit cell are shown for clarity. From [2.16], © 1987 The American Physical Society.

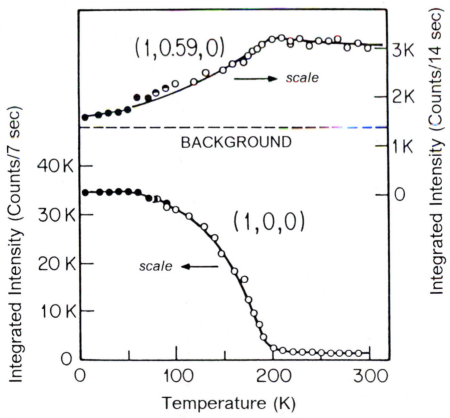

Figure 2.6. Integrated intensities of the (100) 3-D antiferromagnetic Bragg peak and the (1,0.59,0) 2-D quantum spin fluid ridge. The open and filled circles represent separate experiments. From [2.17], © 1987 The American Physical Society.

to localization or to pairing in the strong-coupling limit as proposed by Emery [2.18] and others [2.19]. The resonant valence-band state is also related to the A.F. state [2.20].

From the prevalence of magnetic interactions as primary cause for the occurrence of the high-T_c superconductivity, one would expect the isotope effect to be absent. This, because the latter effect is found when the Cooper pairing is mediated by phonon interaction, as found in most of the metallic superconductors previously known. Indeed, substituting O^{16} by O^{18} in the $YBa_2Cu_3O_{7-\delta}$ compound at AT&T did not reveal a shift in T_c [2.21]. However, substitution experiments in the $La_{2-x}Sr_xCuO_{4-y}$ carried out shortly thereafter did reveal an isotope effect with $0.14 < \beta < 0.35$ [2.22] as compared with the full effect of $\beta = 1/2$ deduced from the weak-coupling formula [1.10]

$$T_c = 1.13\Theta_D \exp-(1/N(E_F)_xV^*), \qquad (2.3)$$

with the Debye temperature $\Theta_D \propto 1/M^{1/2}$ of the reduced mass. Thus in the lanthanum compound, oxygen motion is certainly present. As it is highly unlikely that the mechanism is substantially different in the 123 compound, oxygen motion should also be there. This, because absence of the isotope effect does not necessarily exclude a phonon mechanism, which has to be present if Jahn-Teller polarons participate. Indeed, a subsequent, more accurate experiment did show a weak isotope effect in $YBa_2Cu_3O_{7-\delta}$ with $\Delta T_c \approx 0.3$ to 0.5 K [2.23]. From these results, it appears likely that there is more than one interaction present which leads to the high transition temperatures, the low quasi-2D properties certainly being of relevance.

The x-ray and photoemission studies mentioned earlier had indicated strong correlation effects. Cooper pairing having been ascertained, it was therefore of considerable interest whether the new superconductors were of the strong- or the weak-coupling variety. In the latter case, the gap 2Δ to kT_c ratio is [1.10]

$$\frac{2\Delta}{kT_c} = 3.52, \qquad (2.4)$$

whereas in the former it is larger.

Tunneling experiments have been widely used to determine the gap in the classical superconductors. However, the very short coherence length yields too low values of 2Δ, as will be discussed later [2.24]. Infrared transmission and reflectivity measurements on powders were carried out at quite an early stage. With the availability of $YBa_2Cu_3O_7$ single crystals, powder infrared data are less relevant, but are quoted in the more recent work. An interesting example is the reflectivity study by Schlesinger *et al.* [2.25] of superconducting $YBa_2Cu_3O_7$ and a Drude fit to the nonsuperconducting $YBa_2Cu_3O_{6.5}$ data. From the Mattis-Bardeen enhanced peak in the superconducting state, these authors obtained $2\Delta_{ab}/kT_c \approx 8$, i.e., strong coupling in the Cu-O planes, see Figure 2.7. NMR relaxation experiments by Mali *et al.* [2.26], although not yet completely analyzed, yield two gaps with ratios 4.3 and 9.3, respectively i.e., the latter in the range of the infrared data.

NMR relaxation experiments were among the first at the time to prove the

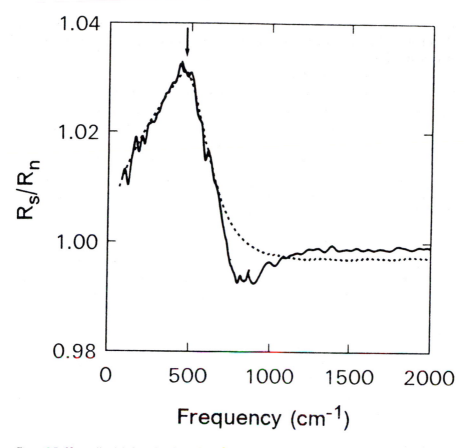

Figure 2.7. Normalized infrared reflectivity of a single crystal of $YBa_2Cu_3O_7$ and fitted Mattis-Bardeen form of $\delta(\omega)$ (dotted line) in the superconducting state. The arrow shows the peak occurring at $2\Delta \sim 480$ cm^{-1}, hence $2\Delta \approx 8kT_c$ with $T_c = 92$ K [2.25]. Courtesy of Z. Schlesinger *et al.*

existence of a gap [2.27]. They also appear to be important for the new class of superconductors. Zero-field nuclear spin lattice relaxation measurements of [139]La in $La_{1.8}Sr_{0.2}CuO_{4-\delta}$ below T_c behave like $1/T_1 \propto \exp(-\Delta/kT)$, see Figure 2.8, with activation energy $\Delta = 1.1$ meV at low temperatures $kT \ll 2\Delta$ due to a $T_c = 38$ K. A ratio of $2\Delta/kT_c = 7.1$ was obtained [2.28]. Therefore, strong coupling appears to be also present in the La compound. The value of Δ probably has to be attributed to the gap parallel to the planes. In fact, it could be shown that infrared reflectivity data on powders measure the gap along the c-axis, and a ratio of $2\Delta_c/kT_c \approx 2.5$ was given [2.29]. Thus the coupling between the planes would be weak. Such a substantial anisotropic property was not previously found in other superconductors.

From the first measurements of resistivity as a function of magnetic field, the slopes dH_{c2}/dT near T_c could be obtained, and from them very high critical fields at low temperatures were extrapolated. From the many works published, we quote that of Decroux *et al.* [2.30], also because this was the first paper to

Physics 1987

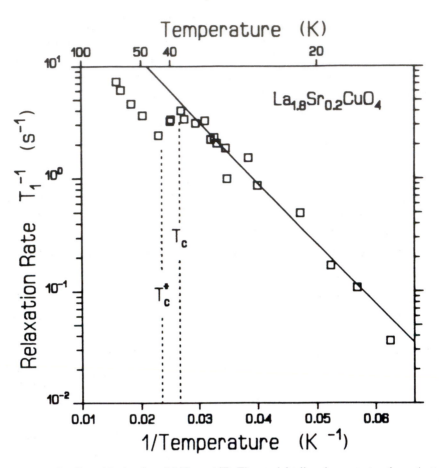

Figure 2.8. Semilogarithmic plot of $1/T_1$ vs $1/T$. The straight line demonstrates the activated behavior $1/T_1 \sim \exp[-\Delta/kT]$ for $T \ll T_c^*$. An activation energy of $\Delta/k = 135$ K is obtained from this graph. From [2.28], © Les Editions de Physique 1988.

report a specific-heat plateau at T_c. The group at the University of Geneva found $dH_{c2}/dT = -2.5$ T/K, yielding an extrapolated $H_{c2}(T = 0) = 64$ T.

From the well-known formula the critical field in type-II superconductors,

$$H_{c2} = \frac{\phi_0}{2\pi\xi^2},\tag{2.5}$$

one calculates that the coherence length ξ is of the order of the lattice distances. Actually the coherence lengths evaluated have become smaller; recent results on single crystals by IBM's group in Yorktown Heights [2.31] and the Stanford group on expitaxial layers [2.32] are of the order $\xi_c \approx 3 - 4\ \Omega$ for the coherence length parallel to c and $\xi_{ab} \approx 20 - 30\Omega$ perpendicular to c.

Such short coherence lengths could be expected when one considers the relation of ξ with the gap and the Fermi energy E_F. Weisskopf [2.33] deduced

$$\xi \approx \frac{E_F}{\Delta}d\tag{2.6}$$

from the Heisenberg uncertainty principle. In Eq. (2.6), d is the screening length, which one can assume to be of the order of a unit-cell distance. The ratio E_F/Δ is near unity owing to the large Δ and the small E_F, the latter resulting from the low carrier density and the sizeable electron mass. Therefore in oxides, ξ is considerably smaller than in metals. Because Δ is anisotropic, so is ξ. The comparable size of E_F and Δ indicates that most of the carriers participate in the superconductivity of the new oxides for temperatures $T < T_c$, in contrast to the classic superconductors, where $E_F \gg \Delta \approx 1.7\ kT_c$.

The short coherence lengths in the layered copper-oxide superconductors are important theoretically, experimentally and applicationwise: The short ξ's and carrier concentrations of the order of $n = 10^{21}/cm^3$ make one wonder whether boson-condensation approaches are not more appropriate, i.e., real-space Cooper pairing in contrast to the wave-vector space pairing of classical BCS theory [1.10], which applies so well for metals with large ξ's and concentrations n. Actually, Schafroth [2.34] back in 1955 was the first to work out a superconductivity theory with boson condensation. Referring to Chakraverti's phase diagram in Figure 1.2 [1.11], one may regard the metal superconducting phase line as BCS with weak coupling, and the superconducting insulator boundary for large coupling constants λ as the Schafroth line.

The short coherence lengths induce considerable weakening of the pair potential at surfaces and interfaces, as emphasized by Deutscher and Müller [2.24]. Using an expression for the "extrapolation length" b [2.35] for the boundary condition at the superconducting-insulator interface, the $\Delta(x)$ profile was deduced as shown in Figure 2.9 for $T \lesssim T_c$ and $T \ll T_c$.

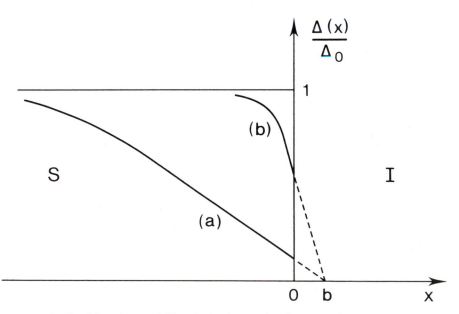

Figure 2.9. Profile of the pair potential in a short-coherence-length superconductor near a superconductor-insulator boundary. Curve a: $T \lesssim T_c$; curve b: $T \ll T_c$. From [2.24], © 1987 The American Physical Society.

Analogous behavior of $\Delta(x)$ will also be present at superconducting-normal (SN) interfaces. Thus, the depressed order parameter involving experiments of SIS and SNS will result in tunneling characteristics [2.24] with a reduced value of the Δ observed. In consequence, such experiments are less suitable than infrared and NMR to detemine Δ, and actually lead to erroneous conclusions regarding gapless superconductivity also in point-contact spectroscopy [2.36]. $YBa_2Cu_3O_7$ undergoes a tetragonal-to-orthorhombic phase transition near 700 C. Thus upon cooling, (110) twin boundaries are formed, separating the orthorhombic domains and inducing *intragrain* Josephson or weak-link junctions. These junctions form a network dividing the crystallites into Josephson-coupled domains, with possibility of fluxon trapping as well. Therefore even single crystals can form a superconducting glass *in the presence* of a sizeable magnetic field.

The basic Hamiltonian regarding the phases is [1.22]

$$\beta H = -\sum J_{ij} \cos(\phi_i - \phi_j - A_{ij}). \qquad (2.7)$$

Here J_{ij} is the Josephson coupling constant between domains. The phase factors $A_{ij} = K_{ij}H$ introduce randomness for $H \neq 0$ because K_{ij} is a random geometric factor. A review of the superconducting glass state has recently appeared [2.37].

The first experimental evidence indicating the presence of superconducting glassy behavior was deduced from field-cooled and zero-field-cooled magnetization data [1.24, 2.38] in $La_{2-x}Ba_xCuO_4$ ceramics. In addition to the twin-boundary-induced *intragrain* junctions, such a material also has junctions resulting from the *intergrain* boundaries. The latter J_{ij}'s are *much weaker* and uncouple at lower magnetic fields and currents J_c. Consequently, the critical currents observed in the ceramics are more of the order of 10^3 to 10^4 A/cm^2 [2.39], whereas those in epitaxial layers [2.40] and single crystals [2.1] are of the order of 10^6 to 10^7 A/cm^2 [2.41]. The latter work, carried out by two IBM groups, is a major breakthrough in the field.

The decay length of the superconducting wave functions at SNS and SIS junctions are both of the order of $\xi(0)$. This entails an anomalous temperature dependence of $J_c \propto (T-T_c)^2$. Such behavior is seen in the mid-temperature range for $J_c(T)$ in the $YBa_2Cu_3O_{7-\delta}$ epilayer on $SrTiO_3$ of Figure 2.10 [2.40]. Such critical currents are acceptable for thin-film applications at 77 K for low magnetic fields (Figure 2.10), whereas in the ceramics much lower J_c's require substantial inventiveness or, perhaps better still, a new type of high-T_c superconductor that should exist.

The geometrical critical magnetic field H^*_{c1} is of the order of [1.22]

$$H^*_{c1} = \phi_0/2S, \qquad (2.8)$$

where S is the projected area of the superconducting loop with uniform phase. In single crystals, the S of domains is of the order of $S = 100 \, \mu m^2$, whereas that of grains in ceramics is $S = 1 - 10 \, \mu m^2$. In agreement with Eq. (2.8), H^*_{c1} is of

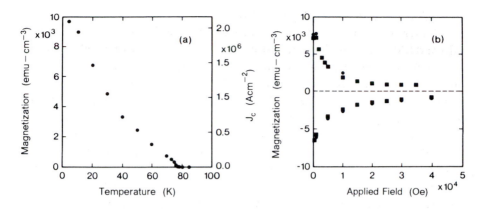

Figure 2.10. (a) The volume magnetization vs increasing temperature for an epitaxial sample. The y-axis scale on the right was obtained with the use of the Bean formula $J_c = 30$ M/d and with the mean radius of the sample of d = 0.14 cm. (b) The volume magnetization vs applied field at 4.2 K for two samples. From [2.40], © 1987 The American Physical Society.

the order of 0.5 Gauss for the penetration of H into twinned crystals, and 5 to 100 Gauss to disrupt intergranular nets in ceramics.

Since the publication on the existence of this new class of materials, the interest and work have far exceeded the expectations of the laureates, whose aim was primarily to show that oxides could "do better" in superconductivity than metals and alloys. Due to this frenzy, progress on the experimental side has been rapid and is expected to continue. This will also assist in finding new compounds, with T_c's reaching at least 130 K (Figure 2.4). Quantitative theoretical models are expected in the not too distant future, first perhaps phenomenological ones. On this rapidly growing tree of research, separate branches are becoming strong, such as glassy aspects, growth techniques for single crystals, epitaxial films, and preparation of ceramics, the latter two being of crucial importance for applications. The former will dominate the small-current microelectronics field, while the latter will have to be mastered in the large-current field. Here the hopes are for energy transport, and large magnetic-field applications for example in beam bending in accelerators and plasma containment in fusion.

References Part 2

[2.1]   Cooper, L. N. (1956) Phys. Rev. *104*, 1189.
[2.2]   Estève, D., Martinis, J. M., Urbina, C., Devoret, M. H., Collin, G., Monod, P., Ribault, M., and Revcolevschi, A. (1987) Europhys. Lett. *3*, 1237.
[2.3]   Shapiro, S. (1963) Phys. Rev. Lett, *11*, 80.
[2.4]   Josephson, B. D. (1962) Phys. Lett, *1*, 251.
[2.5]   London, F. (1950) *Superfluids,* Vol. I. (John Wiley and Sons, Inc., New York), p. 152.
[2.6]   Gough, C. E., Colclough, M. S., Forgan, E. M., Jordan, R. G., Keene, M., Muirhead, C. M., Rae, A. I. M., Thomas, N., Abell, J. S., and Sutton, S. (1987) Nature (London) *326*, 855.
[2.7]   Steglich, F., Bredl, C. D., De Boer, F. R., Lang M., Rauchschwalbe, U., Rietschel, H., Schefzyk, R., Sparn, G., and Stewart, G. R. (1987) Physica Scripta *T19*, 253.
[2.8]   Fujimori, A., Takayama-Muromachi, E., Uchida, Y., and Okai, B. (1987) Phys. Rev. B *35*, 8814.
[2.9]   Bianconi, A., Castellano, A. C., De Santis, M., Delogu, P., Gargano, A., and Giorgi, R. (1987) Solid State Commun. *63*, 1135; Bianconi, A., Castellano, A. C., De Santis, M., Politis, C., Marcelli, A., Mobilio, S., and Savoia, A. (1987) Z. Phys. B *67*, 307.
[2.10]   Thiry, P., Rossi, G., Petroff, Y., Revcolevschi, A., and Jegoudez, J. (1988) Europhys. Lett, *5*, 55.
[2.11]   Nücker, M., Fink. J., Fuggle, J. C., Durham, P. J., and Temmerman, W. M. (1988) Phys. Rev. B *37*, 5158.
[2.12]   Steiner, P., Hüfner, S., Kinsinger, V., Sander, I., Siegwart, B., Schmitt, H., Schulz, R., Junk, S., Schwitzgebel, G., Gold, A., Politis, C., Müller, H. P., Hoppe, R., Kemmler-Sack, S., and Kunz, C. (1988) Z. Phys. B *69*, 449.
[2.13]   Shafer, M. W., Penney, T., and Olson, B. L. (1987) Phys. Rev. B *36*, 4047.
[2.14]   Tarascon, J. M., McKinnon, W. R., Greene, L. H., Hull, G. W., and Vogel, E. M. (1987) Phys. Rev. B *36*, 226; see also Bagley, B. G., Greene, L. H., Tarascon, J. M., and Hull, G. W. (1987) Appl. Phys. Lett. *51*, 622.
[2.15]   Greene, R. L., Maletta, H., Plaskett, T. S., Bednorz, J. G., and Müller, K. A. (1987) Solid State Commun, *63*, 379.
[2.16]   Vaknin, D., Sinha, S. K., Moncton, D. E., Johnston, D. C., Newsam, J. M., Safinya, C. R., and King, H. E., Jr. (1987) Phys. Rev. Lett. *58*, 2802.
[2.17]   Shirane, G., Endoh, Y., Birgeneau, R. J., Kastner, M. A., Hidaka, Y., Oda, M., Suzuki, M., and Murakami, T. (1987) Phys. Rev. Lett. *59*, 1613.
[2.18]   Emery, V. J. (1987) Phys. Rev. Lett. *58*, 2794; *idem* (1987) Nature (London) *328*, 756.
[2.19]   Hirsch, J. E. (1987) Phys. Rev. Lett. *59*, 228; Cyrot, M. (1987) Solid State Commun. *62*, 821, and to be published.
[2.20]   Anderson, P. W. (1987) Science *235*, 1196; Anderson, P. W., Baskaran, G., Zou, Z., and Hsu, T. (1987) Phys. Rev. Lett. *58*, 2790.
[2.21]   Batlogg, B., Cava, R. J., Jayaraman, A., Van Dover, R. B., Kourouklis, G. A., Sunshine, S., Murphy, D. W., Rupp, L. W., Chen, H. S., White, A., Short, K. T., Mujsce, A. M., and Rietman, E. A. (1987) Phys. Rev. Lett. *58*, 2333; Bourne, L. C., Crommite, M. F., Zettl, A., zur Loye, H.-C., Keller, S. W., Leary, K. L., Stacy, A. M., Chang, K. J., Cohen, M. L., and Morris, D. E. (1987) *ibid*, 2337.
[2.22]   Faltens, T. A., Ham, W. K., Keller, S. W., Leary, K. J., Michaels, J. N., Stacy, A. M., zur Loye, H.-C., Morris, D. E., Barbee III, T. W., Bourne, L. C., Cohen, M. L., Hoen, S., and Zettl, A. (1987) Phys. Rev. Lett. *59*, 915; Batlogg, B., Kourouklis, G., Weber, W., Cava, R. J., Jayaraman, A., White, A. E., Short, K. T., Rupp, L. W., and Rietman, E. A. (1987) *ibid*, 912.

[2.23] Leary, K. J., zur Loye, H.-C., Keller, S. W., Faltens, T. A., Ham, W. K., Michaels, J. N., and Stacy, A. M. (1987) Phys. Rev. Lett. *59*, 1236.

[2.24] Deutscher, G. and Müller, K. A. (1987) Phys. Rev. Lett. *59*, 1745.

[2.25] Schlesinger, Z., Collins, R. T., Kaiser, D. L., and Holtzberg, F., (1987) Phys. Rev. Lett. *59*, 1958.

[2.26] Mali, M., Brinkmann, D., Pauli, L., Roos, J., Zimmermann, H., and Hulliger, J. (1987) Phys. Lett. A *124*, 112, and references therein. It should be noted that analysis of the relaxation data $1/T_c \propto T^3$ is also possible, as one expects in the presence of anisotropic gaps.

[2.27] Hebel, L. C. and Slichter, C. P. (1959) Phys. Rev. *113*, 1504.

[2.28] Seidel, H., Hentsch, F., Mehring, M., Bednorz, J. G., and Müller, K. A. (1988) Europhys. Lett. *5*, 647.

[2.29] Schlesinger, Z., Collins, R. T., and Shafer, M. W. (1987) Phys. Rev. B *35*, 7232.

[2.30] Decroux, M., Junod, A., Bezinge, A., Cattani, D., Cors, J., Jorda, J. L. Stettler, A., François, M., Yvon, K., Fischer, Ø., and Muller, J. (1987) Europhys. Lett. *3*, 1035.

[2.31] Worthington, T. K., Gallagher, W. J., and Dinger, T. R. (1987) Phys. Rev. Lett. *59*, 1160.

[2.32] Kapitulnic, A., Beasley, M. R., Castellani, C., and Di Castro, C. (1988) Phys. Rev. B *37*, 537.

[2.33] Weisskopf, V. F. (1979) "The Formation of Cooper Pairs and the Nature of Superconducting Currents," CERN, Geneva, Memorandum No. CERN 79-12, Theoretical Studies Division, Dec. 21, 1979.

[2.34] Schafroth, M. R., (1954) Phys. Rev. *96*, 1149 and 1442; *idem* (1955) Phys. Rev. *100*, 463.

[2.35] de Gennes, P. G. (1968) *Superconductivity of Metals and Alloys* (Benjamin, New York) p. 229.

[2.36] Yanson, I. K., Rybal'chenko, L. F., Fisun, V. V., Bobrov, N. L., Obolenskii, M. A., Brandt, N. B., Moshchalkov, V. V., Tret'yakov, Yu. D., Kaul', A. R., and Graboi, I. É. (1987) Fiz. Nizk. Temp. *13*, 557 [Sov. J. Low Temp. Phys. *13*, 315 (1987)]

[2.37] Müller, K. A., Blazey, K. W., Bednorz, J. G., and Takashige, M. (1987) Physica *148 B*, 149.

[2.38] Razavi, F. S., Koffyberg, F. P., and Mitrović, B. (1987) Phys. Rev. B *35*, 5323.

[2.39] Kwak, J. F., Venturini, E. L., Ginley, D. S., and Fu, W. (1987) *Novel Superconductivity. Proceedings of the International Workshop on Novel Mechanisms in Superconductivity*, editors, S. A. Wolf and V. Z. Kresin (Plenum Press, New York) p. 983.

[2.40] Chaudhari, P., Koch, R. H., Laibowitz, R. B., McGuire, T. R., and Gambino, R. J. (1987) Phys. Rev. Lett. *58*, 2684.

[2.41] Dinger, T. R., Worthington, T. K., Gallagher, W. J., and Sandstrom, R. L. (1987) Phys. Rev. Lett. *58*, 2687.

Physics 1988

LEON M LEDERMAN, MELVIN SCHWARTZ and JACK STEINBERGER

for the neutrino beam method and the demonstration of the doublet structure of the leptons through the discovery of the muon neutrino

THE NOBEL PRIZE IN PHYSICS

Speech by Professor Gösta Ekspong of the Royal Swedish Academy of Sciences.
Translation from the Swedish text.

Your Majesties, Your Royal Highnesses, Ladies and Gentlemen,

The Royal Swedish Academy of Sciences has decided to award this year's Nobel Prize in Physics jointly to Dr Leon Lederman, Dr Melvin Schwartz and Dr Jack Steinberger. The citation has the following wording, "for the neutrino beam method and the demonstration of the doublet structure of the leptons through the discovery of the muon neutrino".

The neutrino figures in George Gamow's entertaining book "Mr Tompkins Explores the Atom", written in the 1940's. Gamow describes how Mr Tompkins in a dream visits a woodcarvers shop, where the building blocks of the elements — protons, neutrons and electrons — are stored in separate caskets. Mr Tompkins sees many unusual things, but above all a carefully closed, but apparently empty casket labelled: "NEUTRINOS, Handle with care and don't let out". The woodcarver does not know whether there is anything inside. The friend, who had presented the casket to him, must have been Wolfgang Pauli, Nobel Laureate in Physics in 1945, who proposed the existence of the neutrino in the early 1930's.

The neutrino is electrically neutral and almost or totally massless — hence the name. It cannot be seen and it interacts only weakly with atoms. It travels with the speed of light or nearly so. It is impossible to completely stop a beam of neutrinos. To do so would require a wall of several hundred thousands of steel blocks stacked in depth one after the other, each with a thickness corresponding to the distance from here to the sun.

Our sun is a source of neutrinos, which are copiously produced in its hot central region. They pass through the whole sun without much difficulty. Every square centimeter on Earth is bombarded by many billion solar neutrinos every second and they pass straight through the Earth without leaving a noticeable mark. The neutrinos are — if I may say so — "lazy", they do almost nothing but steal energy, which they carry away.

The great achievement of the Nobel prize winners was to put the "lazy" neutrinos to work. Lederman, Schwartz and Steinberger are famous for several other important discoveries concerning elementary particles. At the time of the neutrino experiment they were associated with Columbia University in New York. They and their co-workers designed the world's first beam of neutrinos at the Brookhaven National Laboratory, using its large proton accelerator as a source. Their neutrinos had considerably more energy than usual, because they were produced from the decay in flight of fast moving mesons. Such neutrinos are much more apt to interact with matter and the collisions with atomic particles become much more interesting. Although a neutrino collision is a rare event it can be spectacular at high energy — and very informative.

In their pioneering experiment the prizewinners dealt with a total of about 10^{14}, i.e. a hundred thousand billion neutrinos. To catch just a few dozen collisions from all these, the research team invented and built a huge, sophisticated detector with the weight of 10 tons. Other unwanted particles in the beam had to be prevented from entering the detector. An enormous 13-meter thick steel wall served this purpose. To save time and money, the wall material was taken from scrapped battleships. Unwanted particles came also from the outside in the form of cosmic ray muons. Various tricks were used to prevent these muons from playing a role as false neutrinos. The first neutrino beam experiment was a bold endeavor, which proved successful. The method has since been much used as a tool for investigating the weak force and the quark structure of matter. It has also been used to investigate the neutrino itself.

At the time of the prizewinners' experiment, physicists were puzzled by the fact that a possible, alternative decay of the muon particle did not happen. No known law forbade it, and there is a general principle which says that a process must occur unless it is explicitly forbidden by law. The mystery was solved when the prizewinners' team discovered that Mother Nature provides two completely different species of neutrino, as had been suggested by a theoretical analysis. The old type of neutrino is paired with and may be transformed into an electron, the newly discovered type of neutrino is similarly paired with the muon. The two pairs constitute two separate lepton families, which never mix with each other. Thus, a new law of Nature had been discovered.

Cosmologists and physicists alike want to know how many different lepton families, i.e. how many neutrino species there are in Nature. Present ideas about the birth and early evolution of our universe cannot tolerate more than four. A third is already on the books. One of the goals of the experimental program at the large LEP accelerator ring at CERN, which will be ready to start operation next summer, is to give a precise answer as to the number of neutrino species and thus the exact number of lepton families in the universe.

Professors Dr Lederman, Dr Schwartz and Dr Steinberger,
You started a bold new line of research, which gave rich fruit from the beginning by establishing the existence of a second neutrino. Furthermore, problems which could not even be formulated at the time of your experiment, have been successfully elucidated in later experiments using your method. The pairing of the leptons, which you discovered, is also of much wider applicability than could be foreseen at the time and is now an indispensible ingredient in the standard model for quarks and leptons.

On behalf of the Royal Swedish Academy of Sciences, I have the privilege and the great honour to extend to you our warmest congratulations. May I now ask you to receive the 1988 Nobel Prize in physics from the hands of His Majesty the King.

MELVIN SCHWARTZ

Having been born in 1932, at the peak of the great depression, I grew up in difficult times. My parents worked extraordinarily hard to give us economic stability but at the same time they managed to instill in me two qualities which became the foundation of my personal and professional life. One is an unbounded sense of optimism; the other is a strong feeling as to the importance of using one's mind for the betterment of mankind.

My interest in Physics really began at the age of 12 when I entered the Bronx High School of Science in New York. That school has become famous for the large number of outstanding individuals it has produced including among them four Nobel Laureates in Physics. The four years I spent there were certainly among the most exciting and stimulating of my life, mostly because of the interaction with other students having similar background, interest and ability. It's rather amazing how important the interaction with the one's peers can be at that age in determining one's direction and success in life.

Upon graduating from high school the path to follow was fairly obvious. The Columbia Physics Department at that time was unmatched by any in the world. Largely a product of the late Professor I.I. Rabi, it was a department which was to provide the ambiance for six Nobel Prize pieces of work in widely diverse fields during the next thirteen years. And, in addition, it was the host for a period of time to another half dozen or so future Nobel Laureates either as students or as post-doctoral researchers. I know of no other institution either before or since that has come close to that record.

Thus, it was that I became an undergraduate at Columbia in 1949, to stay there through my graduate years and take up a faculty position as Assistant Professor in 1958. I became an Associate Professor in 1960 and a Professor in 1963.

In order for me to put my life into perspective, I must mention four individuals who have given it meaning, direction and focus. Foremost among these is my wife Marilyn whom I married 35 years ago and who has provided the one most enduring thread throughout these years. Without her constant encouragement and enthusiasm there would have been far less meaning to my life. The second is of course Jack Steinberger. Jack was my teacher, my mentor and my closest colleague during my years at Columbia. Whatever taste and judgement I have ever had in the field of Particle Physics came from Jack. Third of course is T.D. Lee. He was the inspirer of this experiment and the person who has served as a constant sounding board for any ideas I have had. He has also become, I am proud to say, a dear personal friend. And finally, my close collaborator Leon Lederman. If there is any one person who has served as the sparkplug for high energy physics in the U.S. it has been Leon. I am proud to have been his collaborator.

In 1966, after having spent 17 years at Columbia, I decided to move West to Stanford, where a new accelerator was just being completed. During the ensuing years I was involved in two major research efforts. The first of these investigated the charge asymmetry in the decay of the long-lived neutral kaon. The second of these, which was quite unique, succeeded in producing and detecting relativistic hydrogen-like atoms each made up of a pion and a muon.

During the 1970's, lured in part by the new industrial revolution in "Silicon Valley" I decided to try my hand at a totally new adventure. Digital Pathways, Inc. of which I am currently the Chief Executive Officer is a company dedicated to the secure management of data communications. Although it is difficult to predict the future I still have all the optimism that I had back when I first grew up in New York — life can be a marvelous adventure.

(added in 1991): A new change in my career occurred in February 1991 when I became Associate Director, High Energy and Nuclear Physics, at Brookhaven National Laboratory.

THE FIRST HIGH ENERGY NEUTRINO EXPERIMENT

Nobel Lecture, December 8, 1988

by

MELVIN SCHWARTZ

Digital Pathways, Inc., 201 Ravendale Avenue, Mountain View, CA 94303, USA.

In the first part of my lecture I would like to tell you a bit about the state of knowledge in the field of Elementary Particle Physics as the decade of the 1960's began with particular emphasis upon the Weak Interactions. In the second part I will cover the planning, implementation and analysis of the first high energy neutrino experiment. My colleagues, Jack Steinberger and Leon Lederman, will discuss the evolution of the field of high energy neutrino physics beyond this first experiment and the significance of this effort when seen in the context of today's view of elementary particle structure.

I. HISTORICAL REVIEW

By the year 1960 the interaction of elementary particles had been classified into four basic strengths. The weakest of these, the gravitational interaction does not play a significant role in the laboratory study of elementary particles and will be ignored. The others are:

1. *Strong Interactions*

This class covers the interactions among so-called hadrons. Among these hadrons are the neutrons and protons that we are all familiar with along with the pions and other mesons that serve to tie them together into nuclei. Obviously, the interaction that ties two protons into a nucleus must overcome the electrostatic repulsion which tends to push them apart. The strong interactions are short range, typically acting over a distance of 10^{-13} cm, but at that distance are some two orders of magnitude stronger than electromagnetic interactions.

In general, as presently understood, hadrons are combinations of the most elementary strongly interacting particles, called quarks. You will hear more about them later.

2. *Electromagnetic Interactions*

You are all familiar with electromagnetic interactions from your daily experience. Like charges repel one another. Opposite charges attract. The earth acts like a giant magnet. Indeed matter itself is held together by the electromagnetic interactions among electrons and nuclei. With the exception of the neutrinos, all elementary particles have electromagnetic interactions either through charge, or magnetic property, or the ability to directly interact with charge or magnetic moment. In 1960, the only known elementary particles apart from the hadrons were the three leptons — electron, muon and neutrino with some suspicion that there might be two types of neutrinos. Both the electron and muon are electromagnetically interacting.

3. *Weak Interactions*

Early in the century it was discovered that some nuclei are unstable against decay into residual nuclei and electrons or positrons. There were two important characteristics of these so-called decays.

a. They were "slow". That is to say, the lifetimes of the decaying nuclei corresponded to an interaction which is much weaker than that characteristic of electromagnetism.

b. Energy and momentum were missing.

If one examined the spectrum of the electrons which were emitted, then it was clear that to preserve energy, momentum and angular momentum in the decay it was necessary that there be another decay product present. That decay product needed to be of nearly zero mass and have half integral spin. This observation was first made by Pauli. Fermi later gave it the name of neutrino.

The development of the Fermi theory of weak interactions in fact made the neutrino's properties even more specific. The neutrino has a spin of $1/2$ and a very low probability of interacting in matter. The predicted cross-section for the interaction of a decay neutrino with nucleons is about 10^{-43} cm^2. Thus, one of these neutrinos would on the average pass through a light year of lead without doing anything.

The β-decay reactions can be simply written as:

$$Z \to (Z+1) + e^- + \bar{\nu}$$
$$Z \to (Z-1) + e^+ + \nu$$

By the failure to detect neutrino-less double β decay, namely the process $Z \to (Z+2) + e^- + e^-$, it was established that the neutrino and anti-neutrino were indeed different particles. In the 1950's, by means of a series of experiments associated with the discovery of parity violation it was also established that the neutrinos and anti-neutrinos were produced in a state of complete longitudinal polarization or helicity, with the neutrinos being left-handed and anti-neutrinos right-handed.

In the 1940's and 1950's, a number of other weak interactions were discovered. The pion, mentioned earlier as the hadron which serves to hold the nucleus together, can be produced in a free state. Its mass is about 273 times the electron mass and it decays in about 2.5×10^{-8} seconds into a muon and a particle with neutrino-like properties. The muon in turn exhibits all of the properties of a heavy electron with a mass of about 207 times the electron mass. It decays in about 2.2×10^{-6} seconds into an electron and two neutrinos. The presumed reactions, when they were discovered, were written as:

$$\pi^+ \to \mu^+ + \nu$$
$$\pi^- \to \mu^- + \bar{\nu}$$
$$\mu^+ \to e^+ + \nu + \bar{\nu}$$

It was also known by 1960 that these decays were parity violating and that the neutrinos here had the same helicity as the neutrinos emitted in β decay.

Needless to say, there was a general acceptance in 1959 that the neutrinos associated with β decay were the same particles as those associated with pion and muon decay. The only hint that this may not be so came from a paper by G. Feinberg in 1958 in which he showed that the decay $\mu \to e + \gamma$ should occur with a branching ratio of about 10^{-4}, if a charged intermediate boson (W) moderated the weak interaction. Inasmuch as the experimental limit was much lower ($\sim 10^{-8}$) this paper was thought of as a proof that there was no intermediate boson. Feinberg did point out, however, that a boson might still exist if the muon neutrino and the electron neutrino were different.

One final historical note with respect to neutrinos. In the mid-nineteen fifties Cowen and Reines in an extremely difficult pioneering experiment were able to make a direct observation of the interaction of neutrinos in matter. They used a reactor in which a large number of $\bar{\nu}$ are produced and observed the reaction $\bar{\nu} + p \to n + e^+$. The cross section observed was consistent with that which was required by the theory.

II. CONCEPTION, PLANNING AND IMPLEMENTATION OF THE EXPERIMENT

The first conception of the experiment was in late 1959. The Columbia University Physics Department had a tradition of a coffee hour at which the latest problems in the world of Physics came under intense discussions. At one of these Professor T. D. Lee was leading such a discussion of the possibilities for investigating weak interactions at high energies. A number of experiments were considered and rejected as not feasible. As the meeting broke up there was some sense of frustration as to what could ever be done to disentangle the high energy weak interactions from the rest of what takes place when energetic particles are allowed to collide with targets. The only ray of hope was the expectation that the cross sections characteristic of the weak interactions increased as the square of the center of mass energy at least until such time as an intermediate boson or other damping mechanism took hold.

That evening the key notion came to me — perhaps the neutrinos from pion decay could be produced in sufficient numbers to allow us to use them in an experiment. A quick "back of the envelope" calculation indicated the feasibility of doing this at one or another of the accelerators under construction or being planned at that time. I called T. D. Lee at home with the news and his enthusiasm was overwhelming. The next day planning for the experiment began in earnest. Meanwhile Lee and Yang began a study of what could be learned from the experiment and what the detailed cross-sections were.

Not long after this point we became aware that Bruno Pontecorvo had also come up with many of the same ideas as we had. He had written up a proposed experiment with neutrinos from stopped pions, but he had also discussed the possibilities of using energetic pions at a conference in the Soviet Union. His overall contribution to the field of neutrino physics was certainly major.

Leon Lederman, Jack Steinberger, Jean-Marc Gaillard and I spent a great deal of time trying to decide on an ideal neutrino detector. Our first choice, if it were feasible, would have been a large Freon bubble chamber that Jack Steinberger had built. (In the end that would have given about a factor of 10 fewer events at the Brookhaven A.G.S. than the spark chamber which we did use. Hence it was not used in this experiment).

Fortunately for us, the spark chamber was invented at just about that time. Gaillard, Lederman and I drove down to Princeton to see one at Cronin's laboratory. It was small, but the idea was clearly the right one. The three of us decided to build the experiment around a ten ton spark chamber design.

In the summer of 1960, Lee and Yang again had a major impact on our thinking. They pointed out that it was essentially impossible to explain the absence of the decay $\mu \rightarrow e + \gamma$ without positing two types of neutrinos. Their argument as presented in the 1960 Rochester Conference was more or less as follows:

1. The simple four-fermion point model which explains low energy weak interactions leads to a cross-section increasing as the square of the center of mass energy.

2. At the same time, a point interaction must of necessity be S-wave and thus the cross-section cannot exceed $\lambda^2/4\pi$ without violating unitarity. This violation would take place at about 300 GeV.

3. Thus, there must be a mechanism which damps the total cross-section before the energy reaches 300 GeV. This mechanism would imply a "size" to the interaction region which would in turn imply charges and currents which would couple to photons. This coupling would lead to the reaction $\mu \rightarrow e + \gamma$ through the diagram.

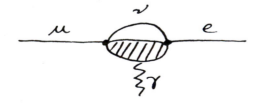

4. The anticipated branding ratio for $\mu \rightarrow e + \gamma$ should not differ appreciably from 10^{-5}. The fact that the branching ratio was known to be less than 10^{-8} was then *strong evidence* for the two-neutrino hypothesis.

With these observations in mind the experiment became highly motivated toward investigating the question of whether $\nu_\mu = \nu_e$. If there were only one type of neutrino then the theory predicted that there should be equal numbers of muons and electrons produced. If there were two types of neutrinos then the production of electrons and muons should be different. Indeed, if one followed the Lee-Yang argument for the absence of $\mu \rightarrow e + \gamma$ then the muon neutrino should produce *no* electrons at all.

We now come to the design of the experiment. The people involved in the effort were Gordon Danby, Jean-Marc Gaillard, Konstantin Goulianos, Nariman Mistry along with Leon Lederman, Jack Steinberger and myself. The facility used to produce the pions was the newly completed Alternate Gradient Synchrotron (A.G.S.) at the Brookhaven National Laboratory. Although the maximum energy of the accelerator was 30 GeV, it was necessary to run it at 15 GeV in order to minimize the background from energetic muons.

Pions were produced by means of collisions between the internal proton beam and a beryllium target at the end of a 3-meter straight section (see Figure 1). The detector was set at an angle of 7.5° to the proton direction behind a 13.5-meter steel wall made of the deck-plates of a dismantled cruiser. Additional concrete and lead were placed as shown.

To minimize the amount of cosmic ray background it was important to minimize the fraction of time during which the beam was actually hitting the target. Any so-called "events" which occurred outside of that window could then be excluded as not being due to machine induced high energy radiation.

The A.G.S. at 15 GeV operator at a repetition rate of one pulse per 1.2 seconds. The beam RF structure consisted of 20 ns bursts every 220 ns. The beam itself was deflected onto the target over the course of 20–30 μs for each cycle of the machine. Thus, the target was actually being bombarded for only 2×10^{-6} sec. for each second of real time.

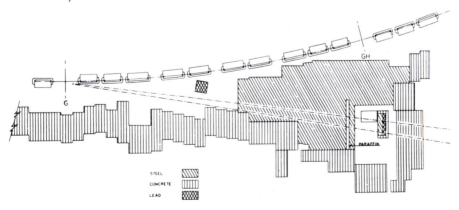

Figure 1. Plan view of the A.G.S. neutrino experiment.

In order to make effective use of this beam structure it was necessary to gate the detector on the bursts of pions which occurred when the target was actually being struck. This was done by means of a 30 ns time window which was triggered through the use of a Cerenkov counter in front of the shielding wall. Phasing of the Cerenkov counter relative to the detector was accomplished by raising the A.G.S. energy and allowing muons to penetrate the shield.

Incidentally, this tight timing also served to exclude 90 % of the background induced by slow neutrons.

The rate of production of pions and kaons was well known at the time and it was quite straightforward to calculate the anticipated neutrino flux. In Figure 2 we present an energy spectrum of the neutrino flux for a 15 GeV proton beam making use of both pion and kaon decay. It is clear that kaon decay is a major contributor for neutrino energies greater than about 1.2 GeV. (These neutrinos come from the reaction $K^{\pm} \rightarrow \mu^{\pm} + (\frac{\nu}{\bar{\nu}})$).

Needless to say, the main shielding wall is thick enough to suppress all strongly interacting particles. Indeed, the only hadrons that were expected to emerge from that wall were due to neutrino interactions in the last meter or so. Muons entering the wall with up to 17 GeV would have been stopped by ionization loss. The only serious background was due to neutrons leaking through the concrete floor; these were effectively eliminated in the second half the experiment.

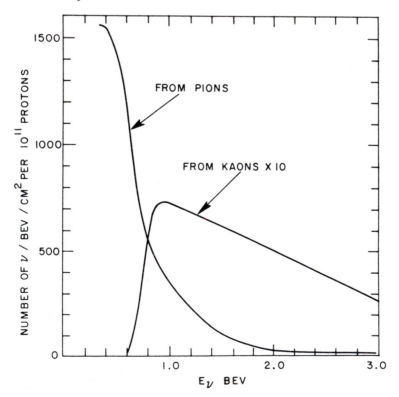

Figure 2. Energy spectrum of neutrinos as expected for A.G.S. running at 15 GeV.

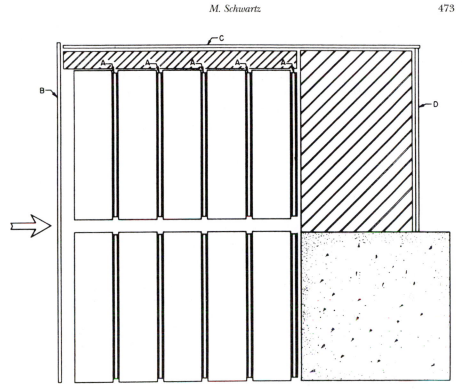

Figure 3. Spark chamber and counter arrangement. This is the front view with neutrinos entering on the left. A are the trigger counters. B, C and D are used in anti-coincidence.

Figure 4. A photograph of the chambers and counters.

The spark chamber is shown in Figure 3 and 4. It consisted of ten modules, each of 9 aluminum plates, 44 in. x 44 in. x 1 in. thick separated by 3/8 in. Lucite spacers. Anticoincidence counters covered the front, top and rear of the assembly, as shown, to reduce the effect of cosmic rays and muons which penetrate the shielding wall. Forty triggering counters were inserted between modules and at the end of the assembly. Each triggering counter consisted of two sheets of scintillator separated by 3/4 in. of aluminum. The scintillators were put in eletronic coincidence.

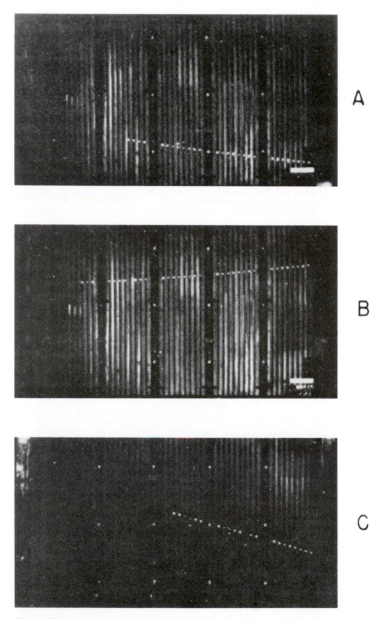

Figure 5. Some typical single muon events.

Events were selected for further study if they originated within a fiducial volume which excluded the first two plates, two inches at top and bottom and four inches at front and rear of the assembly. Single track events also needed to stay within the fiducial volume if extrapolated back for two gaps. Single tracks were not accepted unless their production angle relative to the neutrino direction was less than 60°.

A total of 113 events were found which satisfied these criteria. Of these, 49 were very short single tracks. All but three of these appeared in the first half of the experiment before the shielding was improved and they were considered to be background. In retrospect, some of these were presumably neutral current events, but at the time it was impossible to distinguish them from neutron induced interactions due to leakage over and under the shield.

The remaining events included the following categories.

a) 34 "single muons" of more than 300 MeV/c of visible momentum. Some of these are illustrated in Figure 5. Among them are some with one or two extraneous sparks at the vertex, presumably from nuclear recoils.

b) 22 "vertex" events. Some of these show substantial energy release. These events are presumably muons accompanied by pions in the collision. (See Figure 6)

c) 8 "shower" candidates. Of these 6 were selected so that their potential range, had they been muons would correspond to more than the 300 MeV/c. These were the only candidates for single electrons in the experiment. We will consider them in detail shortly.

It was quite simple to demonstrate that the 56 events in categories (a) and (b) were almost all of neutrino origin.

By running the experiment with the accelerator off and triggering on cosmic rays it was possible to place a limit of 5 ± 1 on the total number of the single muon events which could be due to such background. Indeed, the slight asymmetry in Figure 7 is consistent with this hypothesis.

It was simple to demonstrate that these events were not neutron induced. Referring to Figure 7 we see how they tend to point toward the target through the main body of steel shielding. No more than 10^{-4} events should have arisen from neutrons penetrating the shield (other than from neutrino induced events in the last foot of the shield itself). Indeed, removing four feet of steel from the front would have increased the event rate by a factor of 100; no such increase was seen. Futhermore, if the events were neutron induced they would have clustered toward the first chambers. In fact they were uniformly spread throughout the detector subject only to the 300 MeV/c requirement.

The evidence that the single particle tracks were primarily due to muons was based on the absence of interactions. If these tracks were pions we would have expected 8 interactions. Indeed, even if all of the stopping tracks were considered to be interacting, it would still lead to the conclusion that the mean free path of these tracks was 4 times that expected for hadrons.

As a final check on the origin of these events we effectively replaced four feet of the shield by an equivalent amount as close as possible to the beryllium target. This reduced the decay distance by a factor of 8. The rate of events decreased from $1.46 \pm .02$ to 0.3 ± 0.2 per 10^{16} incident protons.

All of the above arguments convinced us that we were in fact looking at neutrino induced events and that 29 of the 34 single track events were muons produced by neutrinos (The other five being background due to cosmic rays). It is these events that will form the basis of our arguments as to

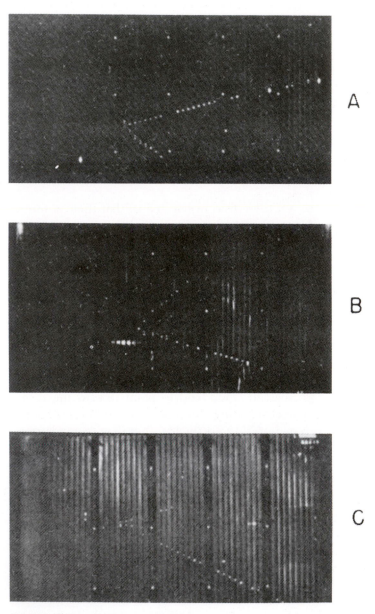

Figure 6. Some typical "vertex" events.

the identity of v_μ and v_e. But, first we must see what electrons would look like in passing through our spark chambers. An electron will on the average radiate half of its energy in about four of the aluminum plates. This will lead to gammas which will in turn convert to other electron—positron pairs. The net result is called a "shower". Typically an electron shower shows a number of sparks in each gap between plates. The total number of sparks in the shower increases roughly linearly with electron energy in 400 MeV region.

In order to calibrate the chambers we exposed them to a beam of 400 GeV electrons at the Brookhaven Cosmotron (See Figure 8). We noted that the triggering system was 67 % efficient with respect to these electrons. We then plotted the spark distribution as shown in Figure 9 for a sample of 2/3 × 29 expected showers. The 6 "shower" events were also plotted. Clearly, the difference between the expected distribution, had there been only one neutrino, and the observed distribution was substantial. We concluded that $v_\mu \neq v_e$.

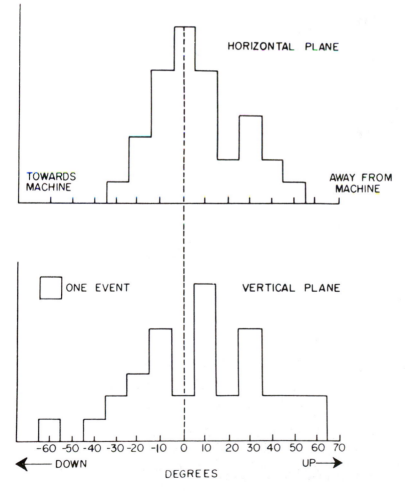

Figure 7. Projected angular distribution of the single track events. The neutrino direction is taken as zero degrees.

As a further point, we compared the expected rate of neutrino events with that predicted by the Fermi theory and found agreement within 30 %.

The results of the experiment were described in an article in Physical Review Letters Volume 9, pp. 36–44 (1962).

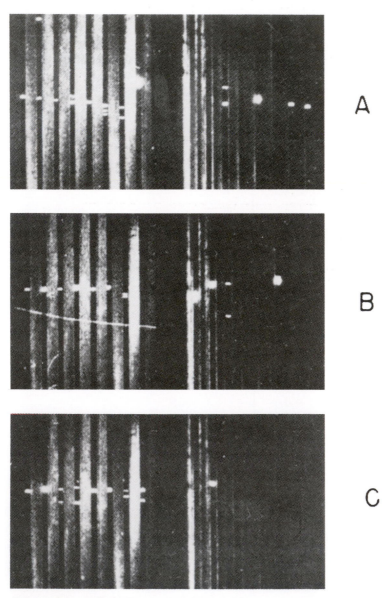

Figure 8. Typical 400 MeV/c electrons from the Cosmotron calibration run.

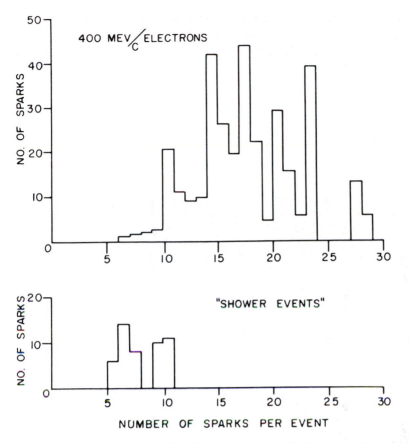

Figure 9. Spark distribution for 400 MeV/c electrons normalized to expected number of showers should be $\nu_\mu \neq \nu_e$. Also shown are the observed "shower" events.

Jack Steinberger

JACK STEINBERGER

I was born in Bad Kissingen (Franconia) in 1921. At that time my father, Ludwig, was 45 years old. He was one of twelve children of a rural 'Vieh-händler' (small-time cattle dealer). Since the age of eighteen he had been cantor and religious teacher for the little Jewish community, a job he still held when he emigrated in 1938. He had been a bachelor until he returned from four years of service in the German Army in the first World War. My mother was born in Nuremberg to a hop merchant, and was fifteen years the younger. Unusual for her time, she had the benefit of a college education and supplemented the meagre income with English and French lessons, mostly to the tourists which provided the economy of the spa. The childhood I shared with my two brothers was simple; Germany was living through the post-war depression.

Things took a dramatic turn when I was entering my teens. I remember Nazi election propaganda posters showing a hateful Jewish face with crooked nose, and the inscription "Die Juden sind unser Unglück", as well as torchlight parades of SA storm troops singing "Wenn's Juden Blut vom Messer fliesst, dann geht's noch mal so gut". In 1933, the Nazis came to power and the more systematic persecution of the Jews followed quickly. Laws were enacted which excluded Jewish children from higher education in public schools. When, in 1934, the American Jewish charities offered to find homes for 300 German refugee children, my father applied for my older brother and myself. We were on the SS Washington, bound for New York, Christmas 1934.

I owe the deepest gratitude to Barnett Faroll, the owner of a grain brokerage house on the Chicago Board of Trade, who took me into his house, parented my high-school education, and made it possible also for my parents and younger brother to come in 1938 and so to escape the holocaust. New Trier Township High School on the well-to-do Chicago North Shore, enjoyed a national reputation, and, with a swimming pool, athletic fields, cafeteria, as well as excellent teachers, offered horizons unimaginable to the young emigrant from a small German town.

The reunited family settled down in Chicago. We were helped to acquire a small delicatessen store which was the basis of a very marginal income, but we were used to a simple life, so this was no problem. I was able to continue my education for two years at the Armour Institute of Technology (now the Illinois Institute of Technology) where I studied chemical engineering. I was a good student, but these were the hard times of the depression, my scholarship came to an end, and it was necessary to work to supplement the family income.

The experience of trying to find a job as a twenty-year-old boy without connections was the most depressing I was ever to face. I tried to find any job in a chemical laboratory: I would present myself, fill out forms, and have the door closed hopelessly behind me. Finally through a benefactor of my older brother, I was accepted to wash chemical apparatus in a pharmaceutical laboratory, G.D. Searl and Co., at eighteen dollars a week. In the evenings I studied chemistry at the University of Chicago, the weekends I helped in the family store.

The next year, with the help of a scholarship from the University of Chicago, I could again attend day classes, so that in 1942 I could finish an undergraduate degree in chemistry.

On 7 December 1942, Japan attacked the United States at Pearl Harbor. I joined the Army and was sent to the MIT radiation laboratory after a few months of introduction to electromagnetic wave theory in a special course, given for Army personnel at the University of Chicago. My only previous contact with physics had been the sophomore introductory course at Armour. The radiation laboratory was engaged in the development of radar bomb sights; I was assigned to the antenna group. Among the outstanding physicists in the laboratory were Ed Purcell and Julian Schwinger. The two years there offered me the opportunity to take some basic courses in physics.

After Germany surrendered in 1945, I spent some months on active duty in the Army, but was released after the Japanese surrender, to continue my studies at the University of Chicago. It was a wonderful atmosphere, both between professors and students and also among the students. The professors to whom I owe the greatest gratitude are Enrico Fermi, W. Zachariasen, Edward Teller and Gregor Wentzel. The courses of Fermi were gems of simplicity and clarity and he made a great effort to help us become good physicists also outside the regular class-room work, by arranging evening discussions on a widespread series of topics, where he also showed us how to solve problems. Fellow students included Yang, Lee, Goldberger, Rosenbluth, Garwin, Chamberlain, Wolfenstein and Chew. There was a marvellous collaboration, and I feel I learned as much from these fellow students as from the professors.

I would have preferred to do a theoretical thesis, but nothing within reach of my capabilities seemed to offer itself. Fermi then asked me to look into a problem raised in an experiment by Rossi and Sands on stopping cosmic-ray muons. They did not find the expected number of decays. After correcting for geometrical losses there was still a missing factor of two, and I suggested to Sands that this might be due to the fact that the decay electron had less energy than expected in the two-body decay, and that one might test this experimentally. When this idea was not followed, Fermi suggested that I do the experiment, instead of waiting for a theoretical topic to surface. The cosmic-ray experiment required less than a year from its conception to its conclusion, in the end of the summer of 1948. It showed that the muon's is a three-body decay, probably into an electron and two neutrinos, and helped lay the experimental foundation for the concept of a universal weak interaction.

There followed an interlude to try theory again at the Institute for Advanced Study in Princeton, where Oppenheimer had become director. It was a frustrating year: I was no match for Dyson and other young theoreticians assembled there. Towards the end I managed to find a piece of work I could do, on the decay of mesons via intermediate nucleons. I still remember how happy Oppenheimer was to see me come up with something, at last.

In 1949, Gian Carlo Wick, with whom I had done some work on the scattering of polarized neutrons in magnetized iron while still a graduate student at Chicago University, invited me to be his assistant at the University of California in Berkeley. There the experimental possibilities in the Radiation Laboratory, created by E.O. Lawrence, were so great that I reverted easily to my wild state, that is experimentation. During the year there, I had the magnificent opportunity of working on the just completed electron synchrotron of Ed McMillan. It enabled me to do the first experiments on the photoproduction of pions (with A.S. Bishop) to establish the existence of neutral pions (with W.K.H. Panofsky and J. Stellar) as well as to measure the pion mean life (with O. Chamberlain, R.F. Mozley and C. Weigand).

I survived only a year in Berkeley, partly because I declined to sign the anticommunist loyalty oath, and moved on to Columbia University in the summer of 1950. At its Nevis Laboratory, Columbia had just completed a 380 MeV cyclotron; this, for the first time, offered the possibility of experimenting with beams of π mesons. In the next years I exploited these beams to determine the spins and parities of charged and neutral pions, to measure the $\pi^- \pi^0$ mass difference and to study the scattering of charged pions. This work leaned heavily on the collaboration of Profs. D. Bodansky and A.M. Sachs, as well as of several Ph.D. students: R. Durbin, H. Loar, P. Lindenfeld, W. Chinowsky and S. Lokanathan.

These experiments all utilized small scintillator counters. In the early fifties, the bubble-chamber technique was discovered by Don Glaser, and in 1954 three graduate students, J. Leitner, N.P. Samios and M. Schwartz, and myself began to study this technique which had not as yet been exploited to do physics. Our first effort was a 10 cm diameter propane chamber. We made one substantial contribution to the technique, that was the realization of a fast recompression (within $\sim$ 10 ms), so that the bubbles were recompressed before they could grow large and move to the top. This permitted chamber operation at a useful cycling rate. The first bubble-chamber paper to be published was from our experiment at the newly built Brookhaven Cosmotron, using a 15 cm propane chamber without magnetic field. It yielded a number of results on the properties of the new unstable (strange) particles at a previously unattainable level, and so dramatically demonstrated the power of the new technique which was to dominate particle physics for the next dozen years. Only a few months later we published our findings on three events of the type $\Sigma^0 \rightarrow \Lambda^0 + \gamma$, which demonstrated the existence of the Σ^0 hyperon and gave a measure of its mass. This experiment used a new propane chamber, eight times larger in volume, and with a magnetic field. This chamber also introduced the use of more than two stereo cameras, a development which is crucial for the rapid, computerized analysis of events, and has been incorporated into all subsequent bubble chambers.

In the decade which followed, the same collaborators, together with Profs. Plano, Baltay, Franzini, Colley and Prodell, and a number of new students, constructed three more bubble chambers: a 12″ H_2 chamber as well as 30″ propane and H_2 chambers, developed the analysis techniques, and performed a series of experiments to clarify the properties of the new particles. The experiments I remember with the most pleasure are:
— the demonstration of parity violation in Λ decay, 1957;
— the demonstration of the β decay of the pion, 1958;
— the determination of the π^0 parity on the basis of angular correlation in the double internal conversion of the γ rays, 1962;
— the determination of the ω and φ decay widths (lifetimes), 1962;
— the determination of the $\Sigma^0 - \Lambda^0$ relative parity, 1963;
— the demonstration of the validity of the $\Delta S = \Delta Q$ rule in K^0 and in hyperon decays, 1964.

This long chain of bubble-chamber experiments, in which I also enjoyed and appreciated the collaboration of two Italian groups, the Bologna group of G. Puppi and the Pisa group of M. Conversi, was interrupted in 1961, in order to perform, at the suggestion of Mel Schwartz, and with G. Danby, J.M. Gaillard, D. Goulianos, L. Lederman and N. Mistri, the first experiment using a high-energy neutrino beam now recognized by the Nobel Prize, and described in the paper of M. Schwartz.

In 1964, CP violation was discovered by Christensen, Cronin, Fitch and Turlay. Soon after I found myself on sabbatical leave at CERN, and proposed, together with Rubbia and others, to look for the interference between K_S^0 and K_L^0 amplitudes in the time dependence of K^0 decay. Such interference was expected in the CP violation explanation of the results of Christensen et al., but not in other explanations which had also been proposed. The experiment was successful, and marked the beginning of a set of experiments to learn more about CP violation, which was to last a decade. The next result was the observation of the small, CP-violating, charge asymmetry in K_L^0 leptonic decay, in 1966. Measurement of the time dependence of this charge asymmetry, following a regenerator, permitted a determination of the regeneration phase; this, together with the earlier interference experiments, yielded, for the first time, the CP-violating phase $\varphi_{\eta+-}$ and, in consequence, as well as the observed magnitudes of the CP-violating amplitudes in the two-pion and the leptonic decays, certain checks of the superweak model. The same experiment also gave a more sensitive check of the $\Delta S = \Delta Q$ rule, an ingredient of the present Standard Model.

In 1968, I joined CERN. Charpak had just invented proportional wire chambers, and this development offered a much more powerful way to study the K^0 decay to which I had become addicted. Two identical detectors were constructed, one at CERN together with Filthuth, Kleinknecht, Wahl, and others, and one at Columbia together with Christensen, Nygren, Carithers and students. The Columbia beam was long, and therefore contained no K_S but only K_L, the CERN beam was short, and therefore contained a mixture of K_S and K_L. It was contaminated by a large flux of Λ^0, and so was also a hyperon beam, permitting the first measurements of Λ^0 cross-sections as well as the Coulomb excitation of Λ^0 to Σ^0, a difficult and interesting experiment carried out chiefly by Steffen and Dydak. The most important result to come from the Columbia experiment was the observation of the rare decay $K_L \rightarrow \mu^+ \mu^-$ with a branching ratio compatible with theoretical predictions based on unitarity. Previously, a Berkeley experiment had searched in vain for this decay and had claimed an upper limit in violation of unitarity. Since unitarity is fundamental to field theory, this result had a certain importance.

The CERN experiment, which extended until 1976, produced a series of precise measurements on the interference of K_S and K_L in the two-pion and leptonic decay modes, thus leading us to obtain highly precise results on the CP-violating parameters in K^0 decay. I believe the experiment was beautiful, and take some pride in it, but the results were all in agreement with the superweak model and so did little towards understanding the origin of CP violation.

In 1972, the K^0 collaboration of CERN, Dortmund and Heidelberg was joined by a group from Saclay, under R. Turlay, to study the possibilities for a neutrino experiment at the CERN SPS then under construction. The CDHS detector, a modular array of magnetized iron disks, scintillation counters and drift chambers, 3.75 m in diameter, 20 m long, and weighing 1 200 t, was designed, constructed, and exposed to different neutrino beams at the SPS during the period 1977 to 1983. It provided a large body of data on the charged-current and neutral-current inclusive reactions in iron, which permitted first of all the clearing away of a number of incorrect results, e.g. the "high-y anomaly" produced at Fermilab, allowed the first precise and correct determination of the Weinberg angle, demonstrated the existence of right-handed neutral currents, provided measurements of the structure functions which gave quantitative support to the quark constituent model of the nucleon, and, through the Q^2 evolution of the structure functions, gave quantitative support to QCD. The study of multimuon events gave quantitative support to the GIM model of the Cabibbo current through its predictions on charm production.

In the CDHS experiment we were about thirty physicists. Since 1983, I have been spokesman for a collaboration of 400 physicists engaged in the design and construction of a detector for the 100 + 100 GeV $e^+ e^-$ Collider, LEP, to be ready at CERN in the beginning of 1989. In the meantime I had also helped to design an experiment to compare CP violation in the charged and neutral two-pion decay of the K_L^0. This experiment was the first to show "direct" CP violation, an important step towards the understanding of CP violation.

In 1986, I retired from CERN and became part-time Professor at the Scuola Normale Superiore in Pisa. However, my chief activity continues as before in my research at CERN.

I am married to Cynthia Alff, my former student and now biologist, and we have two marvelous children, Julia, 14 years old, and John, 11 years old. From an earlier marriage to Joan Beauregard, there are two fine sons, Joseph Ludwig and Richard Ned.

I play the flute, unfortunately not very well, and have enjoyed tennis, mountaineering and sailing, passionately.

EXPERIMENTS WITH
HIGH-ENERGY NEUTRINO BEAMS

Nobel Lecture, December 8, 1988

by

JACK STEINBERGER

CERN, Geneva, Switzerland, and Scuola Normale Superiore, Pisa, Italy

1. INTRODUCTION

High-energy neutrino beams have found intensive and varied application in particle physics experimentation in the last decades. This review is constrained to a few of the most fruitful examples: the discovery of neutral currents, the measurement of the Weinberg angle, the study of weak currents and the consequent test of the electroweak theory, the study of nucleon quark structure, and the testing of quantum chromodynamics (QCD). Other studies such as the production of "prompt" neutrinos, the search for finite neutrino masses and neutrino oscillations, the search for heavy leptons or other new particles, or the measurement of proton and neutron structure functions, elastic and pseudoelastic cross-sections, and other exclusive processes, are not discussed here. Neutrino experiments have been pursued vigorously at the Brookhaven National Laboratory, at Fermilab, and at CERN. It is fair to say that they have made large contributions to our understanding of particle physics.

2. NEUTRINO BEAMS

Present neutrino beams are produced in four steps: i) production of secondary hadrons in the collision of high-energy protons on a fixed target; ii) momentum (charge) selection and focusing of the hadrons; iii) passage of the beam through an (evacuated) decay region, long enough to permit a substantial fraction of the hadrons to decay; iv) absorption of the remaining hadrons and the muons that are produced along with the neutrinos in a shield of adequate thickness. The two-body decays $\pi^{+(-)} \rightarrow \mu^{+(-)} + \nu(\bar{\nu})$ and $K^{+(-)} \rightarrow \mu^{+(-)} + \nu(\bar{\nu})$ account for $\sim 97\%$ of the neutrino flux in present beams. Positive hadrons produce neutrinos, negative hadrons produce antineutrinos. Figures 1a and 1b give an impression of the two hadron beam-forming options that are available, side by side, at CERN: a conventional, so-called narrow-band beam (NBB), and an achromatic, Van der Meer horn-focused, wide-band beam (WBB). The neutrino spectra produced by these two beams are very different, as shown in Figure 2. The

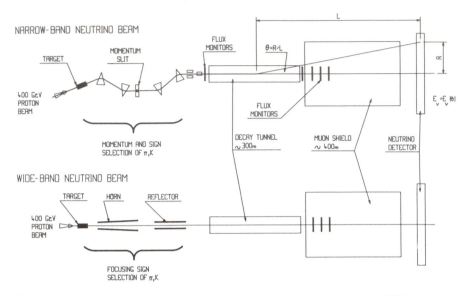

NARROW-BAND NEUTRINO BEAM

MOMENTUM SLIT

TARGET

FLUX MONITORS

$\theta = R \diagup L$

L

R

400 GeV PROTON BEAM

$E_\nu = E_\nu |\theta|$

FLUX MONITORS

MOMENTUM AND SIGN SELECTION OF π,K

DECAY TUNNEL ∼ 300m

MUON SHIELD ∼ 400m

NEUTRINO DETECTOR

WIDE-BAND NEUTRINO BEAM

TARGET   HORN   REFLECTOR

400 GeV PROTON BEAM

FOCUSING SIGN SELECTION OF π,K

Figure 1a Sketch of narrow-band and wide-band neutrino beam layouts at CERN, showing disposition of primary target, focusing elements, decay region, shielding, and monitoring devices.

Figure 1b View of the neutrino beam tunnel at the CERN SPS in 1976, before operations began. The NBB line is seen in the centre; on the right is the pulse transformer for the WBB horn, but the horn itself, destined for the pedestal on the left, is not yet installed. At the far end, the 2.5 m diameter titanium window of the evacuated decay region can be seen.

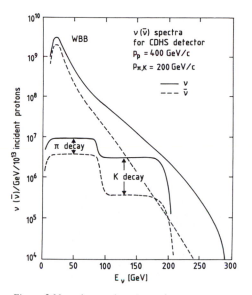

Figure 2 Neutrino and antineutrino energy spectra, calculated for the horn-focused WBB and the more conventional NBB.

WBBs are characterized by high intensity, a steep (generally undesirable) energy fall-off, and a substantial contamination of wrong-"sign" neutrinos. The NBBs have lower intensity, a flat energy dependence in the contribution from each of the two decays, and small wrong-sign background. They also have the important feature that the energy of the neutrino can be known, subject to a twofold π-K dichotomy, if the decay angle is known. In general this can be inferred from the impact parameter of the event in the detector.

3. DETECTORS

The low cross-sections of neutrinos are reflected in two general features of neutrino detectors:

   i) they are massive;
   ii) the target serves also as detector.

In the seventies, the most successful detectors were large bubble chambers. The most splendid of these were the cryogenic devices built at CERN and Fermilab, each with a volume of $\sim$ 15 m^3, in large magnetic fields, and capable of operating with liquid hydrogen, deuterium, or neon. A picture of a typical neutrino event in the CERN chamber is shown in Figure 3. It is an example of the "charged-current" (CC) reaction $\nu + N \rightarrow \mu^- +$ hadrons. However, one of the major discoveries at CERN was made not in this but in a large Freon-filled bubble chamber, affectionately called Gargamelle. The active volume was a cylinder 4.8 m long and 1.9 m in diameter, for a volume of about 13 m^3, inside a magnet producing a field of 2 T. Figure 4 gives some impression of its size.

    The bubble chamber has now been largely replaced by detectors based on electronic detection methods. As an example, I mention here the CDHS

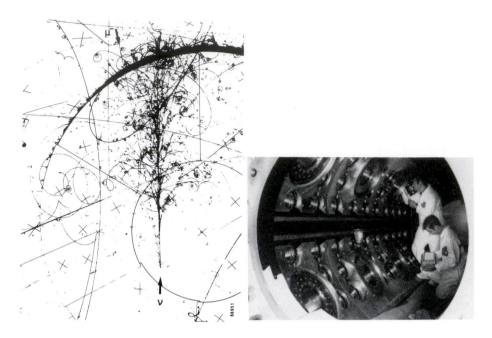

Figure 3 A typical neutrino event as observed in the Big European Bubble Chamber (BEBC) filled with neon at the CERN 450 GeV Super Proton Synchrotron (SPS) accelerator. The muon can be seen on the left. It has been tagged by an external muon identifier. The many-particle hadron shower is to the right.

Figure 4 Preparation of the interior of the 13 m³ bubble chamber Gargamelle, later to be filled with Freon. It is with this detector that the neutral currents were discovered.

(CERN-Dortmund-Heidelberg-Saclay Collaboration) detector used at CERN from 1977 to 1985. It consists of 19 modules made of iron plates 3.75 m in diameter, each with total iron thickness of 75 cm and a weight of ⁓ 65 t. The iron is toroidally magnetized to a field of 1.7 T by means of coils that pass through a hole in the centre.

Interleaved with the 5 cm thick iron plates are scintillator strips, which serve to measure the energy of the secondary hadrons by sampling the ionization. The typical hadron shower is ⁓ 25 cm in radius and ⁓ 1 m long, so the shower dimensions are very small compared with the size of the detector. The muon momenta are determined on the basis of curvature in the magnetic field, with the help of drift chambers inserted between the iron modules. These measure the positions of traversing tracks in three projections. The useful target weight is ⁓ 800 t. Figure 5 shows the CDHS experiment and Figure 6 a typical event of the same type as that shown in Figure 3.

Figure 5 View along the 19 modules of the CDHS electronic neutrino detector at the SPS. The black light-guides and phototubes, which are used to measure the hadron energy, can be seen sticking out of the magnetized iron modules. The hexagonal aluminium structures are the drift chambers that measure the muon trajectories.

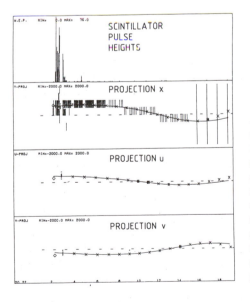

Figure 6 Computer reconstruction of a typical event of the reaction $v + Fe \rightarrow \mu^- + X$. Four views are shown, with the horizontal axis along the beam direction. The top view shows the scintillator pulse height, or hadron energy, and its distribution along the detector. The next view shows the scintillator hits as well as the horizontal wire hits and the reconstructed track in the x projection. The other two views show the wire hits and the reconstructed track for the $\pm 60°$ projections.

4. NEUTRAL CURRENTS

4.1 Discovery

The evolution of the electroweak unified gauge theory in the late sixties and early seventies was a miraculous achievement, but one that had no immediate impact on the majority of particle physicists — certainly not on me — perhaps because it was a theoretical construct which left the existing experimental domain intact. However, it predicted some entirely new phenomena, and of these the neutral weak currents were the first to be discovered. The verification of neutral currents (NCs) established the theory overnight, and subsequent experiments on their detailed structure reinforced this. This observation[1] of neutral weak currents at CERN in 1973 by the Gargamelle group was the first great discovery made at CERN. It was followed 10 years later by the second — also a prediction of the same theory — the intermediate boson.

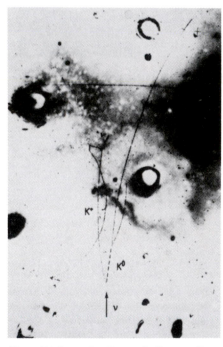

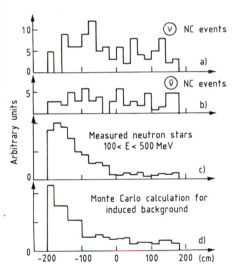

Figure 7 A "muonless" event in Gargamelle. All tracks stop or interact in the chamber. None could be a muon. The neutrino produces one K^+ and one K^0 meson. The K^+ meson interacts in the liquid and then decays. The invisible K^0 meson decays to two pions.

Figure 8 Distribution of the origin of muonless events along the beam direction in Gargamelle. Neutrino events are expected to be uniformly distributed, whereas neutron events should decrease with distance because of their absorption in the Freon. The nuclear mean free path in Freon is about 80 cm. The expected and observed distributions of neutron interactions are shown in the bottom two histograms. The muonless events are consistent with neutrino and inconsistent with neutron origin (Ref. 1).

The bubble chamber, built under the direction of A. Lagarrigue at the École Polytechnique in Paris, was exposed to neutrino and antineutrino WBBs at the CERN 24 GeV proton accelerator. The normal CC reactions,

$$\nu(\bar\nu) + N \rightarrow \mu^-(\mu^+) + \text{hadrons},$$

were found as usual, but NC "muonless" reactions,

$$\nu(\bar\nu) + N \rightarrow \nu(\bar\nu) + \text{hadrons},$$

which had hardly been looked for before — and therefore had not been found — were there as well. Such an event is shown in Figure 7. These events were selected on the basis of no muon candidate among the observed particles. The main experimental challenge was to show that they were not due to stray neutrons in the beam. I myself was a sceptic for a long time, and I lost a bottle or two of good wine on this matter. However, the neutron background would be expected to decrease exponentially along the length of the chamber, roughly with the neutron mean free path in Freon. Instead, the event distribution was flat, as expected for neutrino events (see Fig. 8.). I have never enjoyed paying up a debt more than at the dinner we gave for the winners, very good friends, Jacques Prentki, John Iliopoulos and Henri Epstein.

The ratios of the cross-sections

$$R_\nu = \frac{\sigma_\nu^{NC}}{\sigma_\nu^{CC}} \quad \text{and} \quad R_{\bar\nu} = \frac{\sigma_{\bar\nu}^{NC}}{\sigma_{\bar\nu}^{CC}}$$

are given in the electroweak theory in terms of the Weinberg angle θ_w:

$$R_\nu = \frac{1}{2} - \sin^2\theta_w + (1 + r)\frac{5}{9}\sin^4\theta_w \tag{1}$$

and

$$R_{\bar\nu} = \frac{1}{2} - \sin^2\theta_w + \left(1 + \frac{1}{r}\right)\frac{5}{9}\sin^4\theta_w, \tag{2}$$

where r is the ratio of antineutrino to neutrino CC total cross-sections: $r = \sigma^{CC,\bar\nu}/\sigma^{CC,\nu} = 0.48 \pm 0.02$ experimentally. On the basis of these ratios, the experiment yielded a first measure of $\sin^2\theta_w$ that was not very different from present, more precise determinations. In the same exposure a beautiful example of another NC process, the scattering of an antineutrino on an electron, was also found[2].

4.2 Precision measurement of $\sin^2\theta_w$ and right-handed neutral currents

The higher energies that became available a few years later at Fermilab and CERN made the study of NC processes much easier. The muons of the CC background had now a greater penetration power, which permitted cleaner separation of NC and CC events. Also, with the advent of the higher energies, the advantage in the study of inclusive neutrino scattering had shifted to electronic detection techniques. In the period 1977 to 1985, hadronic NC neutrino scattering was studied extensively by the CDHS

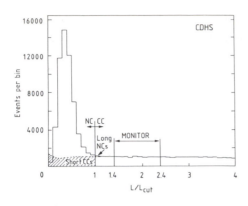

Figure 9 Identification, by the CDHS Collaboration, of NC events by their short event length. The peak at small event lengths is due to NC events. The long tail is due to muonic events, which must be subtracted under the peak to give the NC rate (Ref. 3).

Collaboration at CERN in order to get a more precise value for θ_w[3] and to check the prediction of the electroweak theory for the ratio of right-handed to left-handed NCs[4]. The NC events are selected on the basis of short event length, i.e. the short penetration of the hadronic shower compared with that of the muon of CC events. This is illustrated in Figure 9. A 15 % background of CC events is subtracted. The neutrino NC-to-CC ratio R_ν yielded the most precise value of the weak mixing angle available at present, $\sin^2\theta_w = 0.227 \pm 0.006$. Once $\sin^2\theta_w$ is known, the antineutrino ratio $R_{\bar\nu}$ follows from Eq. (2). Its measurement provided a sensitive test of the electroweak theory, and confirmed it in its simplest form. The presence of right-handed NCs (CCs are purely left-handed) in the amount predicted by the theory could be demonstrated by comparing the hadron energy distributions of the NC and CC processes. The result is shown in Figure 10.

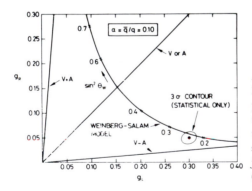

Figure 10 Strengths of the left- and right-handed neutral currents. If the NC were purely left-handed, as is the case for the CC, the experimental point would be expected to fall on the V-A line. The experiment shows a right-handed component, which is just that expected in the electroweak theory (Weinberg-Salam model) (Ref. 4).

4.3 Neutrino-electron scattering

The elastic-scattering reactions of neutrinos on atomic electrons

$$\nu_\mu + e^- \rightarrow \nu_\mu + e^- \qquad \text{and} \qquad \bar\nu_\mu + e^- \rightarrow \bar\nu_\mu + e^-$$

proceed via NCs. They are characterized by small cross-sections — smaller than their hadronic counterparts by the mass ratio m_e/m_p because of the smaller c.m. energies — and, for the same reason, by small electron production angles, $\theta_c \approx \sqrt{m_c/E_v}$. Until now, these angles have not been resolved by the experiments, so only total cross-sections have been measured. The expectations in the electroweak theory are:

$$\sigma^{v,e} = \frac{G_F^2 \, Em_e}{\pi} \left(1 - 4 \sin^2 \theta_w + \frac{16}{3} \sin^4 \theta_w \right)$$

and

$$\sigma^{\bar{v},e} = \frac{G_F^2 \, Em_e}{\pi} \left(\frac{1}{3} - \frac{4}{3} \sin^2 \theta_w + \frac{16}{3} \sin^4 \theta_w \right) .$$

These reactions can also serve to test this theory, and have the advantage that strongly interacting particles are not involved, so that the understanding of strong-interaction corrections is not necessary in the interpretation. They have the experimental disadvantage of low rates and consequent large background. The best results at present are from a BNL experiment[5] using relatively low energy neutrinos, $E \approx 1.5$ GeV, and a 140 t detector entirely composed of many layers of plastic scintillator and drift chambers. The background is subtracted on the basis of the distribution in the production angle of the electron shower (see Figure 11). Instead of comparing the neutrino and antineutrino cross-sections directly with the theory, the authors form the ratio of the two, which is less sensitive to some systematic errors. From this they find the result $\sin^2\theta_w = 0.209 \pm 0.032$. A CERN group reports a similar result, with $\sin^2\theta_w = 0.211 \pm 0.037$. The agreement with other methods of obtaining this angle is an important confirmation of the theory. A massive experiment to improve the precision is currently under way at CERN.

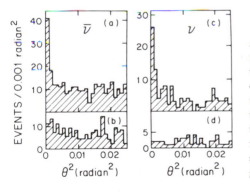

Figure 11 Identification of neutrino-electron events on the basis of their small angle with respect to the beam in the experiment of Ahrens et al.[5]. The peak at small angles in the two top graphs is due to neutrino-electron scattering. The bottom graphs show the flat distribution observed if photons rather than electrons are detected. This shows the angular distribution of background events.

5. NEUTRINO-NUCLEON INCLUSIVE SCATTERING AND THE QUARK STRUCTURE OF HADRONS

5.1 *Phenomenology*

We consider the CC reactions,

$$v + N \rightarrow \mu^- + \text{hadrons} \quad \text{and} \quad \bar{v} + N \rightarrow \mu^+ + \text{hadrons,}$$

independently of the final hadron configuration. This is called the inclusive process. It is assumed that the lepton vertex is described by the vector—

axial vector current of the electroweak theory. Let k be the initial and k' the final lepton energy-momentum four vectors, p that of the incident nucleon, and p' that of the final hadron state:

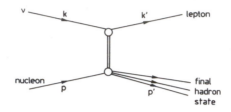

Define the kinematic variables:

$$Q^2 \equiv (k-k')^2 = 4EE' \sin^2 \theta/2,$$
$$\nu \equiv p \cdot Q^2/m_p = E_h - m_p \simeq E_h$$

(the energy of the final-state hadrons in the laboratory system),

$$x \equiv Q^2/2m_p\nu, \ 0 \leqslant x \leqslant 1,$$
$$y \equiv \nu/E \simeq E_h/E, \ 0 \leqslant y \leqslant 1,$$

where E and E' are the energies of the initial neutrino and final muon, respectively, and θ is the angle between these, all in the laboratory system. The cross-sections can be written in terms of three structure functions, each a function of the variables x and Q^2 that characterize the hadronic vertex:

$$\frac{d^2\sigma^{\nu(\bar{\nu})}}{dxdy} = \frac{G^2 \, m_p E_\nu}{2\pi} \{ F_2(x,Q^2) [1 + (1 - y)^2] - y^2 F_L(x,Q^2) \underset{(-)}{\overset{+}{}} xF_3(x,Q^2)$$
$$[1 - (1 - y)^2] \} \ .$$

The functions $F_2(x,Q^2)$, $xF_3(x,Q^2)$, and $F_L(x,Q^2)$ are the three structure functions that express what happens at the hadron vertex. The sum of neutrino and antineutrino cross-sections has the same structure-function dependence as does the cross-section for charged leptons:

$$\frac{d^2\sigma^{l^\pm}}{dxdy} = \frac{2\pi\alpha^2 \, m_p E}{Q^4} \{ F_2^{l^\pm}(x,Q^2) [1 + (1 - y)^2] - y^2 F_L(x,Q^2) \} \ .$$

5.2 Quark structure of the nucleon

In 1969, at the newly completed 2-mile linear electron accelerator at SLAC, it was discovered[7] that in electron-proton collisions, at high momentum transfer, the form factors were independent of Q^2. This so-called "scaling" behaviour is characteristic of "point", or structureless, particles. The interpretation in terms of a composite structure of the protons — that is, protons composed of point-like quarks — was given by Bjorken[8] and Feynman[9].

Neutrinos are projectiles *par excellence* for investigating this structure, in part because of the heavy mass of the intermediate boson, and in part because quarks and antiquarks are scattered differently by neutrinos owing to the V-A character of the weak currents; they can therefore be distinguished in neutrino scattering, whereas in charged-lepton scattering this is not possible. The quark model makes definite predictions for neutrino-hadron scattering, which are beautifully confirmed experimentally. Many of the predictions rest on the fact that now the kinematical variable x takes on a physical meaning: it can be interpreted as the fraction of the nucleon momentum or mass carried by the quark on which the scattering takes place. The neutrino experiments we review here have primarily used iron as the target material. Iron has roughly equal numbers of protons and neutrons. For such nuclei, the cross-sections can be expressed in terms of the total quark and total antiquark distributions in the proton. Let u(x), d(x), s(x), c(x), etc., be the up, down, strange, charm, etc., quark distributions in the proton. The proton contains three "valence" quarks: two up-quarks and one down-quark. In addition, it contains a "sea" of virtual quark-antiquark pairs. The up valence-quark distribution is $u(x) - \bar{u}(x)$, and the down valence-quark distribution is $d(x) - \bar{d}(x)$. The sea quarks and antiquarks have necessarily identical distributions, so that $s(x) = \bar{s}(x)$, $c(x) = \bar{c}(x)$, etc. For the neutron, u and d change roles, but s and c are the same. Let

$$q(x) = u(x) + d(x) + s(x) + c(x) + \ldots$$

and

$$\bar{q}(x) = \bar{u}(x) + \bar{d}(x) + \bar{s}(x) + \bar{c}(x) + \ldots$$

be the total quark and antiquark distributions of the proton, respectively. For spin-1/2 quarks interacting according to the Standard Model, for a target with equal numbers of protons and neutrons, and for $Q^2 \ll m_W^2$ and $m_p \ll E$:

$$\frac{d^2\sigma^\nu}{dxdy} = \frac{G_F^2 Em_p}{\pi} \, x[q(x) + (1-y)^2 \, \bar{q}(x)]$$

and

$$\frac{d^2\sigma^{\bar\nu}}{dxdy} = \frac{G_F^2 Em_p}{\pi} \, x[\bar{q}(x) + (1-y)^2 \, q(x)]$$

Comparison with the expression for the cross-section in terms of structure functions then gives these functions in terms of quark distributions:

$F_2(x,Q^2) = x[q(x) + \bar{q}(x)]$: $q(x) + \bar{q}(x)$ is the total quark + antiquark distribution;

$xF_3(x,Q^2) = x[q(x) - \bar{q}(x)]$: $q(x) - \bar{q}(x)$ is the 'valence'-quark distribution;

$F_L(x,Q^2) = 0$: this is a consequence of the spin-1/2 nature of the quarks.

From these simple expressions for the cross-sections, in terms of quark structure, several tests of the quark model are derived. For the experimental comparisons, we take the CDHS experiments[10]. It should be noted that the measurements in the detector, i.e. the hadron energy and the muon momentum, are just sufficient to define the inclusive process.

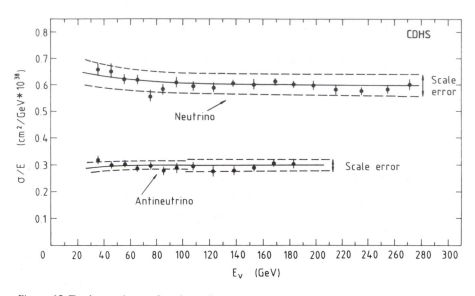

Figure 12 Total neutrino and antineutrino cross-sections per nucleon divided by neutrino energy. The flat 'scaling' behaviour is a consequence of the point-like interaction of the constituents (Ref. 10).

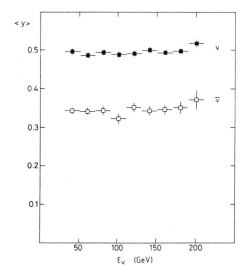

Figure 13 The average of y (the fraction of the neutrino energy transmitted to the final hadron state) as a function of the neutrino energy, for neutrinos and antineutrinos. The uniformity is a consequence of scaling, which in turn is a consequence of the point-like interaction of the quark (Ref. 10).

1) Scaling. The independence of the differential cross-sections with respect to Q^2 is evident everywhere, over a large domain in Q^2. As one example, Figure 12 shows the linearity of the total cross-sections with neutrino energy; as another, Figure 13 shows the uniformity of the average of y with respect to neutrino energy; both examples are consequences of scaling. Small deviations from scaling are observed in the structure functions, as we will see later, but these have their explanation in the strong interactions of the quarks.

2) The y-dependence of the cross-sections. We expect

$$\frac{d\sigma^{\nu}}{dy} + \frac{d\sigma^{\bar{\nu}}}{dy} \propto [1 + (1 - y)^2] \int x[q(x) + \bar{q}(x)] \, dx$$

and

$$\frac{d\sigma^{\nu}}{dy} - \frac{d\sigma^{\bar{\nu}}}{dy} \propto [1 - (1 - y)^2] \int x[q(x) - \bar{q}(x)] \, dx \quad .$$

The agreement with this expectation is quite good, as can be seen from Figure. 14. A corollary of this agreement is that $F_L(x)$ is small. It is found that $\int F_L(x)dx/\int F_2(x)dx \simeq 0.1$. Again, this deviation from the simple quark picture is understood in terms of the strong interactions of the quarks, as we will see later.

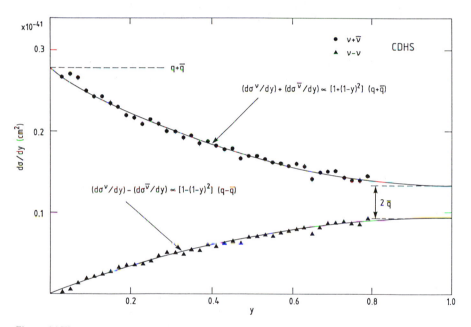

Figure 14 The y-dependence of the sum and the difference of neutrino and antineutrino cross-sections. Spin-$^1/_2$ quarks are expected to have y-dependences $1 + (1-y)^2$ for the sum and $1 - (1-y)^2$ for the difference (Ref. 10).

3) Correspondence between $F_2^{l\,\pm}$ (y) and F_2^{ν} (x). Both are proportional to q(x) + $\bar{q}$(x), and so are expected to have the same x-dependence in the simple quark model. They are related by the factor

$$\frac{F_2^{l\,\pm}(x)}{F_2^{\nu}(x)} = \frac{1}{2}\left[\left(\frac{2}{3}\right)^2 + \left(-\frac{1}{3}\right)^2\right] = \frac{5}{18} .$$

Here $^2/_3$ and $-^1/_3$ are the up- and down-quark electric charges, respectively. The agreement in shape and magnitude, shown in Figure 15, not only supports the quark picture, but also demonstrates the third integral quark electric charges.

4) $\int xF_3(x)\,dx/x = 3$. Since $xF_3(x) = x[q(x) - \bar{q}(x)]$ in the quark model and $q(x) - \bar{q}(x)$ is the valence-quark distribution, this sum rule states that there are three valence quarks in the nucleon. The experimental demonstration is not without problems, because the v and $\bar{v}$ cross-sections are finite as $x \to 0$, and the difference, which is $xF_3(x)$, has a consequent large error at small x, which is divided by x as $x \to 0$. However, all experiments give a value near 3, with typical uncertainties of $\backsim$ 10%.

Together with the charged-lepton inclusive scattering experiments, the neutrino experiments leave no doubt about the validity of the quark picture of nucleon structure. In addition, the neutrino experiments are unique in offering the possibility of measuring independently the quark and antiquark distributions in the nucleon, shown in Figure 15.

If the quarks were the sole nucleon constituents, we would expect $\int F_2\,dx = \int x[q(x) + \bar{q}(x)]\,dx$ to be equal to 1. Experimentally, $\int F_2(x)\,dx = 0.48 \pm 0.02$. We should have expected that some of the nucleon momentum is carried by the gluons, the mesons that bind the quarks. The experimental result is therefore interpreted to mean that gluons account for about half of the nucleon momentum (or mass).

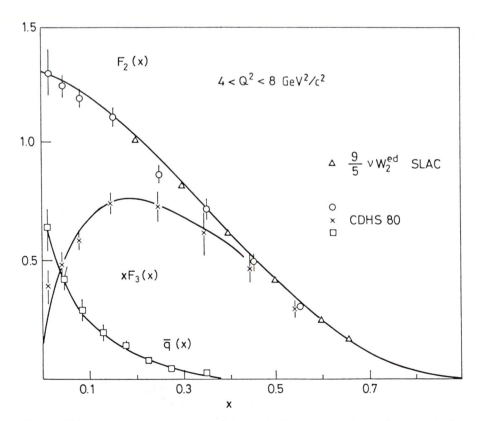

Figure 15 The structure functions $xF_3(x)$, $F_2(x)$, and $\bar{q}(x)$. In the simple quark picture $F_2(x) = x[q(x) + \bar{q}(x)]$ and $xF_3(x) = x[q(x) - \bar{q}(x)]$.

5.3 Neutrino scattering and quantum chromodynamics (QCD)

QCD is the elegant new gauge theory of the interaction of quarks and gluons, which describes the binding of quarks into the hadrons. Deep-inelastic lepton scattering provided a means of testing the predictions of this important theory and gave it its first experimental support. So far, no one has succeeded in calculating low-energy hadronic phenomena such as the wave functions of quarks in hadrons, because of the large coupling constant that frustrates perturbation methods at low energy. At high Q^2, however, the effective coupling constant becomes logarithmically smaller, and perturbation calculations become credible. The theory predicts 'scaling violations' in the form of a 'shrinking' of the structure functions towards smaller x as Q^2 gets larger. This is observed experimentally, as can be seen from Figure 16. In the theory, the 'shrinking' is the consequence of the

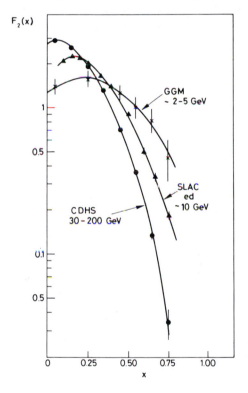

Figure 16 Scaling is only approximately true for the structure functions. Early measurements of $F_2(x)$ in three different energy domains exhibit shrinking, as expected in the QCD theory.

emission of gluons in the scattering process. This emission can be calculated. The Q^2 evolution at sufficiently high Q^2 is therefore quantitatively predicted by the theory. In neutrino experiments, this Q^2 evolution could be measured, and these measurements confirmed the theory and contributed to its acceptance. In the case of xF_3, the theoretical predictions have only one free parameter, the coupling constant α_S. In the case of F_2, the Q^2 evolution is coupled to the gluon distribution $G(x,Q^2)$. The experimental Q^2 evolutions of xF_3 and F_2 in the latest CDHS experiment are shown, together with their QCD fits, in Figsure 17 and 18. The theory fits the data adequately. These fits give a value for the parameter Λ in the running strong-coupling constant,

$$\alpha_s = [6/(33 - 2N_f) \ln (Q^2/\Lambda^2)],$$

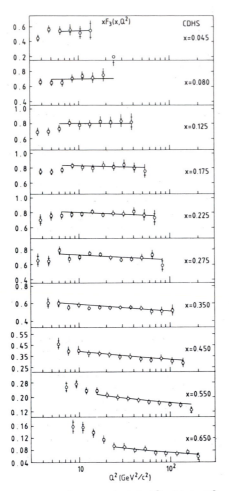 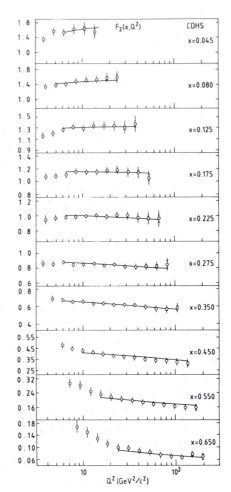

Figure 17 Variation of $xF_3(x,Q^2)$ with ln Q^2 in different x bins. The Q^2 evolution predictions of QCD with Λ = 128 MeV are also shown (Ref. 10).

Figure 18 Variation of $F_2(x,Q^2)$ with ln Q^2. The QCD fit is also shown.

where N_f = number of excited quark flavours ($N_f \simeq 4$ in this experiment), $\Lambda \simeq 100$ MeV. They also give the gluon distribution shown in Figure 19. These QCD comparisons suffer somewhat from the fact that Q^2 is still too low to reduce non-perturbative effects to a negligible level, but the calculable perturbative effects dominate and are confirmed by the experiments. Perturbative QCD also predicts a non-zero longitudinal structure function $F_L(x,Q^2)$ as another consequence of the emission of gluons. This prediction is compared with the CDHS experimental results in Figure 20. Again, the experiment lends support to the theory.

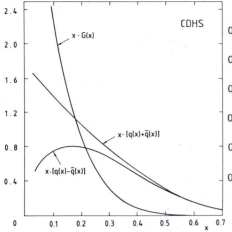

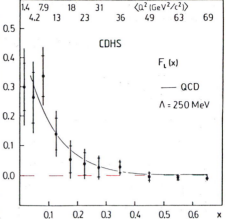

Figure 19 The gluon distribution G(x) derived from the QCD fits to $F_2(x,Q^2)$, $\bar{q}(x,Q^2)$, and $xF_3(x,Q^2)$ (Ref. 10).

Figure 20 The structure function $F_L(x)$ associated with longitudinally polarized intermediate bosons, and the QCD predictions. In the simple quark model, F_L is zero (Ref. 10).

6. NEUTRINO INTERACTIONS, THE GIM WEAK CURRENT, AND THE STRANGE QUARK IN THE NUCLEON

Among the most beautiful results obtained with neutrino beam experiments are those concerning the opposite-sign 'dimuons' first observed at Fermilab[11] and studied in detail in the CDHS experiments[12]. These reactions occur with roughly $1/100$ of the rate of the dominant single-muon events. The experiments are interesting, on the one hand because they confirm the doublet structure of the quark weak current proposed some years ago by Glashow, Iliopoulos and Maiani[13], and which is fundamental to the electroweak theory, and on the other hand because they give such a vivid confirmation of the nucleon quark structure altogether.

The origin of the extra muon was quickly understood as being due to the production of charmed quarks and their subsequent muonic decay. In the GIM model, the charm-producing reactions are

GIM cross-section
proportionality

$$\nu + d \rightarrow \mu^- + c; \ c \rightarrow \mu^+ + \dots \qquad xd(x)\sin^2\theta_C \qquad (3)$$
$$\nu + s \rightarrow \mu^- + c; \ c \rightarrow \mu^+ + \dots \qquad xs(x)\cos^2\theta_C \qquad (4)$$

and

$$\bar{\nu} + \bar{d} \rightarrow \mu^+ + \bar{c}; \ \bar{c} \rightarrow \mu^- + \dots \qquad x\bar{d}(x)\sin^2\theta_C \qquad (5)$$
$$\bar{\nu} + \bar{s} \rightarrow \mu^+ + \bar{c}; \ \bar{c} \rightarrow \mu^- + \dots \qquad x\bar{s}(x)\cos^2\theta_C \qquad (6)$$

The identification of the extra muon events with charm decay is experimentally confirmed in a number of ways:

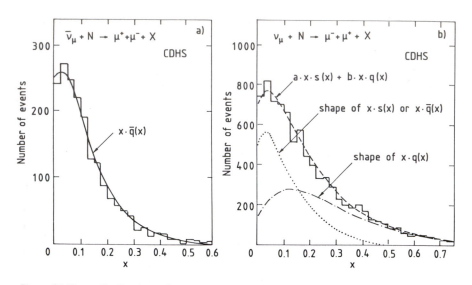

Figure 21 The x-distributions of opposite-sign dimuon events. a) For antineutrinos. The dominant process is $\bar{\nu} + \bar{s} \rightarrow \mu^+ + \bar{c}$. The observed x-distribution is therefore that of the strange sea in the nucleon.

b) For neutrinos. The process is $\nu + s$ or $d \rightarrow \mu^- + c$. The shape allows the determination of the relative contributions of s and d quarks, and therefore the relative coupling constant. This confirmed the GIM prediction (Ref. 12).

   i) opposite-sign muons are produced, like-sign ones are not;

   ii) in general, the extra muon has little energy;

  iii) the extra muon is correlated, as expected, to the direction of the hadron shower, of which the charmed particle is a part.

The GIM paper[13] preceded the experimental discovery of charm by five years. It was proposed because of the theoretical attractiveness of the doublet structure of the weak currents. The predictions were precise. The cross-sections are proportional to $\sin^2\theta_C$ for d and $\bar{d}$ quarks and to $\cos^2\theta_C$ for s and $\bar{s}$ quarks. The Cabibbo angle θ_C was previously known, with $\cos^2\theta_C = 0.97$, close to 1, and $\sin^2\theta_C = 0.05$, very much smaller. Reactions (3) and (4), or (5) and (6), are not experimentally separable since the target nucleon contains both s and d quarks, and the final state is the same. In the antineutrino case, reaction (6) dominates (5) because $\sin^2\theta_C$ is so small. For each event, x and y are measured as for single-muon events. Therefore, the x-distribution for antineutrino dimuon production, shown in Figure 21a, measures the amount and the shape of the strange sea s(x).

In the neutrino reactions, the smallness of $\sin^2\theta_C$ for reaction (3) is very closely compensated by the fact that d(x), containing also valence quarks, is much greater than s(x) of reaction (4). By fitting, it can be seen that the x-distribution in Figure 21b is a roughly equal mixture of s(x) as obtained with the antineutrinos and d(x), previously known from the normal CC reactions. The ratio of the two contributions is a measure of θ_C as it enters the charm production reaction. The Cabibbo angle obtained in this way is found to be equal, within errors, to θ_C measured in strange decays, as proposed in the

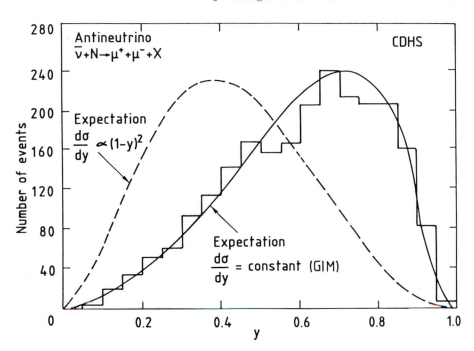

Figure 22 The y-distribution of $\bar{\nu}$-produced dimuons. The acceptance over the y-domain is unfortunately very non-uniform, because of the 5 GeV minimum energy required of each muon. The observed y-distribution agrees with an acceptance-corrected flat y-distribution as predicted by the GIM current, but differs strikingly from the $(1-y)^2$ distribution characteristic of the single-muon antineutrino cross-section (Ref. 12).

GIM hypothesis. Further support of the GIM current is provided by the y-distributions. They reflect the relative helicities of the neutrino and the struck quark: if the two helicities are the same, as is the case for all four charm-producing reactions, the expected y-distribution is flat; if they are opposite, as is the case for instance for $\nu + \bar{q}$ and $\bar{\nu} + q$, the expected distribution is $(1-y)^2$. Both neutrino and antineutrino single-muon reactions are mixtures of the two, as we saw in Figure 14. The contrast is especially strong for antineutrinos, where the experimental single-muon y-distribution is dominated by $(1-y)^2$, whereas the dimuon distribution is flat, as shown in Figure 22, again confirming the GIM picture.

7. CONCLUDING REMARKS

I have given some examples to illustrate the impact of high-energy neutrino research on the particle physics progress of the past years, both in the field of the weak interactions and in that of nucleon structure. How will this develop in the future? I do not know, of course. The increase of proton accelerator energies into the 10 TeV range will certainly permit better QCD tests than those cited above. In general, however, it can be expected that progress in particle physics will depend more and more on colliders, because of their higher centre-of-mass energies. High-energy e-p machines, such as HERA, will permit exploration of inclusive scattering to higher Q^2 domains than will be possible with fixed-target neutrino beams.

However, the fascination with neutrinos and the unanswered questions concerning them — such as their masses — are motivating a broad line of research in astrophysics, accelerator physics, and nuclear physics. One of the first and most important results expected from the two large e^+e^- colliders just coming into operation, the Stanford Linear Collider and the CERN LEP, which will produce lots of Z^0 mesons, is the determination of how many families of leptons and quarks there really are. Are there others besides the three already known? This fundamental question will be answered by determining how often the Z^0 decays to neutrinos, even if the masses of the other members of possible additional families are too large to permit their production at these energies.

REFERENCES

1. F. J. Hasert et al., Phys. Lett. *46B* (1973) 138.
2. F. J. Hasert et al., Phys. Lett. *46B* (1973) 121.
3. M. Holder et al., Phys. Lett. *71B* (1977) 222.
   H. Abramowicz et al., Z. Phys. *C28* (1985) 51 and Phys. Rev. Lett. *57* (1986) 298.
4. H. Holder et al., Phys. Lett. *72B* (1977) 254.
5. L. H. Ahrens et al., Phys. Rev. Lett. *54* (1985) 18.
6. J. Dorenbosch et al., Experimental results on electron-neutrino scattering, submitted to Zeitschrift für Physik C.
7. E. D. Bloom et al., Phys. Rev. Lett. *23* (1969) 930.
   M. L. Breitenbach et al., Phys. Rev. Lett. *23* (1969) 935.
   E. D. Bloom et al., Recent results in inelastic electron scattering, Stanford preprint SLAC-PUB 796 (1970), paper 7b−17 submitted to the 15th Int. Conf. on High-Energy Physics, Kiev, 1970. [See in Proc. talk by R. Wilson (Naukova Dumka, Kiev, 1972), p. 238.]
8. J. D. Bjorken, Phys. Rev. *179* (1969) 1547.
9. R. P. Feynman, Photon-hadron interactions (Benjamin, New York, 1972).
10. J. G. H. De Groot et al., Z. Phys. *C1* (1979) 143.
    H. Abramowicz et al., Z. Phys. *C12* (1982) 289 and *C17* (1983) 283.
    H. Abramowicz et al., A measurement of differential cross-sections and nucleon structure functions in charged-current neutrino interactions on iron, to be submitted to Zeitschrift für Physik C.
11. A. Benvenuti et al., Phys. Rev. Lett. *34* (1975) 419.
12. M. Holder et al., Phys. Lett. *69B* (1977) 377.
    H. Abramowicz et al., Z. Phys. *C15* (1982) 19.
13. S. L. Glashow, J. Iliopoulos and L. Maiani, Phys. Rev. *D2* (1970) 1285.

LEON M. LEDERMAN

New York City in the period of 1922 to 1979 provided the streets, schools, entertainment, culture and ethnic diversity for many future scientists. I was born in New York on July 15, 1922 of immigrant parents. My father, Morris, operated a hand laundry and venerated learning. Brother Paul, six years older, was a tinkerer of unusual skill. I started my schooling in 1927 at PS 92 on Broadway and 95th Street and received my Ph.D. in 1951 about one mile north, at Columbia University. In between there were neighborhood junior and senior high schools and the City College of New York. There I majored in chemistry but fell under the influence of such future physicists as Isaac Halpern and my high school friend, Martin J. Klein. I graduated in 1943 and proceeded promptly to spend three years in the U.S. Army where I rose to the rank of 2nd Lieutenant in the Signal Corps. In September of 1946 I entered the Graduate School of Physics at Columbia, chaired by I. I. Rabi.

The Columbia Physics Department was constructing a 385 MeV Synchrocyclotron at their NEVIS Laboratory, located in Irvington-on-the-Hudson, New York. Construction was aided by the Office of Naval Research and "NEVIS" eventually proved to be an extremely productive laboratory, as judged by physics results and students produced.

I joined that project in 1948 and worked with Professor Eugene T. Booth, the director of the cyclotron project. My thesis assignment was to build a Wilson Cloud Chamber. Rabi invited many experts to Columbia to assist the novice staff in what was, for Columbia, a totally new field. Gilberto Bernardini came from Rome and John Tinlot came from Rossi's group at MIT. Somewhat later, Jack Steinberger was recruited from Berkeley. After receiving my Ph.D. in 1951 I was invited to stay on, which I did, for the next 28 years. Much of my early work on pions was carried out with Tinlot and Bernardini.

In 1958, I was promoted to Professor and took my first sabbatical at CERN where I organized a group to do the "g-2" experiment. This CERN program would continue for about 19 years and involve many CERN physicists (Picasso, Farley, Charpak, Sens, Zichichi, etc.). It was also the initiation of several collaborations in CERN research which continued through the mid-70s.

I became Director of the Nevis Labs in 1961 and held this position until 1978. I have been a guest scientist at many labs but did the bulk of my research at Nevis, Brookhaven, CERN and Fermilab. During my academic career at Columbia (1951–1979) I have had 50 Ph.D. students, 14 are professors of physics, one is a university president and the rest with few exceptions, are physicists at national labs, in government or in industry. None, to my knowledge, is in jail. In 1979, I became Director of the Fermi National Accelerator Laboratory where I supervised the construction and utilization of the first superconducting synchrotron, now the highest energy accelerator in the world.

I have three children with my first wife, Florence Gordon. Daughter Rena is an anthropologist, son Jesse is an investment banker and daughter Rachel a lawyer. I now live with my second wife Ellen at the Fermilab Laboratory in Batavia, Illinois, where we keep horses for riding and chickens for eggs. I have been increasingly involved in development via scientific collaboration with Latin America, with science education for gifted children and with public understanding of science. I helped to found and am on the Board of Trustees of the Illinois Mathematics and Science Academy, a three year residence public school for gifted children in the State of Illinois.

Honors: Leon Lederman is the recipient of fellowships from the Ford, Guggenheim, Ernest Kepton Adams and National Science Foundations. He is a founding member of the High Energy Physics Advisory Panel (to AEC, DOE) and the International Committee on Future Accelerators. He has received the National Medal of Science (1965) and the Wolf Prize for Physics (1982) among many other awards.

Honorary D.Sc's have been awarded to Leon M. Lederman by City College of New York, University of Chicago, Illinois Institute of Technology, Northern Illinois University, Lake Forest College and Carnegie Mellon University.

(added in 1991): I retired from Fermilab in 1989 to join the faculty of the University of Chicago as Professor of Physics. In 1989 I was appointed Science Adviser to the Governor of Illinois. I helped to organize a Teachers' Academy for Mathematics and Science, designed to retrain 20,000 teachers in the Chicago Public Schools in the art of teaching science and mathematics. In 1991 I became President of the American Association for the Advancement of Science.

Honors

D.Sc.'s have been awarded among others by the universities at Pisa, Italy and Guanajuarto, Mexico. Elected to the National Academies of Science in Finland and in Argentina. Serves on thirteen (non-paying) Boards of Directors of museums, schools, science organizations and government agencies.

OBSERVATIONS IN PARTICLE PHYSICS FROM TWO NEUTRINOS TO THE STANDARD MODEL

Nobel Lecture, December 8, 1988

by

LEON M. LEDERMAN

The Fermi National Accelerator Laboratory

I. Introduction

My colleagues Melvin Schwartz and Jack Steinberger and I, sharing the 1988 Nobel Award, were faced with a dilemma. We could, in *Rashomon*-like fashion, each describe the two-neutrino experiment (as it became known) in his own style, with his own recollections, in the totally objective manner of true scientists. Whereas this could be of some interest to sociologists and anthropologists, this definitely would run the risk of inducing boredom and so we decided on a logical division of effort. Dr. Schwartz, having left the field of physics a decade ago, would concentrate on the origins and on the details of the original experiment. Dr. Steinberger would concentrate on the exploitation of neutrino beams, a field in which he has been an outstanding leader for many years. I volunteered to discuss "the rest," a hasty decision which eventually crystallized into a core theme—how the two-neutrino discovery was a crucial early step in assembling the current world view of particle physics which we call "the Standard Model." Obviously, even a "first step" rests on a pre-existing body of knowledge that could also be addressed. My selection of topics will not only be subjective, but it will also be obsessively personal as befits the awesome occasion of this award ceremony.

I will relate a sequence of experiments which eventually, perhaps even tortuously contributed to the Standard Model, that elegant but still incomplete summary of all subnuclear knowledge. This model describes the 12 basic fermion particles, six quarks and six leptons, arranged in three generations and subject to the forces of nature carried by 12-gauge bosons. My own experimental work brought me to such accelerators as the Nevis Synchrocyclotron (SC); the Cosmotron and Alternate Gradient Synchrotron (AGS) at the Brookhaven National Laboratory (BNL); the Berkeley Bevatron and the Princeton-Penn Synchrotron; the (SC), Proton Synchrotron (PS), and Intersecting Storage Ring (ISR) machines at CERN; the Fermilab 400-GeV accelerator; and the electron-positron collider Cornell Electron Storage Rings (CESR) at Cornell. I can only hint of the tremendous creativity which brought these magnificent scientific tools into being.

One must also have some direct experience with the parallel development of instrumentation. This equally bright record made available to me and my colleagues a remarkable evolution of the ability to record particular subnuclear events with ever finer spatial detail and even finer definition in time. My own experience began with Wilson cloud chambers, paused at photographic nuclear emulsions, exploited the advances of the diffusion cloud chamber, graduated to small arrays of scintillation counters, then spark chambers, lead-glass high-resolution Cerenkov counters, scintillation hodoscopes and eventually the increasingly complex arrays of multiwire proportional chambers, calorimeters, ring imaging counters, and scintillators, all operating into electronic data acquisition systems of exquisite complexity.

Experimentalists are often specialists in reactions initiated by particular particles. I have heard it said that there are some physicists, well along in years, who only observe electron collisions! In reviewing my own bibliography, I can recognize distinct periods, not too different from artists' phases, e.g., Picasso's Blue Period. My earliest work was with pions which exploded into the world of physics (in 1947) at about the time I made my quiet entry. Later, I turned to muons mostly to study their properties and to address questions of their curious similarity to electrons, e.g., in order to answer Richard Feynman's question, "Why does the muon weigh?" or Rabi's parallel reaction, "Who ordered that?" Muons, in the intense beams from the AGS, turned out to be a powerful probe of subnuclear happenings not only in rather classical scattering experiments (one muon in, one muon out), but also in a decidedly non-classical technique (no muons in, two muons out). A brief sojourn with neutral kaons preceded the neutrino program, which my colleagues will have discussed in detail. This led finally to studies of collisions with protons of the highest energy possible, in which leptons are produced. This last phase began in 1968 and was still going on in the 1980's.

Accelerators and detection instruments are essentials in particle research, but there also needs to be some kind of guiding philosophy. My own approach was formed by a specific experience as a graduate student.

My thesis research at Columbia University involved the construction of a Wilson cloud chamber designed to be used with the brand new 400-MeV synchrocyclotron under construction at the Nevis Laboratory about 20 miles north of the Columbia campus in New York City.

I. I. Rabi was the Physics Department Chairman, maestro, teacher of us all. He was intensely interested in the new physics that the highest energy accelerator in the world was producing. At one point I described some curious events observed in the chamber which excited Rabi very much. Realizing that the data was very unconvincing, I tried to explain that we were a long way from a definitive measurement. Rabi's comment, "First comes the observation, then comes the measurement," served to clarify for me the fairly sharp distinction between "observation" and "measurement."

Both experimental approaches are necessary to progress in physics. Observations are experiments which open new fields. Measurements are subsequently needed to advance these. Observations may be qualitative and may require an apparatus which sacrifices detail. Measurement is more usually concerned with the full panoply of relevant instruments. And of course, there are blurred boundaries. In the course of the next 30 or so years I have been concerned with measurements of great precision, e.g., the magnetic moment of the muon[1], or the mass, charge and lifetime of the muon[2], measurements of moderate precision like the rho value in muon decay, the elastic scattering of muons[3], or the lifetimes of the lambda and kaon particles[4]. I have also been involved in observations, which are attempts to see entirely new phenomena. These "observations" have, since 1956, been so labelled in the titles of papers, some of which are listed in chronological order in Table I and as references 5 – 11. I selected these because 1) I loved each one; and 2) they were reasonably important in the evolution of particle physics in the amazing period from the 1950's to the 1980's.

TABLE I. MAJOR OBSERVATIONS

- Observation of Long-Lived Neutral V Particles (1956) Ref. 5.
- Observation of the Failure of Conservation of Parity and Charge Conjugation in Meson Decays: The Magnetic Moment of the Free Muon(1957) Ref. 6.
- Observation of the High-Energy Neutrino Reactions and the Existence of Two Kinds of Neutrinos (1962) Ref. 7.
- Observation of Massive Muon Pairs in Hadron Collisions (1970) Ref. 8.
- Observation of π Mesons with Large Transverse Momentum in High-Energy Proton-Proton Collisions (1973) Ref. 9.
- Observation of a Dimuon Resonance at 9.5 GeV in 400-GeV Proton-Nucleus Collisions (1977) Ref. 10.
- Observation of the Upsilon 4-Prime at CESR (1980) Ref. 11.

II. Long-Lived Neutral Kaons Observation of a Long-Lived Neutral V Particle[5]
In 1955, Pais and Gell-Mann[12] noted that the neutral K meson presented a unique situation in particle physics. In contrast to, e.g., the π^0, the K^0 is not identical to its antiparticle, even though they cannot be distinguished by their decay. Using charge-conjugation invariance, the bizarre particle mixture scheme emerges: K^0 and $\bar{K}^0$ are appropriate descriptions of particle states produced with the well-defined quantum number, strangeness, but two other states, K_L and K_S, have well-defined decay properties and lifetimes.

The essence of the theoretical point, given in a Columbia University lecture by Abraham Pais in the spring of 1955, was that there should exist, in equal abundance with the already observed K_S (lifetime 10^{-10} sec), a particle with much longer lifetime, forbidden by C-invariance from decaying, as did K$_S$, into two pions. The clarity of the lecture stimulated what appeared to me to be an equally clear experimental approach, using the cloud chamber which had been invented back in 1896 by the Scottish physicist C.T.R. Wilson. The cloud chamber was first used for making visible the tracks of subatomic particles from nuclear disintegrations in 1911. Supplemented with strong magnetic fields or filled with lead plates, it became the workhorse of cosmic ray and early accelerator research, and was used in many discoveries, e.g., those of the positron, the muon, the lambda, the "θ" (now K$_s$), and K$^+$. As an instrument, it was more biological than physical, subject to poisons, track distortions, and an interminable period of about one minute. To obtain precise momentum and angle measurements with cloud chambers required luck, old-world craftsmanship, and a large, not-to-be-questioned burden of folklore and recipes. Their slow repetition rate was a particular handicap in accelerator science. Donald Glaser's invention of the bubble chamber and Luis Alvarez's rapid exploitation of it offered a superior instrument for the most purposes and by the mid-50's, very few cloud chambers were still operating at accelerators. At Columbia I had some success with the 11″-diameter chamber built at the Nevis Synchrocyclotron for my thesis, a comparison of the lifetimes of negative and positive pions[13]. In a stirring finale to this thesis, I had concluded (wrongly as it turned out) that the equality of lifetimes implied that charge conjugation was invariant in weak interactions!

In its history at Nevis, the cloud chamber produced results on the decay of pions[14], on the mass of the neutrino born in pion decay[15] (enter the muon neutrino; it would be almost a decade before this number was improved), on the scattering of pions[16], including the first suggestions of

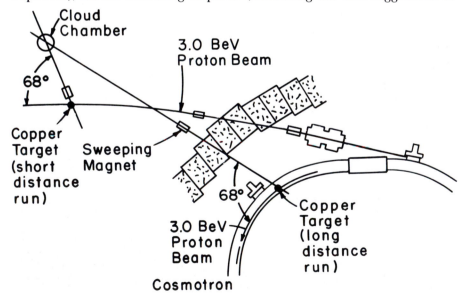

Figure 1. Experimental arrangemants for lifetime study.

strong backward scattering that was later found by E. Fermi to be the indicator of the "3,3" resonance, and on the Coulomb-nuclear interference of π^+ and π^- scattering in carbon. The carbon scattering led to analysis in terms of complex optical-model parameters which now, over 30 years later, are still a dominating subject in medium-energy physics convocations.

When the Cosmotron began operating in BNL about 1953, we had built a 36″-diameter chamber, equipped with a magnetic field of 10,000 gauss, to study the new Λ^0's and θ^0's which were copiously produced by pions of $\backsim 1$ GeV. The chamber seemed ideal to use in a search for long-lived kaons. Figure 1 shows the two arrangements that were eventually used and Figure 2 shows a K_L event in the 36″ cloud chamber. The Cosmotron produced

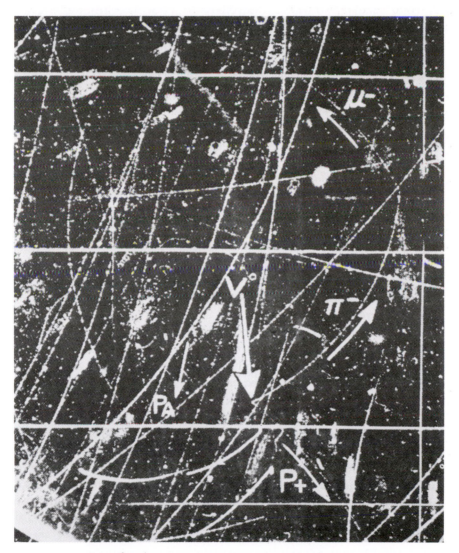

Figure 2. Example of $K^0 \to \pi^+ + \pi^-$ neutral particle. P_+ is shown to be a pion by ionization measurements. P_A is a proton track used in the ionization calibration.

ample quantities of 3-GeV protons and access to targets was particularly convenient because of the magnetic structure of the machine. The trick was to sweep all charged particles away from the chamber and reduce the sensitivity to neutrons by thinning the chamber wall and using helium as chamber gas. By mid-1956, our group of five had established the existence of K_L, and had observed its principal three-body decay modes. Our discussion of alternative interpretations of the "V" events seen in the chamber was exhaustive and definitive. In the next year we measured the lifetime by changing the flight time from target to chamber (both the cloud chamber and the accelerator were immovable). This lifetime, so crudely measured, agrees well with the 1988 handbook value. The K_L was the last discovery made by the now venerable Wilson cloud chamber.

In 1958, we made a careful search of the data for the possibility of a two-body decay mode of K_L. This search was a reflection of the rapid pace of events in the 1956−58 period. Whereas C-invariance was the key argument used by Pais and Gell-Mann to generate the neutral K mixture scheme, the events of 1957 (see below) proved that, in fact, C-invariance was strongly violated in weak decays. Since the predictions turned out to be correct, the improved argument, supplied by Lee, Oehme and Yang[17], replaced C-invariance by CP-invariance, and in fact, also CPT invariance. CP invariance would strictly forbid the decay

$$K_1 \rightarrow \pi^+ + \pi^-$$

and, in our 1958 paper based upon 186 K_L events, we concluded: "... only two events had zero total transverse momentum within errors ... and none of these could be a two-body decay of the K_L^0. An upper limit to $K_L^0 \rightarrow \pi^+ + \pi^-$ was set at 0.6% ... the absence of the two-pion final state is consistent with the predictions of time reversal invariance."

Six years later, at the much more powerful AGS accelerator, V. Fitch and J. Cronin[18], capitalizing on progress in spark chamber detectors, were able to vastly increase the number of observed K_L decays. They found clear evidence for the two-pion decay mode at the level of 0.22 % establishing the fact that CP is, after all, not an absolute symmetry of nature.

The K^0 research eventually provided a major constraint on the Standard Model. On the one hand, it served to refine the properties of the strange quark proposed in 1963 by Gell-Mann. On the other hand, the famous Kobayashi-Maskawa (KM) quark mixing matrix with three generations of quarks was an economical proposal to accommodate the data generated by the K^0 structure *and* the observation of CP violation. Finally, the neutral K-meson problem (essentially the K_S decay modes) led to the next major observation, that of charge conjugation (C) and parity (P) violation and, together, a major advance in the understanding of the weak interactions. In 1988, neutral K research remains a leading component of the fixed-target measurements at Fermilab, BNL, and CERN.

III. Observation of the Failure of Conservation of Parity and Charge Conjugation in Meson Decays[6]

In the summer of 1956 at BNL, Lee and Yang had discussed the puzzle of the K's (θ, τ puzzle) and were led to propose a number of reactions where possible P violation could be tested in weak interactions[19]. At first glance these all seemed quite difficult experimentally, since one was thinking of relatively small effects. Only C. S. Wu, our Columbia colleague, attempted, with her collaborators at the National Bureau of Standards, the difficult problem of polarizing a radioactive source. When, at a Christmas party in 1956, Wu reported that early results indicated large parity-violating effects in the decay of Co^{60}, it became conceivable that the chain of parity violating reactions: $\pi \rightarrow \mu + \nu$ and then $\mu \rightarrow e + 2\nu$ would not reduce the parity violating effect to unobservability. The "effect" here was the asymmetry in the emission of electrons around the incident, stopped, and spinning polarized muon.

Experience in two key areas set in course a series of events which would convert a Friday Chinese-lunch discussion, just after New Year, 1957, into a Tuesday morning major experimental observation. One was that I knew a lot about the way pion and muon beams were formed at the Nevis cyclotron. In 1950, John Tinlot and I had been pondering how to get pions into the cloud chamber. Until that time, external beams of pions were unknown at the existing cyclotrons such as those at Berkeley, Rochester, and Liverpool. We plotted the trajectories of pions produced by 400-MeV protons hitting a target inside the machine, near the outer limit of orbiting protons, and we discovered fringe field focussing. Negative pions would actually emerge from the accelerator into a well-collimated beam. It remained only to invent a target holder and to modify the thick concrete shield so as to "let them out." In about a month, we had achieved the first external pion beam and had seen more pions in the cloud chamber than had ever been seen anywhere.

The second key area had to do with my student, Marcel Weinrich, who had been studying the lifetime of negative muons in various materials. To prepare his beam we had reviewed the process of pions converting to muons by decay-in-flight. What was more subtle, but easy to play back during the 30-minute Friday evening drive from Columbia to Nevis, was that a correlation of the muon spin relative to its CM momentum would, in fact, be preserved in the kinematics of pion decay-in-flight, resulting in a polarized muon beam. One totally unclear issue was whether the muon would retain its polarization as it slowed from $\frown$ 50 MeV to rest in a solid material. Opportunities to pick up an electron and depolarize seemed very large, but I recalled Rabi's dictum: "A spin is a slippery thing" and decided — why not try it?

Preempting Weinrich's apparatus and enlisting Richard Garwin, an expert on spin precession experiments (as well as on almost everything else), we began the Friday night activities which culminated, Tuesday morning, in a 50 standard deviation parity violating asymmetry in the distribution of

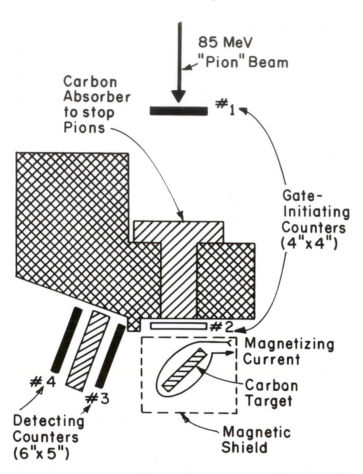

Figure 3. Experimental arrangement. The magnetizing coil was close wound directly on the carbon to provide a uniform vertical field of 79 gauss per ampere.

decay electrons relative to muon spin. Figure 3 shows the very simple arrangement and Fig. 4 shows the data. The following 10 conclusions were contained in the publicatin of our results:

1. The large asymmetry seen in the $\mu^+ \rightarrow e^+ + 2\nu$ decay establishes that the μ^+ beam is strongly polarized.
2. The angular distribution of the electrons is given by
   $$1 + a \cos \theta \text{ where } a = -1/3 \text{ to a precision of 10 \%.}$$
3. In reactions $\pi^+ \rightarrow \mu^+ + \nu$ *and* $\mu^+ \rightarrow e^+ + 2\nu$ parity is not conserved.
4. By a theorem of Lee, Oehme, and Yang, the observed asymmetry proves that invariance under charge conjugation is violated.
5. The g-value of the free μ^+ is found to be $+2.00 \pm 0.10$.
6. The measured g-value and the angular distribution in muon decay lead to the strong probability that the spin of the μ^+ is 1/2.
7. The energy dependence of the observed asymmetry is not strong.
8. Negative muons stopped in carbon show an asymmetry (also peaked backwards) of $a = -1/20$, i.e., about 15 % of that for μ^+.

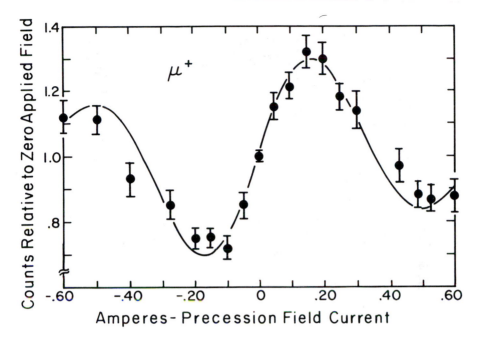

Figure 4. Variation of gated 3−4 counting rate with magnetizing current. The solid curve is computed from an assumed electron angular distribution 1-1/3 cosθ, with counter and gate-width resolution folded in.

9. The magnetic moment of the μ^- bound in carbon is found to be negative and agrees within limited accuracy with that of μ^+.

10. Large asymmetries are found for the e^+ from polarized μ^+ stopped in polyethylene and calcium. Nuclear emulsions yield an asymmetry half that of carbon.

Not bad for a long weekend of work.

This large effect established the two-component neutrinos and this, together with details of the decay parameters as they emerged over the next year, established the V-A structure of the weak interactions. A major crisis emerged from the application of this theory to high energy where the weak cross section threatened to violate unitarity. Theoretical attempts to prevent this catastrophe ran into the absence of evidence for the reaction:

$$\mu \rightarrow e + \gamma$$

The rate calculated by Columbia colleague G. Feinberg[20] was 10^4 times larger than that of the data. This crisis, as perceived by Feinberg, by T. D. Lee, and by Bruno Pontecorvo, provided motivation for the two-neutrino experiment. The stage was also set for increasingly sharp considerations of the intermediate vector boson hypothesis and, indeed, ultimately the electroweak unification.

The 1957 discovery of parity violation in pion and muon decay proved to be a powerful tool for additional research and, indeed, it kept the "pion-factories" at Columbia, Chicago, Liverpool, CERN, and Dubna going for decades, largely pursuing the physics that polarized muons enabled one to do. The earliest application was the precise magnetic resonance measurement of the muon magnetic moment at Nevis in 1957[1]. The high level of precision in such measurements had been unknown to particle physicists who had to learn about precisely measured magnetic fields and spin flipping. A more profound follow-up on this early measurement was the multi-decade obsession at CERN with the g-value of the muon. This measurement provides one of the most exacting tests of Quantum Electrodynamics and is a very strong constraint on the existence of hypothetical particles whose coupling to muons would spoil the current excellent agreement between theory and experiment.

One conclusion of the 1957 parity paper states hopefully that, "... it seems possible that polarized positive and negative muons will become a powerful tool for exploring magnetic fields in nuclei, atoms, and interatomic regions." Today "μSR" (muon spin resonance) has become a widespread tool in solid-state and chemical physics, meriting annual conferences devoted to this technique.

IV. Observation of High-Energy Neutrino Reactions and the Existence of Two Kind of Neutrinos[7]

Since this is the subject of Melvin Schwartz' paper I will not review the details of this research.

The two-neutrino road (a better metaphor would perhaps be; piece of the jigsaw puzzle) to the Standard Model passed through a major milestone with the 1963 quark hypothesis. In its early formulation by both Gell-Mann and George Zweig, three quarks, i.e., a triplet, were believed adequate along the lines of other attempts at constituent explanations (e.g., the Sakata model) of the family groupings of hadrons.

Before the quark hypothesis, a feeling for baryon-lepton symmetry had motivated many theorists, one even opposing the two-neutrino hypothesis before the experiment because "... two types of neutrinos would imply two types of protons." However, after the quark flavor model, Bjorken and Glashow, in 1964[21], transformed the baryon-lepton symmetry idea to quark-lepton symmetry and introduced the name "charm". They predicted the existence of a new family of particles carrying the charm quantum number. This development, and its enlargement by the Glashow, Illiopolis, Maiami (GIM) mechanism in 1970, was another important ingredient in establishing the Standard Model[22].

In GIM, the quark family structure and weak interaction universality explains the absence of strangeness changing neutral weak decays. This is done by assuming a charmed quark counterpart to the second neutrino ν_μ. With the 1974 discovery of the J/ψ at BNL/Stanford Linear Accelerator Center (SLAC) and subsequent experiments establishing the c-quark, the Standard Model, at least with two generations, was experimentally established. Included in this model was the doublet structure of quarks and leptons, e.g., (u,d), (c,s), (e,ν_e), (μ,ν_μ).

The measurements which followed from this observation are given in detail in Jack Steinberger's paper. Major neutrino facilities were established at BNL, CERN, Serpukhov, and Fermilab. Out of these came a rich yield of information on the properties of the weak interaction including neutral as well as charged currents, on the structure functions of quarks and gluons within protons and neutrons, and on the purely leptonic neutrino-electron scattering.

V. Partons and Dynamical Quarks
A. Observation of Dimuons in 30 GeV Proton Collisions[8]

The two-neutrino experiment moved, in its follow-up phase at BNL, to a much more massive detector and into a far more potent neutrino beam. To provide for this, the AGS proton beam was extracted from the accelerator, not at all an easy thing to do because an extraction efficiency of only 95% would leave an unacceptably large amount of radiation in the machine.

However, the ability to take pions off at 0^0 to the beam rather than at the 7^0 of the original experiment, represented a very significant gain in pions, hence in neutrinos. Thus, the second neutrino experiment, now with healthy competition from CERN, could look forward to thousands of events instead of the original 50.

The major motivation was to find the W particle. The weak interaction theory could predict the cross section for any given mass. The W production was

$$\nu_\mu + A \rightarrow W^+ + \mu^- + A*$$
$$\bar{\nu}_\mu + A \rightarrow W^- + \mu^+ + A*.$$

Since W will immediately decay, and often into a charged lepton and neutrino, two opposite-sign leptons appear in the final state at one vertex. Figures 5a, 5b show W candidates. The relatively low energy of the BNL and CERN neutrino beams produced by 30-GeV protons ($\bar{E} \backsim 1$ GeV) made this a relatively insensitive way of searching for W's but both groups were able to set limits

$$M_W > 2 \text{ GeV}.$$

We were then stimulated to try to find W's produced directly with 30-GeV protons, the signature being a high transverse momentum muon emerging from W-decay ($\backsim M_W/2$). The experiment found no large momentum muons and yielded[23] an improved upper limit for the W mass of about 5 GeV which, however, was burdened by theoretical uncertainties of how W's are produced by protons. The technique led, serendipitously, to the opening of a new field of high-energy probes.

To look for W's, the neutrino-producing target was removed and the beam of protons was transported across the former flight path of 22 m (for pions) and buried in the thick neutrino shield. The massive W could show itself by the appearance of high transverse momentum muons. This beam

Figure 5a. Neutrino event with long muon and possible second μ-meson.

Figure 5b. Neutrino event with long muon and possible electron.

dump approach was recognized in 1964 to be sensitive to short-lived neutrino sources[24], e.g., heavy leptons produced by 30 GeV protons. However, the single muon produced by a hypothetical W could also have been a member of a pair produced by a virtual photon. This criticism, pointed out by Y. Yamaguchi and L. Okun[24], presented us with the idea for a new small-distance probe: virtual photons.

We promptly began designing an experiment to look for the virtual photon decay into muon pairs with the hope that the decreasing yield as a function of effective mass of the observed pair is a measure of small-distance

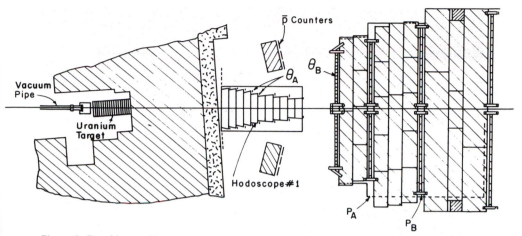

Figure 6. Brookhaven dimuon setup.

physics and that this slope could be interrupted by as yet undiscovered vector mesons. Observation here would be using the illumination of virtual photons whose parameters could be determined from the two-muon final state. In 1967, we organized a relatively simple exploration of the yield of muon pairs from 30-GeV proton collisions. Emilio Zavattini from CERN, Jim Christenson, a graduate of the Fitch-Cronin experiment from Princeton, and Peter Limon, a postdoc from Wisconsin, joined the proposal. Figure 6 shows the apparatus and Figure 7 shows the data. Later we were taught (by Richard Feynman) that this was an *inclusive* experiment:

$$p + U \rightarrow \mu^+ + \mu^- + \text{anything.}$$

The yield of muon pairs decreased rapidly from 1 GeV to the kinematic limit of nearly 6 GeV with the exception of a curious shoulder near 3 GeV. The measurement of muons was by *range* as determined by liquid and plastic scintillation counters interspersed with steel shielding. Each angular bin (there were 18) had four range bins and for two muons this made a total of only 5 000 mass bins into which to sort the data. Multiple scattering in the minimum of 10 feet of steel made finer binning useless. Thus, we could only note that: "Indeed, in the mass region near 3.5 GeV, the observed spectrum may be reproduced by a composite of a resonance and a steeper continuum." This 1968–69 experiment was repeated in 1974 by Aubert et al.[25], with a magnetic spectrometer based upon multiwire proportional chambers. The shoulder was refined by the superior resolution into a towering peak (see Fig. 7 a) called the "J" particle.

Our huge flux of 10^{11} protons/pulse made the experiment very sensitive to small yields and, in fact, signals were recorded at the level of 10^{-12} of the total cross section. A crucial development of this class of super-high-rate experiments was a foolproof way of subtracting accidentals.

The second outcome of this research was its interpretation by S. Drell and T-M Yan. They postulated the production of virtual photons by the annihi-

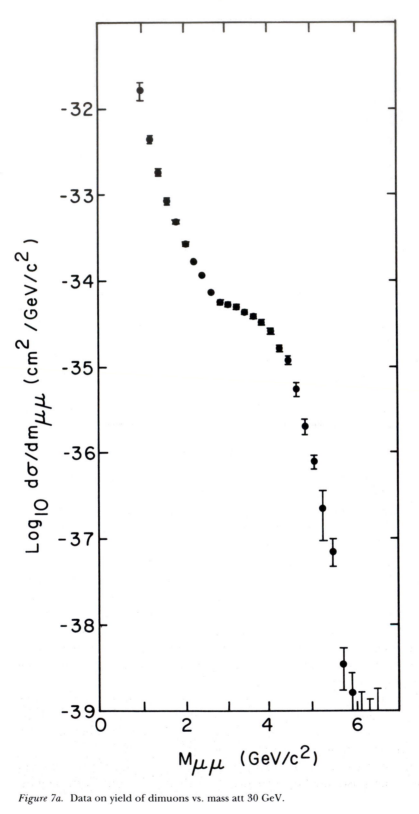

Figure 7a. Data on yield of dimuons vs. mass att 30 GeV.

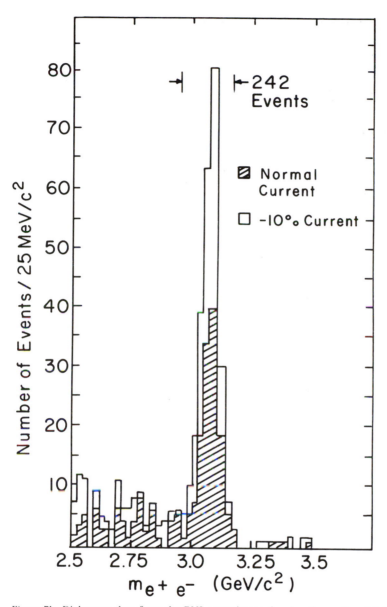

Figure 7b. Dielectric data from the BNL experiment showing the peak at 3.1 GeV which was named "J".

lation of a quark and antiquark in the colliding particles. The application of the now firmly named Drell-Yan process (this is how theorists get all the credit!) in the unraveling of quark dynamics has become increasingly incisive. It lagged behind the deeply inelastic scattering (DIS) analysis by Bjorken and others, in which electrons, muons, and neutrinos were scattered from nucleons with large energy loss. The Drell-Yan process is more dependent upon the strong interaction processes in the initial state and is more subject to the difficult problem of higher-order corrections. However, the dileption kinematics gives direct access to the constituent structure of hadrons with the possibility of experimental control of important parameters of the parton distribution function. Indeed, a very large Drell-Yan industry now flourishes in all the proton accelerators. Drell-Yan processes also allow one to study structure functions of pions, kaons, and antiprotons.

A major consequence of this experimental activity, accompanied by a much greater theoretical flood (our first results stimulated over 100 theoretical papers!), was a parameter-free fit of fairly precise (timelike) data[26] of "two leptons out" to nucleon structure functions determined by probing the nuclear constituent with incident leptons. Some of the most precise data here were collected by the CDHS group of Jack Steinberger and he has covered this in his paper. The agreement of such diverse experiments on the behavior of quark-gluon constituents went a long way toward giving quarks the reality of other elementary particles, despite the confinement restriction.

B.   *Observation of π Mesons with Large Transverse Momentum in High-Energy Proton-Proton Collisions*[9]

The dynamics of quark-parton constituents were first convincingly demonstrated by James Bjorken's analysis and interpretation of the DIS experiments at SLAC in 1970. Feynman's parton approach must, of course, also be mentioned. The Berman-Bjorken-Kogut (BBK) paper[27] became the Bible of hard collisionists. In 1971, the brand new ISR at CERN began operations and experimenters were able to observe head-on collisions of 30-GeV protons on 30-GeV protons. The ISR, as the highest-energy machine in the 1970's, was a superb place to practice observation strategy. Impressed by the power of the dilepton proble at BNL and by its hints of structure, Rodney Cool of Rockefeller University and I co-oped Luigi DiLella from CERN to help us design an approach which would trade luminosity for resolution. Recall that with the "beam dump" philosophy at BNL we had been able to observe dimuon yields as low as 10^{-12} of the total cross section. However, the penalty was a resolution roughly analogous to using the bottom of a Coca-Cola bottle as the lens for a Nikon. The balance of resolution and luminosity would be a crucial element in the increasing power of the dilepton process.

We learned from Carlo Rubbia about the excellent properties of lead glass as an electromagnetic spectrometer. Photons or electrons would multiply in the high Z medium and dissipate all of their energy in a relatively short length. Improved manufacturing techniques had yielded a dense but transparent glass in which Cerenkov light could be efficiently coupled to good quality photomultiplier tubes. The relatively small response of lead glass to pions and kaons compared to electrons and photons is its great advantage. Six months of hard work in Brookhaven test beams gave us a good command of and respect for this technique and its essential weakness, the calibration process.

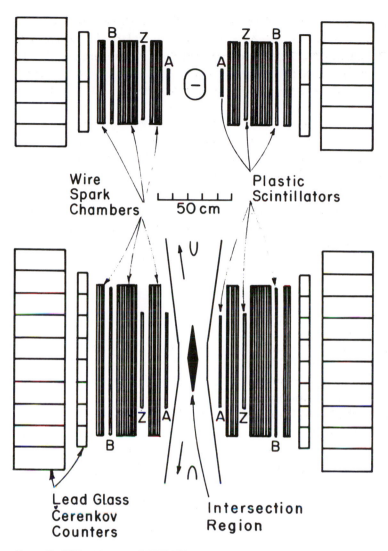

Figure 8. CCR apparatus, CERN ISR.

The idea then was to have two arrays, on opposite sides of the interaction point, each subtending about one steradian of solid angle. Figure 8 shows the CCR apparatus and Figure 9 the data.

The CERN-Columbia-Rockefeller (CCR) team was assembled in 1971 to follow up on the BNL dilepton results, but now electron pairs where the particles of choice and a large lead-glass array was in place around the interaction point of this very first hadron collider. Here again, the discovery of the J/ψ was frustrated by an interesting background that was totally unexpected but, here again, a new technique for probing small distances was discovered—the emission of high transverse momentum hadrons.

Before the ISR research, a handy rule was that hadron production would fall exponentially with transverse momentum. The CCR result had, at a P_t of

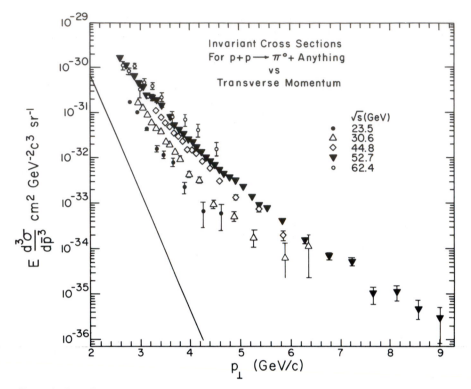

Figure 9. Data from the yield of inclusive π^{o}'s.

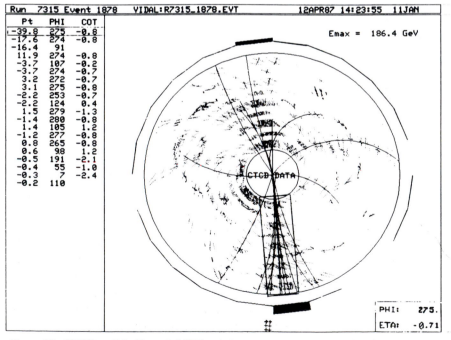

Figure 10. CDF Fermilab dijet at 1.8 TeV.

3 GeV, orders of magnitude higher yield of single π^0's, well detected by the high-resolution lead-glass array. The production rate was observed to be:

$$\backsim P_t^{-8} \text{ at } \sqrt{s} = 62 \text{ GeV}$$

which provided a stringent test of the quark-parton model in the early 70's and QCD some few years later. Other ISR experiments quickly confirmed the CCR result, but only CCR had the quality and quantity of data to provide a phenomenological fit. It turned out that one could eventually go directly from these data to parton-parton (or quark-quark, etc.) hard scattering processes. The study of "single inclusive π^0's, at high P_t" evolved into study of the more typical *jet* structure which now shows up so spectacularly in proton-antiproton collider data. See Figure 10.

Thus, the dilepton adventure using scintillation counters at BNL and the lead-glass exposures to the ISR, initiated independent programs to contribute to the conviction that protons and pions are bound states of confined quarks interacting strongly via the exchange of gluons which are themselves capable of becoming virtual $q\bar{q}$ pairs.

VI. The Third Generation: Observation of a Dimuon Resonance at 9.5 GeV in 400-GeV Proton-Nucleus Collisions[10]

In 1969—1970, the BNL dimuon result had not only stimulated the ISR proposal but also a proposal to the Fermilab (then known as NAL and still a large hole-in-the-ground) to do a high-resolution lepton pair experiment. By the time the machine came on in 1972/3, a single-arm lepton detector had been installed, using the very powerful combination of magnetic measurement *and* lead-glass in order to identify electrons with a pion contamination of $\leq 10^{-5}$. Such rejection is needed when only one particle is involved.

While the study of "direct" electrons fully occupied the Columbia-Fermilab-Stony Brook collaboration in 1974, the J/ψ was being cheerfully discovered at BNL and SLAC. The single-lepton effects turned out to be relatively unfruitful, and the originally proposed pair experiment got underway in 1975. In a series of runs the number of events with pair masses above 4 GeV gradually increased and eventually grew to a few hundred. During this phase hints of resonant peaks appeared and then disappeared. The group was learning how to do those difficult experiments. In early 1977, the key to a vastly improved dilepton experiment was finally discovered. The senior Ph. D.s on the collaboration, Steve Herb, Walter Innes, Charles Brown, and John Yoh, constituted a rare combination of experience, energy, and insight. A new rearrangement of target, shielding, and detector elements concentrated on muon pairs but with hadronic absorption being carried out in beryllium, actually 30 feet of beryllium. The decreased multiple scattering of the surviving muons reduced the mass resolution to 2%, a respectable improvement over the 10—15 % of the 1968 BNL experiment. The filtering of all hadrons permitted over 1000 times as many protons to hit the

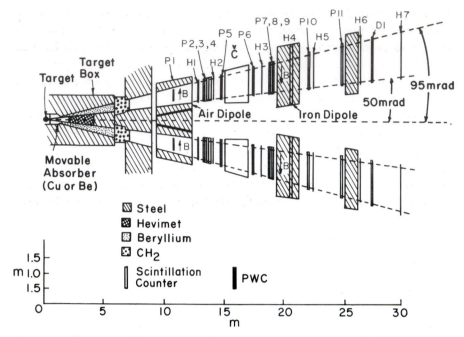

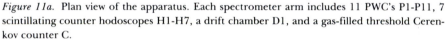

Figure 11a. Plan view of the apparatus. Each spectrometer arm includes 11 PWC's P1-P11, 7 scintillating counter hodoscopes H1-H7, a drift chamber D1, and a gas-filled threshold Cerenkov counter C.

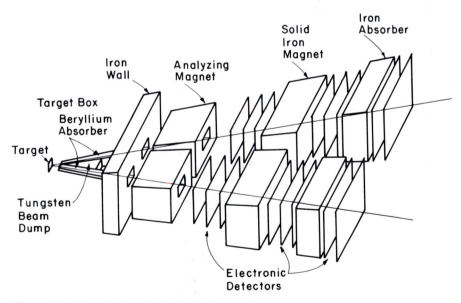

Figure 11b. Schematic sketch of Fermilab dimuon experiment which led to the discovery of the Upsilon particle.

target as compared to open geometry. The compromise between luminosity and resolution was optimized by meticulous attention to the removal of cracks and careful arrangement of the shielding. Recall that this kind of *observation* can call on as many protons as the detector can stand, typically 1 percent of the available protons. The multiwire proportional chambers and triggering scintillators were crowded in towards the target to get maximum acceptance. Muon-ness was certified before and after bending in iron toroids to redetermine the muon momentum and discourage punch-throughs. Figures 11 a, 11 b show the apparatus.

In a month of data taking in the spring of 1977, some 7000 pairs were recorded with masses greater than 4 GeV and a curious, asymmetric, and wide bump appeared to interrupt the Drell-Yan continuum near 9.5 GeV.

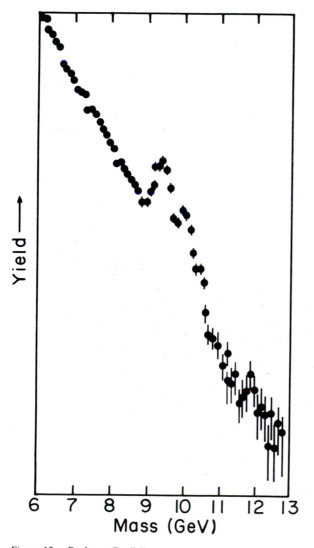

Figure 12a. Peaks on Drell-Yan continuum.

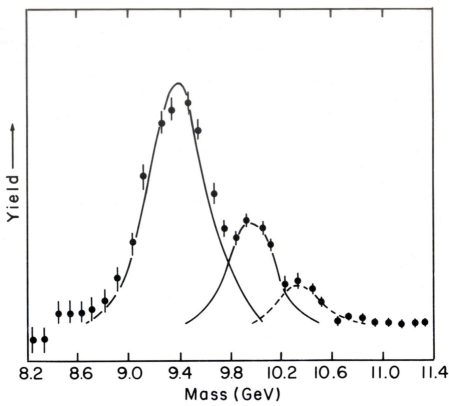

Figure 12b. Peaks with continuum subracted.

With 800 events in the bump, a very clean Drell-Yan continuum under it and practically no background as measured by looking (simultaneously) for same-sign muons, the resonance was absolutely clear. It was named upsilon and a paper was sent off in August of 1977. By September, with 30,000 events, the enhancement was resolved into three clearly separated peaks, the third "peak" being a well-defined shoulder. See Figures 12a, 12b. These states were called Υ, Υ'' and Υ'''. Shortly afterwards, the DORIS accelerator in DESY produced the upsilon in e^+e^- collisions and also served to confirm the only plausible interpretation of the upsilon as a bound state of a new quark b with its antiparticle $\bar{b}$. The Υ'' and Υ''' were the 2S and 3S states of this non-relativistic "atom". In the Standard Model, we had a choice of charge, $+2/3$ (up-like) or $-1/3$ (down-like) for the b-quark. The Fermilab data favored $-1/3$.

Fallout was relatively swift. Taken together with the discovery by Martin Perl and his colleagues[28] of the τ lepton at SLAC slightly earlier, a third generation was added to the Standard Model with the b quark at 5 GeV and the τ-lepton at 2 GeV. This fully confirmed the KM speculation that CP violation may require a third generation. (Clearly we are vastly oversimplifying the theoretical efforts here.)

The b$\bar{\text{b}}$ system was a beautiful addition to c$\bar{\text{c}}$ (charmonium) as a measurement laboratory for the study of potential models for the strong quark-quark force. To get in on the fun, I organized a group from Columbia and Stony Brook to design a lead-glass, sodium iodide spectrometer to be used at the CESR machine, ideally suitable for γ-spectroscopy. This Columbia, Stony Brook collaboration (CUSB) began taking data in 1979 and soon assisted in the identification of the 4S state[11]. The 4S state is especially important because it is above threshold for hadronic decay to B-states, i.e., mesons having one b quark and a lighter antiquark. Follow-up experiments to learn more about the upsilons were also carried out at Fermilab. These used a number of tricks to even further advance the resolving power without losing luminosity — see Figure 13. By now many other states, including p-states, have been identified in this new heavy-quark spectroscopy.

Recent studies of the B-states in electron-positron colliders indicate that the B system may be far richer in physics than the charm equivalent, the D system. B^0's mix like the K^0 and $\bar{K}^0$ particles. Quoting one of CERN's leading phenomenologists, G. Altarelli: "The observation by Argus at DESY of a relatively large amount of $B^0 - \bar{B}^0$ mixing ... was the most important experimental result of the year [1987] in particle physics." There is the strong possibility that CP violation, seen to date *only* in the K^0 system, may possibly be observable in the B^0 system. B-factories, usually high-intensity e^+e^- machines, are being proposed in various labs around the world. The Cornell machine is being upgraded to produce of the order of 10^6 B$\bar{\text{B}}$ pairs a year. Meanwhile the hadron machines are trying hard to solve the very difficult experimental problem of detecting B's (e.g., at the 800-GeV Fermilab fixed target) in a background of 10^6 times as many inelastic collisions. An ambitious detector is being proposed for the Fermilab collider, with the goal of obtaining 10^{10} B$\bar{\text{B}}$ pairs/year. Judging from 1988 activity, measurements in B-physics will play an increasingly important role in particle research over the next decade. The driving force is the recognition that the third generation seems to be needed to account for CP violation. Taken together with baryon non-conservation, CP violation plays a key role in our understanding of the evolution of the universe, including why we are here. For physicists with a less grandiose view, the quark mixing matrix parameters are part of the basis of our Standard Model and b-physics is the key to these crucial parameters.

The third generation still needs a top quark and as we speak here, searches for this are going on now at the CERN S$\bar{\text{p}}$pS machine and at the Fermilab collider.

Both machines are operating at very good intensities averaging $200 - 400$ nb^{-1} per week. The Fermilab machine has a decided advantage of 1.8 TeV as compared to CERN's 0.63 TeV, but everything depends on the quality of data, the wisdom invested in the design of the detectors and, of course, the mass of the top quark. It does seem safe to predict that a paper will soon appear, perhaps entitled: "Observation of the Top Quark."

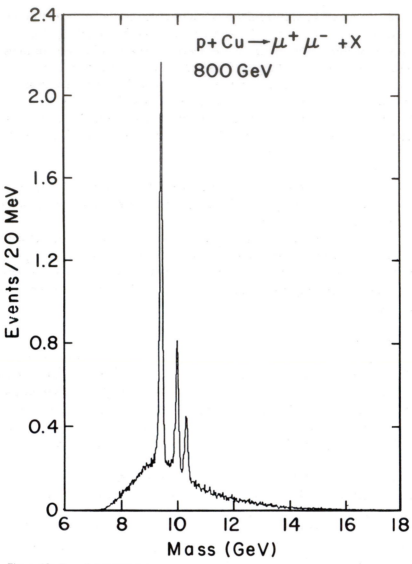

Figure 13. Fermilab E-605 data.

VII. Crucial Issues in Neutrino Physics Today

I conclude this paper with a brief resumé of our ignorance about neutrinos. Neutrino interaction data are in good agreement with electroweak theory of the SM and so they will continue to be used to improve our knowledge of quark structure functions, the crucial Weinberg angle, etc. However, we have not yet seen the ν_τ, we do not know if there is a fourth neutrino, we cannot answer urgent questions about the possibility of neutrino mass, and mixing of different flavors, of the stability of the neutrino, whether it has a magnetic moment, and, finally, the nature of the antineutrino, e.g., whether of the Dirac or Majorana type. What makes all of this intensely interesting are two factors: 1) the astrophysical implications of the answers to these questions are awesome; and 2) the view as expressed by Weinberg that "... neutrino mass illuminates some of the deepest questions in particle physics." This is because, in the Standard Model, with the usual quarks, leptons, and gauge bosons, there is no possible renormalizable interaction that can violate lepton number conservation and give the neutrino a mass. Thus, the observation of mass would very likely be a sign of new physics far beyond the Standard Model, perhaps as far as 10^{15} GeV, the scale of Grand Unification.

A. The Third Neutrino, v_τ

The "three-neutrino" experiment has not been done. Although data from the decay of τ lepton are very strongly suggestive of the existence of v_τ, direct evidence for v_τ has yet to appear.

The technical problem is to move the target as close to the detector as possible but to divert the now unstoppable muons by magnetic sweeping. The flux of v_τ's cannot be predicted with confidence and the shielding configuration is very expensive. This is primarily why the experiment has not yet been done.

B. A Fourth Neutrino?

This question is a shorthand for the issue of the number of generations. Searches for heavier quarks and/or leptons are the *sine qua non* of new accelerators and these have all been negative so far, although the results simply give limits $M_Q > 40$ GeV (same as top quark) and $M_L > 20 - 40$ GeV depending upon the kind of heavy lepton and upon assumptions as to the mass of its accompanying neutrino[29]. Important constraints come from astrophysics where the abundance of helium has been related to the number of low-mass neutrinos[20]. Probably one more low-mass neutrino could still be accommodated within the Big Bang nucleosynthesis arguments. The connection between the cosmological model of creation in the Big Bang and the number of generations in the Standard Model is one of the more romantic episodes in the marriage of particle physics and (early universe) cosmology. In fact, one of the strongest supports of Big Bang cosmology is primordial nucleosynthesis; the cooking of the light elements in the caldron beginning at $t \simeq 1$ sec. The astrophysicists manage to get it right: the abundances of deuterium, helium, and lithium. The key is helium 4; its abundance is a sensitive indicator of the total radiation density at formation time. Contributing to this are all the low-mass, relativistic particles, i.e., photons, electrons, and the three neutrinos plus their antiparticles. Another generation containing a low-mass neutrino would probably not destroy the agreement but it would begin to stretch the agreement. Conclusion: there may be a fourth generation, but a fifth generation which included low-mass particles would provide a major problem for our astrophysical colleagues. Of course, there could be something out there which is outside of the generational structure. One experiment soon to yield results is being carried out at the e^+e^- machines at CERN's Large Electron Positron Collider

(LEP) and the Stanford Linear Collider (SLC) where the width of the Z^0 will give some indication of the number of neutrino pairs into which it can decay. The residual and dominant current interest in the neutrinos comes from astrophysical arguments related to dark matter. This in turn puts the spotlight on the neutrino mass measurements to which we now turn.

C. Neutrino Masses and Oscillation

In the Standard Model, neutrino masses are set to zero and both total lepton number L and lepton flavor number L_i (i = e, μ, τ) are conserved. Neutrino masses "provide a window on the world beyond the SM" and have become one of the outstanding concerns of present-day particle physics. The possibility of oscillation is a statement that $v_\mu \rightarrow v_e$ is not rigorously forbidden as suggested by our two-neutrino experiment. The issue is given great emphasis by the cosmologists, who are increasingly impinging on the orderly development of particle physics (and what a joy that is!) and by the solar neutrino crisis, which has been around for decades. This is the discrepancy between the number of v_e's observed to be coming from the sun and the flux that our best knowledge would predict. The detection of v signals from Supernova 1987a has added to the intensity of interest.

The oscillation possibility was first suggested by B. Pontecorvo in 1967[30]. The neutrino flavor mixing is analogous to the quark mixing as given in the KM matrix. Today, we see many attempts to observe oscillations. These are at the high-energy accelerator labs, at meson factories, at reactors, and indeed in the solar environment. There, the problem is a theoretical one, to understand the lack of neutrinos from the processes that are known to keep the Sun shining. The solar neutrino crisis alone is receiving the attention of at least 14 large experimental groups around the world and many times that number of theorists!

As of this date, no convincing evidence for oscillations or for neutrino masses has been observed. These indirect evidences for mass differences and other experiments which look directly for neutrino masses are summarized by:

$$m(v_e) < \sim 20 \text{ eV}$$
$$m(v_\mu) < 0.25 \text{ MeV}$$
$$m(v_\tau) < 35 \text{ MeV}$$

Oscillation limits are more conventionally given in terms of limits on the mass differences, Δ, and the coupled limits on the phase angle, Θ, that defines the mixing strength. Slowly and inexorably the space on the two-dimensional plot (Δ^2 vs sin 2Θ) is being reduced to the lower left hand corner, although logarithmic scales will encourage experimenters to design ever more sensitive tests.

Cosmologists assure us that we live in a universe whose primary component of mass density is dark (non-luminous) and is presently unidentified. Much of this is probably (they say) non-baryonic and some kind of weakly interacting particle carrying some mass (WIMP) is a likely candidate. The principle of minimum complexity would have these be neutrinos and the condition is $\Sigma m_i \sim 20$ eV (i=e,μ,τ). This brings the ν_τ forward, as emphasized by Harari who proposes as a matter of urgency a renewed search for $\nu_\mu \to \nu_\tau$.

Other experiments occupying the new pion factories (SIN, TRIUMF, and LAMPF) look for (small) violations of lepton flavor conservation via extremely sensitive searches for such reactions as

$$\mu^+ \nrightarrow e^+ \gamma \text{ (again but now at B} \sim 10^{-11}\text{!)}$$
$$\text{and } \mu^+ \nrightarrow e^+ e^+ e^- \text{ (B} < 10^{-12}\text{).}$$

The improvements in experimental techniques and machines conspire to improve these observations by about an order of magnitude every seven years. For completeness we must also list the search for rare decay modes of K-mesons in "kaon factories." Pion, kaon, and B-factories clearly indicate the industrialization of particle physics. The physics objectives of all of these researches are to seek out the tiny influences of presumed new physics which is taking place at the TeV level and higher. For a mature experimenter, these are fun experiments combining the payoff of observations (if and when) with the attention to detail of precise measurements.

To all of the above we should add the new generation structure function research with neutrino beams, probably tagged. Taken together, the 1962 two-neutrino experiment, honored at this meeting, has given rise to a set of activities which, in 1988, continues to play a dominating role in particle physics and its new branches, astrophysics and early universe cosmology.

VIII. Final Comments

I would like to conclude this history of the Standard Model, which is not a history at all. From time to time it follows the main road, e.g., when the two-neutrino experiment pointed to flavor and the generational organization of the Standard Model. More often it takes side trails because my own experiments were down those paths. So we have neglected such milestones as the discovery of neutral currents, the τ lepton, the W and Z bosons, charmonium, etc. We have also been crushingly neglectful of the essential theoretical contributions and blitzed through quarks, color, symmetry-breaking, etc.

However, I regret most not having the space to speak more of the accelerators, the detectors, and the people who brought these to be. The Nevis cyclotron was built under the leadership of Eugene Booth and James Rainwater; the AGS, most successful machine ever, led by Ken Green, Ernest Courant, Stanley Livingston, and Hartland Snyder; Fermilab, of course, by Robert Wilson and his outstanding staff. My own detector experience owes much to Georges Charpak of CERN and William Sippach of Columbia. In neglecting these details I am reminded of my teacher, friend, and thesis professor, Gilberto Bernardini, who, when being shown the Nevis cyclotron's innards, exclaimed: "Just show me where the beam comes out." Finally, I make amends to the theorists who are obviously crucial to the entire enterprise. I have enjoyed and profited from many physicists of the theoretical persuasion, but most especially T. D. Lee, M. Veltman, and J. D. Bjorken.

REFERENCES

1. T. Coffin, R. L. Garwin, L. M. Lederman, S. Penman, A. M. Sachs; Phys. Rev. *106*, 1108 (1957).
2. S. Devons, G. Gidal, L. M. Lederman, G. Shapiro, Phys. Rev. Lett. *5*, 330 (1960). S. Meyer et al. Phys. Rev. *132*, 2963 (1963).
3. R. W. Ellsworth, A. C. Melissinos, J. H. Tinlot, H. von Briesen, Jr., T. Yamanouchi, L. M. Lederman, T. Tannenbaum, R. L., Cool, A. Maschke, Phys. Rev. *165*, 1449 (1968).
   C. P. Sargent, M. Rinehart, L. M. Lederman, K. C. Rogers, Phys. Rev. *99*, 885 (1955).
4. H. Blumenfeld, W. Chinowsky, L. M. Lederman, Nuovo Cimento *8*, 296 (1958).
5. K. Lande, E. T. Booth, J. Impeduglia, L. M. Lederman, Phys. Rev. *103*, 1901 (1956).
6. R. L. Garwin, L. M. Lederman, M. Weinrich, Phys. Rev. *105*, 1415 (1957).
7. G. Danby, J-M. Gaillard, K. Goulianos, L. M. Lederman, N. Mistry, M. Schwartz, J. Steinberger, Phys. Rev. Lett. *9*, 36 (1962).
8. J. H. Christenson, G. S. Hicks, L. M. Lederman, P. J. Limon, B. G. Pope, E. Zavattini, Phys. Rev. Lett. *25*, 1523 (1970).
9. F. W. Büsser, L. Camilleri, L. DiLella, G. Gladding, A. Placci, B. G. Pope, A. M. Smith, J. K. Yoh, E. Zavattini, B. J. Blumenfeld, L. M. Lederman, R. L. Cool, L. Litt, S. L. Segler, Phys. Lett. *46B*, 471 (1973).
10. S. W. Herb, D. C. Hom, L. M. Lederman, J. C. Sens, H. D. Snyder, J. K. Yoh, J. A. Appel, B. C. Brown, C. N. Brown, W. R. Innes, K. Ueno. T. Yamanouchi, A. S. Ito, H. Jostlein, D. M. Kaplan, R. D. Kephart, Phys. Rev. Lett. *39*, 252 (1977).
11. G. Finocchiaro, G. Giannini, J. Lee-Franzini, R. D. Schamberger Jr., M. Siverta, L. J. Spencer, P. M. Tuts, T. Bohringer, F. Costantini, J. Dobbins, P. Franzini, K. Han, S. W. Herb, D. M. Kaplan, L. M. Lederman, G. Mageras, D. Peterson, E. Rice, J. K. Yoh, G. Levman, Phys. Lett. *45*, 222 (1980).
12. M. Gell-Mann, A. Pais, Phys. Rev. *97*, 1387 (1955).
13. L. Lederman, E. T. Booth, H. Byfield, J. Kessler, Phys. Rev. *83*, 685 (1951).
14. L. Lederman, J. Tinlot, E. Booth, Phys. Rev. *81*, (1951).
15. L. M. Lederman, Thesis Columbia U, 1951 (unpublished).
16. H. Byfield, J. Kessler, L. M. Lederman, Phys. Rev. *86*, 17 (1952).
17. T. D. Lee, Reinhard Oehne, C. N. Yang, Phys. Rev. *106*, 340 (1957).
18. J. H. Christenson, J. W. Cronin, V. L. Fitch, R. Turlay, Phys. Rev. Lett. *13*, 138 (1964).
19. T. D. Lee, C. N. Yang, Phys. Rev. *104*, 254 (1956).
20. G. Feinberg, Phys. Rev. *110*, 1482 (1958).
21. J. D. Bjorken, S. L. Glashow, Phys. Lett. *11*, 255 (1964).
22. S. L. Glashow, J. Iliopoulos, L. Maiani, Phys. Rev. *D2*, 1285 (1970).
23. R. Burns, G. Danby, E. Heyman, L. M. Lederman, W. Lee, J. Rettberg, and J. Sunderland, Phys. Rev. Lett. *15*, 830 (1965).
24. Y. Yamaguchi, Nuovo Cimento *43A*, 193 (1966). L. Okun, unpublished (1966).
25. J. J. Aubert, U. Becker, P. J. Biggs, J. Burger, M. Chen, G. Everhart, P. Goldhagen, J. Leong, T. McCorriston, T. G. Rhoades, M. Rohde, Samuel C. C. Ting, Sau Lan Wu, Phys. Rev. Lett. *33*, 1404 (1974).
26. D. M. Kaplan, R. J. Fisk, A. S. Ito, H. Jostlein, J. A. Appel, B. C. Brown, C. N. Brown, W. R. Innes, R. D. Kephart, K. Ueno, T. Yamanouchi, S. W. Herb, D. C.

Hom, L. M. Lederman, J. C. Sens, H. D. Snyder, J. K. Yoh, Phys. Rev. Lett. *40,* 435 (1978) and A. S. Ito et al., Phys. Rev. D *23,* 604 (1981).

27. S. M. Berman, J. D. Bjorken, J. B. Kogut, Phys. Rev. *D4,* 3388 (1971).
28. Martin Perl: Heavy Leptons in 1986. Proceedings of the XXIII International Conference on High Energy Physics, Berkeley, 1986, p. 596.
29. Gary Steigman, David N. Schramm and James E. Gunn, Phys. Lett. *66B,* 202 (1977).
30. B. Pontecorvo, Zh. Eksp. Teor. Fiz. *53,* 1717 (1967) [Soviet Physics JETP *26,* 984 (1968)].

Physics 1989

NORMAN F RAMSEY

for the invention of the separated oscillatory fields method and its use in the hydrogen maser and other atomic clocks

HANS G DEHMELT and WOLFGANG PAUL

for the development of the ion trap technique

THE NOBEL PRIZE IN PHYSICS

Speech by Professor Ingvar Lindgren of the Royal Swedish Academy of Sciences.
Translation from the Swedish text.

Your Majesties, Your Royal Highnesses, Ladies and Gentlemen.

This year's Nobel Prize in Physics is shared between three scientists, Professor *Norman Ramsey*, Harvard University, Professor *Hans Dehmelt*, University of Washington, Seattle, and Professor *Wolfgang Paul*, University of Bonn, for "*contributions of importance for the development of atomic precision spectroscopy.*"

The works of the laureates have led to dramatic advances in the field of precision spectroscopy in recent years. Methods have been developed that form the basis for our present definition of time, and these techniques are applied for such disparate purposes as testing Einstein's general theory of relativity and measuring continental drift.

An atom has certain fixed energy levels, and transition between these levels can take place by means of emission or absorption of electromagnetic radiation, such as light. Transition between closely spaced levels can be induced by means of radio-frequency radiation, and this forms the basis for so-called *resonance methods*. The first method of this kind was introduced by Professor I. Rabi in 1937, and the same basic idea underlies the resonance methods developed later, such as nuclear magnetic resonance (NMR), electron-spin resonance (ESR) and optical pumping.

In Rabi's method a beam of atoms passes through an oscillating field, and if the frequency of that field is right, transition between atomic levels can take place. In 1949 one of this year's laureates, Norman Ramsey, modified this method by introducing *two separate oscillatory fields*. Due to the interaction between these fields, a very sharp interference pattern appears. This discovery made it possible to improve precision by several orders of magnitude, and this started the development towards high-precision spectroscopy.

One important application of Ramsey's method is the *cesium clock*, an atomic clock on which our definition of time has been based since 1967. One second is no longer based on the rotation of the earth or its movement around the sun, but is instead defined as the time interval during which the cesium atom makes a certain number of oscillations. The cesium clock has a margin of error equivalent to one thousandth of a second in three hundred years. Compared with this clock, the earth behaves like a bobbing duck.

The dream of the spectroscopist is to be able to study a single atom or ion under constant conditions for a long period of time. In recent years, this dream has to a large extent been realized. The basic tool is here the *ion trap*, which was introduced in the 1950s by another of this year's laureates,

Wolfgang Paul in Bonn. His technique was further refined by the third laureate, Hans Dehmelt, and his co-workers in Seattle into what is now known as *ion-trap spectroscopy*.

Dehmelt and his associates used this spectroscopy primarily for studying electrons, and in 1973 they succeeded for the first time in observing a *single electron* in an ion trap, and in confining it there for weeks and months. One property of the electron, its magnetic moment, was measured to 12 digits, 11 of which have later been verified theoretically. This represents a most stringent test of the atomic theory known as *quantum electrodynamics* (QED).

In a similar way, Dehmelt and others were later able to trap and study a *single ion,* which represents a true landmark in the history of spectroscopy. The technique is now being used in development of improved atomic clocks, in particular at the National Institute for Standards and Technology (formerly the National Bureau of Standards) in Boulder, Colorado.

Another technique for storing atoms and observing them for a long period of time has been developed by Ramsey and his co-workers at Harvard University, *the hydrogen maser.* This instrument is mainly used as a secondary standard for time and frequency with a higher stability for intermediate times than the cesium clock. It is used, for instance, for the determination of continental drift, using VLBI (Very Long Base Line Interferometry). Here, signals from a radio star are received with radio telescopes on two continents and compared by means of very accurate time settings from two hydrogen masers. Another application is the test of Einstein's general theory of relativity. According to this theory, time elapses faster on the top of a mountain than down in the valley. In order to test this prediction, a hydrogen maser was sent up in a rocket to a height of 10,000 km and its frequency compared with that of another hydrogen maser on the ground. The predicted shift has been verified to one part in ten thousand.

The continued rapid development of the atomic clock can be foreseen in the near future. An accuracy of *one part in one billion billions* is considered realistic. This corresponds to an uncertainty of less than one second since the creation of the universe fifteen billion years ago.

Do we need such accuracy? It is clear that navigation and communication in space require a growing degree of exactness, and existing atomic clocks are already being utilized in these fields to the limit of their capacity. The new technique may be even more important for testing very fundamental principles of physics. Further tests of quantum physics and relativity theory may force us to revise our assumptions about time and space or about the smallest building blocks of matter.

Norman F. Ramsey

NORMAN F. RAMSEY

I was born August 27, 1915 in Washington, D.C. My mother, daughter of German immigrants, had been a mathematics instructor at the University of Kansas. My father, descended from Scottish refugees and a West Point graduate, was an officer in the Army Ordnance Corps. His frequently changing assignments took us from Washington, DC to Topeka, Kansas, to Paris, France, to Picatinny Arsenal near Dover, New Jersey, and to Fort Leavenworth, Kansas. With two of the moves I skipped a grade and, encouraged by my supportive parents and teachers, I graduated from high school with a high academic record at the age of 15.

My early interest in science was stimulated by reading an article on the quantum theory of the atom. But at that time I did not realize that physics could be a profession. My parents presumed that I would try to follow my father's footsteps to West Point, but I was too young to be admitted there. I was offered a scholarship to Kansas University but my parents again moved — this time to New York City. Thus I entered Columbia College in 1931, during the great depression. Though I started in engineering, I soon learned that I wanted a deeper understanding of nature than was then expected of engineers so I shifted to mathematics. By winning yearly competitive mathematics contests, I was honored in my senior year by being given the mathematics teaching assistantship normally reserved for graduate students. At the time I graduated from Columbia in 1935, I discovered that physics was a possible profession and was the field that most excited my curiosity and interest.

Columbia gave me a Kellett Fellowship to Cambridge University, England, where I enrolled as a physics undergraduate. The Cavendish Laboratory in Cambridge was then an exciting world center for physics with a stellar array of physicists: J.J. Thomson, Rutherford, Chadwick, Cockcroft, Eddington, Appleton, Born, Fowler, Bullard, Goldhaber and Dirac. An essay I wrote at Cambridge for my tutor, Maurice Goldhaber, first stimulated my interest in molecular beams and in the possiblity of later doing my Ph. D. research with I. I. Rabi at Columbia.

After receiving from Cambridge my second bachelors degree, I therefore returned to Columbia to do research with Rabi. At the time I arrived Rabi was rather discouraged about the future of molecular beam research, but this discouragement soon vanished when he invented the molecular beam magnetic resonance method which became a potent source for new fundamental discoveries in physics. This invention gave me the unique opportunity to be the first graduate student to work with Rabi and his associates,

Zacharias, Kellogg, Millman and Kusch, in the new field of magnetic resonance and to share in the discovery of the deuteron quadrupole moment.

Following the completion of my Columbia thesis, I went to Washington,
D. C. as a Carnegie Institution Fellow, where I studied neutron-proton and
proton-helium scattering.

In the summer of 1940 I married Elinor Jameson of Brooklyn, New York,
and we went to the University of Illinois with the expectation of spending
the rest of our lives there, but our stay was short lived. World War II was
rampant in Europe and within a few weeks we left for the MIT Radiation
Laboratory. During the next two years I headed the group developing radar
at 3 cm wavelength and then went to Washington as a radar consultant to
the Secretary of War. In 1943 we went to Los Alamos, New Mexico, to work
on the Manhattan Project.

As soon as the war ended I eagerly returned to Columbia University as a
professor and research scientist. Rabi and I immediately set out to revive
the molecular beam laboratory which had been abandoned during the war.
My first graduate student, William Nierenberg, and I measured a number of
nuclear magnetic dipole and electric quadrupole moments and Rabi and I
started two other students, Nafe and Nelson, on a fundamental experiment
to measure accurately the atomic hydrogen hyperfine separation.. During
this period Rabi and I also initiated the actions that led to the establishment
of the Brookhaven National Laboratory on Long Island, New York, where
in 1946 I became the first head of the Physics Department.

In 1947 I moved to Harvard University where I taught for 40 years
except for visiting professorships at Middlebury College, Oxford University, Mt. Holyoke College and the University of Virginia. At Harvard I
established a molecular beam laboratory with the intent of doing accurate
molecular beam magnetic resonance experiments, but I had difficulty in
obtaining magnetic fields of the required uniformity. Inspired by this
failure, I invented the separated oscillatory field method which permitted us
to achieve the desired accuracy with the available magnets. My graduate
students and I then used this method to measure in many different molecules a number of molecular and nuclear properties including nuclear
spins, nuclear magnetic dipole and electric quadrupole moments, rotational
magnetic moments of molecules, spin-rotational interactions, spin-spin interactions, electron distributions in molecules, etc. Although we studied a
wide variety of molecules we concentrated on the diatomic molecules of the
hydrogen isotopes since these molecules were most suitable for comparing
theory and experiment. During this period I also consulted with various
groups that were applying the separated oscillatory field method to atomic
clocks and I analyzed the precautions which must be taken to avoid errors.
Although our original molecular beam research was only with the magnetic
resonance method, we later built a separated oscillatory fields electric
resonance apparatus and used it to study polar molecules.

In an effort to attain even greater accuracy and to do so with atomic
hydrogen, the simplest fundamental atom, Daniel Kleppner, a former stu-

dent, and I invented the atomic hydrogen maser. We then used it for accurate measurements of the hyperfine separations of atomic hydrogen, deuterium and tritium and for determining the extent to which the hyperfine structure was modified by the application of external electric and magnetic fields. We also participated with Robert Vessot and others in converting a hydrogen maser to a clock of unprecedented stability.

While these experiments were being carried out with some of my graduate students, I worked with other students and associates to apply similar precision methods to beams of polarized neutrons. At the Institut Laue-Langevin in Grenoble, France, we measured accurately the magnetic moment of the neutron, set a low limit to the electric dipole moment of the neutron as a test of time reversal symmetry and discovered and measured the parity non-conserving rotations of the spins of neutrons passing through various materials.

Concurrently with my molecular and neutron beam research, I was also teaching and involved with other scientific activities. I was director of the Harvard Cyclotron during its construction and early operation and participated in proton-proton scattering experiments with that cyclotron. I was later chairman of the joint Harvard-MIT committee managing the construction of the 6 GeV Cambridge Electron Accelerator and used that device for various particle physics experiments including electron-proton scattering. For a year and a half I was on leave from Harvard as the first Assistant Secretary General for Science (Science Advisor) in NATO where I initiated the NATO programs for Advanced Study Institutes, Fellowships and Research Grants. For sixteen exciting years I was on leave half time from Harvard as President of Universities Research Association which exercised its management responsibilities for the construction and operation of the Fermilab accelerator through two outstanding laboratory directors, Robert R. Wilson and Leon Lederman.

Although I am primarily an experimental physicist, theoretical physics is my hobby and I have published several theoretical papers including early discussions of parity and time reversal symmetry, the first successful theory of the NMR chemical shifts, theories of nuclear interactions in molecules and the theory of thermodynamics and statistical mechanics at negative absolute temperatures.

I officially retired from Harvard in 1986, but I have remained active in physics. For one year I was a research fellow at the Joint Institute for Laboratory Astrophysics at the University of Colorado and I now periodically revisit JILA as an Adjunct Research Fellow. Subsequent to our year in Colorado, I have been visiting professor at The University of Chicago, Williams College and the University of Michigan. I continue writing and theoretical calculations in my Harvard office and with my collaborators we are continuing our neutron experiments at Grenoble.

After Elinor died in 1983, I married Ellie Welch of Brookline, Massachusetts and we now have a combined family of seven children and six grandchildren. We enjoy downhill and cross country skiing, hiking, bicycling and trekking as well as musical and cultural events.

I have greatly enjoyed my years as a teacher and research physicist and continue to do so. The research collaborations and close friendships with my eighty-four graduate students have given me especially great pleasure. I hope they have learned as much from me as I have from them.

Books:
Experimental Nuclear Physics, with E. Segrè, John Wiley and Sons, Inc. (1953), *Nuclear Moments,* John Wiley and Sons, Inc. (1953), *Molecular Beams,* Oxford University Press (1956 and 1985), and *Quick Calculus,* with D. Kleppner, John Wiley and Sons, Inc. (1965 and 1985).

Honorary D. Sc.:
Case-Western Reserve University, Middlebury College, Oxford University, The Rockefeller University, The University of Chicago, and The University of Sussex.

Honors:
E. O Lawrence Award, 1960; Trustee Carnegie Endowment for International Peace, 1962−86; Davisson-Germer Prize,1974; Trustee of The Rockefeller University, 1977− ; President of the American Physical Society, 1978−79; Chairman Board of Governors of American Institute of Physics, 1980−86; President of United Chapters of Phi Beta Kappa, 1984−88; IEEE Medal of Honor, 1984; Rabi Prize, 1985; Rumford Premium, 1985; Chairman Board of Physics and Astronomy of National Research Council, 1985−1989; Compton Medal, 1986; Oersted Medal, 1988; National Medal of Science, 1988.

(added in 1991):
Doctor of Civil Law (D.C.L.), Oxford University (1990)
D.Sc., University of Houston (1990) and Carleton College (1991)
Foreign Associate, French Academy of Science (1990)

Principal Publications:

1. Magnetic Moments of Proton and Deuteron. Radiofrequency Spectrum of H_2 in Magnetic fields. With J. M. B. Kellogg, I. I. Rabi and J. R. Zacharias, Phys. Rev. *56,* 728 (1939).
2. Electrical Quadrupole Moment of the Deuteron. Radiofrequency Spectra of HD and D_2 Molecules in a Magnetric Field. With J. M. B. Kellogg, I. I. Rabi and J. R. Zacharias, Phys. Rev. *57,* 677 (1940).
3. Rotational Magnetic Moments of H_2, D_2 and HD molecules. Phys. Rev. *58,* 226 (1940).
4. Molecular Beam Resonance Method with Separated Oscillating Fields. Phys. Rev. *78,* 695 (1950).
5. Magnetic Shielding of Nuclei in Molecules. Phys. Rev. *78,* 699 (1950).
6. On the Possibility of Electric Dipole Moments for Elementary Particles and Nuclei. With E. M. Purcell, Phys. Rev. *78,* 807(L) (1950).

7. Nuclear Audiofrequency Spectroscopy by Resonant Heating of the Nuclear Spin System. With R. V. Pound, Phys. Rev. *81*, 278(L) (1951).

8. Proton-Proton Scattering at 105 MeV and 75 MeV. With R. W. Brige and U. E. Kruse, Phys. Rev. *83*, 274 (1951).

9. Theory of Molecular Hydrogen and Deuterium in Magnetic Fields. Phys. Rev. *85*, 60 (1952).

10. Chemical Effects in Nuclear Magnetic Resonance and in Diamagnetic Susceptibility. Phys. Rev. *86*, 243 (1952).

11. Nuclear Radiofrequency Spectra of H_2 and D_2 in High and Low Magnetic Fields. With H. G. Kolsky, T. E. Phipps, and H. B. Silsbee, Phys. Rev. *87*, 395 (1952).

12. Nuclear Radiofrequency Spectra of D_2 and H_2 in Intermediate and Strong Magnetic Fields. With N. J. Harrick, R. G. Barns and P. J. Bray, Phys. Rev. *90*, 260 (1953).

13. Electron Coupled Interations between Nuclear Spins in Molecules. Phys. Rev. *91*, 303 (1953).

14. Use of Rotating Coordinates in Magnetic Resonance Problems. With I. I. Rabi and J. Schwinger, Rev. Mod. Phys. *26*, 167 (1954).

15. Resonance Transitions Induced by Perturbations at Two or More Different Frequencies. Phys. Rev. *100*, 1191 (1955).

16. Thermodynamics and Statistical Mechanics at Negative Absolute Temperatures, Phys. Rev. *103*, 20 (1956).

17. Molecular Beams, Published by Oxford University Press, England (1956).

18. Resonance Experiments in Successive Oscillatory Fields. Rev. Sci. Instr. *28*, 57(L) (1957).

19. Experimental Limit to the Electric Dipole Moment of the Neutron. With J. H. Smith and E. M. Purcell, Phys. Rev. *108*, 120 (1957).

20. Time Reversal, Charge Conjugation, Magnetic Pole Conjugation, and Parity. Phys. Rev. *109*, 225 (1958).

21. Molecular Beam Resonances in Oscillatory Fields of Nonuniform Amplitudes and Phases. Phys. Rev. *109*, 822 (1958).

22. Radiofrequency Spectra of Hydrogen Deuteride in Strong Magnetic Fields. With W. E. Quinn, J. M. Baker, J. T. LaTourrette, Phys. Rev. *112*, 1929 (1958).

23. On the Significance of Potentials in Quantum Theory. With W. H. Furry, Phys. Rev. *118*, 623 (1960).

24. Atomic Hydrogen Maser. With H. M. Goldenberg and D. Kleppner, Phys. Rev. Letters *8*, 361 (1960).

25. Theory of the Hydrogen Maser. With D. Kleppner and H. M. Goldenberg, Phys. Rev. *126*, 603 (1962).

26. Hyperfine Structure of Ground State of Atomic Hydrogen. With S. B. Crampton and D. Kleppner, Phys. Rev. Letters *11*, 338 (1963).

27. Hydrogen Maser Principles and Techniques. With D. Kleppner, H. C. Berg, S. B. Crampton, R. F. C. Vessot, H. E. Peters and J. Vanier, Phys. Rev. *138*, A972 (1965).

28. Measurement of Proton Electromagnetic Form Factors at High Momentum Transfer. With K. W. Chen, J. R. Dunning, Jr., A. A. Cone, J. K. Walker and Richard Wilson, Phys. Rev. *141*, 1267 (1966).

29. Absolute Value of the Proton g Factor. With T. Myint, D. Kleppner and H. G. Robinson, Phys. Rev. Lett. *17*, 405 (1966).

30. Magnetic Resonance Molecular Beam Spectra of Methane. With C. H. Anderson, Phys. Rev. *149*, 14 (1966).

31. Hyperfine Separation of Tritium. With B. S. Mathur, S. B. Crampton, and D. Kleppner, Phys. Rev. *158*, 14 (1967).

32. Measurement of the Hydrogen-Deuterium Atomic Magnetic Moment Ratio and of the Deuterium Hyperfine Frequency. With D. J. Larson and P. A. Valberg, Phys. Rev. Letters *23*, 1369 (1969).
33. Multiple Region Hydrogen Maser with Reduced Wall Shift. With E. E. Uzgiris, Phys. Rev. *A1*, 429 (1970).
34. Molecular Beam Magnetic Resonance Studies of HD and D2. With R. F. Code, Phys. Rev. *A4*, 1945 (1971).
35. Atomic Deuterium Maser With D. J. Wineland, Phys. Rev. *A5*, 821 (1972).
36. The Molecular Zeeman and Hyperfine Spectra of LiH and LiD by Molecular Beam High Resolution Electric Resonance. With Richard R. Freeman, Abram R. Jacobson, and David W. Johnson, J. of Chem. Physics *63*, 2597 (1975).
37. The Tensor Force Between Two Protons at Long Range, Physica *96A*, 285 (1979).
38. Measurement of the Neutron Magnetic Moment. With G. L. Green, W. Mampe, J. M. Pendelbury, K. Smith, W. B. Dress, P. D. Miller and P. Perrin, Phys. Rev. *D20*, 2139 (1979).
39. First Measurement of Parity-Nonconserving Neutron Spin Rotation: The Tin Isotopes. With M. Forte, B. R. Heckel, K. Green, and G. L. Greene, Phys. Rev. Lett. *45*, 2088 (1980).
40. Search for P and T Violations in the Hyperfine Structure of Thallium Fluoride. With D. A. Wilkening and D. J. Larson, Phys. Rev. *A29*, 425 (1984).
41. Search for a Neutron Electric Dipole Moment. With J. M. Pendlebury, et al., Phys. Letters *136B*, 327 (1984).
42. Neutron Magnetic Resonance Experiments. Physica *137B*, 223 (1986).
43. Quantum Mechanics and Precision Measurements, IEEE Transactions on Instrumentation and Measurement *IM36*, 155 (1987).
44. Precise Measurements of Time. American Scientist *76*, 42 (1988).
45. The Electric Dipole Moment of the Neutron. Physica Scripta *T22*, 40 (1988).

EXPERIMENTS WITH SEPARATED OSCILLATORY FIELDS AND HYDROGEN MASERS

Nobel Lecture, December 8, 1989

by

NORMAN F. RAMSEY

Physics Department, Harvard University, Cambridge, MA 02138, USA

I am honored to receive the Nobel Prize, which I feel is also an honor to the physicists and engineers in many countries who have done beautiful experiments using the methods I shall be discussing. In particular, I am grateful to my eighty-four wonderful Ph.D. students and, to Daniel Kleppner and Daniel Larson, who were my close collaborators for a number of years.

THE METHOD OF SUCCESSIVE OSCILLATORY FIELDS

In the summer of 1937 following two years at Cambridge University, I went to Columbia University to work with I. I. Rabi. After I had been there only a few months, Rabi invented[1-4] the molecular beam magnetic resonance method so I had the great good fortune to be the only graduate student to work with Rabi and his colleagues[1-7] on one of the first two experiments to develop and utilize magnetic resonance spectroscopy, for which Rabi received the 1944 Nobel Prize in Physics.

By 1949, I had moved to Harvard University and was looking for a way to make more accurate measurements than were possible with the Rabi method and in so doing I invented the method of separated oscillatory fields.[3-6] In this method the single oscillatory magnetic field in the center of the Rabi device is replaced by two oscillatory fields, one at the entrance and one at the exit of the space in which the properties of the atoms or molecules are studied. As I will discuss, the separated oscillatory fields method has many advantages over the single oscillatory field method and in subsequent years it has been extended to many experiments beyond those of molecular beam magnetic resonance. The device shown in Figure 1 is a molecular beam apparatus embodying successive oscillatory fields that has been used at Harvard for an extensive series of experiments.

Let me now review the successive oscillatory field method, particularly in its original and easiest to explain application — the measurement of nuclear magnetic moments. The extension to more general cases is then straightforward.

Physics 1989

Figure 1. Molecular beam apparatus with separated oscillatory fields. The beams of molecules emerges from a small source aperture in the left third of the apparatus, is focussed there and passes through the middle third in an approximately parallel beam. It is focussed again in the right third to a small detection aperture. The separated oscillatory electric fields at the beginning and end of the middle third of the apparatus produce resonance transitions that reduce the focussing and therefore weaken the detected beam intensity.

The method was initially an improvement on Rabi's resonance method for measuring nuclear magnetic moments, whose principles are illustrated schematically in Figure 2. Consider a classical nucleus with spin angular momentum $\hbar \mathbf{J}$ and magnetic moment $\mu = (\mu/J)\mathbf{J}$. Then in a static magnetic field $\mathbf{H}_o = H_0 \mathbf{k}$, the nucleus, due to the torque on the nuclear angular momentum, will precess like a top about $\mathbf{H}_0$ with the Larmor frequency υ_0 and angular frequency ω_0 given by

$$\omega_0 = 2\pi \, \upsilon_0 = \frac{\mu H_0}{\hbar J} \tag{1}$$

as shown in Figure 3. Consider an additional magnetic field $\mathbf{H}_1$ perpendicular to $\mathbf{H}_0$ and rotating about it with angular frequency ω. Then, if at any time H_1 is perpendicular to the plane of H_0 and J, it will remain perpendicular to it provided $\omega = \omega_0$. In that case, in a coordinate system rotating with H_1, $\mathbf{J}$ will precess about $\mathbf{H}_1$ and the angle ϕ will continuously change in a fashion analogous to the motion of a "sleeping top"; the change of orientation can be detected by allowing the molecular beam containing the magnetic moments to pass through inhomogeneous fields as in Figure 2. If ω is not equal to ω_0, $\mathbf{H}_1$ will not remain perpendicular to $\mathbf{J}$; so ϕ will increase for a short while and then decrease, leading to no net change. In this fashion the Larmor precession frequency ω_0 can be detected by measuring the oscillator frequency ω at which there is maximum reorientation of the angular momentum and hence a maximum change in beam intensity for an apparatus as in Figure 2. This procedure is the basis of the Rabi molecular beam resonance method.

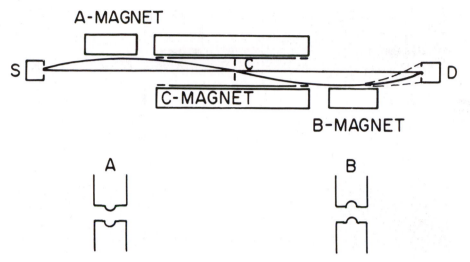

Figure 2. Schematic diagram of a molecular beam magnetic resonance apparatus. A typical molecule which can be detected emerges from the source, is deflected by the inhomogeneous magnetic field A, passes through the collimator and is deflected to the detector by the inhomogeneous magnetic field B. If, however, the oscillatory field in the C region induces a change in the molecular state, the B magnet will provide a different deflection and the beam will follow the dashed lines with a corresponding reduction in detected intensity. In the Rabi method, the oscillatory field is applied uniformly throughout the C region as indicated by the long rf lines F, whereas in the separated oscillatory field method the rf is applied only in the regions E and G.

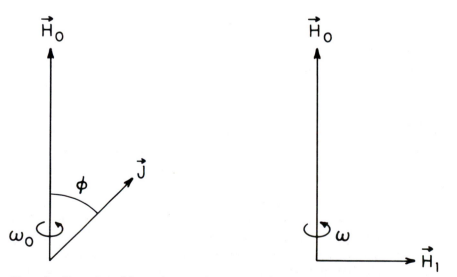

Figure 3. Precession of the nuclear angular momentum **J** (left) and the rotating magnetic field **H**$_1$ (right) in the Rabi method.

The separated oscillatory field method in this application is much the same except that the rotating field H_1 seen by the nucleus is applied initially for a short time τ, the amplitude of H_1 is then reduced to zero for a relatively long time T and then increased to H_1 for a time τ, with phase coherency being preserved for the oscillating fields as shown in Figure 4. This can be done, for example, in the molecular-beam apparatus of Figure 2 in which the molecules first pass through a rotating field region, then a region with no rotating field and finally a region with a second rotating field driven phase coherently by the same oscillator.

If the nuclear spin angular momentum is initially parallel to the fixed field (so that ϕ is equal to zero initially) it is possible to select the magnitude of the rotating field so that ϕ is 90° or $\pi/2$ radians at the end of the first oscillating region. While in the region with no oscillating field, the magnetic moment simply precesses with the Larmor frequency appropriate to the magnetic field in that region. When the magnetic moment enters the second oscillating field region there is again a torque acting to change ϕ . If the frequency of the rotating field is exactly the same as the mean Larmor frequency in the intermediate region there is no relative phase shift between the angular momentum and the rotating field.

Consequently, if the magnitude of the second rotating field and the length of time of its application are equal to those of the first region, the second rotating field has just the same effect as the first one — that is, it increases ϕ by another $\pi/2$, making $\phi = \pi$, corresponding to a complete reversal of the direction of the angular momentum. On the other hand, if the field and the Larmor frequencies are slightly different, so that the relative phase angle between the rotating field vector and the precessing angular momentum is changed by π while the system is passing through the intermediate region, the second oscillating field has just the opposite effect to the first one; the result is that ϕ is returned to zero. If the Larmor frequency and the rotating field frequency differ by an amount such that the relative phase shift in the intermediate region is exactly an integral multiple of 2π, ϕ will again be left at π just as at exact resonance.

In other words if all molecules had the same velocity, the transition probability would be periodic as in Figure 5. However, in a molecular beam resonance experiment one can easily distinguish between exact resonance and the other cases. In the case of exact resonance, the condition for no change in the relative phase of the rotating field and of the precessing angular momentum is independent of the molecular velocity. In the other cases, however, the condition for integral multiple of 2π relative phase shift is velocity dependent, because a slower molecule is in the intermediate region longer and so experiences a greater shift than a faster molecule. Consequently, for the non-resonance peaks, the reorientations of most molecules are incomplete so the magnitudes of the non-resonance peaks are smaller than at exact resonance and one expects a resonance curve similar to that shown in Figure 6, in which the transition probability for a particle of spin $1/2$ is plotted as a function of frequency.

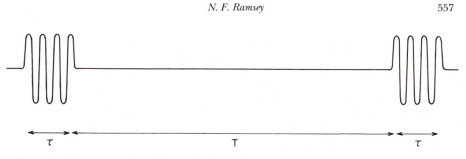

Figure 4. Two separated oscillatory fields, each acting for a time τ, with zero amplitude oscillating field acting for time T. Phase coherency is preserved between the two oscillatory fields so it is as if the oscillation continued, but with zero amplitude for time T.

Although the above description of the method is primarily in terms of classical spins and magnetic moments, the method applies to any quantum mechanical system for which a transition can be induced between two energy states W_i and W_f which are differently focussed. The resonance frequency ω_0 is then given by

$$\omega_0 = (W_i - W_f)/\hbar \tag{2}$$

and one expects a resonance curve similar to that shown in Figure 6, in which the transition probability for a particle of spin $1/2$ is plotted as a function of frequency.

From a quantum-mechanical point of view, the oscillating character of the transition probability in Figures 5 and 6 is the result of the cross term in the calculation of the transition probability from probability amplitudes. Let C_{iif} be the probability amplitude for the nucleus to pass through the first oscillatory field region with the initial state i unchanged but for there to be a transition to state ϕ in the final field, whereas C_{iff} is the amplitude for the alternative path with the transition to the final state ϕ being in the first field with no change in the second. The cross term $C_{iif}C^*_{iff}$ produces an interference pattern and gives the narrow oscillatory pattern of the transition probability shown in the curves of Figures 5 and 6. Alternatively the pattern can in part be interpreted as resulting from the Fourier spectrum of an oscillating field which is on for a time τ, off for T and on again for τ, as in Figure 4,. However, the Fourier interpretation is not fully valid since with finite rotations of J, the problem is a non-linear one. Furthermore, the Fourier interpretation obscures some of the key advantages of the separated oscillatory field method. I have calculated the quantum mechanical transition probabilities[3,6,7,8] and these calculations provide the basis for Figure 6.

The separated oscillatory field method has a number of advantages including the following:

(1) The resonance peaks are only 0.6 as broad as the corresponding ones with the single oscillatory field method. The narrowing is somewhat analogous to the peaks in a two slit optical interference pattern being

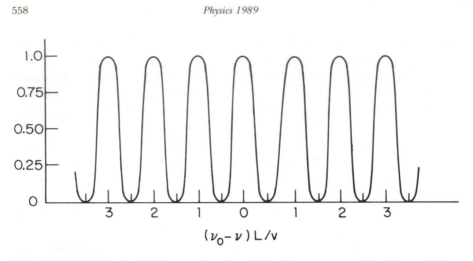

Figure 5. Transition probability as a function of the frequency $v = \omega/2\pi$ that would be observed in a separated oscillatory field experiment if all the molecules in the beam had a single velocity.

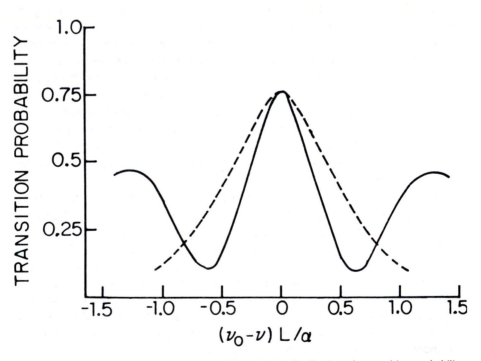

Figure 6. When the molecules have a Maxwellian velocity distribution, the transition probability is as shown by the full line for optimum rotating field amplitude. (L is the distance between oscillating field regions, α is the most probable molecular velocity and v is the oscillatory frequency $= \omega/2\pi$). The dashed line represents the transition probability with the single oscillating field method when the total duration is the same as the time between separated oscillatory field pulses.

narrower than the central diffraction peak of a single wide slit whose width is equal to the separation of the two slits.

(2)   The sharpness of the resonance is not reduced by non-uniformities of the constant field since both from the qualitative description and from the theoretical quantum analysis, it is only the space average value of the energies along the path that enter Eq. (2) and are important.

(3)   The method is more effective and often essential at very high frequencies where the wave length of the radiation used may be comparable to or smaller than the length of the region in which the energy levels are studied.

(4)   Provided there is no unintended phase shift between the two oscillatory fields, first order Doppler shifts and widths are eliminated.

(5)   The method can be applied to study energy levels in a region into which an oscillating field can not be introduced; for example, the Larmor precession frequency of neutrons can be measured while they are inside a magnetized iron block.

(6)   The lines can be narrowed by reducing the amplitude of the rotating field below the optimum, as shown by the dotted curve in Figure 6. The narrowing is the result of the low amplitude favoring slower than average molecules.

(7)   If the atomic state being studied decays spontaneously, the separated oscillatory field method permits the observation of narrower resonances than those anticipated from the lifetime and the Heisenberg uncertainty principle provided the two separated oscillatory fields are sufficiently far apart; only states that survive sufficiently long to reach the second oscillatory field can contribute to the resonance. This method, for example, has been used by Lundeen and others[9] in precise studies of the Lamb shift.

The advantages of the separated oscillatory field method have led to its extensive use in molecular and atomic beam spectroscopy. One of the best known is in atomic cesium standards of frequency and time which will be discussed later.

Although in most respects, the separated oscillatory field method offers advantages over a single oscillatory field, there are sometimes disadvantages. In studying complicated overlapping spectra the subsidiary maxima of Figure 6 can cause confusion. Furthermore, it is sometimes difficult at the required frequency to obtain sufficient oscillatory field strengths with two short oscillatory fields, whereas adequate field strength may be achieved with a weaker, longer oscillatory field. Therefore for most molecular beam resonance experiments, it is best to have both separated oscillatory fields and a single long oscillatory field available so the most suitable method under the circumstances can be used.

As in any high precision experiment, care must be exercised with the separated oscillatory field method to avoid obtaining misleading results. Ordinarily these potential distortions are more easily understood and eliminated with the separated oscillatory field method than are their counterparts in most other high-precision spectroscopy. Nevertheless, the effects are important and require care in high-precision measurements. I have discussed the various effects in detail elsewhere[3,7,8,10] but I will briefly summarize them here.

Variations in the amplitudes of the oscillating fields from their optimum values may markedly change the shape of the resonance, including the replacement of a maximum transition probability by a minimum. However, symmetry about the exact resonance frequency is preserved, so no measurement error need be introduced by such amplitude variations.[7,8]

Variations of the magnitude of the fixed field between, but not in, the oscillatory field regions do not ordinarily distort a molecular beam resonance provided the average transition frequency (Bohr frequency) between the two fields equals the values of the transition frequencies in each of the two oscillatory field regions alone. If this condition is not met, there can be some shift in the resonance frequency.[7,8]

If, in addition to the two energy levels between which transitions are studied, there are other energy levels partially excited by the oscillatory field, there will be a pulling of the resonance frequency as in any spectroscopic study and as analyzed in detail in the literature.[3,7,8]

Even in the case when only two energy levels are involved, the application of additional rotating magnetic fields at frequencies other than the resonance frequency will produce a net shift in the observed resonance frequency, as discussed elsewhere.[3,7,8] A particularly important special case is the effect identified by Bloch and Siegert[11] which occurs when oscillatory rather than rotating magnetic fields are used. Since an oscillatory field can be decomposed into two oppositely rotating fields, the counter-rotating field component automatically acts as such an extraneous rotating field. Another example of an extraneously introduced rotating field is that which results from the motion of an atom through a field H_0 whose direction varies in the region traversed. The theory of the effects of additional rotating fields at arbitrary frequencies has been developed by Ramsey,[7,8,10,12] Winter,[10] Shirley,[13] Code,[12] and Greene.[14]

Unintended relative phase shifts between the two oscillatory field regions will produce a shift in the observed resonance frequency.[13,14,15] This is the most common source of possible error, and care must be taken to avoid it either by eliminating such a phase shift or by determining the shift — say by measurements with the molecular beam passing through the apparatus first in one direction and then in the opposite direction.

A number of extensions to the separated oscillatory field method have been made since its original introduction:

(1) It is often convenient to introduce phase shifts deliberately to modify the resonance shape.[15] As discussed above, unintended phase shifts can cause distortions of the observed resonance, but some distortions are useful. Thus, if the change in transition probability is observed when the relative phase is shifted from $+\pi/2$ to $-\pi/2$ one sees a dispersion curve shape[15] as in Figure 7. A resonance with the shape of Figure 7 provides maximum sensitivity for detecting small shifts in the resonance frequency.

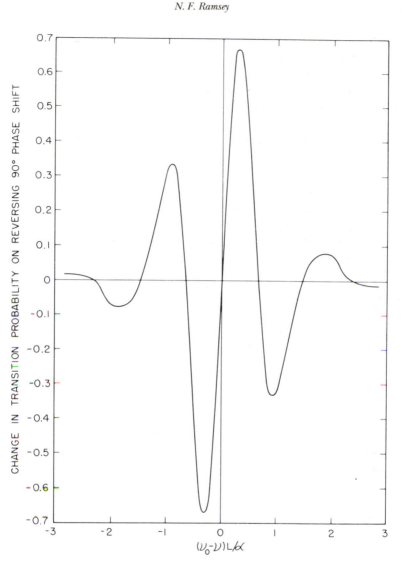

Figure 7. Theoretical change in transition probability on reversing a $\pi/2$ phase shift. At the resonance frequency there is no change in transition probability, but the curve at resonance has the steepest slope.

(2)   For most purposes the highest precision can be obtained with just two oscillatory fields separated by the maximum time, but in some cases it is better to use more than two separated oscillatory fields.[4] The theoretical resonance shapes[7] with two, three, four and infinitely many oscillatory fields are given in Figure 8. The infinitely many oscillatory field case, of course, by definition becomes the same as the single long oscillatory field if the total length of the transition region is kept the same and the infinitely many oscillatory fields fill in the transition region continuously as we assumed in

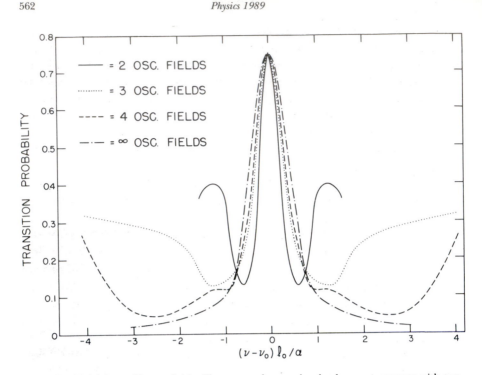

Figure 8. Multiple oscillatory fields. The curves show molecular beam resonances with two, three, four and infinitely many successive oscillating fields. The case with an infinite number of oscillating fields is essentially the same as Rabi's single oscillatory field method.

Figure 8. For many purposes this is the best way to think of the single oscillatory field method, and this point of view makes it apparent that the single oscillatory field method is subjected to complicated versions of all the distortions discussed in the previous section. It is noteworthy that, as the number of oscillatory field regions is increased for the same total length of apparatus, the resonance width is broadened; the narrowest resonance is obtained with just two oscillatory fields separated the maximum distance apart. Despite this advantage, there are valid circumstances for using more than two oscillatory fields. With three oscillatory fields the first and largest side lobe is suppressed, which may help in resolving two nearby resonances; for a larger number of oscillatory fields additional side lobes are suppressed, and in the limiting case of a single oscillatory field there are no side lobes. Another reason for using a large number of successive pulses can be the impossibility of obtaining sufficient power in a single pulse to induce adequate transition probability with a small number of pulses.

(3)   The earliest use of the separated oscillatory field method involved two oscillatory fields separated in space, but it was early realized that the method with modest modifications could be generalized to a method of successive oscillatory fields with the separation being in time, say by the use of coherent pulses.[16]

(4)   If more than two successive oscillatory fields are utilized it is not necessary to the success of the method that they be equally space in time;[4] the only requirement is that the oscillating fields be coherent — as is the case if the oscillatory fields are all derived from a single continuously running oscillator. In particular, the separation of the pulses can even be random,[16] as in the case of the large box hydrogen maser[17] discussed later. The atoms being stimulated to emit move randomly into and out of the cavities with oscillatory fields and spend the intermediate time in the large container with no such fields.

(5)   The full generalization of the successive oscillatory field method is excitation by one or more oscillatory fields that vary arbitrarily with time in both amplitude and phase.[7,8]

(6)   V. F. Ezhov and his colleagues,[6,18] in a neutron-beam experiment, used an inhomogeneous static field in the region of each oscillatory field region such that initially when the oscillatory field is applied conditions are far from resonance. Then, when the resonance condition is slowly approached, the magnetic moment that was originally aligned parallel to H_0 will adiabatically follow the effective magnetic field on a coordinate system rotating with H_0 until at the end of the first oscillatory field region the moment is parallel to H_1. This arrangement has the theoretical advantage that the maximum transition probability can be unity even with a velocity distribution, but the method may be less well adapted to the study of complicated spectra.

(7)   I emphasized earlier that one of the principal sources of error in the separated oscillatory field method is that which arises form uncertainty in the exact value of the relative phase shift in the two oscillatory fields. Jarvis, *et al.*[19] have pointed out that this problem can be overcome with a slight loss in resolution by driving the two cavities at slightly different frequencies so that there is a continual change in the relative phase. In this case the observed resonance pattern will change continuously from absorption to dispersion shape. The envelope of these patterns, however, can be observed and the position of the maximum of the envelope is unaffected by relative phase shifts. Since the envelope is about twice the width of a specific resonance there is some loss of resolution in this method, but in certain cases this loss may be outweighed by the freedom from phase-shift errors.

(8)   The method has been extended to electric as well as magnetic transitions and to optical laser frequencies as well as radio- and microwave-frequencies. The application of the separated oscillatory field method to optical frequencies requires considerable modifications because of the short wave lengths, as pointed out by Blaklanov, Dubetsky and Chebotsev[20] Successful applications of the separated oscillatory field method to lasers have been made by Bergquist,[21] Lee,[21] Hall,[21] Salour,[22] Cohen-Tannoudji,[22] Bordé,[23] Hansch,[24] Chebotayev[25] and many others.[25]

(9)   The method has been extended to neutron beams and to neutrons stored for long times in totally reflecting bottles.

(10)   In a recent beautiful experiment, S. Chu and his associates.[26] have

successfully used the principle of separated oscillatory fields with a fountain of atoms that rises up slowly, passes through an oscillating field region, falls under gravity and passes again through the same oscillatory field region. This fountain experiment was attempted many years ago by J. R. Zacharias and his associates,[3] but it was unsuccessful because of the inadequate number of very slow atoms. Chu and his collaborators used laser cooling[27,28,29] to slow the atoms to a low velocity and obtained a beautifully narrow separated oscillatory fields resonance pattern.

THE ATOMIC HYDROGEN MASER

The atomic hydrogen maser grew out of my attempts to obtain even greater accuracy in atomic beam experiments. By the Heisenberg uncertainty principle (or by the Fourier transform), the width of a resonance in a molecular beam experiment cannot be less than approximately the reciprocal of the time the atom is in the resonance region of the apparatus. For atoms moving through a 1 m long resonance region at 100 m/s this means that the resonance width is about 100 Hz wide. To decrease this width and hence increase the precision of the measurements required an increase in this time. To increase the time by drastically lengthening the apparatus or selecting slower molecules would decrease the already marginal beam intensity or greatly increase the cost of the apparatus. I therefore decided to plan an atomic beam in which the atoms, after passing through the first oscillatory field would enter a storage box with suitably coated walls where they would bounce around for a period of time and then emerge to pass through the second oscillatory field. My Ph.D. student, Daniel Kleppner,[30] undertook the construction of this device as his thesis project. The original configuration required only a few wall collisions and was called a broken atomic beam resonance experiment. Initially the beam was cesium and the wall coating was teflon. The experiment[30] was a partial success in that a separated oscillatory field pattern for an atomic hyperfine transition was obtained, but it was weak and disappeared after a few wall collisions. The results improved markedly when paraffin was used for the wall coating and a hyperfine resonance was eventually obtained after 190 collisions giving a resonance width of 100 Hz, but with the resonance frequency shifted by 150 Hz.

To do much better than this, we decided we would have to use an atom with a lower mass and a lower electric polarizability to reduce the wall interactions. Atomic hydrogen appeared ideal for this purpose, but atomic hydrogen is notoriously difficult to detect. We, therefore, calculated the possibility of detecting the transitions through their effects on the electromagnetic radiations. Townes[31] had a few years earlier made the first successful maser (acronym for microwave amplifier by stimulated emission of radiation) but no one had previously made a maser based on a magnetic dipole moment or on a frequency as low as that of an atomic hyperfine transition. We concluded, however, that if the resonance could be made narrow enough by multiple wall collisions, we should be able to obtain

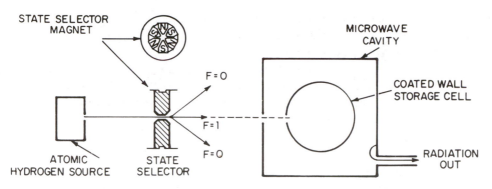

Figure 9. Schematic diagram of atomic hydrogen maser. Only the paths of the m=0 atoms are shown since the m=1 atoms are not involved in the Δm=0 transitions studied.

maser oscillations. The apparatus was designed and constructed by Goldenberg, Kleppner and myself[32] and after a few failures we obtained maser oscillations at the atomic hydrogen hyperfine frequency. Both the proton and the electron have spin angular momenta **I** and **J** as well as magnetic moments. The atomic hyperfine transitions are those for which there is a change of the relative orientation of these two magnetic moments between the initial and final states in Eq. (2). We studied H atoms in the $1^2S_{1/2}$ ground electronic state and mostly observed the transitions (F=1, m=0 → F=0, m=0) where F is the quantum number of the total angular momentum **F = I + J** and m is the associated magnetic quantum number.

The principles of an atomic hydrogen maser are shown schematically in Figure 9. An intense electrical discharge in the source converts commercially available molecular hydrogen (H_2) into atomic hydrogen (H). The atoms emerge from the source into a region that is evacuated to 10^{-6} torr and enter a state selecting magnet which has three north poles alternating in a circle with three south poles. By symmetry, the magnetic field is zero on the axis and increases in magnitude away from the axis. Since the energy of a hydrogen atom in the F=1 m=0 state increases with energy and since mechanical systems are accelerated toward lower potential energy, an atom in F=1 state that is slightly off axis will be accelerated toward the axis, i.e. the F=1 state will be focussed onto the small aperture of the 15 cm diameter storage cell whereas the F=0 state is defocussed. As a result, if the atomic beam flows steadily, the storage bottle in equilibrium will contain more high energy F=1 atoms than low. If these atoms are exposed to microwave radiation at the hyperfine frequency, more atoms are stimulated to go from the higher energy state to the lower one than in the opposite direction. Energy is then released from the atoms and makes the microwave radiation stronger. Thus the device is an amplifier or maser. If the storage cell is placed inside a tuned cavity, an oscillation at the resonance frequency will increase in magnitude until an equilibrium value is reached. At this level the oscillation will continue indefinitely, with the energy to maintain the oscillation coming from the continuing supply of hydrogen atoms in the high energy hyperfine state. The device then becomes a free running maser oscillator at the atomic hyperfine frequency.

The atomic hydrogen maser oscillator has unprecedented high stability due to a combination of favorable features. The atoms typically reside in the storage cell for 10 seconds, which is much longer than in an atomic beam resonance apparatus so the resonance line is much narrower. The atoms are stored at low pressure so they are relatively free and unperturbed while radiating. The first order Doppler shift is removed, since the atoms are exposed to a standing wave and since the average velocity is extremely low for atoms stored for 10 seconds. Masers have very low noise levels, especially when the amplifying elements are isolated atoms. Over periods of several hours the hydrogen maser stability is better than 1×10^{-15}.

The major disadvantage of the hydrogen maser is that the atoms collide with the walls at intervals, changing slightly the hyperfine frequency and giving rise to wall shifts of 1×10^{-11}. However, the wall shifts can be experimentally determined by measurements utilizing storage bottles of two different diameters or with a deformable bulb whose surface to volume ratio can be altered. As in all precision measurements, care must be taken in adjusting and tuning the hydrogen maser to avoid misleading results. These limitations and precautions are discussed in a series of publications by various authors.[32,33,34] The designs of hydrogen masers have been modified in many ways either for special purposes or for increased stability and reliability. For example different hyperfine transitions have been used and masers have been operated in relatively strong magnetic fields. A hydrogen maser has also been operated[17] with a storage bottle that is much larger than the wave length of the stimulating radiation by confining the microwave power to two small cavities so that it functions as a separated oscillatory field device. As shown in Figure 10 the atoms that are stimulated to emit radiation move randomly into and out of the two oscillatory field cavities and spend the intermediate time in the large container where there is no oscillatory field. Due to the larger size of the storage box there are longer storage times and less frequent wall collisions, so the resonances are narrower and the wall shifts are smaller than for a normal hydrogen maser.

PRECISION SPECTROSCOPY

Now that I have discussed extensively the principles of the separated oscillatory field method and of the atomic maser, I shall give some illustrations as to their value. One major category of applications is to precision spectroscopy, especially at radio and microwave frequencies. Another category of applications is to atomic clocks and frequency standards.

It is difficult to summarize the spectroscopic applications since there are so many of them. Many beautiful experiments have been done by a large number of scientists in different countries, including Sweden. I shall, therefore use just a few illustrations from experiments in which I have been personally involved.

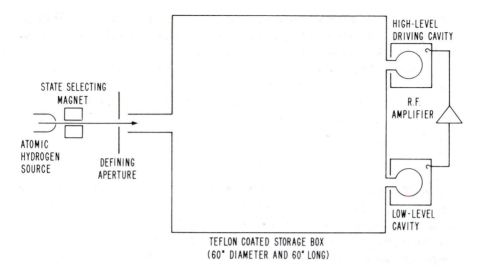

Figure 10. Schematic diagram of a large box hydrogen maser. The two cavities on the right act as two separated oscillating fields with that of the high level cavity being obtained by amplification from the low.

My graduate students have made precision measurements of the radiofrequency spectra of different molecules in various rotational states. For each of these states more than seven different molecular properties can be inferred and thus the variations of the properties with changes in the rotational and vibrational quantum numbers can be determined. These properties include nuclear and rotational magnetic moments, nuclear quadrupole interactions, nuclear spin-spin magnetic interactions, spin rotational interactions, etc. I shall illustrate the accuracy and significance of the measurements with a single example. With both D_2 and LiD we have accurately measured[35,36] the deuteron quadrupole interaction eqQ where e is the proton electric charge, q is the gradient of the molecular electric field at the deuteron and Q is the deuteron quadrupole moment which measures the shape of the deuteron and in particular its departure from spherical symmetry. These measurements were made with a high resolution molecular beam apparatus based on the method of separated oscillatory fields. We found for eqQ the value $+225{,}044 \pm 20$ Hz in D_2 and $+34{,}213 \pm 33$ Hz in LiD. Since q has been calculated[37,38] for each of these quite different molecules, two independent values of Q can be calculated. The results agree to within 1.5% which confirms the validity of the difficult calculation; with it we find $Q = 2.9 \times 10^{-27}$ cm^2.

In an experiment with collaborators[39] at the Institut Laue-Langevin at Grenoble, France, we have used the separated oscillatory field method with a beam of slow neutrons to make an accurate measurement of the neutron magnetic moment and found[37,40] it to be $-1.91304275 \pm 0.00000045$ nuclear magnetons. In a somewhat different experiment with neutrons moving so slowly that they can be bottled for more than 80 s in a suitable storage vessel, we have used the method of successive oscillatory fields with the two coherent radiofrequency pulses being separated in time rather than space. In this manner and as a fundamental test of time reversal symmetry, we have recently set a very low upper limit for the neutron electric dipole moment by finding[41] its value to be $(-3 \pm 5) \times 10^{-26}$ e cm.

The atomic hydrogen maser gives very accurate data on the microwave spectrum of the ground electronic state of the hydrogen atom. The hyperfine frequency Δv for atomic hydrogen has been measured in our laboratory and in a number of other laboratories. The best value[42,43] is

$$\Delta v_H = 1,420,405,751.7667 \pm 0.0009 \text{ Hz}$$

This value agrees with present quantum electromagnetic theory[44] to within the accuracy of the theoretical calculation and can be used to obtain information on the proton structure. Similarly accurate values have been found for atomic deuterium and tritium and the dependence[45] of these results on the strengths of externally applied electric fields have been measured. With a modified form of the hydrogen maser designed to operate at high magnetic fields, the ratio of magnetic moment of the electron to that of the proton is found[40,46] to be -658.210688 ± 0.000006. Incidentally when this result is combined with the beautiful electron measurements from Professor Dehmelt's labortory[40,47,48] we obtain the best values for the free proton magnetic moment in both Bohr and nuclear magnetons.

ATOMIC CLOCKS

In the past 50 years there has been a major revolution in time keeping with accuracy and reproducibility of the best clocks at the end of that period being approximately a million times those at the beginning. This revolution in time keeping and frequency control is due to atomic clocks.

Any clock or frequency standard depends on some regular periodic motion such as the pendulum of the grandfather's clock. In the case of atomic clocks the periodic motion is internal to the atoms and is usually that associated with an atomic hyperfine structure as discussed in the section on atomic hydrogen maser.

In the most widely used atomic clocks, the atom whose internal frequency provides the periodicity is cesium and the usual method of observing it is with a separated oscillatory field magnetic resonance apparatus as in Figure 2. The first commercial cesium beam clock was developed in 1955 by a group led by J. R. Zacharias[4] and in the same year L. Essen and V. L. Parry[4] constructed and operated the first cesium beam apparatus that was extensively used as an actual frequency standard. Subsequently many scientists and engineers throughout the world contributed to the development of atomic clocks, as discussed in greater detail elsewhere.[4]

Cesium atomic clocks now have an accuracy and stability of about 10^{-13} which was so far superior to all previous clocks that in 1967 the internationally adopted definition of the second was changed from one based on motion of the earth around the sun to 9,192,631,770 periods of the cesium atom.

For many purposes even greater stability is required over shorter time intervals. When such stability is needed the hydrogen maser is frequently used with a stability of 10^{-15} over periods of several hours.

Atomic clocks based on the above principles have for a number of years provided clocks of the greatest stability and accuracy and these are sufficiently great that further improvements might seem to neither be desirable nor feasible. But as we shall see in our final section, there are applications that already push atomic clocks to their limits and there are many current developments with great promise for the future. These include improvements to the existing devices, use of higher frequency, use of lasers, electromagnetic traps for storing both ions and atoms, laser cooling, etc.

APPLICATIONS FOR ACCURATE CLOCKS

Accurate atomic clocks are used for so many different purposes that a list of them all is tediously long so I shall here just briefly mention a few that push clock technology to its limit.

In radio astronomy one looks with a parabolic reflector at the radio waves coming from a star just as in optical astronomy one looks with an optical telescope at the light waves coming from a star. Unfortunately, in radio astronomy the wavelength of the radiation is about a million times longer than the wavelength of light. The resolution of the normal radio telescope is therefore about a million times worse since the resolution of a telescope depends on the ratio of the wave length to the telescope aperture. However, if there are two radio telescopes on opposite sides of the earth looking at the same star and if the radio waves entering each are matched in time, it is equivalent to a single telescope whose aperture is the distance between the two telescopes and the resolution of such a combination exceeds that of even the largest single optical telescope. However, to do such precise matching in time each of the two radio telescopes needs a highly stable clock, usually an atomic hydrogen maser.

One of the exciting discoveries in radio astronomy has been the discovery of pulsars, that emit their radiation in short periodic pulses. Precision clocks have been needed to measure the pulsar periods and the changes in the periods with time; these changes sometimes occur smoothly and sometimes abruptly. Of particular interest from the point of view of time measurements, are the millisecond pulsars which have remarkable constancy of period, rivaling the stability of the best atomic clocks.[49] Another millisecond pulsar is part of a rapidly rotating binary star that is slowly changing its period of rotation.[49] This slow change in rotation can be attributed to the loss of energy by the radiation of gravity waves — the first experimental evidence for the existence of gravity waves.

Time and frequency can now be measured so accurately that wherever possible other fundamental measurements are reduced to time or frequency measurements. Thus the unit of length by international agreement has recently been defined as the distance light will travel in a specified time and voltage will soon be represented in terms of frequency measurements.

Accurate clocks have provided important tests of both the special and general theories of relativity. In one experiment, a hydrogen maser was shot in a rocket to a 6,000 mile altitude and its periodic rate changed with speed and altitude just as expected by the special and general theories of relativity.[50] In other experiments, observers have measured the delays predicted by relativity for radio waves passing near the sun.

Precision clocks make possible an entirely new and more accurate navigational system, the global positioning system or GPS. A number of satellites containing accurate atomic clocks transmit signals at specific times so any observer receiving and analyzing the signals from four such satellites can determine his position to within ten yards and the correct time within one hundredth of a millionth of a second (10^{-8} s).

A particularly fascinating navigation feat dependent on accurate clocks was the recent and highly successful tour of the Voyager spacecraft to Neptune. The success of this mission depended upon the ground controllers having accurate knowledge of the position of the Voyager. This was accomplished by having three large radio telescopes at different locations on the earth, each of which transmitted a coded signal to Voyager which in turn transmitted the signals back to the telescopes. The distances from each telescope to Voyager could be determined from the elapsed times and thus Voyager could be located. To achieve the required timing accuracy, two hydrogen masers were located at each telescope. Due to the rotation of the earth in the eight hours required for the electromagnetic wave to travel from the earth to Voyager and back again at the speed of light, the telescope transmitting the signal in some cases had to be different from the one receiving; this placed an additional stringent requirement on the clocks. Thus, the spectacular success of the Voyager mission was depended on the availability of highly stable clocks.

REFERENCES

1. I. I. Rabi, J. R. Zacharias, S. Millman and P. Kusch, Phys. Rev. *53*, 318 (1938) and *55*, 526 (1939).
2. J. M. B. Kellogg, I. I. Rabi, N. F. Ramsey and J. R. Zacharias, Phys. Rev. *55*, 729 (1939); *56*, 728 (1939) and *57*, 677 (1940).
3. N. F. Ramsey, *Molecular Beams*, Oxford Press (1956 and 1985).
4. N. F. Ramsey, *History of Atomic Clocks*, Journal of Research of NBS *88*, 301 (1983). This paper contains an extensive list of references.
5. N. F. Ramsey, Phys. Rev. *76*, 996 (1949) and *78*, 695 (1950).
6. N. F. Ramsey, Physics Today *33* (7), 25 (July 1980).
7. N. F. Ramsey, Phys. Rev. *109*, 822 (1958).
8. N. F. Ramsey, Jour. Phys. et Radium *19*, 809 (1958).
9. S. R. Lundeen, P. E. Jessop and F. M. Pipkin, Phys. Rev. Lett. *34*, 377 and 1368 (1975).
10. N. F. Ramsey, Phys. Rev. *100*, 1191 (1955).
11. F. Bloch and A. Siegert, Phys. Rev. *57*, 522 (1940).
12. R. F. Code and N. F. Ramsey, Phys. Rev. *A4*, 1945 (1971).
13. J. H. Shirley, J. Appl. Phys. *34*, 783 (1963).
14. G. Greene, Phys. Rev. *A18*, 1057 (1970).
15. N. F. Ramsey and H. B. Silsbee, Phys. Rev. *84*, 506 (1951).
16. N. F. Ramsey, Rev. Sci. Inst. *28*, 57 (1957).
17. E. Uzgiris and N. F. Ramsey, Phys. Rev. *A1*, 429 (1970).
18. V. F. Ezhov, S. N. Ivanov, I. M. Lobashov, V. A. Nazarenko, G. D. Porsev, A. P. Serebrov and R. R. Toldaev, Sov. Phys. − JETP *24*, 39 (1976).
19. S. Jarvis, D. J. Wineland and H. Hellwig, J. Appl. Phys. *48*, 5336 (1977).
20. Y. V. Blaklanov, B. V. Dubetsky and V. B. Chebotsev, Appl. Phys. *9*, 171 (1976).
21. J. C. Bergquist, S. A. Lee and J. L. Hall, Phys. Rev. Lett. *38*, 159 (1977) and Laser Spectroscopy *III*, 142 (1978).
22. M. M. Salour, C. Cohen-Tannoudji, Phys. Rev. Lett. *38*, 757 (1977); Laser Spectroscopy *III*, 149 (1978), Appl. Phys. *15*, 119 (1978) and Phys. Rev. *A17*, 614 (1978).
23. C. J. Bordé., C. R. Acad. Sci. Paris *284B*, 101 (1977).
24. T. W. Hansch, Laser Spectroscopy *III*, 149 (1978).
25. V. P. Chebotayev, A. V. Shishayev, B. Y. Yurshin, L. S. Vasilenko, N. M. Dyuba and M. I. Skortsov, Appl. Phys. *15*, 43, 219 and 319 (1987).
26. M. Kasevich, E. Riis, S. Chu and R. S. DeVoe, Phys. Rev. Lett. *63*, 612 (1989).
27. D. Wineland and H. Dehmelt, Bull. Am. Phys. Soc. *18*, 1521 (1973) and *20*, 60, 61, 637 (1975).
28. T. W. Hansch and A. L. Schawlow Opt. Commun. *13*, 68 (1975) and review by V. S. Letokhow, Comments on Atomic and Molecular Physics *6*, 119 (1977).
29. D. J. Wineland and W. M. Itano, Physics Today *40*, (6) 34 (June 1987).
30. D. Kleppner, N. F. Ramsey and P. Fjelstadt, Phys. Rev. Lett. *1*, 232 (1958).
31. J. P. Gordon, H. Z. Geiger and C. H. Townes, Phys. Rev. *95*, 282 (1954) and *99*, 1264 (1955).
32. H. M. Goldenberg, D. Kleppner and N. F. Ramsey, Phys. Rev. Lett. *5*, 361 (1960) and Phys. Rev. *126*, 603 (1962).
33. D. Kleppner, H. C. Berg, S. B. Crampton, N. F. Ramsey, R. F. C. Vessot, H. E. Peters and J. Vanier, Phys. Rev. *138*, A972 (1965).
34. J. M. V. A. Koelman, S. B. Crampton, H. T. C. Luiten and B. J. Verhaar, Phys. Rev. *A38*, 3535 (1988). This paper contains an extended series of references to other papers on hydrogen maser limitations, principles and practices.
35. R. F. Code and N. F. Ramsey, Phys. Rev. *A4*, 1945 (1971).

36. R. R. Freeman, A. R. Jacobson, D. W. Johnson and N. F. Ramsey, Jour. Chem. Phys. *63*, 2597 (1975).

37. R. V. Reid and M. L. Vaida, Phys. Rev. A7, 1841 (1973).

38. K. K. Docken and R. R. Freeman, J. Chem. Phys. *61*, 4217 (1974).

39. G. L. Green, N. F. Ramsey, W. Mampe, J. M. Pendlebury, K. Smith, W. B. Dress, P. D. Miller and P. Perrin, Phys. Rev. D*20*, 2139 (1979).

40. E. R. Cohen and B. Taylor, Rev. Mod. Phys. **59,** 1121 (1987).

41. K. F. Smith, N. Crampin, J. M. Pendlebury, D. J. Richardson, D. Shiers, K. Green, A. I. Kilvington, J. Moir, H. B. Prosper, D. Thompson, N. F. Ramsey, B. R. Heckel, S. K. Lamoreaux, P. Ageron, W. Mampe and A. Steyerl, Phys. Lett. 136B, 327 (1984) and Phys. Lett. *234*, 191 (1990).

42. H. Hellwig, R. F. Vessot, M. Levine, P. W. Zitzewitz, D. W. Allan and D. T. Glaze, IEEE Trans. Instruments and Measurements *IM-19*, 200 (1970).

43. L. Essen, M. J. Donaldson, M. J. Bangham and E. G. Hope, Nature *229*, 110 (1971).

44. G. L. Baldwin and D. R. Yennie, Phys. Rev. *D37*, 498 (1988).

45. P. C. Gibbons and N. F. Ramsey, Phys. Rev. *A5*, 73 (1972).

46. P. F. Winkler, D. Kleppner, T. Myint and F. G. Walther, Phys. Rev. *A5*, 83 (1972) and E. Cohen and B. Taylor, Phys. Lett. B204 (April 1988).

47. R. S. van Dyck, P. B. Schwinberg and H. Dehmelt, Atomic Physics *9*, 53 (1984) (World Scientific, Singapore).

48. R. S. van Dyck, F. L. Moore, D. L. Farnum and P. B. Schwinberg, Bull. Am. Phys. Soc. *31*, 244 (1986) and Atomic Physics 9, 75 (1984) (World Scientific, Singapore)

49. J. Taylor, *et al.*, Nature *277*, 437 (1979) and *315*, 547 (1985).

50. R. F. C. Vessot, *et al.*, Phys. Rev. Lett. *45*, 2081 (1980).

Hans Dehmelt

HANS G. DEHMELT

My father, Georg, had studied law at the Universität Berlin for some years, and in the first World War had been an artillery officer. He was of a philosophical bend of mind and a man of independent opinions. In the depth of the depression he just managed to make a living in real estate. When the family fortunes had shrunk to ownership of a heavily mortgaged apartment building located in an overwhelmingly Communist part of Berlin, it seemed reasonable to move into one of the apartments ourselves as nobody paid any rent. Cannons were deployed on the streets on occasion and the class war had entered the class rooms. After a few bloody noses administered by a burly repeater, I shifted my interests from roaming the streets more towards playing with rudimentary radio receivers and noisy and smelly experiments in my mother's kitchen. In the spring of 1933 my mother, a very energetic lady, saw to it that, at the age of ten, I entered the Gymnasium zum Grauen Kloster, the oldest Latin school in Berlin, which counted Bismarck amongst its Alumni. This involved a stiff entrance examination and I was admitted on a scholarship. My father at that time expressed the opinion that I probably would be happier as a plumber. However, he apparently didn't quite believe this himself. Thus, in years before, he had bought me an erector set and books on the lives of famous inventors and Greek mythology, and when I was ill he had given me the encyclopedia to read. I supplemented the school curriculum with do-it-yourself radio projects until I had hardly any time left for my class work. Only tutoring from my father rescued me from disaster. Reading popular radio books deepened my interest in physics. While physics was taught at the Kloster only in the later grades, in the public library I read books with titles such as "Umsturz im Weltbild der Physik" and learned about the Balmer series and Bohr's energy levels of the hydrogen atom. My teachers at the Kloster were excellent, I remember in particular Dr. Richter, who taught Latin and Greek, and Dr. Splettstoesser, who taught biology and physics. Richter liked to expand on the classical works, which we were reading in class. I spent most of the ample breaks in related intense discussions with a group of classmates, Heppke, Hübner, Landau and Leiser while others engaged in boxing matches. Splettstoesser was a working scientist who spent Summers as a visitor with a marine biology institute on the Adriatic. I jumped a term and graduated in the spring of 1940.

Having received a notice from the draft board, I found it wise to volunteer for the anti-aircraft artillery and a motorized unit. I was not able to serve as a radio man but was assigned to a gun crew and never rose above the rank of senior private. Sent to relieve the German armies at Stalingrad, my battery was extremely lucky to escape the encirclement. A few months later I was even more lucky to be ordered back to Germany to study physics under an army program at the Universität Breslau in 1943. After one year of study, I was sent to the Western Front and captured in the Battle of the Bulge. I spent a year in an American prisoner of war camp in France and was released early in 1946. Supporting myself with the repair and barter of prewar radios, I took up my study of physics again at the Universität Göttingen. Here I attended lectures by Pohl, Richard Becker, Hans Kopfermann and Werner Heisenberg; Max v. Laue and Max Planck attended the physics colloquia. At the funeral of Planck I was chosen to be one of the pall bearers. At the university, I greatly enjoyed repeating the Frank-Hertz experiment, the Millikan oil drop, Zeemann effect, Hull's magnetron, Langmuir's plasma tube and other classic modern physics experiments in an excellent laboratory class run by Wolfgang Paul. In one of his Electricity & Magnetism classes Becker drew a dot on the blackboard and declared "Here is an electron ..." Having heard in another class that the wave function of an electron at rest spreads out over all of space, and having read about ion trapping in radio tubes in my teens set me to wonder how one might realize Becker's localization feat in the laboratory. However, that had to wait a while. In 1948, in Kopfermann's Institute, which was heavily oriented towards hyperfine structure studies, I completed an experimental Diplom-Arbeit (master's thesis) on a Thomson mass spectrograph under Peter Brix. The results were published in "Die photographischen Wirkungen mittelschneller Protonen II," the first paper of which I was a (co)author. Soon thereafter, I began work on my doctoral thesis under Hubert Krüger in the same Institute. Well prepared by a series of excellent Institute seminars on the NMR work of Bloch and of Purcell, we were able to successfully compete with workers at Harvard University. In 1949 we discovered Nuclear Quadrupole Resonance and reported it in our paper "Kernquadrupolfrequenzen in festem Dichloraethylen." My doctoral thesis had the title "Kernquadrupolfrequenzen in kristallinen Jodverbindungen." This work led to an invitation to join Walter Gordy's well known microwave laboratory at Duke University as postdoctoral associate.

At Duke I had the pleasure of making the acquaintance of James Frank, Fritz London, Lothar Nordheim and Hertha Sponer. I advised Hugh Robinson, a graduate student of Gordy's in an NQR experiment, did my own research and also contributed some NMR expertise to an experiment by Bill Fairbank and Gordy on spin statistics in ^{3}He/^{4}He mixtures, gaining some very useful low temperature experience in this brief collaboration. Through Gordy's and Nordheim's good offices I was able to receive a visiting assistant professor appointment at the University of Washington with a charge to advise Edwin Uehling's students during his sabbatical and to do independent research. I had built my first electron impact tube during a brief interlude in 1955 in George Volkoff's laboratory at the University of British Columbia. Prior to that I had attempted a paramagnetic resonance experiment on free atoms in Göttingen and succeeded in doing so at Duke. During seminars at Göttingen on the magnetic resonance techniques of Rabi and of Kastler, it had occurred to me that because of the analogy between an atom and a radio dipole antenna, (a), *alignment* of the atom should show up in its optical absorption cross section, and (b), electron impact should produce *aligned* excited atoms. I put these two ideas to good use in 1956 in Seattle in an experiment entitled "Paramagnetic Resonance Reorientation of Atoms and Ions Aligned by Electron Impact." In this paper I first pointed out the usefulness of *ion trapping for high resolution spectroscopy* and mentioned the 1923 Kingdon trap as a suitable device. This work also brought me into close contact with spin exchange between electron and target atom, which gave me the idea for my 1958 experiment "Spin Resonance of Free Electrons Polarized by Exchange Collisions." However, first I had to learn how to produce polarized atoms, which could then transfer their orientation to trapped electrons. Falling back on buffer gas techniques developed in my 1955 Duke paper "Atomic Phosphorus Paramagnetic Resonance Experiment," I quickly demonstrated in my 1956 Seattle paper "Slow Spin Relaxation of Optically Polarized Sodium Atoms" how to efficiently produce and monitor a polarized atom cloud. Trapping the electrons in a neutralizing ion cloud slowly diffusing in the buffer gas, I was able to carry out the spin resonance experiment. My optical transmission monitoring scheme proved also very useful in the development of rubidium vapor magnetometers and frequency standards by Earl Bell and Arnold Bloom at Varian Associates, in which I acted as a consultant. The rubidium frequency standard is still the least expensive, smallest and most widely used commercial atomic frequency standard. The thesis "Experimental Upper Limit for the Permanent Electric Dipole Moment of Rb^{85} by Optical Pumping Techniques" of my first graduate student, Earl Ensberg, also made use of these novel optical pumping schemes and was finished in 1962. These early results were improved orders of magnitude by my doctoral student Philip Ekstrom in his 1971 thesis "Search for Differential Linear Stark Shift in Cs^{133} and Rb^{85} Using Atomic Light Modulation Oscillators."

I was not satisfied with the plasma trapping scheme used for the electrons and asked my student, Keith Jefferts, to study ion trapping in an electron beam traversing a field free vacuum space between two grids. Also, I began to focus on the magnetron/Penning discharge geometry, which, in the Penning ion gauge, had caught my interest already at Göttingen and at Duke. In their 1955 cyclotron resonance work on photoelectrons in vacuum Franken and Liebes had reported undesirable frequency shifts caused by accidental electron trapping. Their analysis made me realize that in a pure electric quadrupole field the shift would not depend on the location of the electron in the trap. This is an important advantage over many other traps that I decided to exploit. A magnetron trap of this type had been briefly discussed in J. R. Pierce's 1949 book, and I developed a simple description of the axial, magnetron, and cyclotron motions of an electron in it. With the help of the expert glassblower of the Department, Jake Jonson, I built my first high vacuum magnetron trap in 1959 and was soon able to trap electrons for about 10 sec and to detect axial, magnetron and cyclotron resonances. About the same time, my Göttinger colleague, Otto Osberghaus, sent me a research report on the Paul rf ion cage. This trap had very desirable properties for atomic ions and it did not require a magnetic field. Therefore, I asked my student, Fouad Major, to experiment with a simplified cylindrical version of such a trap in the hope that it might be useful in hfs resonance experiments on hydrogenic helium ions. The early results were very encouraging and Jefferts also switched to the Paul trap. In 1962, Jefferts and Major both finished their Doctoral Theses entitled respectively "Alignment of Trapped H_2^+ Molecular Ions by Selective Photodissociation" and "The Orientation of Electrodynamically Contained He^4 Ions." As a continuation of the latter, a new postdoc, Norval Fortson, Major and I published the 1966 paper "Ultrahigh Resolution $\Delta F = 0 \pm 1 \, {}^3He^+$ HFS Spectra by an Ion Storage — Exchange Collision Technique." My own attempts to detect the polarization of the electrons acquired from a polarized beam of alkali atoms in my Penning (magnetron) trap, described in a 1961 research report to the NSF "Spin Resonance of Free Electrons," were not so quickly successful. However in this work I was much impressed by *seeing* the beam of sodium atoms traversing my glass apparatus in the reflected light from a sodium vapor street lamp adapted as illuminating light source. Only a later concerted effort by Gräff and Werth at Bonn, reinforced by Major and Fortson, as visitors, made a similar spin resonance experiment work in 1968.

In the 1966 paper with Fortson and Major, I also proposed to develop an infrared laser based on ions in an rf trap. To this end my student, David Church, completed a thesis in 1969 entitled "Storage and Radiative Cooling of Light Ion Gases in RF Quadrupole Traps." In this work we demonstrated a race-track-shaped trap and cooled the ions by coupling to a resonant LC circuit. In parallel work my student, Stephan Menasian, in 1968, with some help from G. R. Huggett, succeeded in cooling Hg^+ ions in a race-track-trap with a helium buffer gas and in detecting them by optical absorption. Jefferts' research on hfs spectra of H_2^+ was continued in Seattle by my postdoc Charles Richardson and later by Menasian in his 1973 doctoral thesis "High Resolution Study of the $(1, \frac{1}{2}, \frac{1}{2}) - (1, \frac{1}{2}, \frac{3}{2})$ HFS Transition in H_2^+." The resolution in the ${}^3He^+$ hfs work was greatly enhanced in work with my colleague Fortson and my postdoc Hans Schuessler. Realizing in 1961 that precision measurements of the electron magnetic moment would require a large magnetic field and that Becker's electron localization feat might be approximated in a Penning trap, I began to consider other avenues for magnetic resonance experiments. Some success in the electron work, achieved with the help of my new student, Fred Walls, was described in our 1968 paper "'Bolometric' Technique for the RF Spectroscopy of Stored Ions." I reviewed the work on ions and electrons up to 1968 in two articles "Radiofrequency Spectroscopy of Stored Ions."

The able assistance of two postdocs, David Wineland and my former student Phil Ekstrom, made the isolation of a single electron become a reality in 1973 with our paper "Monoelectron Oscillator." Measuring its magnetic moment was another story. At Göttingen in the late forties I had attended a seminar given by Helmut Friedburg, a doctoral Student of Wolfgang Paul, on focussing spins with a magnetic hexapole. This may be viewed as a refinement of the Stern-Gerlach effect. In subsequent discussions with fellow students a rumor of a Stern-Gerlach experiment for electrons was brought up, and also Bohr's and Pauli's thesis that such experiments were impossible in principle. Though it greatly piqued my interest, I could not understand this thesis. Stimulated by a 1927 paper of Brillouin on the subject, I followed another of the guiding principles formulated by Bohr: "In my Institute we take nothing absolutely serious, including this statement." In 1973 I proposed, together with Ekstrom, to monitor spin and cyclotron quantum numbers of the lone electron by means of the "continuous Stern-Gerlach effect" in an abstract "Proposed g-2/δv_z Experiment on Stored Single Electron or Positron." My new post-doc Robert Van Dyck, Philip Ekstrom and myself reported the first such experiment in our 1976 paper "Axial, Magnetron, and Spin-Cyclotron Beat Frequencies Measured on Single Electron Almost at Rest in Free Space (Geonium)." This work also already made use of the important technique of *side band cooling* of the electron. The demonstration of sideband cooling had eluded us in earlier attempts undertaken together with Walls and later with Wineland. Encouraged by the success of the monoelectron oscillator I had also published in 1973 an abstract "Proposed 10^{14} Δv < v Laser Fluorescence Spectroscopy on Tl^+ Mono-Ion Oscillator." Unfortunately, this proposal infuriated one of the agencies funding our research to the degree that they terminated their support almost immediately. I was rescued by a prize from the Humboldt Foundation and an invitation by Gisbert zu Putlitz to initiate the proposed laser spectroscopy project in his Institute at the Universität Heidelberg. As the fruit of these efforts a paper "Localized visible Ba^+ mono-ion oscillator" by Neuhauser, Hohenstatt, Toschek and myself appeared in 1980.

In 1981 Van Dyck, my doctoral student Paul Schwinberg and myself extended the electron work to its antiparticle in our paper "Preliminary Comparison of the Positron and Electron Spin Anomalies" and I reviewed it in an article "Invariant Frequency Ratios in Electron and Positron Geonium Spectra Yield Refined Data on Electron Structure." In 1986 we published a detailed paper "Electron Magnetic Moment from Geonium Spectra: Early Experiments and Background Concepts" and in 1987 our collaboration reported a 4 parts in 10^{12} resolution in the g factor for electron and positron in "New High-Precision Comparison of Electron and Positron g Factors." A very promising scheme to detect cyclotron excitation through the small relativistic mass increase accompanying it was published in a 1985 paper "Observation of Relativistic Bistable Hysteresis in the Cyclotron Motion of a Single Electron" together with my postdoc, Gerald Gabrielse, and William Kells, a visitor from Fermi Lab.

Two years after the Heidelberg pioneering work an individual magnesium ion was isolated in Seattle with my postdoc Warren Nagourney and my student Gary Janik. The latter's thesis bore the title "Laser Cooled Single Ion Spectroscopy of Magnesium and Barium." "Shelved optical electron amplifier: Observation of quantum jumps," was published in 1986 with my colleague Nagourney, and Jon Sandberg, an exceptional undergraduate assistant. The paper introduced a new technique which has made optical spectroscopy on an individual ion possible with record resolution and reproducibility. To date the best resolution has been realized at NIST by a group headed by my former collaborator Wineland. Peter Toschek who had made important contributions to the visible ion work in Heidelberg has built up a thriving laboratory for monoion-spectroscopy at the Universität Hamburg. With Herbert Walther a collaboration almost came off in 1974. Walther, with his large staff and excellent facilities in Munich, has since developed his own expertise in the field and made outstanding contributions to it. Gabrielse, now a full professor at Harvard, has assembled a large group and is trapping and cooling antiprotons at CERN.

In the 1988 paper "A Single Atomic Particle Forever Floating at Rest in Free Space: New Value for Electron Radius" I have surveyed the field and suggested new avenues for its extension. More precise measurements of the g factor of the electron may well be the most promising approach to study its structure. No less important, a trapped individual atomic ion may reveal itself as a timekeeping element of unsurpassed reproducibility. The research effort in Seattle continues on both projects. The National Science Foundation has supported my research since 1958 without interruption. Initially the Army Office of Ordnance Research and the Office of Naval Research did also provide support for many years.

I am married to Diana Dundore, a practicing physician. I have a grown son, Gerd, from an earlier marriage to Irmgard Lassow who is deceased.

I do regular hatha yoga exercises, enjoy waltzing, hiking in the foothills, reading, listening to classical music, and watching ballet performances.

SELECTED PUBLICATIONS

"Die photographischen Wirkungen mittelschneller Protonen II", P. Brix and H. Dehmelt, Z. Physik *126*, 728 (1949)

"Kernquadrupolfrequenzen in festem Dichloraethylen", H. Dehmelt and H. Krueger, Naturwissenschaften *37*, 111 (1950)

"Nuclear Quadrupole Resonance", H. Dehmelt, Am. J. Phys. *22*, 110 (1954)

"Atomic Phosphorus Paramagnetic Resonance Experiment", H. Dehmelt, Phys. Rev. *99*, 527 (1955)

"Paramagnetic Resonance Reorientation of Atoms and Ions Aligned by Electron Impact" H. Dehmelt, Phys. Rev. *103*, 1125 (1956)

"Slow Spin Relaxation of Optically Polarized Sodium Atoms", H. Dehmelt, Phys. Rev. *105*, 1487 (1957)

"Modulation of a Light Beam by Precessing Absorbing Atoms" H. Dehmelt, Phys. Rev. *105*, 1924 (1957)

"Spin Resonance of Free Electrons Polarized by Exchange Collisions", H. Dehmelt, Phys. Rev. *109*, 381 (1958)

"Spin Resonance of Free Electrons", H. Dehmelt, 1958–61 Progress Report for NSF Grant NSF-G 5955

"Alignment of the H_2^+ Molecular Ion by Selective Photodissociation", H. Dehmelt and K. Jefferts, Phys. Rev. *125*, 1318 (1962)

"Orientation of He Ions by Exchange Collisions with Cesium Atoms", H. Dehmelt and F. Major, Phys. Rev. Lett. *8*, 213 (1962)

"Ultrahigh Resolution $\Delta F=0, \pm 1$ $^3He^+$ HFS Spectra by an Ion Storage - Exchange Collision Technique", N. Fortson, F. Major and H. Dehmelt, Phys. Rev. Lett. *16*, 221 (1966)

"Radiofrequency Spectroscopy of Stored Ions, H. Dehmelt, Adv. At. Mol. Phys. *3*, 53 (1967) and *5*, 109 (1969)

"Alignment of the H_2^+ Molecular Ion by Selective Photodissociation II: Experiments on the RF Spectrum," Ch. Richardson, K. Jefferts and H. Dehmelt, Phys. Rev. *165*, 80 (1968)

"'Bolometric' Technique for the RF Spectroscopy of Stored Ions", H. Dehmelt and F. Walls, Phys. Rev. Lett. *21*, 127 (1968)

"Radiative Cooling of an Electrodynamically Confined Proton Gas", D. Church and H. Dehmelt, J. Appl. Phys. *40*, 3421 (1969)

"Proposed g-2/δv_z Experiment on Stored Single Electron or Positron", H. Dehmelt and P. Ekstrom, Bull. Am. Phys. Soc. *18*, 727 (1973)

"Monoelectron Oscillator", D. Wineland, P. Ekstrom and H. Dehmelt, Phys. Rev. Lett. *31*, 1279 (1973)

"Proposed 10^{14} $\Delta v < v$ Laser Fluorescence Spectroscopy on Tl^+ Mono-Ion Oscillator", H. Dehmelt, Bull. Am. Phys. Soc. *18*, 1521 (1973)

"Principles of the Stored Ion Calorimeter" D. Wineland and H. Dehmelt, J. Appl. Phys. *46*, 919 (1975)

"Proposed 10^{14} $\Delta v < v$ Laser Fluorescence Spectroscopy on Tl^+ Mono-Ion Oscillator II (spontaneous quantum jumps)", H. Dehmelt, Bull. Am. Phys. Soc. *20*, 60 (1975)

"Proposed 10^{14} $\Delta v < v$ Laser Fluorescence Spectroscopy on Tl^+ Mono-Ion Oscillator III (side band cooling)", D. Wineland and H. Dehmelt, Bull. Am. Phys. Soc. *20*, 637 (1975)

"Axial, Magnetron, Cyclotron and Spin-Cyclotron Beat Frequencies Measured on Single Electron Almost at Rest in Free Space (Geonium)", Van Dyck, Jr., R. S., Ekstrom, P., and Dehmelt, H., Nature *262*, 776 (1976)

"Entropy Reduction by Motional Side Band Excitation", Dehmelt, H., Nature *262*, 777 (1976)

"A Progress Report on the g-2 Resonance Experiments", H. Dehmelt, in *Atomic Masses and Fundamental Constants*, Volume 5 (eds. J. H. Sanders, and A. H. Wapstra), p. 499. Plenum New York, 1976

"Precise Measurement of Axial, Magnetron, Cyclotron and Spin-Cyclotron Beat Frequencies on an Isolated 1-meV Electron", Van Dyck, Jr., R. S., Ekstrom, P., and Dehmelt, H., Phys. Rev. Lett. *38*, 310 (1977)

"Electron Magnetic Moment from Geonium Spectra", Van Dyck, Jr., R. S., Schwinberg, P. B. & Dehmelt, H. G., in *New Frontiers in High Energy Physics* (Eds. B. Kursunoglu, A. Perlmutter, and L. Scott), Plenum New York, 1978

"Optical Sideband Cooling of Visible Atom Cloud Confined in Parabolic Well", Neuhauser, W., Hohenstatt, M., Toschek, P. E., and Dehmelt, H. G., Phys. Rev. Lett. *41*, 233 (1978)

"Single Elementary Particle at Rest in Free Space I–IV", Dehmelt, H., Van Dyck, Jr., R. S., Schwinberg, P. B., Gabrielse, G., Bull. Am. Phys. Soc. *24*, 757 (1979)

"Localized visible Ba^+ mono-ion oscillator", Neuhauser, W., Hohenstatt, M., Toschek, P. E., and Dehmelt, H. G., Phys. Rev. *A22*, 1137 (1980)

Physics 1989

"Preliminary Comparison of the Positron and Electron Spin Anomalies", P. B. Schwinberg, R. S. Van Dyck, Jr., and H. G. Dehmelt, Phys. Rev. Lett. *47*, 1679 (1981)

"Invariant Frequency Ratios in Electron and Positron Geonium Spectra Yield Refined Data on Electron Structure", Hans Dehmelt, in *Atomic Physics 7*, D. Kleppner & F. Pipkin Eds., Plenum, New York, 1981

"Mono-Ion Oscillator as Potential Ultimate Laser Frequency Standard", Hans Dehmelt, IEEE Transactions on Instrumentation & Measurement, *IM-31*, 83 (1982)

"Stored Ion Spectroscopy", Hans Dehmelt, in *Advances in Laser spectroscopy*, F. T. Arecchi, F. Strumia & H. Walther, Eds., Plenum, New York, 1983

"Geonium Spectra and the Finer Structure of the Electron", R. Van Dyck, P. Schwinberg, G. Gabrielse & Hans Dehmelt, Bulletin of Magnetic Resonance *4*, 107 (1983)

"g-Factor of Electron Centered in Symmetric Cavity", Hans Dehmelt, Proc. Natl. Acad. Sci. USA *81*, 8037 (1984); Erratum ibidem *82*, 6366 (1985)

"Observation of Relativistic Bistable Hysteresis in the Cyclotron Motion of a Single Electron", G. Gabrielse, H. Dehmelt & W. Kells, Phys. Rev. Letters *54*, 537 (1985).

"Doppler-Free Optical Spectroscopy on the Ba^+ Mono-Ion Oscillator", G. Janik, W. Nagourney, H. Dehmelt, J. Opt. Soc. Am. *B 2*, 1251−1257 (1985)

"Single Atomic Particle at Rest in Free Space: New Value for Electron Radius", Hans Dehmelt, Annales de Physique (Paris) *10*, 777−795 (1985)

"Observation of Inhibited Spontaneous Emission", G. Gabrielse and H. Dehmelt, Phys. Rev. Lett. *55*, 67 (1985)

"Electron Magnetic Moment from Geonium Spectra: Early Experiments and Background Concepts", Van Dyck, Jr., R. S., Schwinberg, P. B. & Dehmelt, H. G., Phys. Rev. *D 34*, 722 (1986)

"Continuous Stern Gerlach Effect: Principle and idealized apparatus", Hans Dehmelt, Proc. Natl. Acad. Sci. USA *83*, 2291 (1986), and *83*, 3074 (1986)

"Shelved optical electron amplifier: Observation of quantum jumps", Warren Nagourney, Jon Sandberg, and Hans Dehmelt, Phys. Rev. Letters *56*, 2797 (1986)

"New High Precision Comparison of Electron/Positron g-Factors", Van Dyck, Jr., R. S., Schwinberg, P. B. & Dehmelt, H. G., Phys. Rev. Letters *59*, 26 (1987)

"Single Atomic Particle at Rest in Free Space: Shift-Free Suppression of the Natural Line Width?", Hans Dehmelt, in *Laser Spectroscopy VIII*, S. Svanberg and W. Persson editors, 1987 (Springer, New York)

"Single Atomic Particle Forever Floating at Rest in Free Space: New Value for Electron Radius", Hans Dehmelt, Physica Scripta *T22*, 102 (1988)

"New Continuous Stern Gerlach Effect and a Hint of 'The' Elementary Particle", Hans Dehmelt, Z. Phys. D *10*, 127−134 (1988)

"Coherent Spectroscopy on a Single Atomic System at Rest in Free Space III", Hans Dehmelt, in *Frequency Standards and Metrology*, A. de Marchi Ed. (Springer, New York, 1989), p. 15

"Triton, .. electron, .. cosmon ..: An infinite regression? Hans Dehmelt, Proc. Natl. Acad. Sci. USA *86*, 8618−8619 (1989)

"Miniature Paul-Straubel ion trap with well-defined deep potential well", Nan Yu, Hans Dehmelt, and Warren Nagourney, Proc. Natl. Acad. Sci. USA *86*, 5672 (1989)

EXPERIMENTS WITH AN ISOLATED SUBATOMIC PARTICLE AT REST

Nobel Lecture, December 8, 1989

by

HANS G. DEHMELT

Department of Physics, University of Washington, Seattle, WA 98195, USA

> *"You know, it would be sufficient to really understand the electron."*
> *Albert Einstein*

The 5th century B.C. philosopher's Democritus' smallest conceivable indivisible entity, the a-tomon (the un-cuttable), is a most powerful but not an immutable concept. By 1920 it had already metamorphosed twice: from something similar to a molecule, say a slippery atomon of water, to Mendeleyev's chemist's atom and later to electron and to proton, both particles originally assumed to be of small but finite size. With the rise of Dirac's theory of the electron in the late twenties their size shrunk to mathematically zero. Everybody "knew" then that electron and proton were indivisible Dirac point particles with radius R = 0 and gyromagnetic ratio g = 2.00. The first hint of cuttability or at least compositeness of the proton came from Stern's 1933 measurement of proton magnetism in a Stern-Gerlach molecular beam apparatus. However this was not realized at the time. He found for its normalized dimensionless gyromagnetic ratio not g = 2 but

$$g = (\mu/A)(2M/q) \approx 5,$$

where μ, A, M, q are respectively magnetic moment, angular momentum, mass and charge of the particle. For comparison the obviously composite $^4He^+$ ion, also with spin ½, according to the above formula has the $| g |$ value 14700, much larger than the Dirac value 2. Also, along with this large $| g |$ value went a size of this atomic ion about 4 orders of magnitude larger than an α-particle. And indeed, with Hofstadter's high energy electron scattering experiments in the fifties the proton radius grew again to R = 0.86 × 10^{-15} m. Similar later work at still higher energies found 3 quarks inside the "indivisible" proton. Today everybody "knows" the *electron* is an indivisible atomon, a Dirac point particle with radius R = 0 and g = 2.00.... But is it? Like the proton, it could be a composite object. History may well repeat itself. This puts a high premium on precise measurements of the g factor of the electron.

GEONIUM SPECTROSCOPY

The metastable pseudo-atom geonium (Van Dyck et al. 1978 and 1986) has been expressly synthesized for studies of the electron g factor under optimal conditions. It consists of an individual electron permanently confined in an ultrahigh vacuum Penning trap at 4K. The trap employs a homogeneous magnetic field $B_0 = 5T$ and a weak electric quadrupole field. The latter is produced by hyperbolic electrodes, a positive ring and two negative caps spaced $2Z_0 = 8$ mm apart, see Fig. 1. The potential, with A a constant, is given by

$$\phi(xyz) = A(x^2 + y^2 - 2z^2),$$

with an axial potential well depth

$$D = e[\phi(000) - \phi(00Z_0)] = 2eAZ_0^2 = 5eV.$$

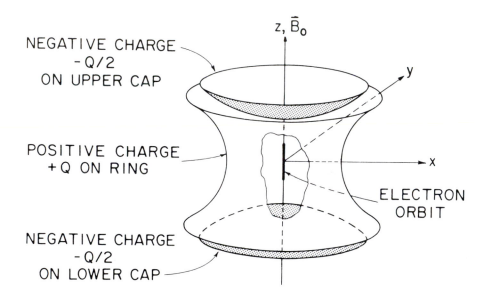

Figure 1. Penning trap. The simplest motion of an electron in the trap is along its symmetry axis, along a magnetic field line. Each time it comes too close to one of the negatively charged caps it turns around. The resulting harmonic oscillation took place at about 60 MHz in our trap. Reproduced from (Dehmelt 1983) with permission, copyright Plenum Press.

The trapping is mostly magnetic. The large magnetic field dominates the motion in the geonium atom. The energy levels of this atom shown in Figure 2 reflect the cyclotron motion, at frequency $v_c = eB_0/2\pi m = 141$ GHz, the spin precession, at $v_c \approx v_c$, the anomaly or g-2 frequency $v_a = v_s - v_c = 164$ MHz, the axial oscillation, at $v_z = 60$ MHz, and the magnetron or drift motion at frequency $v_m = 13$ kHz. The electron is continuously monitored by exciting the v_z-oscillation and detecting via radio the 10^8-fold enhanced spontaneous 60 MHz emission. A corresponding signal appears in Figure 3.

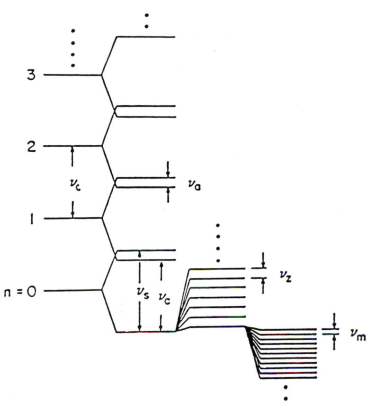

Figure 2. Energy levels of geonium. Each of the cyclotron levels labeled n is split first by the spin — magnetic field interaction. The resulting sublevels are further split into the oscillator levels and finally the manifold of magnetron levels extending downwards. Reproduced from (Van Dyck et al. 1978) with permission, copyright Plenum Press.

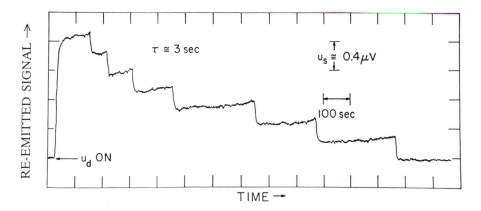

Figure 3. Rf signal produced by trapped electron. When the electron is driven by an axial rf field, it emits a 60 MHz signal, which was picked up by a radio receiver. The signal shown was for a very strong drive and an initially injected bunch of 7 electrons. One electron after the other was randomly "boiled" out of the trap until finally only a single one is left. By somewhat reducing the drive power, this last electron could be observed indefinitely. Reproduced from (Wineland et al. 1973) with permission, copyright American Institute of Physics.

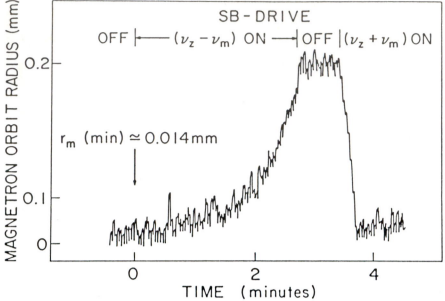

Figure 4. Side-band "cooling" of the magnetron motion at v_m. By driving the axial motion not on resonance at v_z but on the lower side-band at v_z-v_m, it is possible to force the metastable magnetron motion to provide the energy balance hv_m, and thereby expand the magnetron orbit radius. Conversely, an axial drive at v_z + v_m shrinks the radius. The roles of upper and lower side-bands are reversed here from the case of a particle in a well where the energy increases with amplitude because the magnetron motion is metastable and the total energy of this motion *decreases* with radius. Reproduced from (Van Dyck et al. 1978) with permission, copyright Plenum Press.

Side band cooling has made continuous confinement in the trap center of an electron for 10 months (Gabrielse et al. 1985) possible. This process makes the electron absorb rf photons deficient in energy and supply the balance from energy stored in the electron motion to be cooled. The corresponding shrinking of the radius of the magnetron motion is displayed in Figure 4. Extended into the optical region, the cooling scheme is most convincingly demonstrated in Figure 5. The transitions of primary interest at v_c, v_a, v_m are much more difficult to detect than the v_z oscillation. Nevertheless the task may be accomplished by means of the continuous Stern-Gerlach effect (Dehmelt 1988a), in which the geonium atom itself is made to work as a 10^8-fold amplifier. In the scheme a single v_a-photon of only $\approx 1\mu eV$ energy gates the absorption of ≈ 100 eV of rf power at v_z. The continuous effect uses an inhomogeneous magnetic field in a similar way as the classic one. However, the field takes now the form of a very weak Lawrence cyclotron trap or magnetic bottle shown in Figure 6. The bottle adds a minute monitoring well, only

$$D_m = (m + n + \tfrac{1}{2})\, 0.1\mu eV$$

Figure 5. Visible blue (charged) barium atom Astrid at rest in center of Paul trap photographed in natural color. The photograph strikingly demonstrates the close localization, < 1 μm, attainable with geonium techniques. Stray light from the lasers focussed on the ion also illuminates the ring electrode of the tiny rf trap of about 1 mm internal diameter. Reproduced from (Dehmelt 1988) with permission, copyright the Royal Swedish Academy of Sciences.

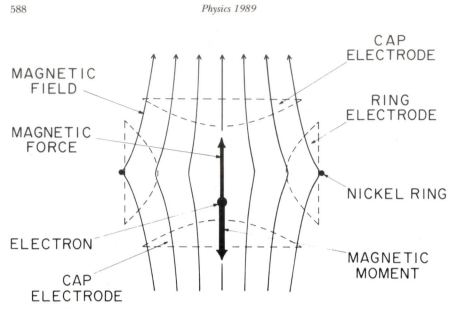

Figure 6. Weak magnetic bottle for continuous Stern-Gerlach effect. When in the lowest cyclotron and magnetron level the electron forms a 1 µm long wave packet, 30 nm in diameter, which may oscillate undistorted in the axial electric potential well. The inhomogeneous field of the auxiliary magnetic bottle produces a minute spin-dependent restoring force that causes the axial frequency v_z for spin ↑ and ↓ to differ by a small but detectable value. Reproduced from (Dehmelt 1988a) with permission, copyright Springer Verlag.

deep, to the axial well of large electrostatic depth D = 5 eV, with m, n respectively denoting spin and cyclotron quantum numbers. Thus jumps in m or n show up as jumps in v_z,

$$v_z = v_{z0} + (m + n + \tfrac{1}{2})\delta,$$

with δ = 1.2 Hz in our experiments, and v_{z0} the axial frequency of a hypothetical electron without magnetic moment. Random jumps in m, n occur, when spin or cyclotron resonances are excited. Figure 6A shows an early example of a series of such jumps in m or spin flips. For the spin spontaneous transitions are totally negligible. Standard text books discuss

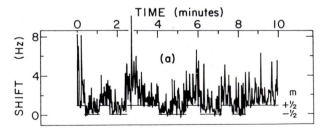

Figure 6A. Spin flips recorded by means of the continuous Stern-Gerlach effect. The random jumps in the base line indicate jumps in m at a rate of about 1/minute when the spin resonance is excited. The upwards spikes or "cyclotron grass" are explained by expected rapid random thermal excitation and spontaneous decay of cyclotron levels with an average value $<n> \approx 1.2$. Adapted from (Van Dyck et al. 1977) with permission, copyright American Institute of Physics.

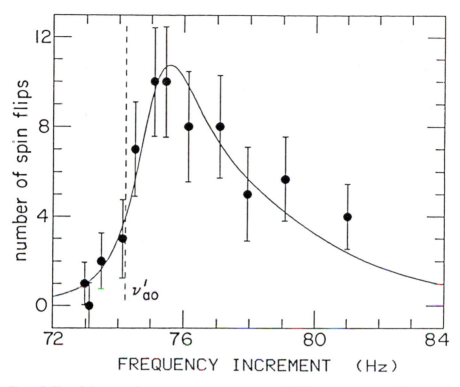

Figure 7. Plot of electron spin resonance in geonium near 141 GHz. A magnetic radiofrequency field causes random jumps in the spin quantum number. As the frequency of the exciting field is stepped through the resonance in small increments, the number of spin flips occurring in a fixed observation period of about ½ hour are counted and then plotted vs frequency. (Actually the 141 GHz field flipping the spin is produced by the cyclotron motion of the electron through an inhomogeneous magnetic rf field at $v_s - v_c = 164$ MHz.) Reproduced from (Van Dyck et al. 1987) with permission, copyright American Institute of Physics.

transitions between two sharp levels induced by a broad electromagnetic spectrum $\rho(v)$: The transition rate from either level is the same and is proportional to the spectral power density $\rho(v_s)$ of the radiation field at the transition frequency v_s. Ergo, the average dwell times in either level are the same, compare Fig. 6A. In the geonium experiments the frequency of the weak rf field is sharp, but the spin resonance is broadened and has a shape $G_s(v)$. One may convince oneself that moving the sharp frequency of the rf field upwards over the broad spin resonance should produce the same results as moving a broad rf field of spectral shape $\rho(v) \propto G_s(v)$ downwards over a sharp spin resonance: The rate of all spin flips or jumps in m in either direction counted in the experiment is proportional to $G_s(v)$. To obtain the plot of $G_s(v)$ in Fig. 7 the frequency of the rf field was increased in small steps, and at each step spin flips were counted for a fixed period of about ½ hour. From our v_s, v_c data for electron and positron (Van Dyck et al. 1987) we have determined

$$\tfrac{1}{2}g^{exp} = v_s/v_c = 1.001\ 159\ 652\ 188(4),$$

the same for particle and anti-particle. The error in their difference is only half as large. Heroic quantum electro-dynamical calculations (Kinoshita 1988) have now yielded for the shift of the g factor of a point electron associated with turning on its interaction with the electromagnetic radiation field

$$\frac{1}{2}(g^{point}-2) = \frac{1}{2}\Delta g^{\ KINOSHITA} = 0.001\ 159\ 652\ 133(29).$$

In the calculations $\Delta g^{KINOSHITA}$ is expressed as a power series in α/π. Kinoshita has critically evaluated the experimental α input data on which he must rely. He warns that the error in his above result, which is dominated by the error in α, may be underestimated. Muonic, hadronic and other small contributions to g amount to less than about 4×10^{-12} and have been included in the shift. Kinoshita's result may be used to correct the experimental g value and find

$$g = g^{exp} - \Delta g^{KINOSHITA1} = 2 + 11(6)x10^{-11}.$$

ELECTRON RADIUS R?

Extrapolation from known to unknown phenomena is a time-honored approach in all the sciences. Thus from known g, and R values of other near-Dirac particles and our *measured* g value of the electron I attempt to extrapolate a value for its radius. Stimulated by 1980 theoretical work of Brodsky & Drell, I (1989a) have plotted $|g-2| = R/\lambda_C$ in Figure 8 for the helium3 nucleus, triton, proton, and electron. Here λ_C is the Compton wavelength of the respective particle. The plausible relation given by Brodsky and Drell (1980) for the simplest composite theoretical model of the electron,

$$|g - 2| = R/\lambda_C, \text{ or}$$

$$|g - g_{DIRAC}| = R/\lambda_C$$

fits the admittedly sparse data surprisingly well. Even for such a very different spin $\frac{1}{2}$ structure as the atomic ion $^4He^+$ composed of an α-particle and an electron the data point does not fall too far off the full line. Intersection in Figure 8 of this line with the line $|g-2| = 1.1x10^{-10}$ for the Seattle g data yields for the electron the extrapolated point shown and with $\lambda_C = 0.39 \times 10^{-10}$ cm an electron radius

$$R \approx 10^{-20} \text{ cm.}$$

The row of X's reflects the data range defined by the uncertainty in the Seattle g data and the upper limit $R < 10^{-17}$ cm determined in high energy collision experiments. It appears that this combination of current data is not in harmony with electron structure models assuming special symmetries that predict the quadratic relation $|g-2| \approx (R/\lambda_C)^2$ shown by the dashed line. This favors the linear relation used in the above extrapolation of R for the electron. Thus, the electron may have *size and structure!*

If one feels that the excess g value $11(6) \times 10^{-11}$ measured is not signifi-

NEAR–DIRAC PARTICLE DATA

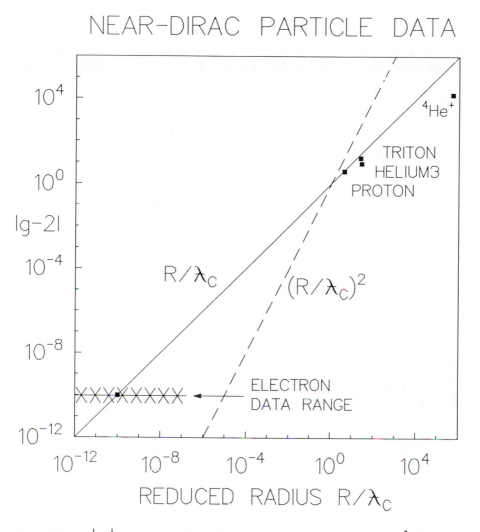

Figure 8. Plot of $|g\text{-}2|$ values, with radiative shifts removed, vs reduced rms radius R/λ_C for near-Dirac particles. The full line $|g\text{-}2| = R/\lambda_C$ predicted by the simplest theoretical model provides a surprisingly good fit to the data points for proton, triton and helium3 nucleus. It may be used to obtain a new radius value for the *physical* electron from its intersection with the line $|g\text{-}2| = 1.1 \times 10^{-10}$ representing the Seattle electron g data. The data are much less well fitted by the relation $|g\text{-}2| = (R/\lambda_C)^2$, which is shown for comparison in the dashed line. The atomic ion $^4\text{He}^+$ is definitely *not* a near-Dirac particle, but even its data point does not fall too far off the full line. Adapted from (Dehmelt 1990) with permission, copyright American Institute of Physics.

cant because of its large relative error then, the value $R \approx 10^{-20}$ cm given here still constitutes an important new upper limit. Changing the point of view, the close agreement of g^{point} with g^{exp} provides the most stringent experimental test of the fundamental theory of Quantum Electrodynamics in which $R = 0$ is assumed. Furthermore the near-identity of the g values measured for electron and positron in Seattle constitutes the most severe test of the CPT theorem or mirror symmetry of a *charged* particle pair.

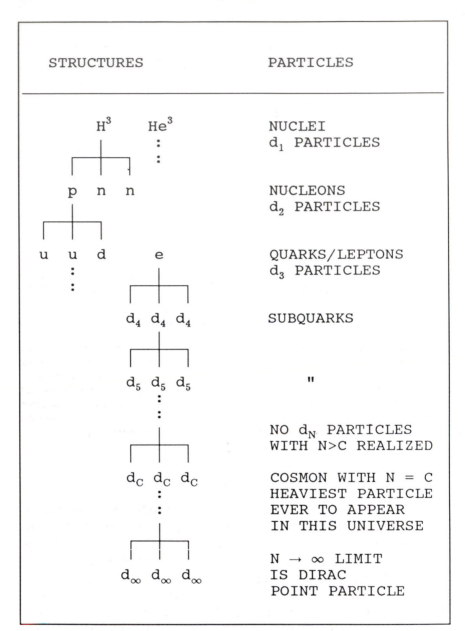

| STRUCTURES | PARTICLES |

$H^3 \quad He^3$

NUCLEI
d_1 PARTICLES

p   n   n

NUCLEONS
d_2 PARTICLES

u   u   d   e

QUARKS/LEPTONS
d_3 PARTICLES

$d_4 \ d_4 \ d_4$

SUBQUARKS

$d_5 \ d_5 \ d_5$

"

NO d_N PARTICLES
WITH N>C REALIZED

$d_C \ d_C \ d_C$

COSMON WITH N = C
HEAVIEST PARTICLE
EVER TO APPEAR
IN THIS UNIVERSE

$d_\infty \ d_\infty \ d_\infty$

N → ∞ LIMIT
IS DIRAC
POINT PARTICLE

Figure 9. Triton model of near-Dirac particles. Reproduced from (Dehmelt 1989b) with permission, copyright the National Academy of Sciences of the USA.

LEMAÎTRE'S "L'ATOME PRIMITIF" REVISITED — A SPECULATION

Beginning 1974 Salam and others have proposed composite electron and quark models (Lyons 1983). On the strength of these proposals and with an eye on Figure 8, I view the electron as the third approximation of a Dirac particle, d_3 for short, and as composed of three fourth-approximation Dirac

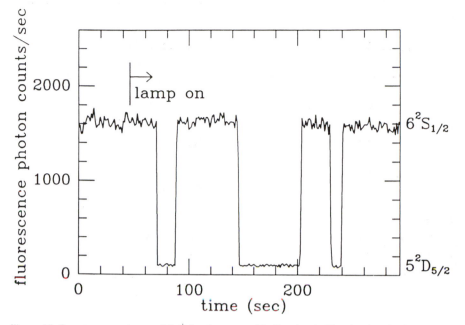

Figure 10. Spontaneous decay of Ba$^+$ ion in metastable $D_{5/2}$-level. Illuminating the ion with a laser tuned close to its resonance line produces strong resonance fluorescence and an easily detectable photon count of 1600 photons/sec. When later an auxiliary, weak Ba$^+$ spectral lamp is turned on the ion is randomly transported into the metastable $D_{5/2}$ level of 30 sec lifetime and becomes invisible. After dwelling in this shelving level for 30 sec on the average, it drops down to the $S_{1/2}$ ground state *spontaneously* and becomes visible again. This cycle then repeats. Reproduced from (Nagourney et al. 1986) with permission, copyright American Institute of Physics.

or d_4 particles. The situation is taken to be quite similar to that previously encountered in the triton and proton subatomic particles, respectively assumed to be of type d_1 and d_2. In more detail, three d_4 subquarks of huge mass m_4 in a deep square well make up the electron in this working hypothesis. However, their mass $3m_4$ is almost completely compensated by strong binding to yield a total relativistic mass equal to the observed mass m_e of the electron. Figure 8 may even suggest a more speculative extrapolation: The e-constituents, in the infinite regression $N \rightarrow \infty$ — proposed in Figure 9, have ever more massive, ever smaller sub-sub-.... constituents d_N. However, these higher order subquarks are realized only up to the "cosmon" with $N = C$, the most massive particle ever to appear in this universe. At the beginning of the universe, a lone bound cosmon-anticosmon pair or life time-broadened cosmonium atom state of near-zero total relativistic mass/energy was created from Vilenkin's (1984) metastable "nothing" state of zero relativistic energy in a spontaneous quantum jump of cosmic rarity. Similar, though much more frequent, quantum jumps that have recently been observed in a trapped Ba$^+$ ion are shown in Figure 10. In this case the system also jumps spontaneously from a state (ion in metastable $D_{5/2}$ level plus no photon) to a new state (ion in $S_{1/2}$ ground level plus photon) of the same total energy. The "cosmonium atom" introduced here is merely a modernized version* of Lemaître's "l'atome primitif" or world-atom whose explosive radioactive decay created the universe. At the beginning of the world the short-lived cosmonium atom decayed into an early gravitation-dominated standard big bang state that eventually developed into a state, in which again rest mass energy, kinetic and Newtonian gravitational potential energy add up to *zero* (see formula 8 of Jordan 1937). The electron is a much more complex particle than the cosmon. It is composed of 3^{C-3} cosmon-like d_C's, but only two particles of this type formed the cosmonium world-atom from which sprang the universe. In closing, I should like to cite a line from *William Blake,*

"To see a world in a grain of sand $- - -$"

and allude to a possible parallel

$-$ to see worlds in an electron $-$

---

* This is by no means the first modernization attempt. M. Goldhaber has kindly brought it to my attention that he had introduced a different "cosmon" already in 1956 in his paper "Speculations on Cosmogony," SCIENCE *124,* 218.

REFERENCES

Brodsky, S. J., and Drell, S. D., "Anomalous Magnetic Moment and Limits on Fermion Substructure," Phys. Rev. D *22*, 2236 (1980).

Dehmelt, H. (1983) "Stored Ion Spectroscopy", in *Advances in Laser Spectroscopy*, F. T. Arecchi, F. Strumia & H. Walther, Eds., Plenum, New York.

Dehmelt, H. (1988a) "Single Atomic Particle Forever Floating at Rest in Free Space: New Value for Electron Radius," Physica Scripta *T22*, 102.

Dehmelt, H. (1988b) "New Continuous Stern Gerlach Effect and a Hint of 'The' Elementary Particle," Z. Phys. D *10*, 127−134.

Dehmelt, H. (1989a) "Geonium Spectra * Electron Radius * Cosmon" in *High Energy Spin Physics*, 8th International Symposium, K. Heller, Ed. (AIP Conference Proceedings No. 187, New York) p. 319.

Dehmelt, H. (1989b) "Triton,..electron,..cosmon..: An infinite regression?," Proc. Natl. Acad. Sci. USA *86*, 8618−8619.

Dehmelt, H. (1990) "Less is more: Experiments with an Individual Atomic Particle at Rest in Free Space" Am. J. Phys., *58*, 17.

Gabrielse, G., Dehmelt, H., and Kells, W. (1985) "Observation of a Relativistic, Bistable Hysteresis in the Cyclotron Motion of a Single electron" Phys. Rev. Letters *54*, 537.

Jordan, P., (1937) "Die physikalischen Weltkonstanten" Naturwissenschaften *25*, 513.

Kinoshita, T. (1988) "Fine-Structure Constant Derived from Quantum Electrodynamics," Metrologia *25*, 233.

Lemaître, G. (1950) *THE PRIMEVAL ATOM* (Van Nostrand, New York) p. 77.

Lyons, L. (1983) "An Introduction to the Possible Substructure of Quarks and Leptons," Progress in Particle and Nuclear Physics *10*, 227, see references cited herein.

Nagourney, W., Sandberg, J., and Dehmelt, H. (1986) "Shelved optical electron amplifier: Observation of quantum jumps." Phys. Rev. Letters *56*, 2797.

Van Dyck, Jr., R. S., Ekstrom, P., and Dehmelt, H. (1977) "Precise Measurement of Axial, Magnetron, and Spin-Cyclotron Beat Frequencies on an Isolated 1-meV Electron", Phys. Rev. Lett. *38*, 310

Van Dyck, Jr., R. S., Schwinberg, P. B. & Dehmelt, H. G. (1978) "Electron Magnetic Moment from Geonium Spectra," in New Frontiers in High Energy Physics (Eds. B. Kursunoglu, A. Perlmutter, and L. Scott), Plenum New York.

Van Dyck, Jr., R. S., Schwinberg, P. B. & Dehmelt, H. G. (1986) "Electron Magnetic Moment from Geonium Spectra: Early Experiments and Background Concepts," Phys. Rev. D *34*, 722.

Van Dyck, Jr., R. S., Schwinberg, P. B. & Dehmelt, H. G. (1987) "New High Precision Comparison of Electron/Positron g-Factors," Phys. Rev. Letters *59*, 26.

Vilenkin, A. (1984) "Quantum Creation of Universe," Phys. Rev. D *30*, 509−515.

Wineland, D., Ekstrom, P., and Dehmelt, H. (1973) "Monoelectron Oscillator," Phys. Rev. Lett. *31*, 1297.

Wolfgang Paul

WOLFGANG PAUL

I was born on August 10, 1913 in Lorenzkirch, a small village in Saxony, as the fourth child of Theodor and Elisabeth Paul, née Ruppel. All in all we were six children. Both parents were descendants from Lutheran ministers in several generations. I grew up in München where my father has been a professor for pharmaceutic chemistry at the university. He had studied chemistry and medicin having been a research student in Leipzig with Wilhem Ostwald, the Nobel Laureate 1909. So I became familiar with the life of a scientist in a chemical laboratory quite early. Unfortunately, my father died when I was still a school boy at the age of fifteen years. But my interest in sciences was awaken, even my parents were very much in favour of a humanistic education. After finishing the gymnasium in München with 9 years of latin and 6 years of ancient greek, history and philosophy, I decided to become a physicist. The great theoretical physicist, Arnold Sommerfeld, an University colleague of my late father, advised me to begin with an apprenticeship in precision mechanics. Afterwards, in the fall 1932, I commenced my studies at the Technische Hochschule München. Listening to the very inspiring physics lectures by Jonathan Zenneck with lots of demonstrations — 6 full hours a week — I felt being on the right track.

After my first examination in 1934 I turned to the Technische Hochschule in Berlin. I was lucky in finding in Hans Kopfermann a teacher with a feeling for the essentials in physics but also a very liberal man, who had taken a fatherly interest in me. He, a former Ph.D. student of James Franck, had just returned from a three years stay at the Niels Bohr Institute in Copenhagen, working in the field of hyperfine spectroscopy and nuclear moments. All in all I worked 16 years with him.

As a theorist Richard Becker taught at the TH Berlin whom I met later at the University of Göttingen again. Both men had the strongest influence on my scientific thinking. But it was not only the scientific aspect. In the Germany of these days just as important was the human and the political attitude. And I am still a little bit proud having been accepted by these sensitive men in this respect. Here are the roots for my later engagement in the anti-nuclear weapon discussion and for having signed the declaration of the so-called "Göttinger Eighteen" in 1957 with its important consequences in German politics.

In 1937 after my diploma exam with Hans Geiger as examinator I followed Kopfermann to the University of Kiel where he had just been appointed Professor Ordinarius. For my doctor thesis I had chosen the determination of the nuclear moments of Beryllium from the hyperfine spectrum. I developed an atomic beam light source to minimize the Doppler effect. But just before the decisive measurements I was drawn to the air force a few days before the war started. Fortunately, a few month later I got a leave of absence to finish my thesis and to take my doctor exam at the TH Berlin. In 1940 I was exempted from military service. I joined again the group around Kopfermann which 2 years later moved to Göttingen. There in 1944 I became Privatdozent at the University.

In these years I worked in mass spectrometry and isotope separation together with W. Walcher. When we heard of the development of the betatron by D. Kerst in the United States and also of a similar development by Gund at the Siemens company, Kopfermann saw immediately that scattering experiments with high energy electrons would enable the study of the charge structure of nuclei. He convinced me to turn to this new very promising field of physics and I soon participated in the first test measurements at the 6 MeV betatron at the Siemens laboratory. Later after the war we succeeded in getting this accelerator to Göttingen.

But due to the restriction in physics research imposed by the military government I turned for a few years my interest to radiobiology and cancer therapy by electrons in collaboration with my colleague G. Schubert from the medical faculty.

Besides we performed some scattering experiments and studied first the electric disintegration of the deuteron, and not to forget for the first time we measured the Lamb shift in the He-spectrum with optical methods.

In 1952 I was appointed Professor at the University of Bonn and Director of the Physics Institute, with very good students waiting for a thesis advisor. I was very lucky that my best young collaborators followed me, O. Osberghaus, H. Ehrenberg, H. G. Bennewitz, G. Knop, and H. Steinwedel as a "house theoretician". Here we started new activities: molecular beam physics, mass spectrometry and high energy electron physics. It was a scanty period after the war. But in order to become in a few years competitive with the well advanced physics abroad we tried to develop new methods and instruments in all our research.

In this period these focusing methods in molecular beam physics with quadrupole and sextupole lenses having already started in Göttingen with H. Friedburg, were further developed and enabled new types of experiments. The quadrupole mass spectrometer and the ion trap were conceived and studied in many respects by research students. And with the generous support of the Deutsche Forschungsgemeinschaft we have built a 500 MeV *electron* synchrotron, the first in Europe working according to the new principle of strong focusing. It was followed in 1965 by a synchroton for 2500 MeV. My colleagues H. Ehrenberg, R. H. Althoff and G. Knop were sharing this success with me.

In recent years my interest turned to neutron physics with a new device, a magnetic storage ring for neutrons.

U. Trinks and K.J. Kügler and later my two sons Lorenz and Stephan, joined me in our experiments with stored neutrons at the ILL in Grenoble.

My experience in accelerator physics brought me in close contact to CERN. I served there from the very early days on as an advisor. Having spent the year 1959 in Genève I became director of the nuclear physics division for the years 1964−67. I was for several years member and later chairman of the Scientific Policy Committee and for many years scientific delegate of Germany in the CERN Council. For a short period I was chairman of ECFA, the European Committee for Future Accelerators.

Together with my friends W. Jentschke and W. Walcher in 1957 we started the German National Laboratory DESY in Hamburg which I joined as chairman of the directorate 1970−73. For several years I was chairman of its scientific council. In the same positions I served in the first years of the Kernforschungsanlage Jülich.

In 1970 I spent some weeks as Morris Loeb lecturer at Harvard University. 1978 I was lecturing as distinguished scientist at the FERMI Institute of the University of Chicago and in a similar position at the University of Tokyo. Since 1981 I am Professur Emeritus at the Bonn University.

In the past decades of recovery of German Universities and Physics research I was engaged in many advisory bodies. I have served as a referee and later as member of senate to the Deutsche Forschungsgemeinschaft. I was member and chairman of several committees: for reforming the university structure and for research planning of the federal government.

Ten years ago I was elected President of the Alexander von Humboldt Foundation which since 130 years fosters the international collaboration among scientists all over the world in the universal spirit of its patron Humboldt.

I was married for 36 years to the late Liselotte Paul, née Hirsche. She shared with me the depressing period during and after the war and due to her optimistic view of life she gave me strength and independence for my profession. Four children were born to us, two daughters, Jutta and Regine, an historian of art and a pharmacist, and two sons, Lorenz and Stephan, both being physicists. Since 1979 I am married to Dr. Doris Walch-Paul, teaching medieval literature at the University of Bonn.

Memberships and Distinctions

Member:

Deutsche Akademie der Naturforscher "Leopoldina"
Akademie der Wissenschaften in Düsseldorf, Heidelberg und Göttingen
Orden Pour le Mérite für Wissenschaft und Künste, Vice chancelor for the Sciences
Honarary member of DESY, Hamburg
Honarary member of KFA Jülich

Distinctions:

Grosses Verdienstkreuz mit Stern der Bundesrepublik Deutschland
Dr. fil. h.c. University Uppsala
Dr.rer.nat.h.c. Technische Hochschule Aachen
Robert-Wichard-Pohl-Preis der Deutschen Physikalischen Gesellschaft
Goldmedal of the Academy of Sciences in Prague

ELECTROMAGNETIC TRAPS FOR CHARGED AND NEUTRAL PARTICLES

Nobel Lecture, December 8, 1989

by

WOLFGANG PAUL

Physikalisches Institut der Universität Bonn, Nussallee 12, D-5300 Bonn, F.R.G.

Experimental physics is the art of observing the structure of matter and of detecting the dynamic processes within it. But in order to understand the extremely complicated behaviour of natural processes as an interplay of a few constituents governed by as few as possible fundamental forces and laws, one has to measure the properties of the relevant constituents and their interaction as precisely as possible. And as all processes in nature are interwoven one must separate and study them individually. It is the skill of the experimentalist to carry out clear experiments in order to get answers to his questions undisturbed by undesired effects and it is his ingenuity to improve the art of measuring to ever higher precision. There are many examples in physics showing that higher precision revealed new phenomena, inspired new ideas or confirmed or dethroned well established theories. On the other hand new experimental techniques conceived to answer special questions in one field of physics became very fruitful in other fields too, be it in chemistry, biology or engineering. In awarding the Nobel prize to my colleagues Norman Ramsey, Hans Dehmelt and me for new experimental methods the Swedish Academy indicates her appreciation for the aphorism the Göttingen physicist Georg Christoph Lichtenberg wrote two hundred years ago in his notebook "one has to do something new in order to see something new". On the same page Lichtenberg said: "I think it is a sad situation in all our chemistry that we are unable to suspend the constituents of matter free".

Today the subject of my lecture will be the suspension of such constituents of matter or in other words, about traps for free charged and neutral particles without material walls. Such traps permit the observation of isolated particles, even of a single one, over a long period of time and therefore according to Heisenberg's uncertainty principle enable us to measure their properties with extremely high accuracy.

In particular, the possibility to observe individual trapped particles opens up a new dimension in atomic measurements. Until few years ago all measurements were performed on an ensemble of particles. Therefore, the measured value — for example, the transition probability between two eigenstates of an atom — is a value averaged over many particles. Tacitly one assumes that all atoms show exactly the same statistical behaviour if one attributes the result to the single atom. On a trapped single atom, however, one can observe its interaction with a radiation field and its own statistical behaviour alone.

The idea of building traps grew out of molecular beam physics, mass spectrometry and particle accelerator physics I was involved in during the first decade of my career as a physicist more than 30 years ago. In these years (1950—55) we had learned that plane electric and magnetic multipole fields are able to focus particles in two dimensions acting on the magnetic or electric dipole moment of the particles. Lenses for atomic and molecular beams [1,2,3] were conceived and realized improving considerably the molecular beam method for spectroscopy or for state selection. The lenses found application as well to the ammonia as to the hydrogen maser [4].

The question "What happens if one injects charged particles, ions or electrons, in such multipole fields" led to the development of the linear quadrupole mass spectrometer. It employs not only the focusing and defocusing forces of a high frequency electric quadrupole field acting on ions but also exploits the stability properties of their equations of motion in analogy to the principle of strong focusing for accelerators which had just been conceived.

If one extends the rules of two-dimensional focusing to three dimensions one posseses all ingredients for particle traps.

As already mentioned the physics or the particle dynamics in such focusing devices is very closely related to that of accelerators or storage rings for nuclear or particle physics. In fact, multipole fields were used in molecular beam physics first. But the two fields have complementary goals: the storage of particles, even of a single one, of extremely low energy down to the micro-electron volt region on the one side, and of as many as possible of extremely high energy on the other. Today we will deal with the low energy part. At first I will talk about the physics of dynamic stabilization of ions in two- and three-dimensional radio frequency quadrupole fields, the quadrupole mass spectrometer and the ion trap. In a second part I shall report on trapping of neutral particles with emphasis on an experiment with magnetically stored neutrons.

As in most cases in physics, especially in experimental physics, the achievements are not the achievements of a single person, even if he contributed in posing the problems and the basic ideas in solving them. All the experiments I am awarded for were done together with research students or young colleagues in mutual inspiration. In particular, I have to mention H. Friedburg and H. G. Bennewitz, C.H. Schlier and P. Toschek in the field of molecular beam physics, and in conceiving and realizing the linear quadrupole spectrometer and the r.f. ion trap H. Steinwedel, O. Osberghaus and especially the late Erhard Fischer. Later H.P. Reinhard, U. v. Zahn and F. v. Busch played an important role in developing this field.

Focusing and Trapping of particles

What are the principles of focusing and trapping particles? Particles are elastically bound to an axis or a coordinate in space if a binding force acts on them which increases linearly with their distance r

$$F = -cr.$$

In other words if they move in a parabolic potential

$$\Phi \sim (ax^2 + \beta y^2 + \gamma z^2)$$

The tools appropriate to generate such fields of force to bind charged particles or neutrals with a dipole moment are electric or magnetic multi-pole fields. In such configurations the field strength, or the potential respectively increases according to a power law and shows the desired symmetry. Generally if *m* is the number of "poles" or the order of symmetry the potential is given by

$$\Phi \sim r^{m/1}\cos(m/2 \cdot \varphi).$$

For a quadrupole $m = 4$ it gives $\Phi \sim r^2 \cos 2\varphi$, and for a sextupole $m = 6$ one gets $\Phi \sim r^3 \cos 3\varphi$ corresponding to a field strength increasing with r and r^2 respectively.

Trapping of charged particles in 2- and 3-dimensional quadrupole fields

In the electric quadrupole field the potential is quadratic in the cartesian coordinates.

$$\Phi = \frac{\Phi_0}{2r^2}\left(ax^2 + \beta y^2 + \gamma z^2\right) \tag{1}$$

The Laplace condition $\Delta\Phi = 0$ imposes the condition $a + \beta + \gamma = 0$ There are two simple ways to satisfy this condition.

a) $a = 1 = -\gamma$, $\beta = 0$ results in the two-dimensional field

$$\Phi = \frac{\Phi_0}{2r_0^2}\left(x^2 - z^2\right) \tag{2}$$

b) $a = \beta = 1$, $\gamma = -2$ generates the three-dimensional configuration, in cylindrical coordinates

$$\Phi = \frac{\Phi_0(r^2 - 2z^2)}{r_0^2 + 2z_0^2} \quad \text{with } 2z_0^2 = r_0^2. \tag{3}$$

The two-dimensional quadrupole or the mass filter [5,6]

Configuration a) is generated by 4 hyperbolically shaped electrodes linearly extended in the y-direction as is shown in Fig. 1. The potential on the electrodes is $\pm\Phi_0/2$ if one applies the voltage Φ_0 between the electrode pairs. The field strength is given by

$$E_x = -\Phi_0/r_0^2 \cdot x \quad , \quad E_z = \Phi_0/r_0^2 \cdot z \quad , \quad E_y = 0$$

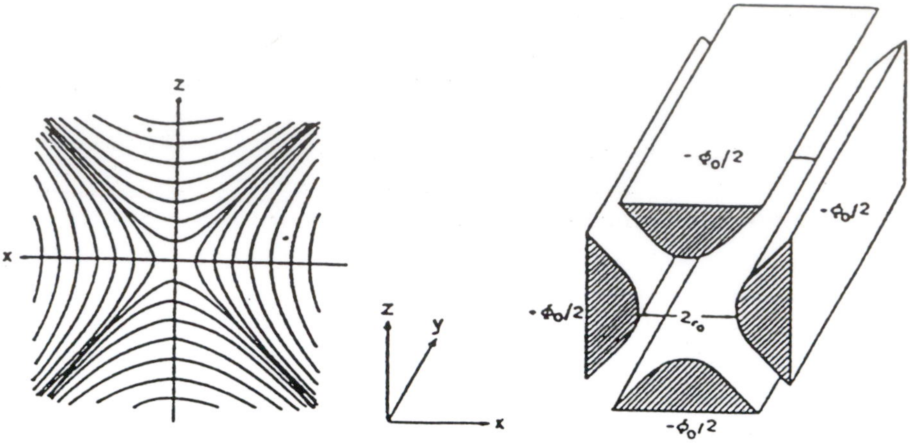

Figure 1. a) Equipotential lines for a plane quadrupole fild, b) the electrodes structure for the mass filter.

If one injects ions in the y-direction it is obvious that for a constant voltage Φ_0 the ions will perform harmonic oscillations in the the x-y-plane but due to the opposite sign in the field E_z their amplitude in the z-direction will increase exponentially. The particles are defocused and will be lost by hitting the electrodes.

This behaviour can be avoided if the applied voltage is periodic. Due to the periodic change of the sign of the electric force one gets focusing and defocusing in both the x- and z-directions alternating in time. If the applied voltage is given by a dc voltage U plus an r.f. voltage V with the driving frequency ω

$$\Phi_0 = U + V \cos\omega t$$

the equations of motion are

$$\ddot{x} + \frac{e}{mr_0^2} (U + V \cos\omega t) \, x = 0$$

$$\ddot{z} - \frac{e}{mr_0^2} (U + V \cos\omega t) \, z = 0 \qquad (4)$$

At first sight one expects that the time-dependent term of the force cancels out in the time average. But this would be true only in a homogenous field. In a periodic inhomogenous field, like the quadrupole field there is a small average force left, which is always in the direction of the lower field, in our case toward the center. Therefore, certain conditions exist that enable the ions to traverse the quadrupole field without hitting the electrodes, i.e. their motion around the y-axis is stable with limited amplitudes in x- and z-directions. We learned these rules from the theory of the Mathieu equations, as this type of differential equation is called.

In dimensionless parameters these equations are written

$$\frac{d^2x}{d\tau^2} + (a_x + 2q_x \cos 2\tau) \, x = 0$$

$$\frac{d^2z}{d\tau^2} + (a_z + 2q_z \cos 2\tau) \, z = 0 \qquad (5)$$

By comparison with equation (4) one gets

$$a_x = -a_z = \frac{4eU}{mr_0^2\omega^2} \; , \; q_x = -q_z = \frac{2eV}{mr_0^2\omega^2} \; , \; \tau = \frac{\omega t}{2}. \qquad (6)$$

The Mathieu equation has two types of solution.
1. stable motion: the particles oscillate in the x-z-plane with limited amplitudes. They pass the quadrupole field in y-direction without hitting the electrodes.

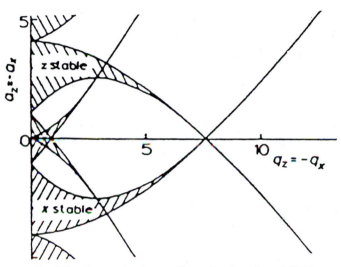

Figure 2. The overall stability diagram for the two-dimensional quadrupole field.

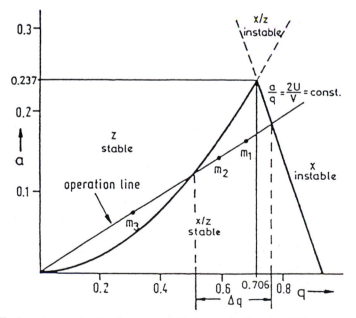

Figure 3. The lowest region for simultaneous stability in x- and z-direction. All ion masses lie on the operation line, $m_1 > m_1$.

2. unstable motion: the amplitudes grow exponentialy in x, z or in both directions. The particles will be lost.

Whether stability exists depends only on the parameters a and q and not on the initial parameters of the ion motion, e.g. their velocity. Therefore, in an a-q-map there are regions of stability and instability (Fig.2). Only the overlapping region for x and z stability is of interest for our problem. The most relevant region $0 < a, q < 1$ is plotted in Fig. 3. The motion is stable in x and z only within the triangle.

For fixed values r_0 ω, U and V all ions with the same m/e have the same operating point in the stability diagram. Since a/q is equal to 2U/V and does not depend on m, all masses lie along the operating line a/q = const. On the q axis (a = 0, no d.c. voltage) one has stability from $0 < q < q_{max} = 0.92$ with the consequence that all masses between $\infty > m > m_{min}$ have stable orbits. In this case the quadrupole field works as a high pass mass filter. The mass range Δm becomes narrower with increasing dc voltage U i.e. with a steeper operating line and approaches $\Delta m = 0$, if the line goes through the tip of the stability region. The bandwidth in this case is given only by the fluctuation of the field parameters. If one changes U and V simultaneously and proportionally in such a way that a/q remains sonstant, one brings the ions of the various masses successively in the stability region scanning through the mass spectrum in this way. Thus the quadrupole works as a mass spectrometer.

A schematic view of such a mass spectrometer is given in Fig. 4. In Figs. 5a,b. the first mass spectra obtained in 1954 are shown [6]. Clearly one sees the influence of the d.c. voltage U on the resolving power.

In quite a number of theses the performance and application of such

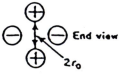

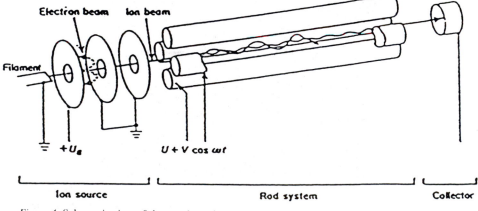

Figure 4. Schematic view of the quadrupole mass spectrometer or mass filter.

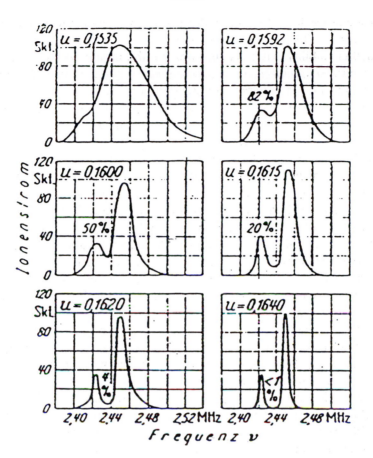

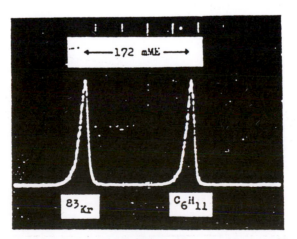

Figure 5. a) Very first mass spectrum of Rubidium. Mass scanning was achieved by periodic variation of the driving frequency v. Parameter: $u = \frac{U}{V}$, at $u = 0.164$ ^{85}Rb and ^{87}Rb are fully resolved. b) Mass doublet $^{83}Kr - C_6H_{11}$. Resolving power $m/\Delta m = 6500$ [9].

instruments was investigated at Bonn University [7,8,9]. We studied the influence of geometrical and electrical imperfections giving rise to higher multipole terms in the field. A very long instrument ($l = 6$ m) for high precision mass measurements was built achieving an accuracy of $2 \cdot 10^{-7}$ in determining mass ratios at a resolving power $\frac{m}{\Delta m} = 16\,000$. Very small ones were used in rockets to measure atomic abundances in the high atmosphere. In another experiment we succeeded in separating isotopes in amounts of milligrams using a resonance method to shake single masses out of an intense ion beam guided in the quadrupole.

In recent decades the r.f. quadrupole whether as mass spectrometer or beam guide due to its versatility and technical simplicity has found broad applications in many fields of science and technology. It became a kind of standard instrument and its properties were treated extensively in the literature [10].

The Ion Trap

Already at the very beginning of our thinking about dynamic stabilization of ions we were aware of the possibility using it for trapping ions in a three-dimensional field. We called such a device "Ionenkäfig"[11,12,13]. Nowadays the word "ion trap" is preferred.

The potential configuration in the ion trap has been given in eq. (3). This configuration is generated by an hyperbolically shaped ring and two hyperbolic rotationally symmetric caps as it is shown schematically in Fig. 6a. Fig. 6b gives the view of the first realized trap in 1954.

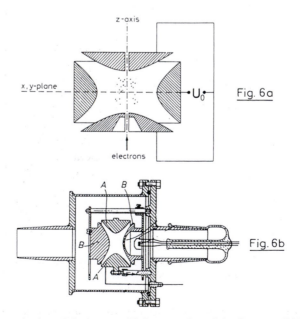

Figure 6. a) Schematic view of the ion trap. b) Cross section of the first trap (1955).

If one brings ions into the trap, which is easily achieved by ionizing inside a low pressure gas by electrons passing through the volume, they perform the same forced motions as in the two-dimensional case. The only difference is that the field in z-direction is stronger by a factor 2. Again a periodic field is needed for the stabilization of the orbits. If the voltage $\Phi_0 = U + V \cos \omega t$ is applied between the caps and the ring electrode the equations of motion are represented by the same Mathieu functions of eq.(5). The relevant parameters for the r motion correspond to those in the x-direction in the plane field case. Only the z parameters are changed by a factor 2.

Accordingly, the region of stability in the a-q-map for the trap has a different shape as is shown in Fig. 7. Again the mass range of the storable ions (i.e. ions in the stable region) can be chosen by the slope of the operation line $a/q = 2U/V$. Starting with operating parameters in the tip of the stable region one can trap ions of a single mass number. By lowering the d.c. voltage one brings the ions near the q-axis where their motions are much more stable.

For many applications it is necessary to know the frequency spectrum of the oscillating ions. From mathematics we learn that the motion of the ions can be described as a slow (secular) oscillation with the fundamental fre-

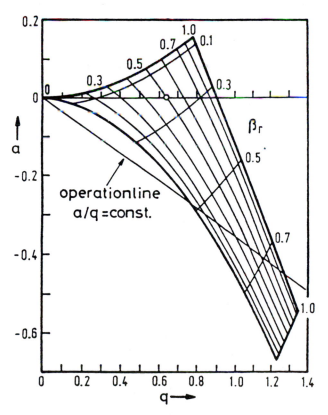

Figure 7. The lowest region for stability in the ion trap. On the lines inside the stability region β_z and β_r resp. are constant.

quencies $\omega_{r,z} = \beta_{r,z} \cdot \omega/2$ modulated with a micromotion, a much faster oscillation of the driving frequency ω if one neglects higher harmonics. The frequency determining factor β is a function only of the Mathieu parameters a and q and therefore mass dependent. Its value varies between 0 and 1; lines of equal β are drawn in Fig. 7.

Due to the stronger field the frequency ω_z of the secular motion becomes twice ω_r. The ratio ω/ω_z is a criterion for the stability. Ratios of 10:1 are easily achieved and therefore the displacement by the micromotion averages out over a period of the secular motion.

The dynamic stabilization in the trap can easily be demonstrated in a mechanical analogue device. In the trap the equipotential lines form a saddle surface as is shown in Fig. 8. We have machined such a surface on a round disc. If one puts a small steel ball on it, then it will roll down: its position is unstable. But if one let the disk rotate with the right frequency

Potential in the Ion Trap

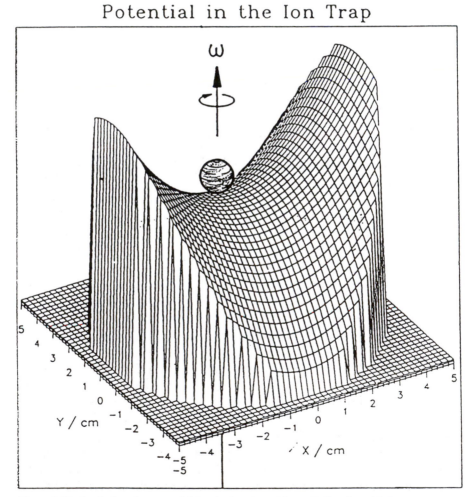

Figure 8. Mechanical analogue model for the ion trap with steelball as "particle".

appropriate to the potential parameters and the mass of the ball (in our case a few turns/s) the ball becomes stable, makes small oscillations and can be kept in position over a long time. Even if one adds a second or a third ball they stay near the center of the disc. The only condition is that the related Mathieu parameter q be in the permitted range. I brought the device with me. It is made out of plexiglas which allows demonstration of the particle motions with the overhead projector.

This behaviour gives us a hint of the physics of the dynamic stabilization. The ions oscillating in the r- and z-directions to first approximation harmonically, behave as if they are moving in a pseudo potential well quadratic in the coordinates. From their frequencies ω_r and ω_z we can calculate the depth of this well for both directions. It is related to the amplitude V of the driving voltage and to the parameters a and q. Without any d.c. voltage the depth is given by $D_z = (q/8) \, V$, in the r-direction it is half of this. As in practice V amounts to a few hundred volts the potential depth is of the order of 10 Volts. The width of the well is given by the geometric dimensions. The resulting configuration of the pseudo potential [14] is therefore given by

$$\Phi = D \frac{(r^2 + 4z^2)}{r_0^2 + 2z_0^2}.$$

Cooling process

As mentioned, the depth of the relevant pseudo-potential in the trap is of the order of a few volts. Accordingly the permitted kinetic energy of the stored ions is of the same magnitude and the amplitude of the oscillations can reach the geometrical dimensions of the trap. But for many applications one needs particles of much lower energy well concentrated in the center of the trap. Especially for precise spectroscopic measurements it is desirable to have extremely low velocities to get rid of the Doppler effect and an eventual Stark effect, caused by the electric field. It becomes necessary to cool the ions. Relatively rough methods of cooling are the use of a cold buffer gas or the damping of the oscillations by an external electric circuit. The most effective method is the laser induced sideband fluorescence developed by Wineland and Dehmelt [15].

In 1959 Wuerker et al. [16] performed an experiment trapping small charged Aluminium particles ($\varnothing \sim mm$) in the quadrupole trap. The necessary driving frequency was around 50 Hz accordingly. They studied all the eigenfrequencies and took photographs of the particle orbits; see Figs. 9a, b. After they have damped the motion with a buffer gas they observed that the randomly moving particles arranged themselves in a regular pattern. They formed a crystal.

In recent years one has succeeded in observing optically single trapped ions by laser resonance fluorescence [17]. Walther et al., using a high resolution image intensifier observed the pseudo-crystallization of ions in the trap after cooling the ions with laser light. The ions are moving to such

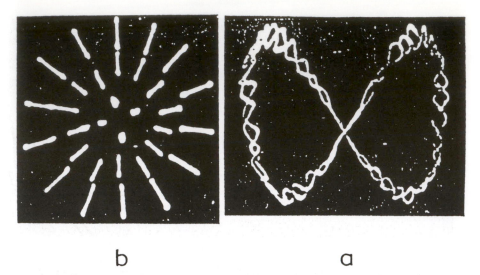

b   a

Figure 9. a) Photomicrograph of a Lissajous orbit in the r-z-plane of a single charged particle of Aluminium powder. The micro motion is visible. b) Pattern of "condensed" Al particles [16].

positions where the repulsive Coulomb force is compensated by the focusing forces in the trap and the energy of the ensemble has a minimum. Figs. 10a, b show such a pattern with 7 ions. Their distance is of the order of a few micrometers. These observations opened a new field of research [18].

The Ion Trap as Mass Spectrometer

As mentioned the ions perform oscillations in the trap with frequencies ω_r and ω_z which at fixed field parameters are determined by the mass of the ion. This enables a mass selective detection of the stored ions. If one connects the cap electrodes with an active r.f. circuit with the eigenfrequency Ω, in the case of resonance $\Omega = \omega_z$, the amplitude of the oscillations increases linearly with time. The ions hit the cap or leave the field through a bore hole and can easily be detected by an electron multiplier device. By modulating the ion frequency determining voltage V in a sawtooth mode one brings the ions of the various masses one after the other into resonance, scanning the mass spectrum. Fig. 11 shows the first spectrum of this kind achieved by Rettinghaus [19].

The same effect with a faster increase of the amplitude is achieved if one inserts a small band of instability into the stability diagram . It can be generated by superimposing on the driving voltage $V \cos \omega t$ a small additional rf voltage, e.g. with frequency $\omega/2$, or by adding a higher multipole term to the potential configuration [5b,20].

In summary the ion trap works as ion source and mass spectrometer at the same time. It became the most sensitive mass analyzer available as only a few ions are necessary for detection. Its theory and performance is reviewed in detail by R.E. March [21].

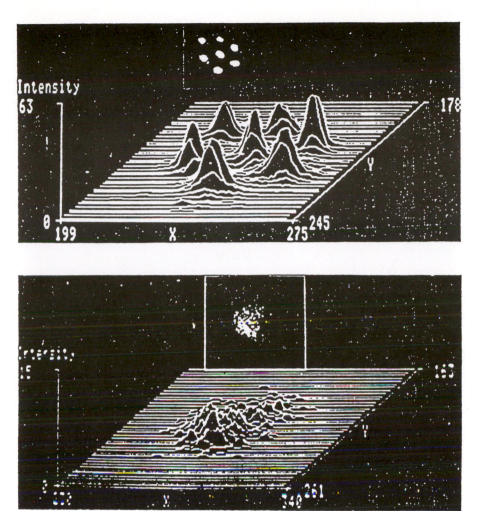

Figure 10. a) Pseudo crystal of 7 magnesium ions. Particle distance 23 μm. b) The same trapped particles at "higher temperature". The crystal has melted [18].

The Penning Trap

If one applies to the quadrupole trap only a d.c. voltage in such a polarity that the ions perform stable oscillations in the z-direction with the frequency $\omega_z^2 = \frac{2eU'}{mr_0^2}$ the ions are unstable in the x-y-plane, since the field is directed outwards. Applying a magnetic field in the axial direction, the z-motion remains unchanged but the ions perform a cyclotron motion ω in the x-y-plane. It is generated by the Lorentz force F_L directed towards the center. This force is partially compensated by the radial electric force $F_r = \frac{eU'}{r_0^2} r$. As long as the magnetic force is much larger than the electric one, stability exists in the r-y-plane as well. No r.f. field is needed. The resulting rotation frequency calculates to

$$\omega = \omega_c - \frac{\omega_z^2}{2\omega}.$$

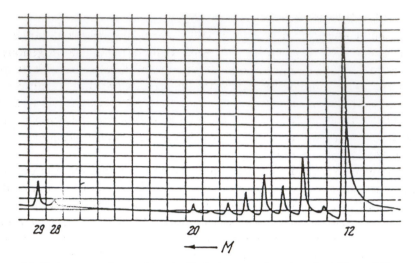

Figure 11. First mass spectrum achieved with the ion trap. Gas: air at $2 \cdot 10^{-9}$ torr [19]

It is slightly smaller than the undisturbed cyclotron frequency eB/m. The difference is due to the magnetron frequency

$$\omega_M = \frac{\omega_z^2}{2\omega}$$

which is independent of the particle mass.

The Penning trap [22], as this device is called, is of advantage if magnetic properties of particles have to be measured, as for example Zeeman transitions in spectroscopic experiments, or cyclotron frequencies for a very precise comparison of masses as are performed e.g. by G. Werth. The most spectacular application the trap has found in the experiments of G. Gräff [23] and H. Dehmelt for measuring the anomalous magnetic moment of the electron. It was brought by Dehmelt [24] to an admirable precision by observing only a single electron stored for many months.

Traps for neutral particles

In the last examination I had to pass as a young man I was asked if it would be possible to confine neutrons in a bottle in order to prove if they are radioactive. This question, at that time only to be answered with "no", pursued me for many years until I could have had replied: Yes, by means of a magnetic bottle. It took 30 years until by the development of super-conducting magnets its realization became feasible.

Using the example of such a bottle I would like to demonstrate the principle of confining neutral particles. Again the basis is our early work on focusing neutral atoms and molecules having a dipole moment by means of multipole fields making use of their Zeeman or Stark effect to first and second order [1,2,3]. Both effects can be used for trapping. Until now only magnetic traps were realized for atoms and neutrons. Particularly, B. Martin, U. Trinks, and K. J. Kügler contributed to their development with great enthusiasm.

The principle of magnetic bottles

The potential energy U of a particle with a permanent magnetic moment μ in a magnetic field is given by $U = -\mu B$. If the field is inhomogenous it corresponds to a force $F = grad(\mu B)$. In the case of the neutron with its spin $\hbar/2$ only two spin directions relative to the field are permitted. Therefore, its magnetic moment can be oriented only parallel or antiparallel to B. In the parallel position the particles are drawn into the field and in the opposite orientation they are repelled. This permits their confinement to a volume with magnetic walls.

The appropriate field configuration to bind the particles harmonically is in this case a magnetic sextupole field. As I have pointed out such a field B increases with r^2, $B = \frac{B_0}{r_0^2} \cdot r^2$ and the gradient $\frac{\delta B}{\delta r}$ with r respectively.

In such a field neutrons with orientation $\mu \uparrow\uparrow B$ satisfy the confining condition as their potential energy $U = + \mu B \sim r^2$ and the restoring force $\mu grad B = -cr$ is always oriented towards the center. They oscillate in the field with the frequency $\omega^2 = \frac{2\mu B_0}{mr_0^2}$. Particles with $\mu \uparrow\downarrow B$ are defocused and leave the field. This is valid only as long as the spin orientation is conserved. Of course, in the sextupole the direction of the magnetic field changes with the azimuth but as long as the particle motion is not too fast the spin follows the field direction adiabatically conserving the magnetic quantum state. This behaviour permits the use of a magnetic field constant in time in contrast to the charged particle in an ion trap.

An ideal linear sextupole in the x-z-plane is generated by six hyperbolically shaped magnetic poles of alternating polarity extended in y-direction, as shown in Figs. 12a, b. It might be approximated by six straight current leads

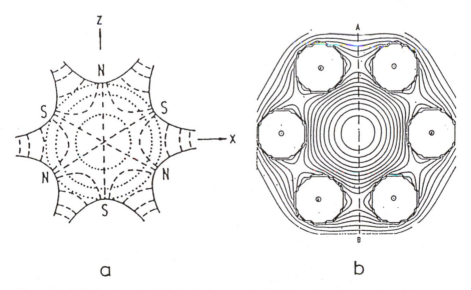

a   b

Figure 12. a) Ideal sextupole field. Dashed: magnetic field lines, dotted: lines of equal magnetic potential, B = const. b) Linear sextupole made of 6 straight current leads with alternating current direction.

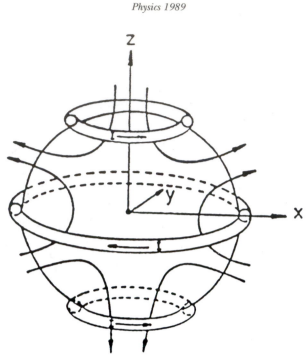

Figure 13. Sextupole sphere.

with alternating current directions arranged in a hexagon. Such a configuration works as a lense for particles moving along the y-axis.

There are two possibilities to achieve a <u>closed storage volume:</u> a sextupole
sphere and a sextupole torus. We have realized and studied both.

The spherically symmetric field is generated by three ring currents in an
arrangement shown in Fig. 13. The field B increases in all directions with r^2
and has its maximum value B_0 at the radius r_0 of the sphere. Using superconducting current leads we achieved $B_0 = 3T$ in a sphere with a radius of 5 cm.
But due to the low magnetic moment of the neutron $\mu = 6 \cdot 10^{-8} \ eV/T$ the
potential depth μB_0 is only $1.8 \cdot 10^{-7} \ eV$ and hence the highest velocity of
storable neutrons is only $v_{max} = 6 m/s$. Due to their stronger moment for Na
atoms these values are $2.2 \cdot 10^{-4} \ eV$ and 37 m/s, respectively.

The main problem with such a closed configuration is the filling process,
especially the cooling inside. However, in 1975 in a test experiment we
succeeded in observing a storage time of 3 s for sodium atoms evaporated
inside the bottle with its Helium cooled walls [25]. But the breakthrough in
confining atoms was achieved by W. D. Phillip and H.J. Metcalf using the
modern technique of Laser cooling [26].

The problem of storing neutrons becomes easier if one uses a linear
sextupole field bent to a closed torus with a radius R as is shown in Fig. 14.
The magnetic field in the torus volume is unchanged $B = \frac{B_0}{r_0^2} \cdot r^2$ and has
no component in azimuthal direction. The neutrons move in a circular orbit
with radius R_S if the centrifugal force is compensated by the magnetic force

$$F_c = \frac{mv_\varphi^2}{R_S} = \mu \left. \frac{\delta B}{\delta r} \right]_{R_S} .$$

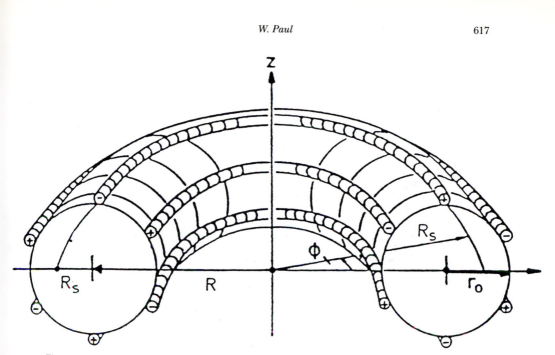

Figure 14. Sextupole torus. R_s orbit of circulating neutrons.

In such a ring the permitted neutron energy is limited by

$$E_{max} = \mu \cdot B_0 \left(\frac{R}{r_0} + 1 \right).$$

It is increased by a factor $(\frac{R}{r_0} + 1)$ compared to the case of the sextupole sphere. As the neutrons have not only an azimuthal velocity but also components in r and z directions they are oscillating around the circular orbit.

But this toroidal configuration has not only the advantage of accepting higher neutron velocities, it also permits an easy injection of the neutrons in the ring from the inside. The neutrons are not only moving in the magnetic potential well but they also experience the centrifugal barrier. Accordingly, one can lower the magnetic wall on the inside by omitting the two inward current leads. The resulting superposition of the magnetic and the centrifugal potential still provides a potential well with its minimum at the beam orbit. But there is no barrier for the inflected neutrons.

It is obvious, that the toroidal trap in principle works analogous to the storage rings for high energy charged particles. In many respects the same problems of instabilities of the particle orbits by resonance phenomena exist, causing the loss of the particles. But also new problems arise like, e.g. undesired spin flips or the influence of the gravitational force. In accelerator physics one has learned to overcome such problems by shaping the magnetic field by employing the proper multipole components. This technique is also appropriate in case of the neutron storage ring. The use of the magnetic force $\mu \cdot gradB$ instead of the Lorentz force being proportional to B just requires multipole terms of one order higher. Quadrupoles for

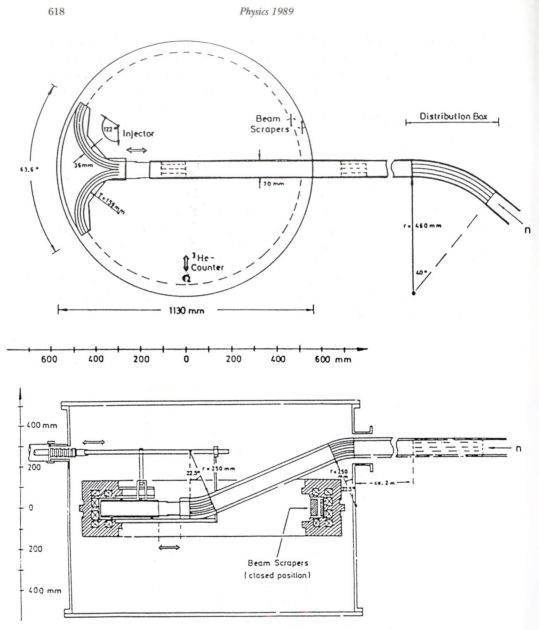

Figure 15. Schematic top and side view of the neutron storage ring experiment.

focusing have to be replaced by sextupoles and e.g. octupoles for stabilization of the orbits by decapoles.

In the seventies we have designed and constructed such a magnetic
storage ring with a diameter of the orbits of 1 m. The achieved usable field
of 3.5 T permits the confinement of neutrons in the velocity range of $5-20$
m/s corresponding to a kinetic energy up to $2 \cdot 10^{-6}$ eV. The neutrons are
injected tangentially into the ring by a neutron guide with totally reflecting
walls. The inflector can be moved mechanically into the storage volume and
shortly afterwards be withdrawn.

The experimental set up is shown in Fig. 15. A detailed description of the storage ring, its theory and performance is given in [27]. In 1978 in a first experiment we have tested the instrument at the Grenoble high flux reactor. We could observe neutrons stored up to 20 min after injection by moving a neutron counter through the confined beam after a preset time. As by the detection process the neutrons are lost, one has to refill the ring starting a new measurement. But due to the relatively low flux of neutrons in the acceptable velocity range, their number was too low to make relevant measurements with it.

In a recent experiment [28] at a new neutron beam with a flux improved by a factor 40 we could observe neutrons up to 90 min, i.e. roughly 6 times the decay time of the neutron due to radioactive decay. Fig. 16 shows the measured profile of the neutron beam circulating inside the magnetic gap. Measuring carefully the number of stored neutrons as a function of time we could determine the lifetime to $\tau = 877 \pm 10$ s (Fig. 17).

The analysis of our measurements lets us conclude that the intrinsic storage time of the ring for neutrons is at least about one day. It shows that we had understood the relevant problems in its design.

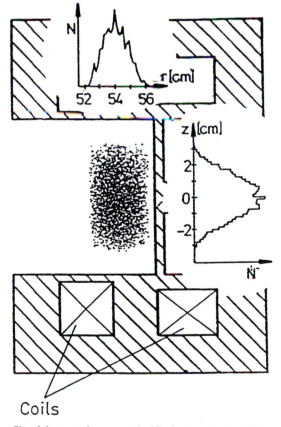

Figure 16. Beam profile of the stored neutrons inside the magnet gap 400 sec after injection.

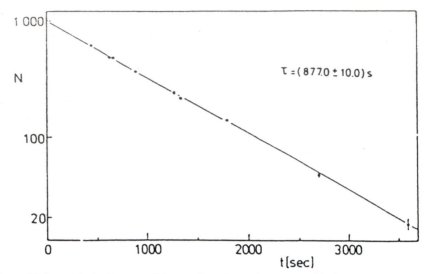

Figure 17. Logarithmic decrease of the number of stored neutrons with time.

The storage ring as a balance

This very reproducible performance permitted another interesting experiment. As I explained the neutrons are elastically bound to the symmetry plane of the magnetic field. Due to their low magnetic moment the restoring force is of the order of the gravitational force. Hence it follows that the weight of the neutron stretches the magnetic spring the particle is hanging on; the equilibrium center of the oscillating neutrons is shifted downwards. The shift z_0 is given by the balance $mg = \mu gradB$. One needs a gradient $\frac{\delta B}{\delta z}$ = 173 *Gauss/cm* for compensating the weight. As the gradient in the ring increases with z and is proportional to the magnetic current I one calculates the shift z_0 to

$$z_0 = const.mg/I.$$

It amounts in our case to $z_0 = 1.2$ *mm* at the highest magnet current $I = 200$ A and 4.8 *mm* at 50 A accordingly.

By moving a thin neutron counter through the storage volume we could measure the profile of the circulating neutron beam and its position in the magnet. Driving alternating the counter downwards and upwards in many measuring runs we determined z_0 as a function of the magnet current.

The result is shown in Fig. 18. The measured data taken with different experimental parameters are following the predicted line. A detailed analysis gives for the gravitational mass of the neutron the value

$$m_g = (1.63 \pm 0.06) \cdot 10^{-24} \, g.$$

It agrees within 4 % with the well known inertial mass.

Thus the magnetic storage ring represents a balance with a sensitivity of 10^{-25} *g*. It is only achieved because the much higher electric forces play no role at all.

I am convinced that the magnetic bottles developed in our laboratory as described will be useful and fruitful instruments for many other experiments in the future as the Ion Trap has already proved.

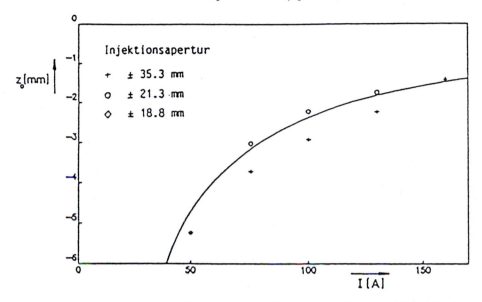

Figure 18. Downward shift of the equilibrium center of the neutron orbits due to the weight of the neutron as function of the magnetic current.

REFERENCES

[1] H. Friedburg and W. Paul, Naturwissenschaft 38, 159 (1951).
[2] H. G. Bennewitz and W. Paul, Z. f. Physik 139, 489 (1954).
[3] H. G. Bennewitz and W. Paul, Z. f. Physik, 141, 6 (1955).
[4] C. H. Townes, Proc. Nat. Acad. of Science, 80, 7679 (1983).
[5] a) W. Paul and H. Steinwedel, Z. f. Naturforschung 8a, 448 (1953); b) German Patent Nr. 944 900; USA Patent 2939958.
[6] W. Paul and M. Raether, Z. f. Physik, 140, 262 (1955).
[7] W. Paul, H. P. Reinhardt, and U. v. Zahn, Z. f. Physik 152, 143 (1958).
[8] F. v. Busch and W. Paul, Z. f. Physik, 164, 581 (1961).
[9] U. v. Zahn, Z. f. Physik, 168, 129 (1962).
[10] P. H. Dawson: *Quadrupole Mass Spectrometry and its Application,* Elsevier, Amsterdam 1976.
[11] W. Paul, O. Osberghaus, and E. Fischer, Forsch.Berichte des Wirtschaftsministeriums Nordrhein-Westfalen Nr. 415 (1958).
[12] K. Berkling, Thesis Bonn 1956.
[13] E. Fischer, Zeitschrift f. Physik 156, 1 (1959).
[14] H. Dehmelt, Adv. in Atom and Molec. Phys., Vol. 3 (1967).
[15] D. J. Wineland and H. Dehmelt, Bull. Am. Phys. Soc., 20, 637 (1975).
[16] R. F. Wuerker and R.V. Langmuir, Appl. Phys. 30, 342 (1959).
[17] W. Neuhauser, M. Hohenstett, P. Toschek and A. Dehmelt, Phys. Rev. A22, 1137 (1980).
[18] F. Dietrich, E. Chen, J. W. Quint and H. Walter, Phys. Rev. Lett. 59, 2931 (1987).
[19] G. Rettinghaus, Zeitschrift Angew. Physik, 22, 321 (1967).
[20] F. v. Busch and W. Paul, Z. f. Physik, 165, 580 (1961).
[21] R. E. March and R. J. Hughes, "*Quadrupole Storage Mass Spectrometry*", John Wiley, New York 1989.
[22] F. M. Penning, Physica 3, 873 (1936).
[23] G. Gräff, E. Klempt and G. Werth, Zeitschrift f. Physik 222, 201 (1969).
[24] R. S. van Dyck, P. B. Schwinberg, H. G. Dehmelt, Phys. Lett. 38, 310 (1977).
[25] B. Martin, Thesis Bonn University 1975.
[26] A. L. Migdal, J. Prodan, W. D. Phillips, Th. H. Bergmann, and H.J. Metcalf, Phys. Rev. Lett. 54, 2596 (1985).
[27] K. J. Kügler, W. Paul, and U. Trinks, Nucl. Instrument. Methods A 228, 240 (1985).
[28] W. Paul, F. Anton, L. Paul, S. Paul, and W. Mampe, Z. f. Physik C 45, 25 (1989).

Physics 1990

JEROME I FRIEDMAN, HENRY W KENDALL and RICHARD E TAYLOR

for their pioneering investigations concerning deep inelastic scattering of electrons on protons and bound neutrons, which have been of essential importance for the development of the quark model in particle physics

THE NOBEL PRIZE IN PHYSICS

Speech by Professor Cecilia Jarlskog of the Royal Swedish Academy of Sciences.
Translation from the Swedish text.

Your Majesties, Your Royal Highnesses, Ladies and Gentlemen,

One of the most important tasks of physics is to provide us with a clearer picture of the world we live in. We know that the observable universe is much larger than any of us could imagine and is even, perhaps, no more than just an island in an ocean of universes. But the creation also has another unfathomable frontier—that towards smaller and smaller constituents: molecules, atoms and elementary particles.

It is the business of science to probe elementary particles as well as the most remote galaxies, collecting facts and deciphering relationships at all levels of creation. The amount of information increases rapidly and without understanding can become overwhelming. Such confusion prevailed at the end of 1950s. At the deepest level of the microscopic world were the electron, the proton and the neutron, particles which for years had been considered to be the fundamental building blocks of matter. However, they were no longer alone but were accompanied by many newly discovered particles. The special roles of the proton and the neutron are evident— among other things they are responsible for more than 99 percent of our weight. But what roles did the other particles play? Where had nature's elegance and beauty gone? Was there a hidden order not yet discovered by man?

There could be order but only at the price of postulating an additional, deeper level in nature—perhaps the ultimate level—consisting of only a few building blocks. Such an idea had been advanced and the new building blocks were called "quarks"—a word borrowed by the 1969 Nobel Prizewinner in Physics, Murray Gell-Mann, from "Finnegans Wake," for most of us an incomprehensible masterpiece by the great Irish novelist James Joyce. But the quark hypothesis was not alone. There was, for example, a model called "nuclear democracy" where no particle had the right to call itself elementary. All particles were equally fundamental and consisted of each other.

This year's Laureates lit a torch in this darkness. They and their co-workers examined the proton (and later on also the neutron) under a microscope—not an ordinary one, but a 2 mile-long electron accelerator built by Wolfgang K.H. Panofsky at Stanford, California. They did not anticipate anything fundamentally new: similar experiments, albeit at lower energies, had found that the proton behaved like a soft gelatinous sphere with many excited states, similar to those of atoms and nuclei. Nevertheless, the Laureates decided to go one step further and study the proton under extreme conditions. They looked for the electron undergoing a large deflection, and where the proton, rather than keeping its identity, seized a lot of the collision energy and broke up into a shower of new particles. This so-called "deep inelastic scattering" had generally been considered to be too rare to be worth investigating. But the experiment showed otherwise: deep inelastic scattering was far more frequent than expected, displaying a totally new facet of proton behavior. This result was at first skeptically received: perhaps the moving electron gave off undetected light. But this year's Prizewinners had been thorough and their findings were subsequently confirmed by other experiments.

The interpretation was given primarily by the theorists James D. Bjorken and the late Richard P. Feynman (Feynman stood in this Hall exactly 25 years ago to receive a Nobel Prize for another of his great contributions to physics). The electrons ricocheted off hard point-like objects inside the proton. These were soon shown to be identical with the quarks, thus simplifying the physicist's picture of the world; but the results could not be entirely explained by quarks alone. The Nobel Prize-winning experiment indicated that the proton also contained electrically neutral constituents. These were soon found to be "gluons," particles glueing the quarks together in protons and other particles.

A new rung on the ladder of creation had revealed itself and a new epoch in the history of physics had begun.

Dear Professors Friedman, Kendall and Taylor,

On behalf of the Royal Swedish Academy of Sciences I wish to convey to you our warmest congratulations for having taken us to the land of deep inelastic scattering where the colourful quarks and gluons first revealed themselves. You will now receive the Nobel Prize from the hands of His Majesty the King.

Richard E. Taylor

RICHARD E. TAYLOR

Medicine Hat is a small town in Southwestern Alberta founded just over 100 years ago in a valley where the Canadian Pacific Railway crossed the South Saskatchewan River. I was born there on November 2, 1929 and raised in comfortable if somewhat Spartan circumstances. My father was the son of a Northern Irish carpenter and his Scottish wife who homesteaded on the Canadian prairies; my mother was an American, the daughter of Norwegian immigrants to the northern United States who moved to a farm in Alberta shortly after the first World War. During my early years our family of three was part of a large family clan headed by my Scottish grandmother. I attended schools named after English Generals and Royalty—Kitchener, Connaught, Alexandra.

Although I read quite a bit and found mathematics easy, I was not an outstanding student. In high school I did reasonably well in mathematics and science thanks to some talented and dedicated teachers.

I was nearly ten years old when World War II began. That conflict had a great effect on our town, and on me. In rapid succession the town found itself host to an R.A.F. flight training school, a prisoner of war camp and a military research establishment. The wartime glamor of the military, the sudden infusion of groups of sophisticated and highly-educated people, and new cultural opportunities (the first live symphonic music I ever heard was played by German prisoners of war) all transformed our town and widened the horizons of the young people there. I developed an interest in explosives and blew three fingers off my left hand just before hostilities ended in Europe. The atomic bomb that ended the war later that summer made me intensely aware of physicists and physics.

Higher education was highly prized in the society of a small prairie town and I was expected to continue on to university. After some difficulties over low grades in some high school subjects, I was admitted to the University of Alberta in Edmonton. I registered in a special program emphasizing mathematics and physics and gradually became interested in experimental physics, continuing my studies towards a Masters degree at the same institution. My thesis research was a rather primitive effort to measure double β-decay in an aging Wilson cloud chamber. Between sessions at the University, I spent two summers as a research assistant at the Defense Research Board installation near Medicine Hat working with Dr. E.J. Wiggins, who encouraged me to continue my studies either in eastern Canada or in the United States.

Those were interesting years, and during this time I met, courted and married Rita Bonneau — a partnership which has enriched my life in every way. Together we decided to try California, and I was accepted into the graduate program at Stanford, while she found work teaching in a military school in order to support us both. The first two years at Stanford were exciting beyond description — the Physics Department at Stanford included Felix Bloch, Leonard Schiff, Willis Lamb, Robert Hofstadter, and W.K.H. (Pief) Panofsky who had just arrived from Berkeley. I found that I had to work hard to keep up with my fellow students, but learning physics was great fun in those surroundings. At the end of the second year I joined the High Energy Physics Laboratory where the new linear accelerator was just beginning to do experiments. My thesis work was accomplished there under Prof. Robert F. Mozley, on a rather difficult experiment producing polarized γ-rays from the accelerator beam and then using those γ-rays to study π-meson production.

In 1958 I was invited to join a group of physicists at the Ecole Normale Superieure in Paris who were planning experiments at an accelerator (similar to the linac at Stanford) which was under construction in Orsay. I stayed in France for about three years working on the experimental facilities for the accelerator, and then participated in some electron scattering experiments. My wife began a new career there as a librarian at the Orsay laboratory, a career which was interrupted for a while when our son, Ted, was born in 1960. We returned to the United States in 1961 but a continuing connection to French physics and physicists has been a significant element in my life since that time — including a Doctorate (Honoris Causa) very kindly conferred upon me in 1980 by the Université de Paris-Sud.

Upon our return to the United States, I joined the staff of the Lawrence Berkeley Laboratory at the University of California. After less than a year in Berkeley, I moved back to Stanford where work on the construction of Stanford Linear Accelerator Center (SLAC) was just beginning. At SLAC, I started working on the design of the experimental areas for the new accelerator. By 1963 I had joined the group considering the requirements for electron scattering apparatus in the larger of two experimental areas. I worked closely with Pief Panofsky, and with collaborators from the California Institute of Technology and the Massachusetts Institute of Technology. I spent the next decade helping to build equipment and taking part in various electron scattering experiments, a number of which are the subject of the 1990 Nobel lectures. This was a period of intense activity, but also one of intense enjoyment for me. I was surrounded by people I liked and admired, and deeply involved in experiments which generated interest in laboratories and universities all over the world. I count myself extremely fortunate to have been at SLAC at that time.

I became a member of the SLAC faculty in 1968. In 1971, I was awarded a Guggenheim fellowship and spent an interesting sabbatical year at CERN, where I was impressed by the great progress that European science had made in the decade since I had worked in France.

Well before my trip to CERN, colleagues in the group at SLAC had become interested in testing some of the invariance properties of the electromagnetic interaction, a field which would absorb our efforts for most of the 1970s. When Charles Prescott joined the group in 1970, he began a serious study of ways to test parity conservation in the interaction between an electron and a nucleon. The electroweak theories of Weinberg and Salam predicted levels of nonconservation that looked extremely hard to measure. We attempted an experiment with the existing Yale polarized source, but the measurements did not reach the desired level of sensitivity. I was not very encouraging to my colleagues who wished to pursue the experiment to higher levels of accuracy. After the theoretical work of Veltman and 't Hooft and the discovery of neutral currents at CERN (during the year I was there) and at NAL (now Fermilab), the interest in experiments on parity conservation greatly intensified. In 1975 a new method for producing polarized electrons was discovered by a group in Colorado which included E. L. Garwin of SLAC. In 1978, after building a source for the linac based on the new method, we were able to demonstrate a violation of parity in close agreement with the electroweak predictions.

After the parity experiments, our group presented two proposals for large experimental facilities at PEP, the e^+e^- collider then being built at SLAC. Both those proposals were rejected. The group was finally successful in proposing a relatively small PEP detector, but I did not take part in that experiment.

In 1981, I received an Alexander von Humboldt award which allowed me to spend most of the 1981−82 academic year at DESY in Hamburg. In 1982 I returned to SLAC as Associate Director for Research, a post I held until 1986 when I resigned to return to research. Since that time I have spent quite a bit of time in Europe and I am presently playing a very small role in the H_1 detector preparations at HERA.

DEEP INELASTIC SCATTERING:
THE EARLY YEARS

Nobel Lecture, December 8, 1990

RICHARD E. TAYLOR

Stanford Linear Accelerator Center, Stanford, California, USA

FOREWORD

Soon after the 1990 Nobel Prize in Physics was announced Henry Kendall, Jerry Friedman and I agreed that we would each describe a part of the deep inelastic experiments in our Nobel lectures. The division we agreed upon was roughly chronological. I would cover the early times, describing some of the work that led to the establishment of the Stanford Linear Accelerator Center where the experiments were performed, followed by a brief account of the construction of the experimental apparatus used in the experiments and the commissioning of the spectrometer facility in early elastic scattering experiments at the Center.

In a second paper, Professor Kendall was to describe the inelastic experiments and the important observation of scale invariance which was found in the early electron-proton data.

In a final paper, Professor Friedman was to describe some of the later experiments at SLAC along with experiments performed by others using muon and neutrino beams, and how these experiments, along with advances in theory, led to widespread acceptance of the quark model as the best description of the structure of the nucleon.

This paper is, therefore, part of a set and should be read in conjunction with the lectures of H. W. Kendall[1] and J. I. Friedman.[2]

There were many individuals who made essential contributions to this work. Our acknowledgements to a number of them are given in Reference 3.

*

Forty years of electron scattering experiments have had a significant impact on the understanding of the basic components of matter. Progress in experimental high energy physics is often directly coupled to improvements in accelerator technology and experimental apparatus. The electron scattering experiments, including the deep inelastic experiments cited this year by the Royal Swedish Academy of Sciences, provide examples of this sort of progress. Experiments made possible by increasing electron energy and intensity, along with increasingly sophisticated detectors have continued to shed light on the structure of nuclei and nucleons over the years. Much additional information has come from experiments using secondary beams of muons and neutrinos from proton accelerators.

Scattering experiments can trace their roots back to the α-particle experiments[4] in Rutherford's laboratory which led to the hypothesis of the nuclear atom.[5] The α-sources used at that time emitted electrons as well as α-particles, but the electron momentum was too small to penetrate beyond the electron cloud of the target atoms, and electron scattering was just an annoying background in those experiments.

Following the landmark experiments of Franck and Hertz[6] on the interaction of electrons with the atoms of various gases, electron scattering was used extensively to investigate the electronic configurations of atoms. Later, after higher energy electrons became available from accelerators, interest in their use as probes of the nucleus increased. Rose[7] gave the first modern treatment of the subject in 1948, followed by Schiff,[8] who was exploring possible experiments for the new electron linear accelerator at Stanford. Schiff stressed the importance of e-p measurements which could probe the structure of the proton itself using the known electromagnetic interaction. Soon after, Rosenbluth[9] calculated the probability that an electron of energy E_0 will scatter through an angle θ in an elastic collision with a proton — corresponding to the following idealized experimental set up:

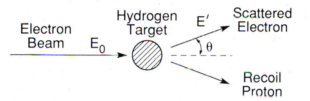

The energy E' of the scattered electron is *less* than the incident energy E_0 because energy is transferred to the recoil proton (of mass M):

$$E' = \frac{E_0}{1 + \frac{2E_0}{M} \sin^2 \theta/2}$$

The square of the four momentum transfer, Q^2, is a measure of the ability to probe structure in the proton. The uncertainty principle limits the spatial definition of the scattering process to $\sim \hbar/Q$ so Q^2, (and therefore E_0) must be large in order to resolve small structures.

$$Q^2 = 4E_0E' \sin^2 \theta/2$$

When only the scattered electron is detected the elastic differential cross section, $d\sigma/d\Omega$, obtained by Rosenbluth is a simple expression, quite similar to the original Rutherford scattering formula:

$$\frac{d\sigma}{d\Omega} = \frac{\alpha^2}{4E_0^2 \sin^4 \theta/2} \cdot \cos^2 \theta/2 \cdot \frac{E'}{E_0} \left[\frac{G_E^2 + \tau G_M^2}{1 + \tau} + 2\tau G_M^2 \tan^2 \theta/2 \right],$$

where

$$\tau = Q^2/4M^2$$

G_E and G_M are form factors describing the distributions of charge and magnetic moment respectively. They are functions of only the momentum transfer, Q^2.

$$G_E = G_E(Q^2) \quad G_M = G_M(Q^2)$$

$$G_E(0) = 1 \qquad G_M(0) = \mu_p.$$

where μ_p is the magnetic moment of the proton (in units of $\hbar$). If the charge and magnetic moment distributions are small compared with $\hbar/Q$, then G_E and G_M will not vary as Q^2 changes, but if the size of those distributions is comparable with $\hbar/Q$ then the G's will decrease with increasing Q^2.

Hanson, Lyman and Scott[10] were the first to observe elastic electron scattering from a nucleus using a 15.7 MeV external beam from the 22 MeV betatron at Illinois. They were studying the scattering of electrons by electrons and observed two peaks in the energy spectrum of the scattered electrons (Fig 1).

In 1953, the commissioning of the first half of the new Mark III linac in the High Energy Physics Laboratory (HEPL) at Stanford provided an external electron beam of unprecedented intensity at energies up to 225 MeV. Complementing this advance in accelerator technology, Hofstadter and his collaborators constructed a quasi-permanent scattering facility (Fig. 2) based on a 180° magnetic spectrometer (radius of bending = 18 inches). The spectrometer could be rotated about the target to measure different scattering angles, and the excitation of the magnet could be varied to

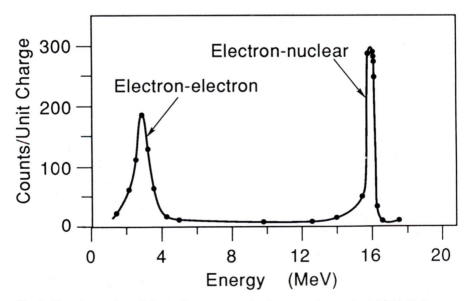

Fig. 1. First observation of elastic electron scattering from a nucleus, using 15.7 MeV electrons from the Illinois betatron, scattered at 10°.

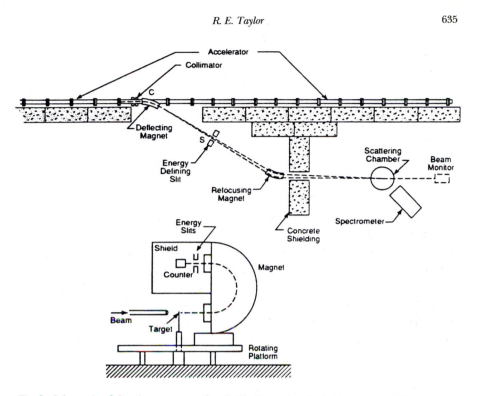

Fig. 2. Schematic of the electron scattering facility located at the halfway point of the Mark III linear accelerator at the High Energy Physics Laboratory at Stanford. The central orbit in the spectrometer has a radius of 18 inches.

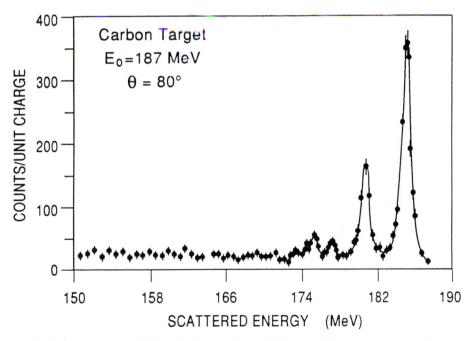

Fig. 3. Energy spectrum of 187 MeV electrons scattered through 80° by a carbon target, using the apparatus in Figure 2.

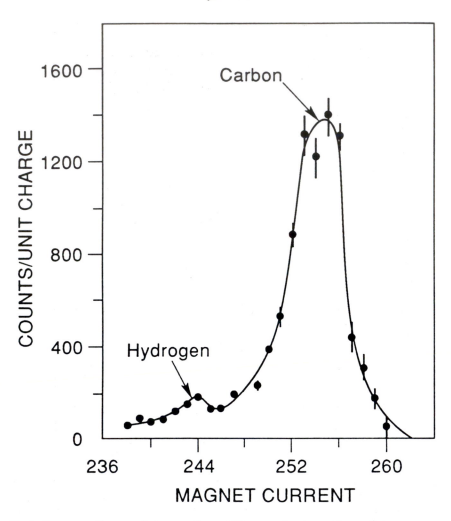

Fig. 4. Spectrum of scattered electrons from a CH_2 target showing evidence of electron-proton scattering, circa 1954.

change the energy of the electrons detected. This apparatus was used for a series of experiments with only minor modifications.

Nuclear scattering was easy to observe with this apparatus. At small angles, the "elastic peak" was the most prominent feature of the energy spectrum of the scattered electrons, although scattering with transitions to excited nuclear states was also evident[11] (Fig. 3). From the behavior of the elastic scattering cross sections at the various beam energies and various scattering angles, Hofstadter and his collaborators were able to measure the size and some simple shape parameters for many nuclides.

In 1953, this facility furnished the first evidence of elastic scattering from the proton, using a polyethylene target[12] as shown in Fig. 4. A hydrogen gas target was then constructed in order to reduce the backgrounds under the elastic peak, and in 1955, Hofstadter and McAllister[13] presented data showing that the form factors in the Rosenbluth cross section were less than

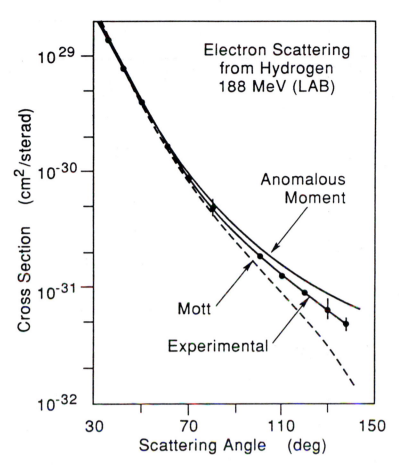

Fig. 5. Elastic electron scattering cross sections from hydrogen compared with the Mott scattering formula (electrons scattered from a particle with unit charge and no magnetic moment) and with the Rosenbluth cross section for a point proton with an anomalous magnetic moment. The data falls between the curves, showing that magnetic scattering is occurring but also indicating that the scattering is less than would be expected from a point proton.

unity (Fig. 5) — and were decreasing with increasing momentum transfer. They gave an estimate of $(0.7 \pm 0.2) \times 10^{-13}$ cm for the size of the proton.

In 1955, new end station facilities at HEPL were commissioned, doubling the energy available for scattering experiments. Beams from the full length of the linac were available in the new area, reaching energies of 550 MeV (Fig. 6). A new spectrometer facility was installed by Hofstadter's group with a magnet of twice the bending radius (36 inches) of the spectrometer in use at the halfway station. A liquid hydrogen target was constructed and installed. This equipment was a considerable improvement (Fig. 7) and a large effort was focused on scattering from hydrogen.[14] A graph of the measured form factors is shown in Fig. 8, which shows data for various values of Q^2 compared with a model proton with a "size" of 0.8×10^{-13} cm.

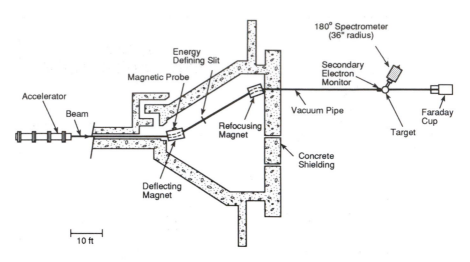

Fig. 6. Layout of the beam line and the 36 inch spectrometer in the End Station of the High Energy Physics Laboratory. This facility was used for electron scattering experiments for more than a decade by R. Hofstadter and his collaborators. (A 72 inch spectrometer was added in 1960 to analyze scattered electrons to an energy of 1000 MeV.)

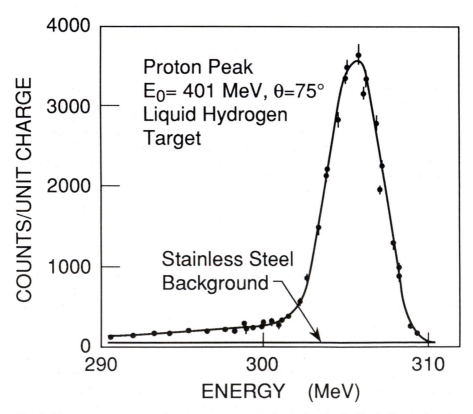

Fig. 7. Electron-proton scattering energy spectrum taken using the facility in Figure 6 and a liquid hydrogen target. The stainless steel container for the liquid hydrogen contributes very little background. The radiative tail of the elastic peak is clearly evident on the low energy side of the peak.

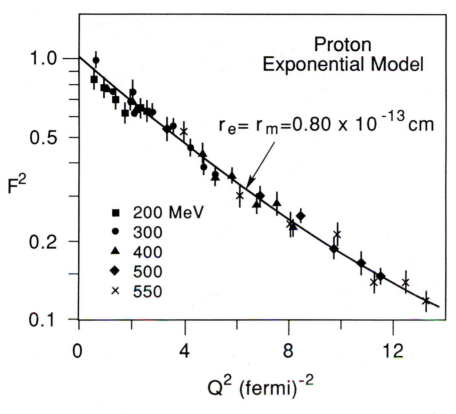

Fig. 8. The proton form factor for various energies and momentum transfers as measured in early experiments using the 36 inch spectrometer facility at HEPL. The value of F^2 was calculated from the original Rosenbluth formula which defined form factors $F_1 (Q^2)$ and $F_2 (Q^2)$. F_1 corresponds to the form factor for a Dirac (spin $\frac{1}{2}$) proton, and F_2 to the form factor for the anomalous magnetic moment. In the analysis of the data it was assumed that $F_1 = F_2$.

At higher values of Q^2 it became evident that $F_1 \neq F_2$, but rather that $G_E = G_M/\mu_p$ for the proton, and the use of the G's then became universal. ($G_M = F_1 + KF_2$ and $G_E \backsimeq F_1$ for small values of Q^2.) The curve shown in the figure was based on a model assuming exponentially falling distributions of charge and magnetic moment, each with a root mean square radius of 0.8×10^{-13} cm (1 Fermi $= 10^{-13}$ cm, 1 (Fermi)$^{-2} = 0.0388$ GeV^2)

These experiments mark the beginning of the search for sub-structure in the proton. They showed persuasively that the proton was not a point, but an extended structure. This fundamental discovery was rapidly accepted by the physics community. It was generally assumed that there was a connection between spatial extent and structure, although I don't think anyone was seriously questioning the "elementary" character of the proton at that time. The available electron energies were not yet high enough for the exploration of inelastic scattering from the proton, and only elastic experiments provided clues about proton structure for the next several years.

The new facility was also used to measure scattering from deuterium, in order to extract information about the neutron. The form factor for elastic scattering from the loosely bound deuterium nucleus falls off extremely

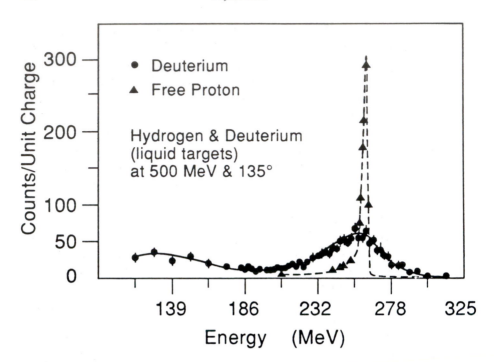

Fig. 9. A comparison of the scattering of electrons from the proton and the quasi-elastic scattering from the individual nucleons in deuterium. The elastic scattering from the deuterium nucleus would occur at an energy above the highest energy shown on the graph and would be negligible in comparison with the cross-sections illustrated here. The quasi-elastic scattering from either the proton or the neutron in deuterium is spread out over a wider range of energies than the scattering from the free proton because of the momentum spread of the nucleons in the deuterium nucleus.

rapidly with increasing momentum transfer, so the neutron was studied via quasi-elastic scattering—scattering from either the proton or the neutron, which together form the deuterium nucleus. The quasi-elastic scattering reaches a maximum near the location of the peak for electron-proton scattering, since the scattering takes place off a single nucleon and the recoil energy is largely determined by the mass of that nucleon (Fig. 9). One also observes the effects of the motion of the nucleons in deuterons, and one result is a measurement of the nucleon's momentum distribution in the deuterium nucleus.

The great success of the scattering program at HEPL had three consequences: Scattering experiments became more popular at existing electron synchrotrons, new synchrotrons were planned for higher energies, and discussions began at Stanford about a much larger linear accelerator—two miles long and powered by one thousand klystrons!

After more than a year of discussions and calculations, the physicists and engineers of the High Energy Physics Laboratory prepared the first proposal for a two-mile linear accelerator to be built at Stanford.[15] E.L. Ginzton, W.K.H. Panofsky and R.B. Neal directed the design effort, and Panofsky and Neal went on to direct the construction of what came to be called the Stanford Linear Accelerator Center (SLAC) — surely one of the great engineering achievements of the early 1960s.[16] The new machine was a bold extrapolation of existing techniques. The design was conservative in the sense that working prototypes of all the machine components were in hand, but a formidable challenge because of the increase in scale. The investigation of the structure of the proton and neutron was a major objective of the new machine. The 20 GeV energy of the accelerator made both elastic and inelastic scattering experiments possible in a new range of values of Q^2, and presented our collaboration with a golden opportunity to pursue the studies of nucleon structure.

When it was proposed, the two-mile linac was the largest and most expensive project ever in high energy physics. Up until that time the field had been dominated by proton accelerators, and electron machines had been relatively small and few in number. Electrons were catching up and, in parallel with the Stanford linac, two large electron synchrotrons were proposed and built: the Cambridge Electron Accelerator (CEA) and the Deutches Electronen Synchrotron (DESY) in Hamburg, with peak energies of 5 and 6 GeV respectively. The establishment of SLAC in 1960 would eventually bring electron physics into direct competition with the largest proton accelerators of the time, the Brookhaven AGS and the CERN PS, both of which were already under construction in the late 1950s. The new electron accelerators would make available many opportunities for physicists.

The new linear accelerator consisted of two miles of accelerating waveguide, mounted in a tunnel buried 25 feet underground. In the initial phase, the waveguide was powered by two hundred and forty 20 − 30 MW

Fig. 10. Cut-away illustration of the two mile Stanford Linear Accelerator, showing the accelerator wave-guide buried 25 feet below the surface and the klystron gallery at ground level. Each klystron feeds 40 feet of accelerator wave-guide through penetrations connecting the accelerator housing with the klystron gallery.

Fig. 11. Aerial view of the SLAC site. On the left are the experimental areas fed by beam lines from the accelerator. On the right is the campus area where offices, laboratories, and shops are located. The scattering experiments were performed in the large shielded building just to the left of center near the bottom of the picture. The structure crossing the accelerator is a super-highway which was under construction at the time this picture was taken.

klystrons housed in a building at ground level. The accelerator was sited in the hills behind Stanford on University land, and was probably the last of the university-based high energy physics accelerators in the U.S. (Figures 10 and 11).

The design parameters of the new machine — 20 GeV in energy and average currents in the neighborhood of 100 μA — presented many new problems for experiments. Two experimental areas (called End Stations in Figure 12) were developed initially — one heavily shielded area, where secondary beams of hadrons and muons could be brought out to various detectors, and a second area for electron and photon beam experiments. The "beam switchyard" connected each area to the accelerator with a magnetic beam transport system which defined the momentum spread of each beam to better than 0.2%, was achromatic and isochronous (in order to preserve the RF time structure of the beam). The transport systems were fed by a system of pulsed magnets, so that a given accelerator pulse could be directed into either of the two experimental areas. Unavoidable beam losses in the system would lead to high levels of radioactivity, and to challenging thermal design problems at the expected levels of beam currents. The design of this "switchyard" area was fairly well fixed by the end of 1963, along with the specifications for the heavily shielded end station buildings (see Ref. 16).

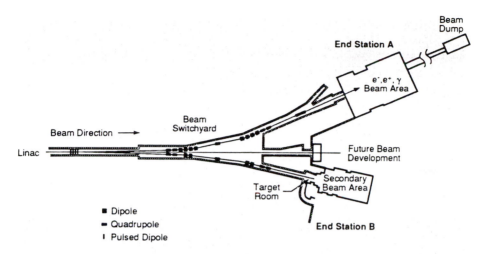

Fig. 12. Layout of the SLAC experimental areas and the beam switchyard.

The experimental area which was to be devoted to electron scattering and photoproduction experiments using the primary beam had to satisfy the experimental needs of several groups of experimenters. The challenge was to build apparatus which would allow rapid and efficient data collection in the new energy region which was being made available. The operating costs of the new accelerator (not to mention the depreciation on the capital costs of over 100 million dollars) would be many thousands of dollars per day, so it was important to balance costs in such a way that the experiments would give good value—a spectrometer with small solid angle would be cheaper, but might take much longer to make a given measurement. The major costs in this area would be for large magnetic spectrometers and shielding, and so some of the smaller components could be developed to a much more sophisticated level than had been possible at the smaller laboratories, while still adding only a small percentage to the overall costs.

Although half a decade had passed since the original proposal for SLAC, the basic physics aims remained much the same. The most effective technique still appeared to be the detection of a single particle from a given interaction. (The duty factor [i.e., the percentage of on-time] was low for the linac — the klystrons were pulsed for approximately two microseconds, at a rate of 360 times per second. This resulted in high instantaneous rates during the short pulses, and made coincidence experiments difficult.) The overall experimental design required instruments which would determine the energy and angle of a particle coming from a target placed in the beam of electrons. Magnetic spectrometers were still the most effective way to accomplish this, but they would be large and cumbersome devices at these energies.

The resolution in energy, ΔE, had to be much better than $m_\pi/E_{max} \sim 0.7\%$ in order to separate reactions that differed in the number of pions emitted. Since the energy of particles from a given reaction is a very steep function of angle, it was also necessary to measure the angle of scattering to high accuracy ($\sim$ 0.15 mrad). Practical spectrometers have angular acceptances much greater than the required resolution in angle, so the optics and the detectors had to be arranged in such a way that the true angle of scattering was determined along with the energy.

There were many discussions about the most effective design for the facilities. Records are sparse, but there are indications of frank and earnest discussions. There was a suggestion that a single 2 GeV spectrometer could cover most of the interesting electron scattering experiments, while others were suggesting that a complex system with a high energy forward spectrometer combined with a huge solenoidal detector in the backward direction was the right way to go.

In the Spring of 1964, I found myself gradually being elected to a position of responsibility for the design and engineering of the facilities in End Station A (as the larger of the two experimental areas was called). This was not an enviable position, since there was little agreement about what should be done, and most of the people involved clearly outranked me.

The sub-group interested in electron scattering experiments was pretty well convinced that a spectrometer of $8-10$ GeV maximum energy with a solid angle $\geq$ 1 milli-steradian would be capable of an extensive program of scattering measurements. By bending in the vertical plane, measurements of scattering angle and momentum could be separated at the location of the detectors. Preliminary designs for such a device had been proposed and had already influenced the layout of the end station, which by this time was in an advanced state of design. The spectrometer incorporated a vertical bend of $\sim$ 30°, with focusing provided by separate quadrupoles preceding and following the bend (Fig. 13, elevation). The magnetic design of the spectrometer involved a lot of computation, but proceeded smoothly. After taking practical and financial constraints into account, the top momentum was fixed at 8 GeV and the solid angle at 1.0 milli-steradians.

In order to cover a range of scattering angles it was our intention to build the spectrometer so that it could be rotated around the target from an external control room (Fig. 13, plan). We needed frames which would hold hundreds of tons of magnets and counters in precise alignment while they were moved about the end station.

It was about this time that we began to assemble a team of engineers and draftsmen to translate the requirements into designs for working hardware. The group began the detailed design of the 8 GeV spectrometer components, while the debate continued about the rest of the complex.

By the middle of 1964 the utility of a forward-angle spectrometer which would analyze particles with a maximum momentum of 20 GeV was no longer questioned. Successful photoproduction experiments were being carried out at energies up to 5 GeV at the CEA electron synchrotron, and extending the energy of these measurements would obviously be a productive program for SLAC. Also, if the electric form factor of the proton, G_E,

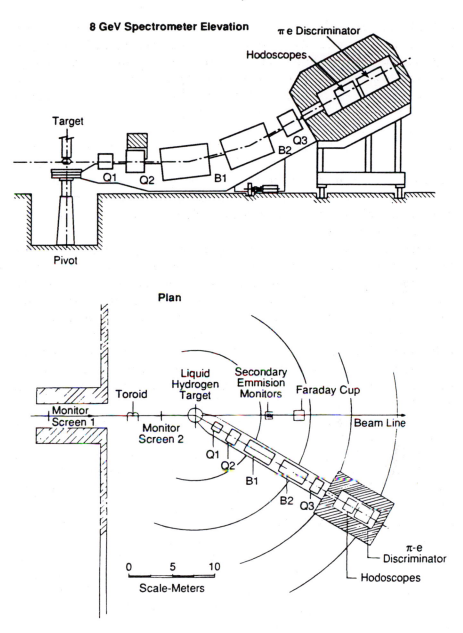

8 GeV Spectrometer Elevation

π e Discriminator

Hodoscopes

Target

Q1 Q2 B1 B2 Q3

Pivot

Plan

Liquid Hydrogen Target

Secondary Emmision Monitors

Faraday Cup

Toroid

Monitor Screen 1

Monitor Screen 2

Beam Line

Q1 Q2 B1 B2 Q3

π-e Discriminator

Hodoscopes

0   5   10

Scale-Meters

Fig. 13. Schematic drawings of the 8 GeV spectrometer. Five magnets (two bending magnets, (B), and three quadrupoles, (Q)) direct scattered particles into the detectors which are mounted in a heavily shielded enclosure. The whole assembly rides on the rails and can be pivoted about the target to change the angle of scattering of the detected electrons.

was to be measured, small-angle scattering experiments would be required.

Scaling up the 8 GeV spectrometer to 20 GeV (and keeping the resolution at 0.1%) would have required very large vertical displacements. Some attempts were made to design a big pit in the end station to accommodate

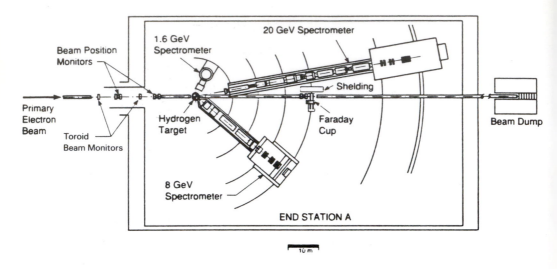

Fig. 14. Layout of spectrometers in End Station A. All three spectrometers can be rotated about the pivot. The 20 GeV spectrometer can be operated from about $1\frac{1}{2}°$ to $25°$, the 8 GeV from about $12°$ to over $90°$. The 1.6 GeV spectrometer coverage is from $\backsim 50° - 150°$.

such a system which would bend downward, but it looked very awkward from a mechanical viewpoint. An ingenious solution was proposed by Panofsky and Coward, in which horizontal bending could be used while preserving orthogonal momentum and angle measurements at the focus. This proposal seemed complicated to me, and I resisted adopting the design. Finally, I was rescued by K. Brown's calculation of aberrations in this device, which he found to be unacceptably large. Shortly thereafter, Brown and Richter proposed a relatively simple spectrometer with a central crossover which allowed vertical bending, but kept the vertical height within bounds. A simple system of sextupoles was required to correct aberrations in the system. Once proposed, this design was accepted by all, and final layout of the spectrometers in the end station was soon accomplished (Fig. 14).

The two large groups at SLAC were not very interested in measurements in the backward direction at the time, but D. Ritson of the Stanford Physics Department saw an opportunity to continue his HEPL program of photo-production measurements at higher energies, and proposed the construction of a 1.5 GeV, $90°$ spectrometer at large angles, a proposal which was accepted by the laboratory after a short delay, and the spectrometer was added to the facility.

With the magnetic design of the two large spectrometers fixed, design and construction of the facility began in earnest. The building of the facility was a joint effort of the SLAC-MIT-CIT group, the SLAC photoproduction group under B. Richter and the Stanford group interested in the 1.6 GeV spectrometer led by D. Ritson. The facility consisted of several parts.

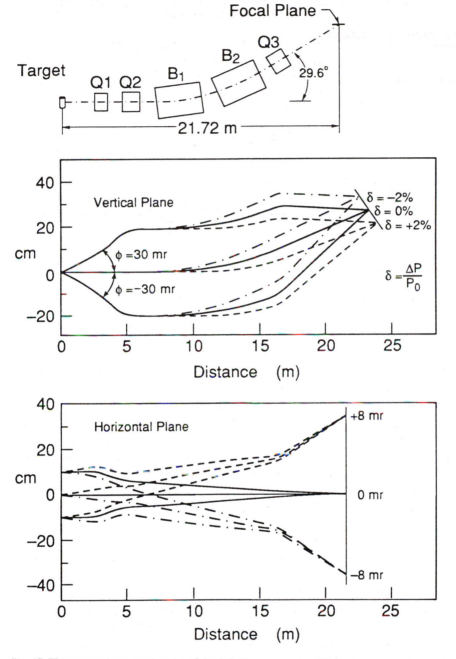

Fig. 15. The magnet layout and optics of the 8 GeV spectrometer. The arrangement of magnet is shown at the top of the figure. In the vertical plane the focusing is "point to point" and momenta are dispersed along the focal plane. In the horizontal, the focusing is parallel to point and angles are dispersed along the θ focal plane. (mr = milli-radian)

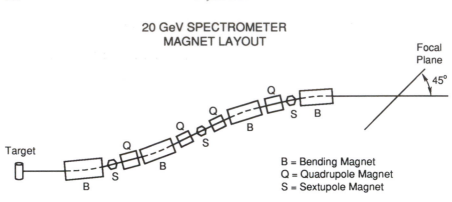

Fig. 16. Magnetic system for the 20 GeV spectrometer. With a momentum focus at the central sextupole, the final two bending magnets add to the momentum dispersion, even though the direction of bending is opposite to that in the first two bending magnets. The three sextupoles are used to adjust the angle of the focal plane to a convenient value.

The 8 GeV spectrometer used five magnetic elements—three quadrupoles and two bending magnets (Fig. 15). It had point-to-point focusing in the vertical plane (the plane in which momentum is dispersed). A detector hodoscope in the p-focal plane defined the differential momentum, Δp. In the horizontal plane (scattering plane), the spectrometer gave parallel-to-point focusing, allowing the use of a long target. A second hodoscope in the θ-focal plane determined the scattering angle. The p- and θ-focal planes were located close to each other, but were not coincident.

The 20 GeV spectrometer used eleven magnetic elements—four bending magnets, four quadrupoles, and three sextupoles—to produce very similar conditions at the p- and θ-focal planes (Fig. 16). An added feature was the extra p-focus in the middle of the magnetic system. A slit at this point could be used to control the $\Delta p/p$ band-pass of the instrument. A system of counters similar to those in the 8 GeV spectrometer was mounted in the shielding hut.

The 1.6 GeV spectrometer had only a single magnetic element (Fig. 17). Focusing was achieved by rotation of the pole tips out of the normal to the central orbit. Some sextupole fields were built into the pole faces to control aberrations.

The liquid hydrogen targets for the facility were of the condensation type. In these devices a separate target cell was in thermal contact with a reservoir of liquid hydrogen at atmospheric pressure. Gaseous hydrogen (or deuterium) introduced into the target cell at greater than atmospheric pressure would condense to the liquid phase.

The first target built for the facility was very simple in concept and used convection in the target cell to transfer the heat generated by the passage of the beam to the reservoir. It turned out that this mechanism was not effective at high beam power levels, and that, as a result, intense beams caused fluctuations in the liquid density. Targets were then built that used

1.6 GeV SPECTROMETER MAGNET

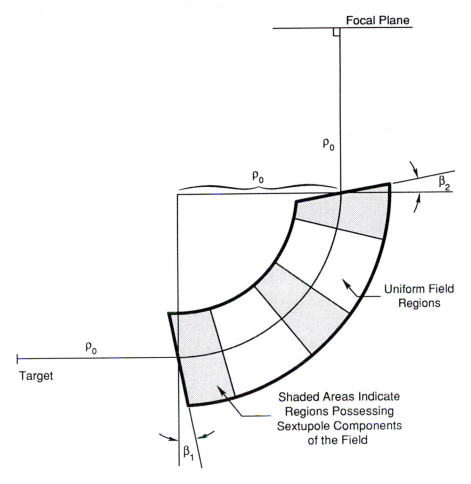

Fig. 17. Schematic of the 1.6 GeV spectrometer. Focusing is achieved by rotated pole tips (angles β_1 and β_2), and sextupoles are built into the pole faces to adjust the focal plane to be at right angles to the central ray.

forced circulation by a fan to keep the liquid in the target cell in closer thermal contact with the reservoir. Schematics of both targets are shown in Fig. 18. (Even the circulating targets had some problems at very high beam currents.)

The accuracy to which cross sections can be measured is directly related to the accuracy with which the incident beam intensity can be measured. The primary standard for the early experiments was a Faraday cup (Fig. 19a) in which 20 GeV electrons were stopped, and the resulting charge measured with an accurate current integrator. The Faraday cup could not be used with the full beam power of the linac because of thermal limitations, but it was used to calibrate other monitors at low repetition rates.

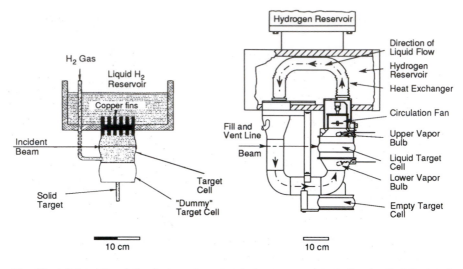

Fig. 18. a) Schematic of the first condensation hydrogen target built for the End Station A facility. The target could be displaced vertically to put either the dummy target or the solid targets on the beamline.

b) Schematic of a condensation target with forced circulation of the condensed hydrogen. As in (a) the target could be displaced vertically so that other targets could be placed in the beam line.

A new toroid monitor was specifically developed for the End Station A experiments. The principle of operation is illustrated in Fig. 19b. The beam acted as the primary winding of a toroidal transformer. Passage of a beam pulse through the toroid set up an oscillation, and the amplitude of that oscillation was sampled after a certain fixed interval. The sampling and subsequent readout of the signal determined the final accuracy of the monitor. The readout was carefully engineered by the SLAC electronics group, and as experience with this device increased, it became the absolute standard for beam current measurements, though often cross checked against the Faraday cup.

In addition to the beam monitors, there were various collimators and screens along the beam line, and a high-power beam dump buried in a hill a hundred meters or so behind the end station. An impressive cable plant connected the spectrometer detectors to the electronics in the "counting house" high above the end station floor.

I wish I had the skills to recreate for you the three years of intense activity that went into translating the paper plans of 1964 into the instruments which began to do physics in early 1967. The problems in procuring the precision magnets, the construction of the giant frames to hold the magnets and to support the massive shields for the detectors, the laying of the rails to extraordinary tolerances — all these and many other problems were attacked with drive and dedication by the mechanical engineering group. Even the professional crews hired to install large parts of the apparatus became

Faraday Cup

Toroid Charge Monitor

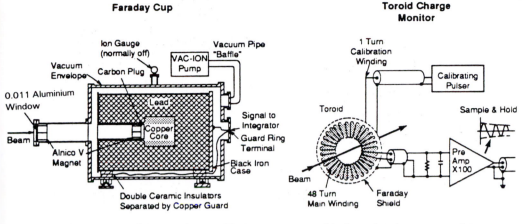

Fig. 19. a) Drawing of the Faraday cup. The beam was stopped in the carbon-copper core of the cup, and the lead absorbed γ-rays created in the shower. The Alnico magnets deflected low energy electrons coming from the window so that they did not reach the cup, and those from the core did not escape from it.

b) Schematic of the toroidal transformer monitor. The beam acted as the primary winding of the ferrite core. A beam pulse caused a "ringing" of a damped LC circuit, the amplitude of which was read out after three quarters of a cycle.

infected with the enthusiasm of the engineers. I lived in mortal fear that a union steward would drop in unannounced and find a millwright (steel worker) building a wooden scaffold, while a carpenter was operating the crane. Figure 20 is a view of the experimental area with the completed 8 and 20 GeV spectrometers in place.

Fig. 20. Photograph of the 8 and 20 GeV spectrometers in End Station A.

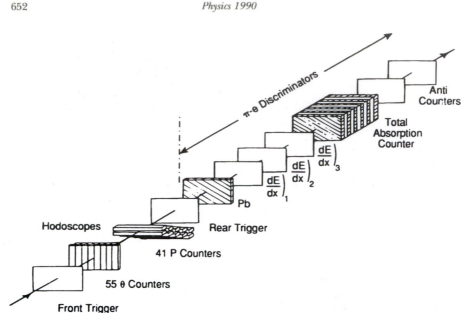

Fig. 21. Schematic drawing of the counter system inside the 8 GeV shielding hut.

The 8 GeV detectors were designed and built at MIT (Fig. 21). Two large scintillation counters acted as trigger counters, signalling the passage of charged particles through the counter system. Two multi-element scintillation counter hodoscopes (mounted between the trigger counters) defined the position of the track in the horizontal (θ) and vertical (p) directions. The hodoscopes each consisted of two layers of overlapping counters, so that each double hit defined the position to half a counter width. The location of the hits together with the angle and energy setting of the spectrometer defined the angle of scattering to ± 0.15 milli-radians and the momentum of the scattered particle to $\pm 0.05\%$. Following the system of hodoscopes was a set of counters used to distinguish electrons from pions. The principal element was a total absorption lead-lucite shower counter. The pulse height threshold was set to be more than 99% efficient for electrons. In the elastic scattering experiments this counter alone was enough to ensure a pure electron signal, but for inelastic scattering, pion backgrounds increased and the use of the dE/dx counters was sometimes necessary. These counters measured the energy loss in a scintillator for particles which had passed through one radiation length of lead. Electrons will often shower in the radiator, giving large pulse height in the counters. In most cases pions will not shower, giving an almost independent indication of their identity. By the time of the first inelastic scattering experiments using the 8 GeV spectrometer, a gas Čerenkov counter had been added in front of the trigger counter as a further tool for particle discrimination. The dE/dx system was used only for the lowest secondary energies where the pion-electron ratios were large. The 20 GeV spectrometer's counter system (Fig.

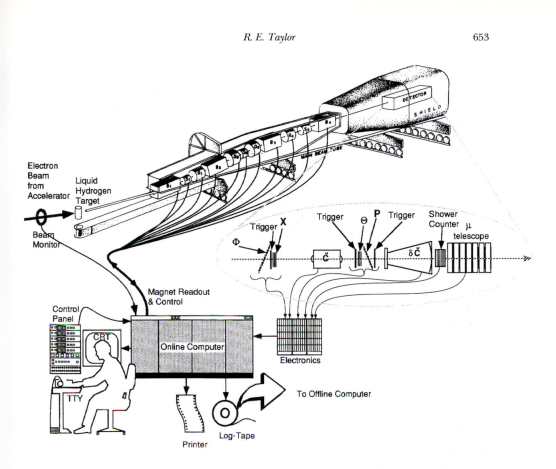

Fig. 22. Schematic of the 20 GeV spectrometer indicating the various computer control and read-out functions. Also shown is a schematic of the 20 GeV counter system. Particle identification in the spectrometer was somewhat more complex than for the 8 GeV instrument, partly because of the higher energies involved, but also because it was sometimes desirable to identify π mesons in a large electron background in the 20 GeV spectrometer.

22) was similar to that in the 8 GeV spectrometer, with the addition of a differential gas Čerenkov counter, and extra sets of hodoscopes which determined the angle of scatter outside the horizontal plane (φ hodoscope) and the position of the scattering center along the beam line (x hodoscope). The MIT group also took responsibility for much of the counting electronics, photo tube power supplies, etc., and were of great assistance to the electrical engineers in the SLAC group who installed the electronics and interfaced the on-line computer.

One innovation by the collaboration was the extensive use of on-line computation in the experiment. While not the first experiment to be equipped with an on-line computer, the degree of computer control was ambitious for the time. We purchased a fairly powerful mainframe, dedicated to only one experiment at a time. A lot of work was done on both software and hardware, so that the effort to set up and operate a given experiment was greatly reduced. The on-line analysis of a fraction of the

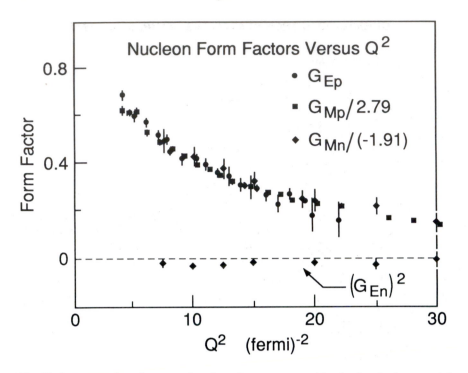

Fig. 23. Summary of results on nuclear form factors presented by the Stanford group at the 1965 "International Symposium on Electron and Photon Interactions at High Energies". (A momentum transfer of 1 GeV^2 is equivalent to 26 Fermis^{-2}.)

increasing data was a powerful way to check on the progress of the experiments (Fig. 22).

In the summer of 1966 there was a call for proposals to use the beam at SLAC. The accelerator was nearing completion, and some early tests of the accelerator with beam were being done with considerable success. Although the initial programs in End Station A were built into the design of the facility, it was now necessary to parcel out beam time and arrange the sequence of experiments for the first year of operation. The Cal Tech-MIT-SLAC collaboration prepared a proposal that consisted of three parts:

a. Elastic electron-proton scattering measurements (8 GeV spectrometer)

b. Inelastic electron-protron scattering measurements (20 GeV spectrometer)

c. Comparison of positron and electron scattering cross sections (8 GeV spectrometer)

It is clear from the proposal that the elastic experiment was the focus of interest at this juncture. "We expect that most members of the groups in the collaboration will be involved in the e-p elastic scattering experiment, and that the other experiments will be done by subgroups."

During the construction of SLAC and the experimental facilities a lot of progress had been made on the measurements of nucleon form factors at other laboratories. The program at HEPL had continued to produce a great

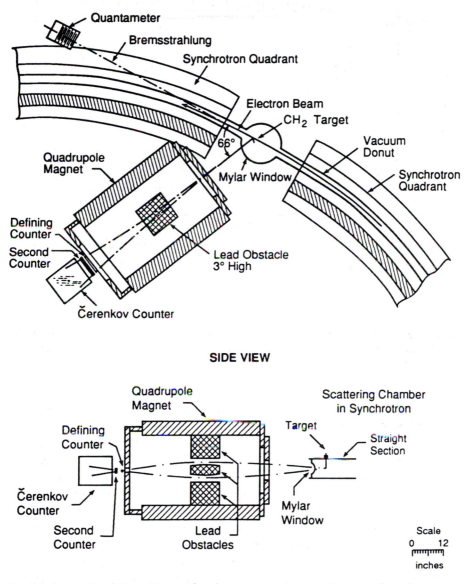

CORNELL ELECTRON SCATTERING SETUP, CIRCA 1960

Fig. 24. Schematic of the equipment for electron scattering experiments at Cornell around 1960. These experiments used a quadrupole spectrometer to analyze electrons scattered from an internal target in the electron synchrotron. The target is mounted away from the normal orbit in the accelerator, and the beam is slowly moved onto the target after acceleration.

deal of new data using the facilities in the end station of the Mark III accelerator. A new spectrometer with a bending radius of 72 inches had been added to accommodate the increased energy available from the accelerator. Extensive results on both the proton and the deuteron were generated and reported[17] (Fig. 23).

At over 1 GeV, the Cornell electron synchrotron was the highest energy electron machine in the world for a few years in the early 1960s. Experimenters there made a series of measurements on CH_2 targets, using a quadrupole spectrometer of novel design[18] (Fig. 24) and a new type of γ-ray monitor.[19] The results from Cornell started a trend toward the use of the electric and magnetic form factors[20] (G_E and G_M), rather than one form factor for a spin 1/2 (Dirac) proton and a second for the anomalous magnetic moment of the proton.

The linear accelerator at Orsay had begun operations in 1959 and by the following year there was an active program of both nucleon and nuclear scattering. The emphasis shifted to colliding beam experiments in later years, but many scattering experiments were done in the intermediate energy stations of that accelerator with beams of up to 750 MeV.

Electrons had become a big success in high energy physics and a new high energy electron synchrotron was approved and built at Harvard. The Cambridge Electron Accelerator was built jointly by Harvard and MIT and came into operation in 1962 with a peak energy of 5 GeV. A program of electron scattering experiments using internal targets was soon in operation. The new accelerator opened up a new range of Q^2 for scattering experiments and several different experimental setups were used to measure the proton and neutron form factors. The higher Q^2 proton measurements fell very close to values expected from a straightforward extrapolation of the data at lower energies. The results[21] were summed up (somewhat later) by Richard Wilson in the words "The peach has no pit." These results were the first evidence that the old core model of the proton was unlikely to be correct (Fig. 25).

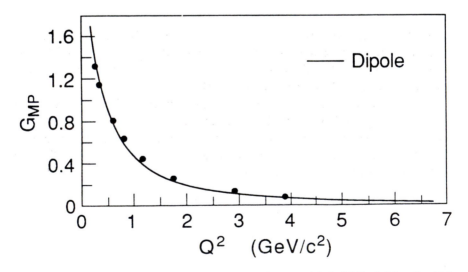

Fig. 25. G_M for the proton from data taken at CEA. The curve labelled "dipole" is a fit which originated in the late 1950's when the maximum measured Q^2 was limited to less than 1 GeV². It has the form $G_M = \mu_p/(1 + Q^2/.71 \text{ GeV}^2)^2$ and is in qualitative agreement with the CEA data at higher Q^2, though the fit is not very good in the statistical sense.

**DESY INTERNAL TARGET
SCATTERING FACILITY**

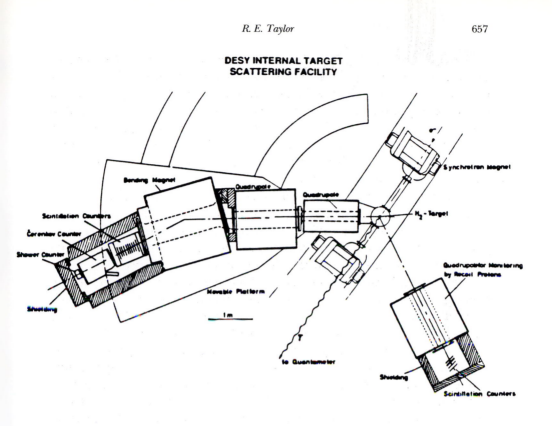

Fig. 26. Layout of the spectrometer setup for internal target electron scattering experiments at DESY. Later on, the same set-up was used to detect electron-proton coincidences in elastic scattering (in order to reduce backgrounds).

A slightly larger synchrotron was built in Hamburg, Germany at about the same time. DESY came into operation in 1964 with a peak energy of 6 GeV. An extensive series of nucleon scattering measurements, using both internal targets[22] (Fig. 26) and external beams[23] (Fig. 27), was undertaken.

With both CEA and DESY operating, the amount of elastic scattering data at high Q^2 (which essentially measures G_M) increased rapidly in both quantity and accuracy. The data continued to follow the so-called dipole model to a good approximation. By the Hamburg conference in 1965 there were no dissenters from the view that

$$G_{Ep} = \frac{G_{Mp}}{\mu_p} = \frac{G_{Mn}}{\mu_n},$$

$$G_{En} \cong 0 \text{ at large } Q^2,$$

and

$$G_{Ep}(Q^2) \cong \left(\frac{1}{1 + \dfrac{Q^2}{0.71 \text{ GeV}^2}} \right)^2 \text{ up to } Q^2 \sim 10 \text{ GeV}^2$$

**DESY EXTERNAL BEAM
SPECTROMETER FACILITY**

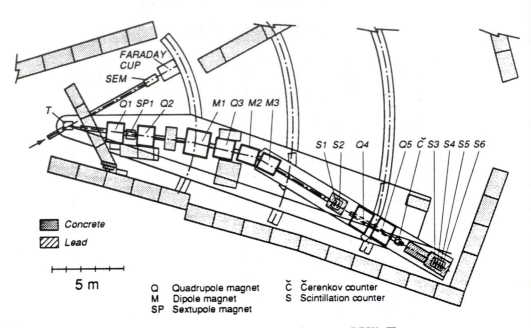

Fig. 27. Setup for external beam scattering experiments at DESY. The spectrometer was articulated between the magnets M_2 and M_3. By varying the bending in M_2 and M_3, lines of constant "missing mass" could be adjusted to a given slope at S_1 for different scattered energies.

SLAC was expected to test this formulation in the new range of Q^2 (Fig. 28) made available with 20 GeV electrons. Questions of interest concerned the evidence for a nucleon core and the validity of the dipole description of the form factor in the extended range of Q^2 available at the new accelerator. The cherished picture of a "real proton" surrounded by a meson cloud was already in pretty serious trouble, but more tests for a small core were outlined in the SLAC proposal . Other questions were related to particular models of behavior for the form factors which are not of great interest today.

Our SLAC proposal demanded certain specifications for the beams to be used in the experiment, which were within the design specifications of SLAC, but which were nonetheless very difficult to meet, given the fact that the accelerator was just being commissioned. Operating the accelerator for the initial scattering experiments was a challenging experience for the crew of accelerator operators, and many of them have indelible memories of those times.

The proposed experiment on elastic scattering aimed at measurements of the cross section at momentum transfers of 16 GeV² and beyond, even in the very first round of experimentation. There was an extensive discussion in the proposal about running at angles and energies in a manner which ·

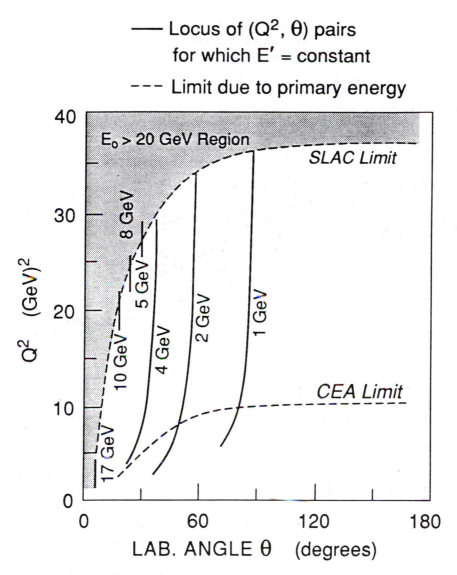

—— Locus of (Q^2, θ) pairs
for which E' = constant

‑‑‑ Limit due to primary energy

Fig. 28. Plot of elastic kinematics showing the extra kinematic region made available at SLAC for spectrometers of different maximum energies (above 4 GeV, only the maximum Q^2 is indicated to avoid confusion on the graph).

would result in an efficient separation of G_E and G_M. Possible backgrounds were considered, and it was expected that they would be negligible. Radiative corrections to elastic scattering were expected to reach up to 30% for our apparatus and incoming energies of 20 GeV. These corrections arose from two related but physically distinct processes:

1. Electrons passing through the target and the target windows might emit radiation as a result of interactions with individual atoms (real bremsstrahlung) and thereby suffer an energy loss.

2. Scattered electrons might emit radiation in the scattering process itself ("wide-angle bremsstrahlung"). The effects of wide-angle bremsstrahlung were first discussed by Schwinger[24] in 1949 and have been the subject of increasingly sophisticated calculations over the years.

In some cases the energy of the emitted radiation (in either reaction) was sufficient to affect the kinematics of the scattering to such an extent that the measuring apparatus would no longer "recognize" the interaction. For example, if sufficient (radiative) energy were lost in an elastic scatter, the energy of the scattered electron might fall below the range that the apparatus defined as the "elastic peak."

The emission of radiation gives rise to the characteristic "radiative tail" in the energy spectrum of elastically scattered electrons as shown schematically in Figure 29. The cross section measured by detecting the electrons in a certain energy range will be smaller than expected because some particles will be lost. It is customary to correct experimental cross sections for these losses — removing the dependence of the final cross section on the energy resolution of the apparatus.

A simple (first order) correction formula illustrates how such a correction might be applied.

$$\frac{d\sigma}{d\Omega}\bigg|_{exp.} = (1 - \delta_s)\, e - \delta_r \frac{d\sigma}{d\Omega}$$

where the wide-angle bremsstrahlung correction δ_s is

$$\delta_s = \frac{2a}{\pi}\left\{\left(1 - ln\,\frac{Q^2}{m_e^2}\right) ln\,\frac{\Delta E}{E}\right\}$$

m_e = mass of electron.
ΔE = energy resolution or acceptance
E = incident energy (assumes $E_s \backsim E$)

and the real bremsstrahlung correction δ_r is

$$\delta_r = -\frac{t}{ln2}\, ln\,\frac{\Delta E}{E}$$

t = thickness of target in
radiation lengths

As long as the corrections can be calculated to sufficient accuracy, they are innocuous in elastic scattering, and determination of elastic form factors is straightforward.

Our proposal included a possible run plan for measuring G_E and G_M to values of Q^2 exceeding 15 GeV². (At the higher Q^2 one finds an upper bound on G_E, rather than a measure of its value.) The program was expected to take about 350 hours of beam time, and a first run of 200 hours was suggested, after which the requests would be updated using measured quantities, rather than estimates. This experiment was to be the first carried out with the new facility.

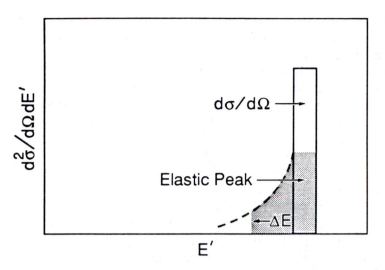

Fig. 29. Radiative effects in elastic scattering. In the absence of radiative effects, all elastic scatters would be found in the box labelled $d\sigma/d\Omega$ (the width of which depends on resolution in the incoming beam and the detection apparatus). Radiative processes result in energy losses for some scattered electrons, and so some electrons will be found in a "tail" on the low energy side of the peak. A measurement of the electrons in the shaded region results in a cross section which is somewhat smaller than $d\sigma/d\Omega$. This smaller $(d\sigma/d\Omega)_{meas}$ can be corrected for radiative losses to determine $d\sigma/d\Omega$.

The second part of the proposal concerned the measurement of inelastic scattering from the proton. Inelastic scattering from the nucleon had a much shorter history than elastic scattering so there was much less guidance for the design of that part of our proposal.

Inelastic scattering from *nuclei* was a common feature of the early scattering data at HEPL. The excitation of nuclear levels and the quasi-elastic scattering from the constituent protons and neutrons of a nucleus were observed in the earliest experiments. The excitation of nuclear levels in carbon could be seen in the data of Fig. 4, for example. Quasi-elastic scattering became more evident as momentum transfer was increased. Fig. 30 shows scattering from the same target as in Fig. 4, and at approximately the same incident energy, but at a scattering angle of 135°. A comparison of the two figures illustrates the growth in the fraction of quasi-elastic scattering as the angle (and therefore the momentum transfer) is increased. When the electrons scatter through 135°, the elastic peak is very small and the pattern of level excitation has changed because the different multipole transitions have different angular dependences. The most prominent feature of the spectrum is the broad quasi-elastic peak in Fig. 30 due to scattering from individual protons and neutrons. The width of the peak reflects the Fermi momentum of the nucleons in the nucleus.

The earliest experiments on the inelastic scattering of electrons from the proton itself were carried out by Panofsky and co-workers at HEPL in the second half of the 1950's.[25,26,27] The early experiments were comparisons

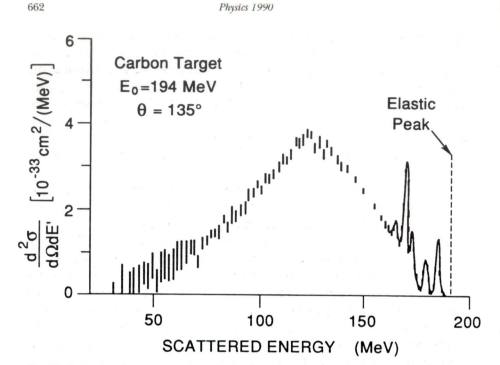

Fig. 30. Spectrum of electrons scattered inelastically from carbon. The excitation of nuclear levels is evident. The large, broad peak between 100 and 150 MeV is due to quasi-elastic scattering from the individual neutrons and protons that make up the carbon nucleus.

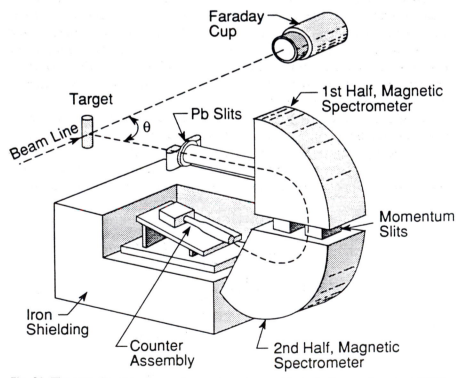

Fig. 31. The zero dispersion magnetic spectrometer used in inelastic experiments at HEPL. Splitting the magnet allowed the insertion of momentum defining slits in the middle of the bend.

of photo- and electroproduction of positive pions in lithium and (later) hydrogen targets. Those experiments checked the calculation of the electromagnetic fields that accompany a relativistic electron, but added little to the knowledge of meson dynamics beyond that which was known from photoproduction (because the dominant contribution to the electroproduction came from virtual photons with very small values of Q). The authors pointed out that observing the scattered *electrons* at a large angle (rather than the pions) might lead to more interesting results, and the next experiment was of that kind.

A new magnetic spectrometer was commissioned at HEPL at about this time,[28] and was used for these experiments (Fig. 31). Panofsky and Allton[29] made measurements of the inelastic scattering of electrons from hydrogen in the region near the threshold for pion production. The energy of the available electrons was not high enough to reach much beyond the threshold for pion production, but the experiment established that the "tail" of the elastic peak was due to the two (calculable) radiative processes mentioned above. One process was elastic scattering preceded (or followed) by emission of bremsstrahlung in the material of the target; the other was "wide-angle bremsstrahlung" — the emission of a photon in the scattering interaction. The experiment was a quantitative test of calculations of the radiative tail of the elastic peak in the region near pion threshold.

The peak energy of the electrons from the Mark III accelerator was improving steadily during those years, and in 1959 Ohlsen[30] used the 36-inch spectrometer in the Hofstadter group's scattering facility (Fig. 6) to do an experiment similar to the Panofsky-Allton measurement. With increased energy, it was possible to make measurements covering the region of the first $\pi-p$ resonance, and a clear peak was observed at the resonance energy. The experimenters were also able to measure a rough Q^2 dependence of the peak cross section.

In 1962, Hand reported on a similar experiment (using the same spectrometer used by Allton) and the results were discussed in modern notation. In particular, there appears an inelastic equivalent of the Rosenbluth formula containing two form factors which are functions of Q^2 and v, the energy loss suffered by the scattered electron. The measured quantities are E_0, E', and θ:

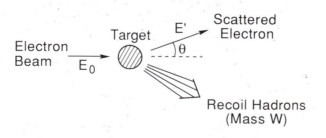

The kinematics of the scattering are described by

$$E' = \frac{E_0 - \dfrac{(W^2 - M^2)}{2M}}{1 + \dfrac{2E_0}{M} \sin^2 \theta/2}$$

W is the mass of the final state of the struck hadron (when $W^2 = M^2$, the elastic kinematics are recovered). The square of the momentum transfer, Q^2,

$$Q^2 = 4E_0E' \sin^2\theta/2$$

the energy loss

$$\nu = E_0 - E'$$

and W^2 are relativistically invariant quantities in the scattering process.

There are two equivalent formulations describing the cross sections which are in current use, one due to Drell and Walecka[31] which is very similar in form to the Rosenbluth expression

$$\frac{d\sigma}{d\Omega dE'} = \frac{\alpha^2}{4E_0^2 \sin^2 \theta/2} \cos^2 \theta/2 \, [W_2 + 2W_1 \tan^2 \theta/2]$$

The structure functions W_1 and W_2 are functions of both the momentum transfer and energy loss, $W_{1,2} \, (Q^2, \nu)$. This is the most general form of the cross section in the (parity conserving) one photon approximation.

Hand[32] popularized a different but equivalent form for the cross section in which one of the form factors reduces to the photoproduction cross section at $Q^2 = 0$

$$\frac{d\sigma}{d\Omega dE'} = \frac{\alpha^2}{4\pi^2} \cdot \frac{(W^2 - M^2) \, E'}{MQ^2E_0 \, (1-\varepsilon)} \left(\sigma_T + \varepsilon\sigma_L \right)$$

where

$$\varepsilon = \frac{1}{[1 + 2 \tan^2 \theta/2 \, (1 + \nu^2/Q^2)]}$$

Again σ_T and σ_L (corresponding to the photo-cross sections for transversely polarized and longitudinally polarized virtual photons respectively) are functions of the momentum transfer and energy loss of the scattered electron, $Q_{L,T} \, (Q^2, \nu)$, with the limiting values at $Q^2 = 0$ of

$$\sigma_T \, (0) = \sigma\gamma_p$$
$$\sigma_L \, (0) = 0$$

These early experiments and the associated theoretical studies developed much of the framework for thinking about inelastic experiments at SLAC. The energy available limited the early experiments on the proton to studies of the $\pi - p$ resonance near 1238 MeV.

An important influence came from the Laboratoire de l'Accelerateur

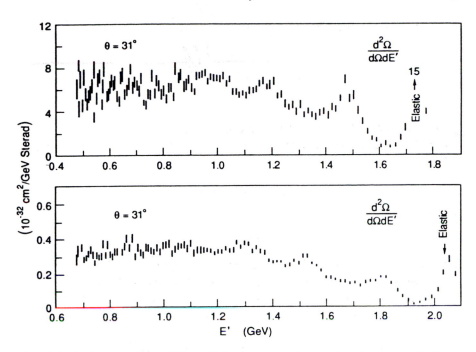

Fig. 32. Inelastic spectra from CEA at 31° for initial energies of 2.4 GeV and 3.0 GeV. Three bumps are clearly evident corresponding to resonance excitations of the proton.

Lineaire in Orsay, where experiments on inelastic electron scattering from nuclei led to the study of radiative processes, and to the determination of radiatively corrected cross sections from inelastic scattering data.

The focus of our thinking about inelastic experiments during the construction period centered on the excitation of resonances and the Q^2 dependences of the "transition form factors" (the nucleon makes a transition from the ground state to the resonant state). We hoped to learn more about each of the observable resonances, and also expected to see new resonances that had not been electroproduced before and even some that had never been observed before in any reaction. Just before the proposal was submitted, data from the CEA was published[33] showing clear evidence of three resonant states excited by inelastic electron scattering. The group at CEA used a quadrupole spectrometer to obtain spectra like those in Figure 32. The background of radiative events is substantial. Very interesting spectra from DESY,[34] showing large non-resonant contributions to the inelastic cross section, would come later, at about the time that the first (inelastic) experiments were starting up at SLAC.

Our proposal was approved in 1966, along with proposals from other groups. The running time for the various parts of our proposal was interleaved with other runs to study photoproduction with the spectrometer facility (and with experiments on a streamer chamber which occupied a building behind End Station A and which used the same beam line).

By January of 1967, the 8 GeV spectrometer was nearly complete and we were beginning preparations for the initial elastic scattering experiment.

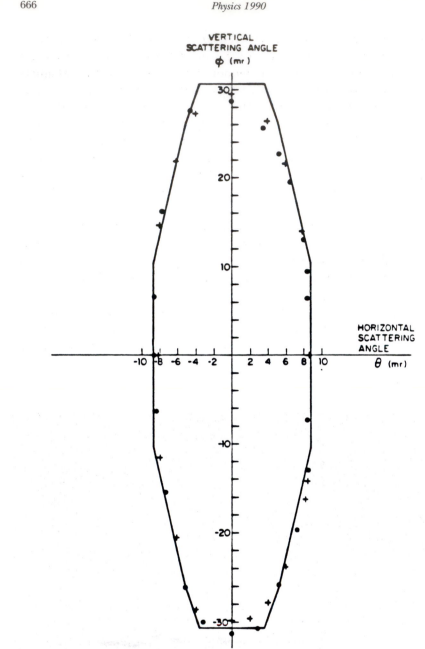

Fig. 33. Angular acceptance of the 8 GeV spectrometer for electrons from the center of the target and with the spectrometer set so that the incoming beam followed the central axis. The points are for two different beam energies (● = 8 GeV , + = 6 GeV). The solid line is the aperture from computer calculations.

The solid angle of the spectrometers entered directly into the calculation of the cross section, and we wanted to check the calculations of the 8 GeV aperture. A special run with beam was planned to study the optics of the spectrometer and the acceptance. The spectrometer was placed at 0° so that the beam entered the spectrometer along the central orbit. The beam energy was adjusted to the setting of the spectrometer and the beam was observed with scintillation screens mounted at the focal planes. Magnets located at the target position steered the beam, tracing out orbits and verifying the optical properties of the spectrometer's magnetic fields. By determining the limiting orbits in the spectrometer the solid angle could be measured. Fig. 33 shows the results for the central momentum case. The agreement with the predictions was quite good, but there were some slight discrepancies with the calculated aperture limits for the extreme rays. After the initial run, lead masks were introduced into the spectrometer to better define the aperture.

Following the optics tests, the counters and shielding were installed along with the hydrogen target and the beam current monitors. By the month of May the first runs of the elastic scattering proposal were underway. The accelerator was operating rather well by this time, though still struggling to meet all of the design specifications.

It is an exciting moment when a new experimental facility is put into operation at a new accelerator, especially when the new accelerator opens up extended new regions of energy for exploration. We were about to use

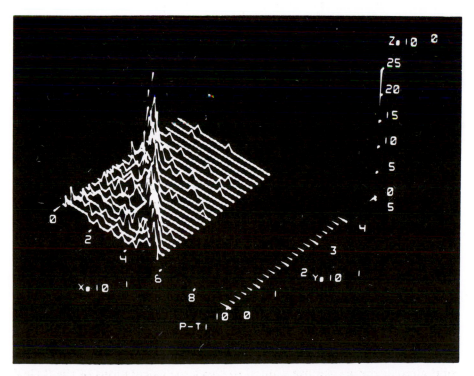

Fig. 34. Computer display of the focal plane location of particles passing through the elements of the p and theta hodoscopes of the 8 GeV spectrometer. The line corresponding to elastic scattering is evident.

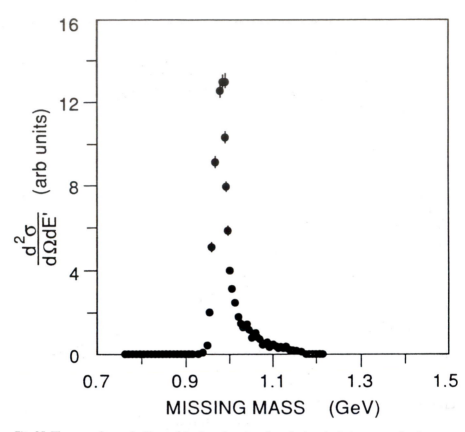

Fig. 35. The same data as in Figure 34, plotted against the calculated missing mass of each event. (The peak is displaced from the mass of the proton at 938 MeV by a slight mismatch in energy calibrations between the switchyard and the spectrometer.)

the biggest physics project ever built to look into places where no one had ever looked before. Nearly a decade of thinking and hard work by hundreds of people would be tested by the events of that evening. Such moments are often spoiled by last minute difficulties, but we were fortunate. Preparations proceeded smoothly, the target was filled with hydrogen and soon the computer was analyzing events. Within a few minutes a respectable elastic peak was showing in the "p-θ" display which sorted events into bins corresponding to the counters hit in the momentum and scattering angle hodoscopes (Fig. 34). The data in this 3-dimensional plot can be converted to a 2-dimensional plot of counts vs. missing mass (Fig. 35) and then to cross sections and form factors. For the next couple of weeks we accumulated data and ran various checks. The system worked well — we could accumulate data fairly rapidly and change both energy and angle from the counting house. The investments made for the sake of efficiency were proving to be valuable, and we were happy with the functioning of our apparatus and the operation of the accelerator.

    A preliminary analysis of the data obtained was made within a few months for presentation at the Electron-Photon Symposium held at SLAC in Au-

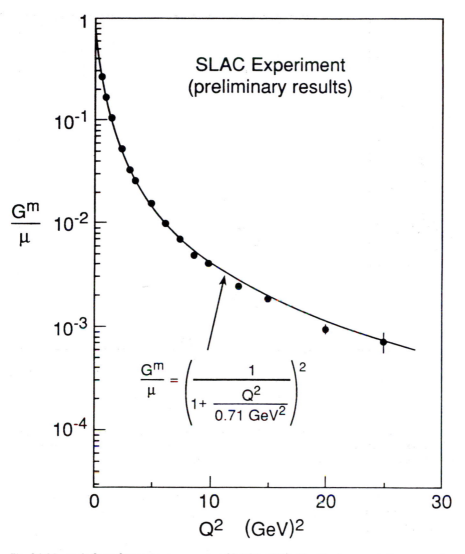

Fig. 36. Magnetic form factor measurement at SLAC in 1967. The dipole curve is the same as in Figure 25, here extended to $Q^2 = 25$ GeV^2. Again, the agreement is imperfect but the curve describes the general behavior of the data quite well.

gust, 1967.[35] The elastic cross sections measured at SLAC behaved in much the same way as those measured at lower energies—falling on the same simple extrapolation of the earlier fits as the CEA and DESY data (Fig. 36). We collected data for G_{Mp} at values of Q^2 up to 25 *GeV*^{2}.

The first opportunity to find something new and unexpected with the spectrometer facility and the SLAC beam had been a disappointment. This is quite normal in experimental physics. Most measurements increment knowledge by just a small amount. Sometimes enough of those small increments eventually result in insights that change our point of view. The sudden observation of unexpected phenomena that result in major new insights is an uncommon event in science. One tries to be ready for such observations, but usually has to be content with adding a small brick of knowledge to the existing edifice. In any case, we had very little time to philosophize over the elastic results because we were busy preparing for the first inelastic scattering experiments. They began in August 1967, using the 20 GeV spectrometer.

In this talk I have tried to point out the importance of advances in accelerators and experimental equipment for the long series of electron scattering experiments at Stanford and elsewhere. The utility of large scale facilities would continue to be demonstrated in later work on nuclear structure with muons and neutrinos at Fermilab and CERN. Large facilities are now commonplace in high energy physics, partially because of the early successes of such facilities in the field of electron scattering.

The Stanford Linear Accelerator and the associated initial complement of experimental equipment were generously supported by U.S. Goverment funding administered by (what is now) the Department of Energy. We were given a chance to build apparatus that was well suited to the opportunities provided by the new linear accelerator. The vast changes in the scale of scientific endeavors during this century have not changed one of the principal preoccupations of the experimental physicist — the building of quality experimental equipment which is matched to the task at hand. In those days the cost-effectiveness of apparatus was considered more important than arbitrary cost-ceilings, and we hope that the physics output of the facilities in End Station A has justified the considerable expense incurred in building them.

In the summer of 1967, SLAC was embarking on a long and productive program of experiments. The story of one of those experiments will be continued in Professor Kendall's lecture.

REFERENCES

1. H. W. Kendall, Deep Inelastic Scattering: Experiments on the Proton and the Observation of Scaling. *Les Prix Nobel 1990*.
2. J. I. Friedman, Deep Inelastic Scattering: Comparisons with the Quark Model. *Les Prix Nobel 1990*.
3. J. I. Friedman, H. W. Kendall, R. E. Taylor, Deep Inelastic Scattering: Acknowledgments. *Les Prix Nobel 1990*.
4. H. Geiger and E. Marsden, Proc. Roy. Soc. **82,** 495 (1909).
5. E. Rutherford, Phil. Mag. **21,** 669 (1911).
6. J. Franck and G. Hertz, Verh. Dtsch. Phys. Ges. **16,** 457 (1914).
7. M. E. Rose, Phys. Rev. **73,** 279 (1948).
8. L. I. Schiff, *Summary of Possible Experiments with a High Energy Linear Electron Accelerator*, SUML—102, Stanford University, Microwave Laboratory, 1949 (unpublished).
9. M. N. Rosenbluth, Phys. Rev. **79,** 615 (1950).
10. E. M. Lyman, A. O. Hanson, and M. B. Scott, Phys. Rev. **84,** 626 (1951).
11. J. H. Fregeau and R. Hofstadter, Phys. Rev. **99,** 1503 (1955).

12. R. Hofstadter, H. R. Fechter and J. A. McIntyre, Phys. Rev. **91,** 422 (1953).
13. R. Hofstadter and R. W. McAllister, Phys. Rev. **98,** 217 (1955).
14. R. Hofstadter, Rev. Mod. Phys. **28,** 214 (1956) and *op. cit.* (This article summarizes the work at HEPL up to 1956 and contains a fairly complete set of references to the early work in the field.)
15. Proposal for a Two Mile Linear Electron Accelerator, Stanford University, April 1957.
16. R. B. Neal, ed., *The Stanford Two Mile Accelerator* (W. A. Benjamin, NY, 1968).
17. C. D. Buchanan *et al., in Proc. 1965 Int. Symp. on Electron and Photon Interactions at High Energies,* Hamburg, Germany, G. Hohler *et al.,* eds. (Hamburg, Germany, Deutsche Phys. Gesellschaft, 1965), pp. 20−42; presented by R. Hofstadter, and *op. cit.*
18. R. R. Wilson *et al.,* Nature **188,** 94 (1960).
19. R. R. Wilson, Nucl. Instrum. **1,** 101 (1957).
20. K. Berkelman *et al.,* Phys. Rev. **130,** 2061 (1965).
21. J. R. Dunning *et al.,* Phys. Rev. Lett. **13,** 631 (1964) and *op. cit.*
22. H. J. Behrend *et al.,* Nuovo Cimento **A38,** 140 (1967) and *op. cit.*
23. W. Bartel *et al.,* Phys. Rev. Lett. **17,** 608 (1966) and *op. cit.*
24. J. Schwinger, Phys. Rev. **75,** 898 (1949).
25. W. K. H. Panofsky, C. Newton, and G. B. Yodh, Phys. Rev. **98,** 751 (1955).
26. W. K. H. Panofsky, W. M. Woodward, and G. B. Yodh, Phys. Rev. **102,** 1392 (1956).
27. G. B. Yodh and W. K. H. Panofsky, Phys. Rev. **105,** 731 (1957).
28. R. A. Alvarez *et al.,* Rev. Sci. Instrum. **31,** 556 (1960).
29. W. K. H. Panofsky and E. A. Allton, Phys. Rev. **110,** 1155 (1958).
30. G. G. Ohlsen, Phys. Rev. **120,** 584 (1960).
31. S. D. Drell and J. D. Walecka, Ann. Phys. (NY) **28,** 18 (1964).
32. L. Hand, Phys. Rev. **129,** 1584 (1963).
33. A. A. Cone *et al.,* Phys. Rev. Lett. **14,** 326 (1965).
34. F. W. Brasse *et al.,* Nuovo Cimento **55,** 679 (1968).
35. R. E. Taylor in *Proc. of the 1967 Int. Symp. on Electron and Photon Interactions at High Energies* (Stanford Linear Accelerator Center, 1967) pp. 70−101.

Henry W. Kendall

HENRY W. KENDALL

I was born on December 9, 1926 in Boston, Massachusetts. My parents were Henry P. Kendall, a Boston businessman, and Evelyn Way Kendall, originally from Canada.

I lived in Boston until the early 1930s when the family — there were five, for by then I had a younger brother and a younger sister — moved to a small town outside Boston, where the three of us grew up and where I still live.

I went briefly to a local grade school but was held back by a reading disability which was cured after I was moved to a school some miles distant. From age 14 to 18, most of the period of World War II, I spent at Deerfield Academy, a college preparatory school. My academic work was poor for I was more interested in non-academic matters and was bored with school work. I had developed — or had been born with — an active curiosity and an intense interest in things mechanical, chemical and electrical and do not remember when I was not fascinated with them and devoted to their exploration. Father was a great encouragement in these projects except when they involved hazards, such as the point, at about age 11, when I embarked on the culture of pathogenic bacteria. He also instilled in both me and my brother a love and respect for the outdoors, especially the mountains and the sea.

I entered the US Merchant Marine Academy in the summer of 1945. I was there, in basic training, when the first atom bombs were exploded over Japan. I was unaware of the human side of these events and only recall a feeling that some of the last secrets of nature had been penetrated and that little would be left to explore. I spent the winter of 1945–46 on a troop transport on the North Atlantic (a most interesting experience), returning to the Academy for advanced training in the spring of 1946. I resigned in October, 1946, to start as a freshman at Amherst College. Although a mathematics major at college, my interest in physics was great and I did undergraduate research and a thesis in that field. But history, English and biology were all most attractive and there was a period, early on, when any one of these might have ended up as the major subject. Non-college enterprises, in the summers particularly, absorbed considerable time. I and a Deerfield friend became interested in diving and two summers were spent in organizing and running a small diving and salvage operation. We wrote our first books after that; one on shallow water diving, another on underwater photography, with a considerable success for both. These activities, mostly self-taught, were a good introduction to two skills very helpful in later experimental work: seeing projects through to successful conclusions and doing them safely.

On the urging of Karl Compton, a family friend and then President of MIT, I applied for, and was accepted at that institution's school of physics in 1950. The years at graduate school were a continuing delight — the first sustained immersion in science at a full professional level. My thesis, carried out under the supervision of Martin Deutsch, was an attempt to measure the Lamb shift in positronium, a transient atom discovered by Deutsch a few years before. The attempt was unsuccessful but it served as a very interesting introduction to electromagnetic interactions and the power of the underlying theory.

The two years after receiving the PhD degree were spent as a National Science Foundation Postdoctoral Fellow at MIT and at Brookhaven National Laboratory, followed by a trip west to join the research group of Robert Hofstadter and the faculty of the Stanford University physics department. Hofstadter was engaged in the study of the proton and neutron structure that was later to bring him the Nobel Prize, work that even at the time was clearly of the greatest interest and importance. The principal facility used in this research was a 300 ft. linear electron accelerator, a precursor to the 2 mile machine at the Stanford Linear Accelerator Center (SLAC), later built in the hills behind the University. Here I met and worked with Jerome Friedman, got to know Richard Taylor, then a graduate student in another group and W. K. H. Panofsky, the driving force behind SLAC. Friedman, Taylor and I were later to join in the long series of measurements on deep inelastic scattering at SLAC.

As in the college years, absorbing non-physics matters claimed a portion of my leisure time: mountaineering and mountain photography. Stanford and the San Francisco Bay area offered a number of skilled climbs as well as Yosemite Valley not far away. After two years of rock and mountain climbing, I was invited on the first of several expeditions to the Andes. Later there have been trips to the Himalayas and the Arctic, with cameras of increasing size to capture some of the astonishing beauty of those remote places. Many of the friends made during those years have remained through life.

After five years at Stanford I moved back to MIT as a member of the faculty. Friedman had gone there a year earlier and we reestablished our collaboration. By 1964, the joint work with Taylor, by then a research group leader at SLAC, was initiated. This collaboration was surely the most enjoyable of any physics I have ever done. It was a pleasure shared by most people in the effort and well recognized at the time. All three of us have remained, up to the present, in the universities we were at then. I have been involved in research in later years, after the SLAC effort wound down in the middle 1970s, at the proton accelerator at Fermilab and since 1981, again at SLAC. The most interesting physics for me has always been the searches for new phenomena or new effects. With colleagues I have searched for limits to quantum electrodynamics, heavy electrons, parity breakdown in electron properties, and other such things. Unfortunately, the ever-growing size, scale, and duration of particle experiments, as well as the much larger collaborations, have made such programs less and less congenial to me over the years, circumstances that disturb many in the physics community.

At the start of the 1960s, troubled by the massive build-up of the superpower's nuclear arsenals, I joined a group of academic scientists advising the U.S. Defense Department. The opportunity to observe the operation of the Defense establishment from the "inside," both in the nuclear weapons area and in the counterinsurgency activities that later expanded to be the U.S. military involvement in South East Asia proved a valuable experience, helpful in later activities in the public domain. It was clear that changing unwise Government policies from inside, especially those the Government is deeply attached to, involves severe, often insurmountable, problems.

In 1969, I was one of a group founding the Union of Concerned Scientists (UCS), and have played a substantial role in its activities in the years hence. UCS is a public interest group, supported by funds raised from the general public, that presses for control of technologies which may be harmful or dangerous. The organization has had an important national role in the controversies over nuclear reactor safety, the wisdom of the US Strategic Defense Initiative, the B2 (Stealth) bomber, and the challenge posed by fossil fuel burning and possible greenhouse warming of the atmosphere, among others. I have been Chairman of the organization since 1974. The activities of the organization are part of a slowly growing interest among scientists to take more responsibility for helping society control the exceedingly powerful technologies that scientific research has spawned. It is hard to conclude that scientists are in the main responsible for the damage and risks that are now so apparent in such areas as environmental matters and nuclear armaments; these have been largely the consequence of governmental and industrial imperatives, both here and abroad. Yet it seems clear that without scientists' participation in the public debates, the chances of great injury to all humanity is much enhanced. In my view, the scientific community has not participated in this effort at a level commensurate with the need, nor with the special responsibilities that scientists ineluctably have in this area.

This expenditure of effort and the sense of responsibility to help achieve control of aberrant technologies which drives it, stems in no small measure from the example set by my Father, who, throughout his life, spent a great deal of time and no small amount of energy on quiet, *pro bono* work. He was not alone among his own friends — nor among his own contemporaries — in this; it has been a tradition in New England of very long standing. In continuing to pursue such objectives, my expectation is that the challenges facing both me and the Union will be made substantially easier by the award of the Nobel Prize. This is perhaps the most attractive part of having gained this exceptional honor.

DEEP INELASTIC SCATTERING: EXPERIMENTS ON THE PROTON AND THE OBSERVATION OF SCALING

Nobel Lecture, December 8, 1990

by

HENRY W. KENDALL

Massachusetts Institute of Technology, Cambridge, Massachusetts, USA

I *Introduction*

A. Overview of the Electron Scattering Program

In late 1967 the first of a long series of experiments on highly inelastic electron scattering was started at the two mile accelerator at the Stanford Linear Accelerator Center (SLAC) using liquid hydrogen and, later, liquid deuterium targets. Carried out by a collaboration from the Massachusetts Institute of Technology (MIT) and SLAC, the object was to look at large energy loss scattering of electrons from the nucleon (the generic name for the proton and neutron), a process soon to be dubbed deep inelastic scattering. Beam energies up to 21 GeV, the highest electron energies then available, and large electron fluxes, made it possible to study the nucleon to very much smaller distances than had previously been possible. Because quantum electrodynamics provides an explicit and well-understood description of the interaction of electrons with charges and magnetic moments, electron scattering had, by 1968, already been shown to be a very powerful probe of the structures of complex nuclei and individual nucleons.

Hofstadter and his collaborators had discovered, by the mid-1960s, that as the momentum transfer in the scattering increased, the scattering cross section dropped sharply relative to that from a point charge. The results showed that nucleons were roughly 10^{-13} cm in size, implying a distributed structure. The earliest MIT-SLAC studies, in which California Institute of Technology physicists also collaborated, looked at elastic electron-proton scattering, later ones at electro-production of nucleon resonances with excitation energies up to less than 2 GeV. Starting in 1967, the MIT-SLAC collaboration employed the higher electron energies made available by the newly completed SLAC accelerator to continue such measurements, before beginning the deep inelastic program.

Results from the inelastic studies arrived swiftly: the momentum transfer dependence of the deep inelastic cross sections was found to be weak, and the deep inelastic form factors — which embodied the information about the proton structure — depended unexpectedly only on a single variable rather than the two allowed by kinematics alone. These results were inconsistent with the current expectations of most physicists at the time. The general belief had been that the nucleon was the extended object found in elastic electron scattering but with the diffuse internal structure seen in pion and proton scattering. The new experimental results suggested point-like constituents but were puzzling because such constituents seemed to contradict well-established beliefs. Intense interest in these results developed in the theoretical community and, in a program of linked experimental and theoretical advances extending over a number of years, the internal constituents were ultimately identified as **quarks,** which had previously been devised in 1964 as an underlying, quasi-abstract scheme to justify a highly successful classification of the then-known hadrons. This identification opened the door to development of a comprehensive field theory of hadrons (the strongly interacting particles), called Quantum Chromodynamics (QCD), that replaced entirely the earlier picture of the nucleons and mesons. QCD in conjunction with electroweak theory, which describes the interactions of leptons and quarks under the influence of the combined weak and electromagnetic fields, constitutes the Standard Model, all of whose predictions, at this writing, are in satisfactory agreement with experiment. The contributions of the MIT-SLAC inelastic scattering program were recognized by the award of the 1990 Nobel Prize in Physics.

B. Organization of lectures

There are three lectures that, taken together, describe the MIT-SLAC experiments. The first, written by R.E.Taylor (Reference 1), sets out the early history of the construction of the two mile accelerator, the proposals made for the construction of the electron scattering facility, the antecedent physics experiments at other laboratories, and the first of our scattering experiments which determined the elastic proton structure form factors. This paper describes the knowledge and beliefs about the nucleon's internal structure in 1968, including the conflicting views on the validity of the quark model and the "bootstrap" models of the nucleon. This is followed by a review of the inelastic scattering program and the series of experiments that were carried out, and the formalism and variables. Radiative corrections are described and then the results of the inelastic electron-proton scattering measurements and the physics picture — the naive parton model — that emerged. The last lecture, by J. I. Friedman (Reference 2), is concerned with the later measurements of inelastic electron-neutron and electron-proton measurements and the details of the physical theory — the constituent quark model — which the experimental scattering results stimulated and subsequently, in conjunction with neutrino studies, confirmed.

II *Nucleon and Hadronic Structure in 1968*

At the time the MIT-SLAC inelastic experiments started in 1968, there was no detailed model of the internal structures of the hadrons. Indeed, the very notion of "internal structure" was foreign to much of the then-current theory. Theory attempted to explain the soft scattering — that is, rapidly

decreasing cross sections as the momentum transfer increased — which was
the predominant characteristic of the high energy hadron-hadron scattering data of the time, as well as the hadron resonances, the bulk of which
were discovered in the late 1950s and 1960s. Quarks had been introduced,
quite successfully, to explain the static properties of the array of hadrons.
Nevertheless, the available information suggested that hadrons were "soft"
inside, and would yield primarily distributions of scattered electrons reflecting diffuse charge and magnetic moment distributions with no underlying
point-like constituents. Quark constituent models were gleams in the eyes of
a small handful of theorists, but had serious problems, then unsolved, which
made them widely unpopular as models for the high energy interactions of
hadrons.

The need to carry out calculations with forces that were known to be very
strong introduced intractable difficulties: perturbation theory, in particular, was totally unjustified. This stimulated renewed attention to S-matrix
theory (Reference 3), an attempt to deal with these problems by consideration of the properties of a matrix that embodied the array of strong
interaction transition amplitudes from all possible initial states to all possible final states.

A.   Theory: Nuclear Democracy

An approach to understanding hadronic interactions, and the large array of
hadronic resonances, was the bootstrap theory (Reference 4), one of several
elaborations of S-matrix theory. It assumed that there were no "fundamental" particles: each was a composite of the others. Sometimes referred to as
"nuclear democracy," the theory was at the opposite pole from constituent
theories.

Regge theory (Reference 5), a very successful phenomenology, was one
elaboration of S-matrix theory which was widely practiced. Based initially on
a new approach to non-relativistic scattering, it was extended to the relativistic S-matrix applicable to high energy scattering (Reference 6). The known
hadrons were classified according to which of several "trajectories" they lay
on. It provided unexpected connections between reactions at high energies
to resonances in the crossed channels, that is, in disconnected sets of states.
For scattering, Regge theory predicted that at high energy, hadron-hadron
scattering cross sections would depend smoothly on s, the square of the
center of mass energy, as $A(s) \sim s^{(\alpha(0))}$, and would fall exponentially with t,
the square of the space-like momentum transfer, as

$$A(t) \sim \exp(\alpha't \ln(s/s_0)).$$

Regge theory led to duality, a special formulation of which was provided by
Veneziano's dual resonance model (Reference 7). These theories still provide the best description of soft, low momentum transfer scattering of pions
and nucleons from nucleons, all that was known in the middle 1960s. There
was a tendency, in this period, to extrapolate these low momentum transfer
results so as to conclude there would be no hard scattering at all.

S-matrix concepts were extended to the electromagnetic processes involving hadrons by the Vector Meson Dominance (VMD) model (Reference 8). According to VMD, when a real or virtual photon interacts with a hadron, the photon transforms, in effect, into one of the low mass vector mesons that has the same quantum numbers as the photon (primarily the rho, omega and phi mesons). In this way electromagnetic amplitudes were related to hadronic collision amplitudes, which could be treated by S-matrix methods. The VMD model was very successful in phenomena involving real photons and many therefore envisaged that VMD would also deal successfully with the virtual photons exchanged in inelastic electron scattering. Naturally, this also led to the expectation that electron scattering would not reveal any underlying structure.

All of these theories, aside from their applications to hadron-hadron scattering and the properties of resonances, had some bearing on nucleon structure as well, and were tested against the early MIT-SLAC results.

B.   Quark Theory of 1964

The quark[1] was born in a 1964 paper by Murray Gell-Mann (Reference 9) and, independently, by George Zweig (Reference 10). For both, the quark (a term Zweig did not use until later) was a means to generate the symmetries of SU(3), the "Eightfold Way," Gell-Mann and Ne'emann's (Reference 11) highly successful 1961 scheme for classifying the hadrons. Combinations of spin 1/2 quarks, with fractional electric charges, and other appropriate quantum numbers, were found to reproduce the multiplet structures of all the observed hadrons. Fractional charges were not necessary but provided the most elegant and economical scheme. Three quarks were required for baryons, later referred to as "valence" quarks, and quark-antiquark pairs for mesons. Indeed the quark picture helped solve some difficulties with the earlier symmetry groupings (Reference 12). The initial successes of the theory stimulated numerous free quark searches. There were attempts to produce them with accelerator beams, studies to see if they were produced in cosmic rays, and searches for "primordial" quarks by Millikan oil drop techniques sensitive to fractional charges. None of these has ever been successful (Reference 13).

C.   Constituent Quark Picture

There were serious problems in having quarks as physical constituents of nucleons and these problems either daunted or repelled the majority of the theoretical community, including some of its most respected members (Reference 14). The idea was distasteful to the S-matrix proponents. The problems were, first, that the failure to produce quarks had no precedent in

---

[1] The word *quork* was invented by Murray Gell-Mann, who later found *quark* in the novel *Finnegan's Wake*, by James Joyce, and adopted what has become the accepted spelling. Joyce apparently employed the word as a corruption of the word *quart*. The author is grateful to Murray Gell-Mann for a discussion clarifying the matter.

physicists' experience. Second, the lack of direct production required the quarks to be very massive, which, for the paired quark configurations of the mesons, meant that the binding had to be very great, a requirement that led to predictions inconsistent with hadron-hadron scattering results. Third, the ways in which they were combined to form the baryons, meant that they could not obey the Pauli exclusion principle, as required for spin one-half particles. Fourth, no fractionally charged objects had ever been unambiguously identified. Such charges were very difficult for many to accept, for the integer character of elementary charges was long established. Enterprising theorists did construct quark theories employing integrally charged quarks, and others contrived ways to circumvent the other objections. Nevertheless, the idea of constituent quarks was not accepted by the bulk of the physics community, while others sought to construct tests that the quark model was expected to fail (Reference 15).

Some theorists persisted, nonetheless. Dalitz (Reference 16) carried out complex calculations to help explain not only splittings *between* hadron multiplets but the splittings *within* them also, using some of the theoretical machinery employed in nuclear spectroscopy calculations. Calculations were carried out on other aspects of hadron dynamics, for example, the successful prediction that Δ^+ decay would be predominantly magnetic dipole (Reference 17). Owing to the theoretical difficulties just discussed, the acceptance of quarks as the basis of this successful phenomenology was not carried over to form a similar basis for high energy scattering.

Gottfried studied electron-proton scattering with a model assuming point quarks, and argued that it would lead to a total cross section (elastic plus inelastic) at fixed momentum transfer, identical to that of a point charge, but he expressed great skepticism that this would be borne out by the forthcoming data (Reference 18). With the exception of Gottfried's work and one by Bjorken stimulated by current algebra, discussed below, all of the published constituent quark calculations were concerned with low energy processes or hadron characteristics rather than high energy interactions. Zweig carried out calculations assuming that quarks were indeed hadron constituents but his ideas were not widely accepted (Reference 19).

Thus, one sees that the tide ran against the constituent quark model in the 60s (Reference 20). One reviewer's summary of the style of the 60s was that "quarks came in handy for coding information but should not be taken seriously as physical objects" (Reference 21). While quite helpful in low energy resonance physics, it was for some "theoretically disreputable," and was felt to be largely peripheral to a description of high energy soft scattering (Reference 22).

D.   Current Algebra

Following his introduction of quarks, Gell-Mann, and others, developed "current algebra," which deals with hadrons under the influence of weak and electromagnetic interactions. Starting with an assumption of free quark fields, he was able to find relations between weak currents that reproduced the current commutators postulated in constructing his earlier hadronic symmetry groups. Current algebra had become very important by 1966. It exploited the concept of *local observables* — the current and charge densities of the weak and electromagnetic interactions. These are field theoretic in character and could only be incorporated into S-matrix cum bootstrap theory by assumptions like VMD. The latter are plausible for moderate momentum transfer, but hardly for transfer large compared to hadron masses. As a consequence, an important and growing part of the theoretical community was thinking in field theoretic terms.

Current algebra also gave rise to a small but vigorous "sum rule" industry. Sum rules are relationships involving weighted integrals over various combinations of cross sections. The predictions of some of these rules were important in confirming the deep inelastic electron and neutrino scattering results, after these became available (Reference 23).

Gell-Mann made clear that he was not suggesting that hadrons were made up of quarks (Reference 24), although he kept open the possibility that they might exist (Reference 25). Nevertheless, current algebra reflected its constituent-quark antecedents, and Bjorken used it to demonstrate that sum rules derived by him and others required large cross sections for these to be satisfied. He then showed that such cross sections arose naturally in a quark constituent model (Reference 26), in analog to models of nuclei composed of constituent protons and neutrons, and also employed it to predict the phenomena of scaling, discussed at length below. Yet Bjorken and others were at a loss to decide how the point-like properties that current algebra appeared to imply were to be accommodated (Reference 27).

E. Theoretical Input to The Scattering Program

In view of the theoretical situation as set out above, there was no consideration that a possible point-like substructure of the nucleon might be observable in electron scattering during the planning and design of the electron scattering facility. Deep inelastic processes were, however, assessed in preparing the proposal submitted to SLAC for construction of the facility (Reference 28). Predictions of the cross sections employed a model assuming off-mass-shell photo-meson production, using photoproduction cross sections combined with elastic scattering structure functions, in what was believed to be the best guide to the yields expected. These were part of extensive calculations, carried out at MIT, designed to find the magnitude of distortions of inelastic spectra arising from photon radiation, necessary in planning the equipment and assessing the difficulty of making radiative corrections. It was found ultimately that these had underpredicted the actual yields by between one and two orders of magnitude.

III *The Scattering Program*

The linear accelerator that provided the electron beam employed in the inelastic scattering experiments was, and remains to the date of this paper, a

Fig. 1. View of the Stanford Linear Accelerator. The electron injector is at the top, the experimental area in lower center. The deep inelastic scattering studies were carried out in End Station A, the largest of the buildings in the experimental area.

device unique among high energy particle accelerators. See Figure 1. An outgrowth of the smaller, 1 GeV accelerator employed by Hofstadter in his studies of the charge and magnetic moment distributions of the nucleon, it relied on advanced klystron technology devised by Stanford scientists and engineers to provide the high levels of microwave power necessary for one-pass acceleration of electrons. Proposed in 1957, approved by the Congress in 1962, its construction was initiated in 1963. It went into operation in 1967, on schedule, having cost $114M (Reference 29).

The experimental collaboration began in 1964. After 1965, R. E. Taylor was head of SLAC Group A with J.I.Friedman and the present author sharing responsibility for the M.I.T. component. A research group from California Institute of Technology joined in the construction cycle and the elastic studies but withdrew before the inelastic work started in order to pursue other interests.

The construction of the facility to be employed in electron scattering was nearly concurrent with the accelerator's construction. This facility was large for its time. A 200 ft. by 125 ft. shielded building housed three magnetic spectrometers with an adjacent "counting house" containing the fast electronics and a computer, also large for its time, where experimenters controlled the equipment and conducted the measurements. See Figure 2a and 2b. The largest spectrometer would focus electrons up to 20 GeV and was employed at scattering angles up to 10^0. A second spectrometer, useful to 8

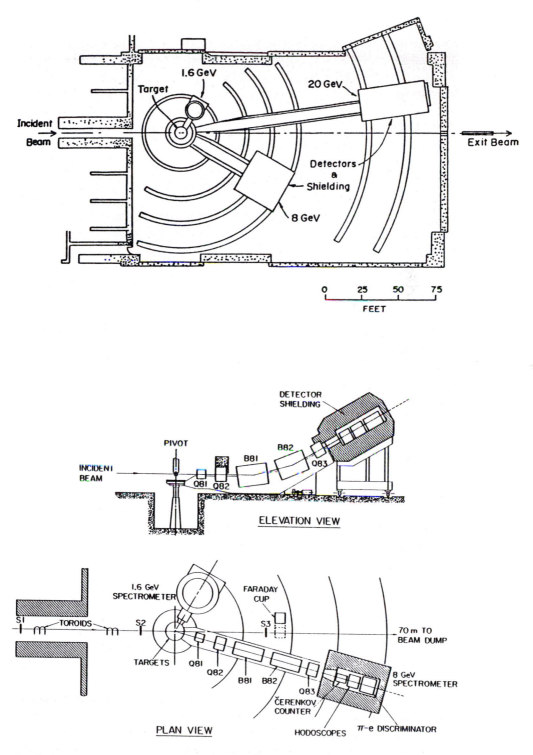

Fig. 2. (a) Plan view of End Station A and the two principal magnetic spectrometers employed for analysis of scattered electrons. (b) Configuration of the 8 GeV spectrometer, employed at scattering angles greater than 12°.

GeV, was used initially out to 34^0, and a third, focusing to 1.6 GeV, constructed for other purposes, was employed in one set of large angle measurements to help determine the uniformity in density of the liquified target gases. The detectors were designed to detect only scattered electrons. The very short duty cycle of the pulsed beam precluded studying the recoil systems in coincidence with the scattered electrons: it would have given rise to unacceptable chance coincidence rates, swamping the signal.

The elastic studies started in early 1967 with the first look at inelastic processes from the proton late the same year. By the spring of 1968, the first inelastic results were at hand. The data were reported at a major scientific meeting in Vienna in August and published in 1969 (Reference 30). Thereafter, a succession of experiments were carried out, most of them, from 1970 on, using both deuterium and hydrogen targets in matched sets of measurements so as to extract neutron scattering cross sections with a minimum of systematic error. These continued well into the 1970s. One set of measurements (Reference 31) studied the atomic-weight dependence of the inelastic scattering, primarily at low momentum transfers, studies that were extended to higher momentum transfers in the early 1980s, and involved extensive reanalysis of earlier MIT-SLAC data on hydrogen, deuterium and other elements (Reference 32).

The collaboration was aware from the outset of the program that there were no accelerators in operation, or planned, that would be able to confirm the entire range of results. The group carried out independent data analyses at MIT and at SLAC to minimize the chance of error. One consequence of the absence of comparable scattering facilities was that the collaboration was never pressed to conclude either data taking or analysis in competitive circumstances. It was possible throughout the program to take the time necessary to complete work thoroughly.

IV   *Scattering Formalism and Radiative Corrections*

A.   Fundamental Processes

The relation between the kinematic variables in elastic scattering, as shown in Figure 3, is:

$$v = E\text{-}E' = q^2/(2M) \qquad q^2 = 2EE'\ (1\text{-}\cos\theta) \tag{1}$$

where E is the initial and E' the final electron energy, θ the laboratory angle of scattering, v the electron energy loss, q the four-momentum transferred to the target nucleon, and M the proton mass.

The cross section for elastic electron-proton scattering has been calculated by Rosenbluth (Reference 33) in first Born approximation, that is, to leading order in $a = 1/137$:

$$\frac{d\sigma}{d\Omega}\ (E) = \sigma_M\ (E) \left\{ \frac{E'}{E} \right\} \left\{ \frac{G_{Ep}^2\ (q^2) + \tau\ G_{Mp}^2\ (q^2)}{1 + \tau} + 2\ \tau\ G_{Mp}^2 \tan^2(\theta/2) \right\} \tag{2}$$

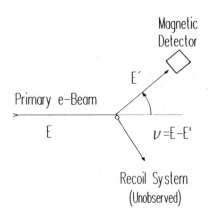

Fig. 3. Scattering kinematics.

where

$$\sigma_M = \frac{4a^2 E'^2}{q^4} \cos^2 \left(\frac{\theta}{2}\right)$$

is the Mott cross section for elastic scattering from a point proton, and

$$\tau = q^2/(4M^2)$$

In these equations, and in what follows, $\hbar = c = 1$, and the electron mass has been neglected. The functions $G_{Ep}(q^2)$ and $G_{Mp}(q^2)$, the electric and magnetic form factors, respectively, describe the time-averaged structure of the proton. In the non-relativistic limit the squares of these functions are the Fourier transforms of the spatial distributions of charge and magnetic moment, respectively. As can be seen from Equation (2), magnetic scattering is dominant at high q^2. Measurements (Reference 34) show that G_{Mp} is roughly described by the "dipole" approximation:

$$G_{Mp}/\mu = 1/(1 + q^2/0.71)^2$$

where q^2 is measured in $(GeV)^2$ and $\mu = 2.79$ is the proton's magnetic moment. Thus, at large q^2 an additional $1/q^8$ dependence beyond that of σ_M is imposed on the elastic scattering cross section as a consequence of the finite size of the proton. This is shown in Figure 4.

In inelastic scattering, energy is imparted to the hadronic system. The invariant or missing mass W is the mass of the final hadronic state. It is given by:

$$W^2 = (2M\nu + M^2 - q^2)$$

When only the electron is observed the composition of the hadronic final state is unknown except for its invariant mass W. On the assumption of one photon exchange (Figure 5), the differential cross section for electron scattering from the nucleon target is related to two structure functions W_1 and W_2 according to (Reference 35):

Elastic Electron-Proton Scattering

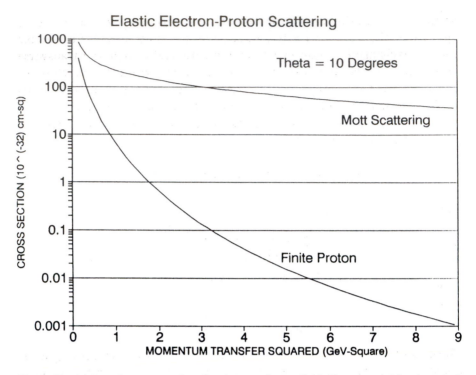

Fig. 4. Elastic scattering cross sections for electrons from a "point" proton and for the actual proton. The differences are attributable to the finite size of the proton.

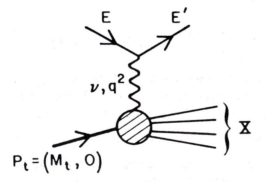

Fig. 5. Feynman diagram for inelastic electron scattering.

$$\frac{d^2\sigma}{d\Omega\,dE'}\,(E,E',\theta) = \sigma_M\left[W_2(\nu,q^2) + 2W_1\,(\nu,q^2)\tan^2(\theta/2)\right] \tag{3}$$

This expression is the analog of the Rosenbluth cross section given above. The structure functions W_1 and W_2 are similarly defined by Equation (3) for the proton, deuteron, or neutron; they summarize all the information about the structure of the target particles obtainable by scattering unpolarized electrons from an unpolarized target.

Within the single-photon-exchange approximation, one may view inelastic electron scattering as photoproduction by "virtual" photons. Here, as opposed to photoproduction by real photons, the photon mass q^2 is variable and the exchanged photon may have a longitudinal as well as a transverse polarization. If the final state hadrons are not observed, the interference between these two components averages to zero, and the differential cross section for inelastic electron scattering is related to the total cross sections for absorption of transverse, σ_T, and longitudinal, σ_L, virtual photons according to (Reference 36)

$$\frac{d^2\sigma}{d\Omega \, dE'}(E,E',\theta) = \Gamma \left[\sigma_T(v,q^2) + \varepsilon\,\sigma_L(v,q^2)\right] \tag{4}$$

where

$$\Gamma = \frac{a}{4\pi^2}\,\frac{KE'}{q^2\,E}\left(\frac{2}{1-\varepsilon}\right)$$

$$\varepsilon = [1 + 2\,(1 + v^2/q^2)\,\tan^2(\theta/2)]^{-1}$$

and

$$K = (W^2 - M^2)/(2M)$$

The quantity Γ is the flux of transverse virtual photons and ε is the degree of longitudinal polarization. The cross sections σ_T and σ_L are related to the structure functions W_1 and W_2 by

$$W_1(v,q^2) = \frac{K}{4\pi^2 a}\,\sigma_T(v,q^2)$$

$$W_2(v,q^2) = \frac{K}{4\pi^2 a}\left(\frac{q^2}{q^2 + v^2}\right)[\sigma_T(v,q^2) + \sigma_L(v,q^2)] \tag{5}$$

In the limit $q^2 \to 0$, gauge invariance requires that $\sigma_L \to 0$ and $\sigma_T \to \sigma_\gamma(v)$, where $\sigma_\gamma(v)$ is the photoproduction cross section for real photons. The quantity R, defined as the ratio σ_L/σ_T is related to the structure functions by

$$R(v,q_2) \equiv \sigma_L/\sigma_T = (W_2/W_1)(1 + v^2/q^2) - 1 \tag{6}$$

A separate determination of the two inelastic structure functions W_1 and W_2 (or, equivalently, σ_L and σ_T) requires values of the differential cross section at several values of the angle θ for fixed v and q^2. According to Equation (4), σ_L is the slope and σ_T is the intercept of a linear fit to the quantity Σ where:

$$\Sigma = \frac{1}{\Gamma}\,\frac{d^2\sigma}{d\Omega \, dE'}(v,q^2,\theta)$$

The structure functions W_1 and W_2 are then directly calculable from Eq. (5). Alternatively, one can extract W_1 and W_2 from a single differential cross-section measurement by inserting a particular functional form for R in the equations

$$W_1 = \frac{1}{\sigma_M} \frac{d^2\sigma}{d\Omega \, dE'} \left[(1 + R) \left(\frac{q^2}{q^2 + v^2} \right) + 2 \tan^2 \left(\frac{\theta}{2} \right) \right]^{-1}$$

$$W_2 = \frac{1}{\sigma_M} \frac{d^2\sigma}{d\Omega \, dE'} \left[1 + \left(\frac{2}{1 + R} \right) \left(\frac{q^2 + v^2}{q^2} \right) \tan^2 \left(\frac{\theta}{2} \right) \right]^{-1}$$

(7)

Equations (5) through (7) apply equally well for the proton, deuteron, or neutron.

In practice, it was convenient to determine values of σ_L and σ_T from straight line fits to differential cross sections as functions of ε. R was determined from the values of σ_L and σ_T, and W_1 and W_2 were, as shown above, determined from R.

B.   Scale Invariance and Scaling Variables.

By investigating models that satisfied current algebra, Bjorken (Reference 37) had conjectured that in the limit of q^2 and v approaching infinity, with the ratio $\omega = 2Mv/q^2$ held fixed, the two quantities vW_2 and W_1 become functions of ω only. That is:

$$2MW_1 (v, q^2) = F_1 (\omega)$$
$$vW_2 (v, q^2) = F_2 (\omega)$$

It is this property that is referred to as "scaling" in the variable ω in the "Bjorken limit." The variable $x = 1/\omega$ came into use soon after the first inelastic measurements; we will use both in this paper.

Since W_1 and W_2 are related by

$$vW_2/W_1 = (1 + R)/(1/v + \omega/(2M))$$

it can be seen that scaling in W_1 accompanies scaling in vW_2 only if R has the proper functional form to make the right hand side of the equation a function of ω. In the Bjorken limit, it is evident that the ratio vW_2/W_1 will scale if R is constant or is a function of ω only.

C.   Radiative Corrections

Radiative corrections must be applied to the measured cross sections to eliminate the effects of the radiation of photons by electrons which occurs during the nucleon scattering itself and during traversals of material before and after scattering. These corrections also remove higher order electrodynamic contributions to the electron-photon vertex and the photon propagator. Radiative corrections as extensive as were required in the proposed scattering program had been little studied previously (Reference 38). Friedman (Reference 39), in 1959 had calculated the elements of the required "triangle," discussed in more detail below, in carrying out corrections to the inelastic scattering of 175 MeV electrons from deuterium. Isabelle and Kendall (Reference 40), studying the inelastic scattering of electrons of energy up to 245 MeV from Bi[209] in 1962, had measured inelastic spectra over a number of triangles and had developed the computer procedures necessary to permit computation of the corrections. These studies provided

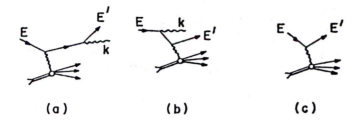

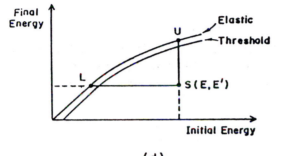

(d)

Fig. 6. Diagrams showing radiation in electron scattering (a) after exchange of a virtual photon (b) before exchange of a virtual photon. Figure (6c) is the diagram with radiative effects removed. Figure (6d) is the kinematic plane relevant to the radiative corrections program. The text contains a further discussion of corrections procedures. A "triangle" as discussed in the text is formed by points L, U, and S.

confidence that the procedures were tractable and the resulting errors of acceptable magnitude.

The largest correction has to be made for the radiation during scattering, described by diagrams (a) and (b) in Figure 6. A photon of energy k is emitted in (a) after the virtual photon is exchanged, and in (b) before the exchange. Diagram (c) is the cross section which is to be recovered after appropriate corrections for (a) and (b) have been made. A measured cross section at fixed E, E', and θ will have contributions from (a) and (b) for all values of k which are kinematically allowed. The lowest value of k is zero, and the largest occurs in (b) for elastic scattering of the virtual electron from the target particle. Thus, to correct a measured cross section at given values of E and E', one must know the cross section over a range of incident and scattered energies.

To an excellent approximation, the information necessary to correct a cross section at an angle θ may all be gathered at the same value of θ. Diagram (d) of Figure 6 shows the kinematic range in E and E' of cross sections which can contribute by radiative processes to the fundamental cross section sought at point S, for fixed θ. The range is the same for contributions from bremsstrahlung processes of the incident and scattered electrons. For single hard photon emission, the cross section at point S will have contributions from elastic scattering at points U and L, and from inelastic scattering along the lines SL and SU, starting at inelastic threshold. If two or more photons are radiated, contributions can arise from line LU and the inelastic region bounded by lines SL and SU. The cross sections needed for these corrections must themselves have been corrected for radiative effects. However, if uncorrected cross sections are available over the whole of the "triangle" LUS, then a one-pass radiative correction procedure may be employed, assuming the peaking approximation (Reference 41), which will produce the approximately corrected cross sections over the entire triangle, including the point S.

The application of radiative corrections required the solution of another difficulty, as it was generally not possible to take measurements sufficiently closely spaced in the E-E' plane to apply them directly. Typically five to ten spectra, each for a different E, were taken to determine the cross sections over a "triangle." Interpolation methods had to be developed to supply the missing cross sections and had to be tested to show that they were not the source of unexpected error. Figure 7 shows the triangles, and the locations of the spectra, for data taken in one of the experiments in the program.

In the procedures that were employed, the radiative tails from elastic electron-proton scattering were subtracted from the measured spectra before the interpolations were carried out. In the MIT-SLAC radiative correction procedures, the radiative tails from elastic scattering were calculated using the formula of Tsai (Reference 42), which is exact to lowest order in α. The calculation of the tail included the effects of radiative energy degradation of the incident and final electrons, the contributions of multiple photon processes, and radiation from the recoiling proton. After the subtraction of the elastic peak's radiative tail, the inelastic radiative tails were removed in a one-pass unfolding procedure as outlined above. The particular form of the peaking approximation used was determined from a fit to an exact calculation of the inelastic tail to lowest order which incorporated a model that approximated the experimental cross sections. One set of formulas and procedures are described by Miller *et al.* (Reference 43) and were employed in the SLAC analysis. The measured cross sections were also corrected in a separate analysis, carried out at MIT, using a somewhat different set of approximations (Reference 44). Comparisons of the two gave corrected cross sections which agreed to within a few percent. Reference 45 contains a complete description of the MIT radiative corrections procedures that were applied, the cross checks that were carried out, and the assessment of errors arising both from the radiative corrections and from other sources of uncertainty in the experiment. Figure 8 shows the relative magnitude of the radiative corrections as a function of W for a typical spectrum with a hydrogen target. While radiative corrections were the largest corrections to the data, and involved a considerable amount of computation, they were understood to a confidence level of 5% to 10% and did not significantly increase the total error in the measurements.

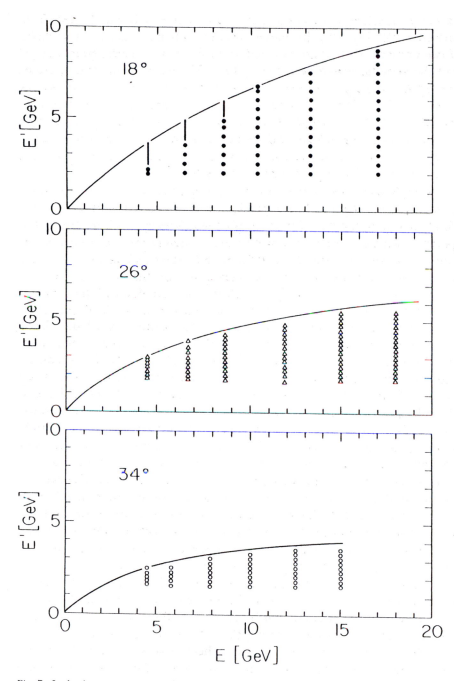

Fig. 7. Inelastic measurements: where spectra were taken to determine "triangles" employed in making radiative corrections for three angles selected for some of the later experiments. The solid curves represent the kinematics of elastic electron-proton scattering.

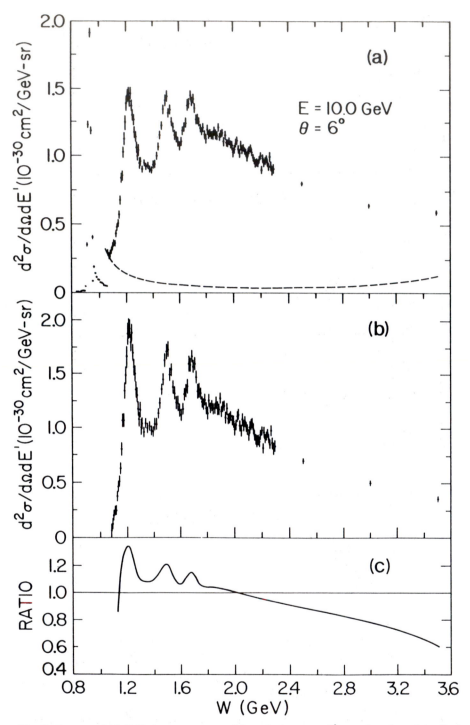

Fig. 8. Spectra of 10 GeV electrons scattered from hydrogen at 6°, as a function of the final hadronic state energy W. Figure (8a) shows the spectrum before radiative corrections. The elastic peak has been reduced in scale by a factor of 8.5. The computed radiative "tail" from the elastic peak is shown. Figure (8b) shows the same spectrum with the elastic peak's tail subtracted and inelastic corrections applied. Figure (8c) shows the ratio of the inelastic spectrum before, to the spectrum after, radiative corrections.

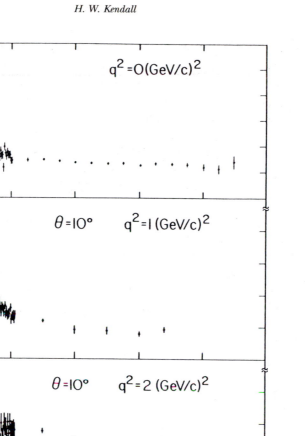

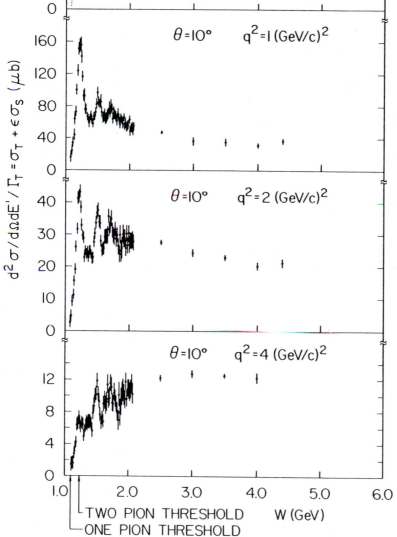

Fig. 9. Spectra of electrons scattered from hydrogen at q^2 up to 4 (GeV/c)^2. The curve for $q^2 = 0$ represents an extrapolation to $q^2 = 0$ of electron scattering data acquired at $\theta = 1.5°$. Elastic peaks have been subtracted and radiative corrections have been applied.

V   *Electron Proton Scattering: Results.*

The scattered electron spectra observed in the experiments had a number of features whose prominence depended on the initial and final electron energies and the scattering angle. At low q^2 both the elastic peak and resonance excitations were large, with little background from non-resonant continuum scattering either in the resonance region or at higher missing masses. As q^2 increased, the elastic and resonance cross sections decreased rapidly, with the continuum scattering becoming more and more dominant. Figure 9 shows four spectra of differing q^2. Data points taken at the elastic peak and in the resonance region were closely spaced in E' so as to allow fits to be made to the resonance yields, but much larger steps were employed for larger excitation energies.

Figures 10a and 10b show visual fits to spectra over a wide range in energy and scattering angle (including one spectrum from the accelerator at the Deutsches Electronen Synchrotron (DESY)), illustrating the points discussed above.

Two features of the non-resonant inelastic scattering that appeared in the first continuum measurements were unexpected. The first was a quite weak q^2 dependence of the scattering at constant W. Examples for $W = 2.0$ and $W = 3.0$ GeV, taken from data of the first experiment, are shown in Figure 11 as a function of q^2. For comparison the q^2 dependence of elastic scattering is shown also.

The second feature was the phenomenon of scaling. During the analysis of the inelastic data, J. D. Bjorken suggested a study to determine if νW_2 was a function of ω alone. Figure 12a shows the earliest data so studied: W_2, for six values of q^2, as a function of ν. Figure 12b shows $F_2 = \nu W_2$ for 10 values of q^2, plotted against ω. Because R was at that time unknown, F_2 was shown for the limiting assumptions, $R = 0$ and $R = \infty$. It was immediately clear that the Bjorken scaling hypothesis was, to a good approximation, correct. This author, who was carrying out this part of the analysis at the time, recalls wondering how Balmer may have felt when he saw, for the first time, the striking agreement of the formula that bears his name with the measured wavelengths of the atomic spectra of hydrogen.

More data showed that, at least in the first regions studied and within sometimes large errors, scaling held nearly quantitatively. As we shall see, scaling holds over a substantial portion of the ranges of ν and q^2 that have been studied. Indeed the earliest inelastic e-p experiments (Reference 30) showed that approximate scaling behavior occurs already at surprisingly non-asymptotic values of $q^2 \geq 1.0$ GeV^2 and $W \geq 2.6$ GeV.

The question quickly arose as to whether there were other scaling variables that converged to ω in the Bjorken limit, and that provided scaling behavior over a larger region in ν and q^2 than did the use of ω. Several were proposed (Reference 46) before the advent of QCD, but because this theory predicts small departures from scaling, the search for such variables was abandoned soon after.

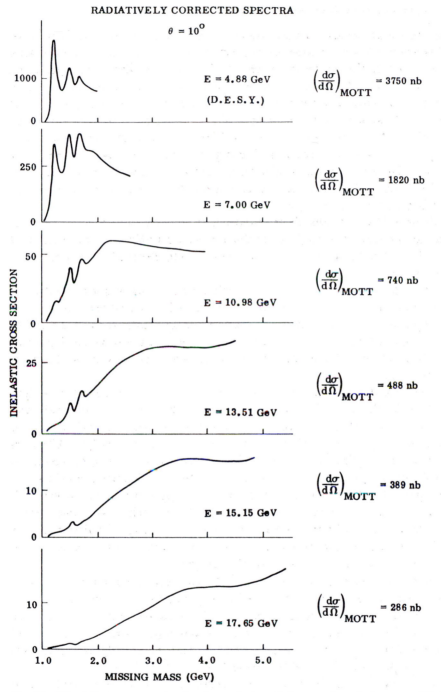

RADIATIVELY CORRECTED SPECTRA

$\theta = 10^{\circ}$

E = 4.88 GeV
(D.E.S.Y.)

$\left(\dfrac{d\sigma}{d\Omega}\right)_{MOTT} = 3750$ nb

E = 7.00 GeV

$\left(\dfrac{d\sigma}{d\Omega}\right)_{MOTT} = 1820$ nb

E = 10.98 GeV

$\left(\dfrac{d\sigma}{d\Omega}\right)_{MOTT} = 740$ nb

E = 13.51 GeV

$\left(\dfrac{d\sigma}{d\Omega}\right)_{MOTT} = 488$ nb

E = 15.15 GeV

$\left(\dfrac{d\sigma}{d\Omega}\right)_{MOTT} = 389$ nb

E = 17.65 GeV

$\left(\dfrac{d\sigma}{d\Omega}\right)_{MOTT} = 286$ nb

INELASTIC CROSS SECTION

MISSING MASS (GeV)

Fig. 10 a. Visual fits to spectra showing the scattering of electrons from hydrogen at 10° for primary energies, *E*, from 4.88 GeV to 17.5 GeV. The elastic peaks have been subtracted and radiative corrections applied. The cross sections are expressed in nanobarns per GeV per steradian. The spectrum for *E* = 4.88 GeV was taken at DESY; W. Bartel, et al., Phys. Lett., **B28** 148 (1968).

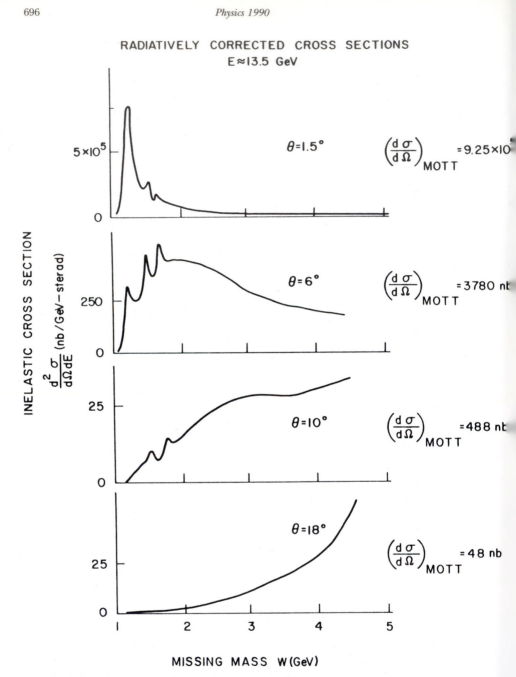

Fig. 10b. Visual fits to spectra showing the scattering of electrons from hydrogen at a primary energy E of approximately 13.5 GeV, for scattering angles from 1.5° to 18°. The 1.5° curve is taken from MIT-SLAC data used to obtain photoabsorption cross sections.

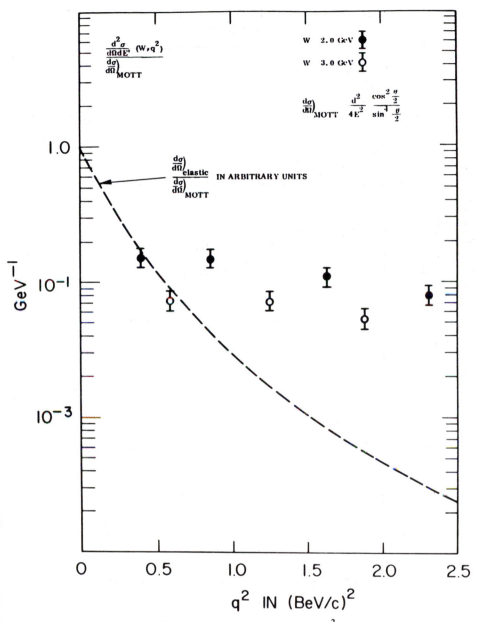

Fig. 11. Inelastic data for $W = 2$ and 3 GeV as a function of q^2. This was one of the earliest examples of the relatively large cross sections and weak q^2 dependence that were later found to characterize the deep inelastic scattering and which suggested point-like nucleon constituents. The q^2 dependence of elastic scattering is shown also; these cross sections have been divided by σ_M

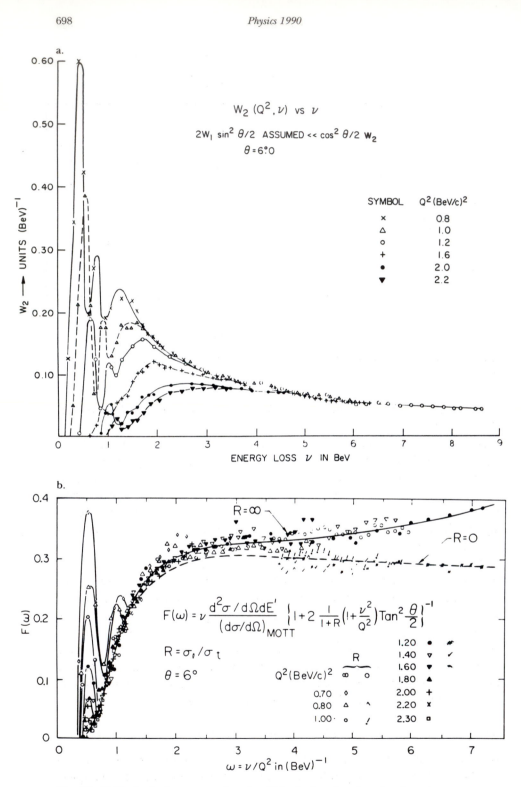

Fig. 12. (a) The inelastic structure function $W_2(\nu, q^2)$ plotted against the electron energy loss ν.
(b) The quantity $F_1 = \nu\, W_2(\omega)$. The "nesting" of the data observed here was the first evidence of
scaling. The figure is discussed further in the text.

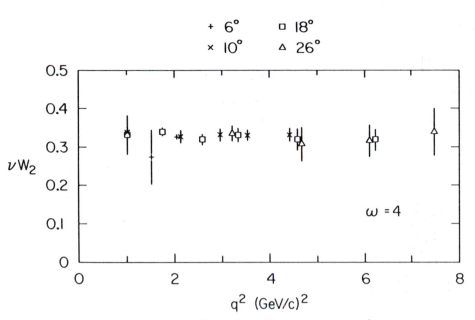

Fig. 13. An early observation of scaling: νW_2 for the proton as a function of q^2 for W > 2 GeV, at $\omega = 4$.

Figure 13 shows early data on νW_2, for $\omega = 4$, as a function of q^2. Within the errors there was no q^2 dependence.

A more complex separation procedure was required to determine R and the structure functions, as discussed above. The kinematic region in $q^2 - W^2$ space available for the separation is shown in Figure 14. This figure also shows the 75 kinematic points where, after the majority of the experiments were complete, separations had been made. Figure 15 displays sample least-square fits to Σ (ν,q^2,θ) vs ε (ν,q^2,θ), as defined earlier, in comparison with data, from which σ_L and σ_T and then R, were found.

A rough evaluation of scaling is provided by, for example, inspecting a plot of the data taken by the collaboration on νW_2 against x as shown in Figure 16. These data, to a fair approximation, describe a single function of x. Some deviations, referred to as scale breaking, are observed. They are more easily inspected by displaying the q^2 dependence of the structure functions. Figure 17 shows separated values of $2MW_1$ and νW_2 from data taken late in the program, plotted against q^2 for a series of constant values of x. With extended kinematic coverage and with smaller experimental errors, sizeable scale breaking was observed in the data.

VI *Theoretical Implications of the Electron-Proton Inelastic Scattering Data.*

As noted earlier, the discovery, during the first inelastic proton measurements, of the weak q^2 dependence of the structure function νW_2, coupled with the scaling concept inferred from current algebra and its roots in the quark theory, at once suggested new possibilities concerning nucleon structure. At the 1968 Vienna Meeting, where the results were made public for

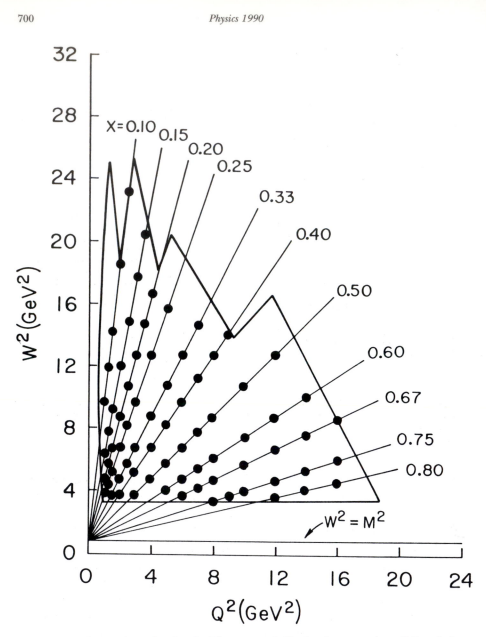

Fig. 14. The kinematic region in $q^2 - W^2$ space available for the extraction of R and the structure functions. Separations were made at the 75 kinematic points (v, q^2) shown.

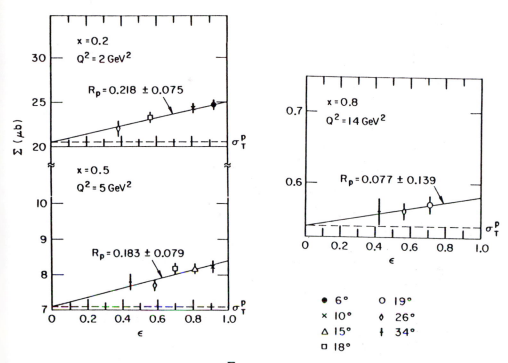

Fig. 15. Sample least-square fits to Σ vs $\mathcal{E}$ in comparison with data from the proton. The quantities R and σ_T were available from the fitting parameters, and from them σ_L was determined.

the first time, the rapporteur, W. K. H. Panofsky, summed up the conclusions (Reference 47): "Therefore theoretical speculations are focussed on the possibility that these data might give evidence on the behavior of point-like, charged structures within the nucleon."

Theoretical interest at SLAC in the implications of the inelastic scattering increased substantially after an August 1968 visit by R. P. Feynman. He had been trying to understand hadron-hadron interactions at high energy assuming constituents he referred to as *partons*. On becoming aware of the inelastic electron scattering data, he immediately saw in partons an explanation both of scaling and the weak q^2 dependence. In his initial formulation (Reference 48), now called the naive parton theory, he assumed that the proton was composed of point-like partons, from which the electrons scattered incoherently. The model assumed an infinite momentum frame of reference, in which the relativistic time dilation slowed down the motion of the constituents. The transverse momentum was neglected, a simplification relaxed in later elaborations. The partons were assumed not to interact with one another while the virtual photon was exchanged: the impulse approximation of quantum mechanics. Thus, in this theory, electrons scattered from constituents that were "free," and therefore the scattering reflected the properties and motions of the constituents. This assumption of a near-vanishing of the parton-parton interaction during lepton scattering, in the

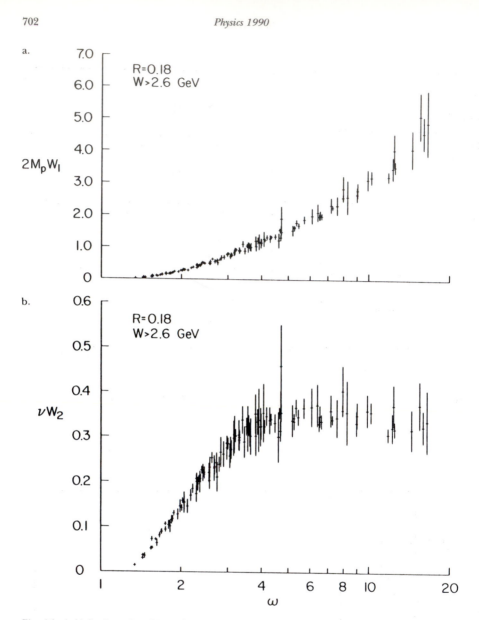

Fig. 16. (a,b) Scaling: $F_1 = 2MW_1 (\omega)$ vs ω, and $F_2 = \nu W_2 (\omega)$ vs ω, for the proton.

Bjorken limit, was subsequently shown to be a consequence of QCD known as *asymptotic freedom*. Feynman came to Stanford again, in October 1968, and gave the first public talk on his parton theory, stimulating much of the theoretical work which ultimately led to the identification of his partons with quarks.

In November 1968, Curt Callan and David Gross (Reference 49) showed that R, given in Equation (6), depended on the spins of the constituents in a parton model and that its kinematic variation constituted an important test

a.

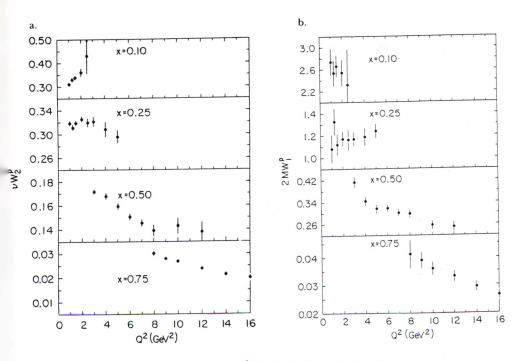

Fig. 17. (a,b) F_1 and F_2 as functions of q^2, for fixed values of x, for the proton.

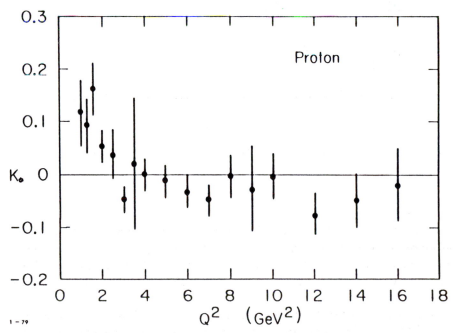

Fig. 18. The Callan-Gross relation: K_0 vs q^2, where K_0 is defined in the text. These results established the spin of the partons as 1/2.

of such models. For spin $1/2$, R was expected to be small, and, for the naive parton model, where the constituents are assumed unbound in the Bjorken limit, $R = q^2/v^2$ (ie, $F_2 = xF_1$). More generally, for spin $1/2$ partons, $R = g(x)(q^2/v^2)$. This is equivalent to the scaling of vR.

Spin zero or one partons led to the prediction $R \neq 0$ in the Bjorken limit, and would indicate that the proton cloud contains elementary bosons. Small values of R were found in the experiment and these were totally incompatible with the predictions of Vector Meson Dominance. Later theoretical studies (Reference 50) showed that deviations from the general Callan-Gross rule would be expected at low x and low q^2. A direct evaluation of the Callan-Gross relation for the naive parton model may be found from

$$K_0 = F_2/(xF_1) - 1$$

which vanishes when the relation is satisfied. K_0 is shown in Figure 18, as a function of q^2. Aside from the expected deviations at low q^2, K_0 is consistent with zero, establishing the parton spin as $1/2$.

VII   *Epilogue*

After the initial inelastic measurements were completed, deuteron studies were initiated to make neutron structure functions accessible. Experiments were made over a greater angular range and statistical, radiative, and systematic errors were reduced. The structure functions for the neutron were found to differ from the proton's. Vector Meson Dominance was abandoned and by 1972 all diffractive models, and nuclear democracy, were found to be inconsistent with the experimental results. Increasingly detailed parton calculations and sum rule comparisons, now focussing on quark constituents, required sea quarks — virtual quark-antiquark pairs — in the nucleon, and, later, gluons — neutral bosons that provided the inter-quark binding.

On the theoretical front, a special class of theories was found that could incorporate asymptotic freedom and yet was compatible with the binding necessary to have stable nucleons. Neutrino measurements confirmed the spin $1/2$ assignment for partons and that they had fractional, rather than integral electric charge. The number of "valence" quarks was found to be 3, consistent with the original 1964 assumptions.

By 1973, the picture of the nucleon had clarified to such an extent that it became possible to construct a comprehensive theory of quarks and gluons and their strong interactions: QCD. This theory was built on the concept of "color," whose introduction years before (Reference 51) made the nucleons' multi-quark wave functions compatible with the Pauli principle, and, on the assumption that only "color-neutral" states exist in nature, explained the absence of all unobserved multi-quark configurations (such as quark-quark and quark-quark-antiquark) in the known array of hadrons. Furthermore, as noted earlier, QCD was shown to be asymptotically free (Reference 52).

By that year the quark-parton model, as it was usually called, satisfactorily explained electron-nucleon and neutrino-nucleon interactions, and provided a rough explanation for the very high energy "hard" nucleon-nucleon scattering that had only recently been observed. The experimenters were seeing quark-quark collisions.

By the end of the decade, the fate of quarks recoiling within the nucleon in high energy collisions had been understood; for example, after quark pair production in electron-positron colliders, they materialized as back-to-back jets composed of ordinary hadrons (mainly pions), with the angular distributions characteristic of spin 1/2 objects. Gluon-jet enhancement of quark jets was predicted and then observed, having the appropriate angular distributions for the spin 1 they were assigned within QCD. Theorists had also begun to deal, with some success, with the problem of how quarks remained confined in stable hadrons.

Quantum Chromodynamics describes the strong interactions of the hadrons and so can account, in principle at least, for their ground state properties as well as hadron-hadron scattering. The hadronic weak and electromagnetic interactions are well described by electroweak theory, itself developed in the late 1960s. The picture of the nucleon, and the other hadrons, as diffuse, structureless objects was gone for good, replaced by a successful, nearly complete theory.

ACKNOWLEDGEMENTS

There were many individuals who made essential contributions to this work. An extensive set of acknowledgements is given in Reference 53.

REFERENCES

1. R.E.Taylor, Deep Inelastic Scattering: The Early Years. *Les Prix Nobel 1990*. Hereafter "Taylor."
2. J. I. Friedman, Deep Inelastic Scattering: Comparisons with the Quark Model. *Les Prix Nobel 1990*. Hereafter "Friedman."
3. S. C. Frautschi, *Regge Poles and S-Matrix Theory*, W. A. Benjamin (1963). Hereafter "Frautschi."
4. G. F. Chew and S. C. Frautschi, Phys. Rev. Lett. **8**, 394 (1961).
5. P. D. B. Collins and E. J. Squires, *Regge Poles in Particle Physics*, Springer-Verlag, Berlin (1968). For a broad review of the strong interaction physics of the period see Martin L. Perl, *High Energy Hadron Physics*, John Wiley & Sons, (New York 1974). See also Frautschi.
6. G. F. Chew, S. C. Frautschi, and S. Mandelstam, Phys. Rev. **126**, 1202 (1962).
7. G. Veneziano, Nuovo Cim. **57A**, 190 (1968). See also J. H. Schwarz, Phys. Rep. **8**, 269 (1973).
8. J.J.Sakurai, Phys. Rev. Lett. **22**, 981 (1969).
9. M. Gell-Mann, Phys. Lett. **8**, 214 (1964).
10. G. Zweig, CERN-8182/Th.401 (Jan. 1964) and CERN-8419/Th.412 (Feb. 1964), both unpublished. Hereafter "Zweig."
11. M. Gell-Mann, C. I. T. Synchrotron Lab. Rep't, CTSL-20 (unpublished) and Y. Ne'eman, Nuc. Phys. **26**, 222 (1961). See also M. Gell-Mann and Y. Ne'eman, *The Eightfold Way*, W. A. Benjamin (1964).

12. The quark model explained why triplet, sextet, and 27-plets of then-current SU(3) were absent of hadrons. With rough quark mass assignments, it could account for observed mass splittings within multiplets, and it provided an understanding of the anomalously long lifetime of the phi meson (discussed later in this paper).

13. Lawrence W. Jones, Rev. Mod. Phys. **49**, 717 (1977).

14. "..we know that...[mesons and baryons] are mostly, if not entirely, made up out of one another. ... The probability that a meson consists of a real quark pair rather than two mesons or a baryon and antibaryon must be quite small." M. Gell-Mann, *Proc.XIIIth Inter. Conf. on High En. Phys.*, Berkeley, California 1967.

15. "Additional data is necessary and very welcome in order to destroy the picture of elementary constituents." J. D. Bjorken. "I think Prof. Bjorken and I constructed the sum rules in the hope of destroying the quark model." Kurt Gottfried. Both quotations from *Proc. 1967 Internat. Symp. on Electron and Photon Interac. at High Energy*, Stanford, California, Sept. 5–9, 1967.

16. R. Dalitz, Session 10, Rapporteur. *Proc. XIIIth Inter. Conf. High En. Phys.*, Berkeley, 1966. (University of California, Berkeley).

17. C. Becchi and G. Morpurgo, Phys. Lett. **17**, 352 (1965).

18. K. Gottfried, Phys. Rev. Lett. **18**, 1174 (1967).

19. Zweig believed from the outset that the nucleon was composed of "physical" quark constituents. This was based primarily on his study of the properties of the phi meson. It did not decay rapidly to $\rho-\pi$ as expected but rather decayed roughly two orders of magnitude slower to kaon-antikaon, whose combined mass was near the threshold for the decay. He saw this as a dynamical effect; one not explainable by selection rules based on the symmetry groups and explainable only by a constituent picture in which the initial quarks would "flow smoothly" into the final state. He was "severely criticized" for his views, in the period before the MIT-SLAC results were obtained. Private communication, February 1991.

20. According to a popular book on the quark search, Zweig, a junior theorist visiting at CERN when he proposed his quark theory, could not get a paper published describing his ideas until the middle 1970s, well after the constituent model was on relatively strong ground; M. Riordan, *The Hunting of the Quark*, Simon and Schuster, New York (1987). His preprints (cf. Zweig) did, however, reach many in the physics community and helped stimulate the early quark searches.

21. A. Pais, *Inward Bound*, Oxford University Press, New York City (1986).

22. "Throughout the 1960s, into the 1970s, papers, reviews, and books were replete with caveats about the existence of quarks." Andrew Pickering, *Constructing Quarks*, University of Chicago Press (Chicago 1984).

23. Further discussion of sum rules and their comparisons with data is to be found in Friedman.

24. "Such particles [quarks] presumably are not real but we may use them in our field theory anyway." Physics **1**, 63 (1964).

25. "Now what is going on? What are these quarks? It is possible that real quarks exist, but if so they have a high threshold for copious production, many BeV; ..." *Proc. XIIIth Inter. Conf. on High En. Phys.*, Berkeley, California, 1967.

26. "We shall find these results [sum rules requiring cross sections of order Rutherford scattering from a point particle] so perspicuous that, by an appeal to history, an interpretation in terms of 'elementary constituents' of the nucleon is suggested." He pointed out that high energy lepton-nucleon scattering could resolve the question of their existence and noted that "it will be of interest to look at very large inelasticity and dispose, without ambiguity, of the model completely." "Current Algebra at Small Distances," lecture given at International School of Physics, "Enrico Fermi," XLI Course, Varenna, Italy, July

1967. SLAC Pub. 338, August 1967 (unpublished) and *Proceedings of the International School of Physics "Enrico Fermi", Course XLI: Selected Topics in Particle Physics*, J. Steinberger, ed. (Academic Press, New York, 1968).

27. T. D. Lee: "I'm certainly not the person to defend the quark models, but it seems to me that at the moment the assumption of an unsubtracted dispersion relation [the subject of the discussion] is as *ad hoc* as the quark model. Therefore, instead of subtracting the quark model one may also subtract the unsubtracted dispersion relation." J. Bjorken: "I certainly agree. I would like to dissociate myself a little bit from this as a test of the quark model. I brought it in mainly as a desperate attempt to interpret the rather striking phenomena of a point-like behavior. One has this very strong inequality on the integral over the scattering cross section. It's only in trying to interpret how that inequality could be satisfied that the quarks were brought in. There may be many other ways of interpreting it." Discussion in *Proc. 1967 Internat. Symp. on Electron and Photon Interac. at High Energy,* Stanford, Sept 5 − 9, 1967.

28. *Proposal for Spectrometer Facilities at SLAC,* submitted by SLAC Groups A and C, and physicists from MIT and CIT. (Stanford, California, undated, unpublished). *Proposal for Initial Electron Scattering Experiments Using the SLAC Spectrometer Facilities:* Proposal 4b "The Electron-Proton Inelastic Scattering Experiment," submitted by the SLAC-MIT-CIT Collaboration, 1 January 1966 (unpublished).

29. *The Stanford Two-Mile Accelerator* R. B. Neal, General Editor, W. A. Benjamin (New York 1968).

30. *14th Int. Conf. High Energy Phys.*, Vienna, Aug., 1968. E. D. Bloom et al., Phys. Rev. Lett. **23**, 930 (1969); M. Breidenbach et al., ibid **23**, 935 (1969).

31. W. R. Ditzler, et al., Phys. Lett. **57B**, 201 (1975).

32. L. W. Whitlow, et al., Phys. Lett. **B250**, 193 (1990) and L. W. Whitlow, SLAC Report 357, March 1990 (unpublished).).

33. M. Rosenbluth, Phys. Rev. **79**, 615 (1950).

34. P. N. Kirk, et al., Phys. Rev. **D8**, 63 (1973).

35. S. D. Drell and J. D. Walecka, Ann. Phys. **28**, 18 (1964).

36. L. Hand, Phys. Rev. **129**, 1834 (1963).

37. J. D. Bjorken, Phys. Rev. **179**, 1547 (1969). Although the conjecture was published after the experimental results established the existence of scaling, the proposal that this might be true was made prior to the measurements as discussed later in the text.

38. J. D. Bjorken, Ann. Phys. **24**, 201 (1963).

39. J. I. Friedman, Phys. Rev. **116**, 1257 (1959).

40. D. Isabelle and H. W. Kendall, Bull. Am. Phys. Soc. **9**, 95 (1964). The final report on the experiment is S. Klawansky et al., Phys. Rev. **C7**, 795 (1973).

41. G. Miller et al., Phys. Rev. **D5**, 528 (1972); L. W. Mo and Y. S. Tsai, Rev. Mod. Phys. **41**, 205 (1969).

42. Y. S. Tsai, *Proc. Nucleon Struct. Conf: Stanford 1963,* p. 221. Stanford Univ. Press (Stanford 1964), edited by R. Hofstadter and L. I. Schiff.

43. G. Miller et al., Phys. Rev. **D5**, 528 (1972).

44. Poucher, J. S. 1971. PhD thesis, MIT, (unpublished).

45. A. Bodek et al., Phys. Rev. **D20**, 1471 (1979).

46. J. I. Friedman and H. W. Kendall, Ann. Rev. Nuc. Sci., **22**, 203 (1972). A portion of this publication is used in the present paper.

47. W.K.H.Panofsky, in 14th *Int. Conf. High Energy Phys.*, Vienna, Aug., 1968, edited by J. Prentki and J. Steinberger (CERN, Geneva), p 23.

48. R. P. Feynman, Phys. Rev. Lett. **23**, 1415 (1969).

49. C. Callan and D. J. Gross, Phys. Rev. Lett. **21**, 311 (1968).

50. R. P. Feynman, *Photon-Hadron Interactions,* W. A. Benjamin (1972).

51. M. Y. Han and Y. Nambu, Phys. Rev. **139B**, 1006 (1965).

52. D. Gross and F. Wilczek, Phys. Rev. Lett. **30**, 1343 (1973), D. Politzer, ibid., 1346.

53. J. I. Friedman, H. W. Kendall, and R. E. Taylor, Deep Inelastic Scattering: Acknowledgements, *Les Prix Nobel 1990*.

Jerome I. Friedman

JEROME I. FRIEDMAN

I was born in Chicago, Illinois on March 28, 1930, the second of two children of Selig and Lillian Friedman, nee Warsaw, who were immigrants from Russia. My father came to the United States in 1913 and later served in the U.S. Army Artillery Corps in World War I. After the war he was employed by the Singer Sewing Machine Co. and later established his own business, repairing and selling used commercial and home sewing machines. My mother arrived in the United States in 1914 on one of the last voyages of the Lusitania. She supported herself until she was married by working in a garment factory. My parents had little formal education, except for courses in English after they arrived in the United States, but were self taught and had wide ranging interests. My father was an avid reader, having interests in science and political history, and our home was filled with books. My mother, who had a lovely singing voice, loved music and, in particular, opera. The education of my brother and myself was of paramount importance to my parents, and in addition to their strong encouragement, they were prepared to make any sacrifice to further our intellectual development. When there were financial difficulties they still managed to provide us with music and art lessons. They greatly respected scholarship in itself, but they also impressed upon us that there were great opportunities available for those who were well educated. I received my primary and secondary education in Chicago. As I very much liked to draw and paint as a child, I entered a special art program in high school, which was very much like being in an art school imbedded in a regular high school curriculum. While I always had some interest in science, I developed a strong interest in physics when I was in high school as a result of reading a short book entitled *Relativity*, by Einstein. It opened a new vista for me and deepened my curiosity about the physical world. Instead of accepting a scholarship to the Art Institute of Chicago Museum School and against the strong advice of my art teacher, I decided to continue my formal education and sought admission to the University of Chicago because of its excellent reputation and because Enrico Fermi taught there. I was fortunate to have been accepted with a full scholarship. As my parents had limited means, my university training would not have been possible without such help. After finishing my requirements in an highly innovative and intellectually stimulating liberal arts program (established by Robert M. Hutchins who was then President of the University), I entered the Physics Department in 1950, receiving a Master's degree in 1953 and a Ph.D. in 1956. It is difficult to convey the sense of excitement that pervaded the Department at that time. Fermi's brilliance, his stimulating, crystal clear lectures that he gave in numerous seminars and courses, the outstanding faculty in the Department, the many notable physicists who frequently came to visit Fermi, and the pioneering investigations of pion proton scattering at the newly constructed cyclotron all combined to create an especially lively atmosphere. I was indeed fortunate to have seen the practice of physics carried out at its "very best" at such an early stage in my development. I also had the great privilege of being supervised by Fermi, and I can remember being overwhelmed with a sense of my good fortune to have been given the opportunity to work for this great man. It was a remarkably stimulating experience that shaped the way I think about physics. My thesis project was an investigation in nuclear emulsion of proton polarization produced in scattering from nuclei at cyclotron energies. The objective was to determine whether the polarization resulted from elastic or inelastic scattering. Professor Fermi tragically died in 1954 after a short illness. What an immense loss it was to all of us. My thesis work was not yet completed, and John Marshall kindly took over my supervision and signed my thesis. After I received my Ph.D., I continued working as a post-doc at the University of Chicago nuclear emulsion laboratory, which was then led by Valentine Telegdi. That year Val Telegdi and I did an emulsion experiment in which we searched for parity violation in muon decay. We were one of the first groups to observe this surprising effect which had been suggested by T.D. Lee and C.N. Yang. Val was not only an excellent mentor but he was instrumental in getting me my first real job with Robert Hofstadter.

In 1957, I joined Hofstadter's group at the High Energy Physics Laboratory at Stanford University as a Research Associate. This was where I learned counter physics and the techniques of electron scattering. While there I did a number of experiments studying elastic and inelastic electron-deuteron scattering. In an experiment to measure a weighted sum-rule for inelastic electron deuteron scattering which was related to the n-p interaction I had to confront the problem of making radiative corrections to inelastic spectra, and I developed a technique which proved to be valuable in my later work. Henry Kendall independently developed a similar technique and later we combined efforts to develop a radiative corrections program for our deep inelastic scattering work at SLAC. It was in Hofstadter's group that I began my long collaboration with Henry Kendall who was also a member of the group. During this period I became acquainted with Richard Taylor, who was just finishing his thesis in another group, and with other future collaborators in the deep inelastic program at SLAC, Dave Coward and Hobey DeStaebler. One of the highlights of this period was attending the wonderfully informal and informative high energy physics seminars in the home of W.K.H. Panofsky, who was Director of the Laboratory.

In 1960, I was hired as a faculty member in the Physics Department of the Massachusetts Institute of Technology. When I arrived I joined David Ritson's research group. A short time later he accepted a position at Stanford University and I inherited a small group. With these resources I soon began working on collaborative effort to measure muon pair production at the Cambridge Electron Accelerator (CEA) in order to test the validity of Quantum Electro-Dynamics. Henry Kendall joined my group in 1961 and we have been collaborators at MIT since that time. The last measurement we did at the CEA was a measurement of the deuteron form factor at the highest momentum transfers that could be reached at that accelerator to get some limits on the size of relativistic effects and meson currents.

In 1963, Henry Kendall and I started a collaboration with W.K.H. Panofsky, Richard Taylor and other physicists from the Stanford Linear Accelerator Center and the California Institute of Technology to develop electron scattering facilities for a physics program at the Stanford Linear Accelerator, a 20 GeV electron linac that was being constructed under the leadership of Panofsky. This required that we both travel between MIT and SLAC on a regular basis. The MIT Physics Department gave us special support by reducing our teaching responsibilities. We soon set up a small MIT group at SLAC and for extended periods of time one of us was always there. We had a rare opportunity. We were part of a group of physicists who were provided a new accelerator, given the support to design and construct optimal experimental facilities, and had the opportunity to participate in the exploration of a new energy range with electrons. From 1967 to about 1975 the MIT and SLAC groups carried out a series of measurements of inelastic electron scattering from the proton and neutron which provided the first direct evidence of the quark sub-structure of the nucleon. It was a very exciting time for all of us. This program is described in detail in the adjoining Physics Nobel Lectures.

As the program at SLAC was nearing completion we joined a collaborative effort at Fermilab involving a number of institutions to build a beam line and a single-arm spectrometer in the Meson Laboratory. During the latter half of the 1970's this collaboration carried out a series of experiments to investigate elastic scattering, Feynman scaling and production mechanisms in inclusive hadron scattering. When this work was completed, our group joined another collaboration to build a large neutrino detector at Fermilab. The objective of this program was to study the weak neutral currents in measurements of inclusive neutrino and anti-neutrino nucleon scattering, which were done in the first half of the 1980's. These investigations confirmed the predictions of the Standard Model.

In 1980, I became Director of the Laboratory for Nuclear Science at MIT and then served as Head of the Physics Department from 1983 to 1988. During the time I was in these administrative positions I managed to maintain a foothold in research, which greatly eased my transition back to full-time teaching and research in 1988. While it was a very interesting period in my life, I was happy to get back to more direct contact with students in the classroom and in my research projects. Currently, our MIT group is participating in the construction of a large detector to study electron-positron annihilations at the Stanford Linear Collider and has also been engaged in design work for a detector for the Superconducting Super Collider, which is now under construction.

Over the years I have served on a number of program and scientific policy advisory committees at various accelerators. I also was a member of the Board of the University Research Association for six years, serving as Vice-President for three years. I am currently a member of the High Energy Advisory Panel for the Department of Energy and also Chairman of the Scientific Policy Committee of the Superconducting Super Collider Laboratory.

Experimental high energy physics research is a group effort. I have been very fortunate to have had outstanding students and colleagues who have made invaluable contributions to the research with which I have been associated. I thank them not only for their contributions, but also for their friendship.

My life has been enhanced by my marriage to Tania Letetsky-Baranovsky who has broadened my horizons and has been an unfaltering source of support. She has endured with cheerful resignation my many absences when I have had to travel to distant particle accelerators. There are four grown children in our family, Ellena, Joel, Martin, and Sandra who pursue their activities in various parts of the country.

With regard to my non-vocational activities, in addition to getting much pleasure from various cultural activities, such as theater, music, ballet, etc., I enjoy painting and study Asian ceramics.

DEEP INELASTIC SCATTERING: COMPARISONS WITH THE QUARK MODEL

Nobel Lecture, December 8, 1990

by

JEROME I. FRIEDMAN

Massachusetts Institute of Technology, Cambridge, Massachusetts, USA.

EARLY RESULTS

In the latter half of 1967 a group of physicists from the Stanford Linear Accelerator Center (SLAC) and the Massachusetts Institute of Technology (MIT) embarked on a program of inelastic electron proton scattering after completing an initial study[1] of elastic scattering with physicists from the California Institute of Technology. This work was done on the newly completed 20 GeV Stanford linear accelerator. The main purpose of the inelastic program was to study the electro-production of resonances as a function of momentum transfer. It was thought that higher mass resonances might become more prominent when excited with virtual photons, and it was our intent to search for these at the very highest masses that could be reached. For completeness we also wanted to look at the inelastic continuum since this was a new energy region which had not been previously explored. The proton resonances that we were able to measure[2] showed no unexpected kinematic behavior. Their transition form factors fell about as rapidly as the elastic proton form factor with increasing values of the four momentum transfer, q. However, we found two surprising features when we investigated the continuum region (now commonly called the deep inelastic region).

(1)   Weak q^2 Dependence

The first unexpected feature of these early results[3] was that the deep inelastic cross-sections showed a weak fall off with increasing q^2. The scattering yields at the larger values of q^2 were between one and two orders of magnitude greater than expected.

The weak momentum transfer dependence of the inelastic cross-sections for excitations well beyond the resonance region is illustrated in Fig. 1. The differential cross section divided by the Mott cross section,[4] σ_{Mott}, is plotted as a function of the square of the four-momentum transfer, $q^2 = 2EE'(1 - \cos\theta)$, for constant values of the invariant mass of the recoiling target system, W, where $W^2 = 2M(E-E') + M^2 - q^2$. The quantity E is the energy of the incident electron, E' is the energy of the final electron, and θ is the scattering angle, all defined in the laboratory system; M is the mass of the proton. The cross section is divided by the Mott cross section in order to remove the major part of the well-known four-momentum transfer depen-

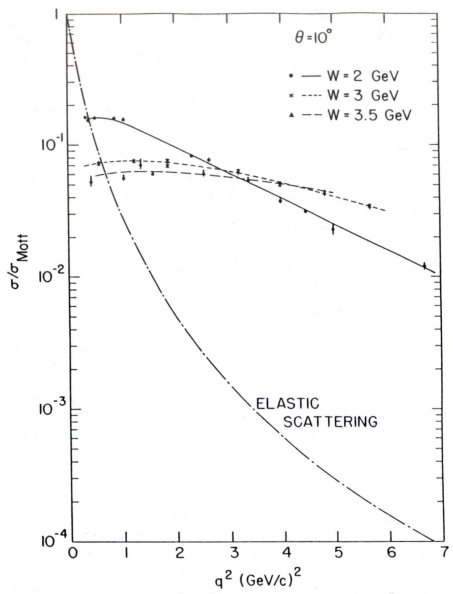

Fig. 1: $(d^2\sigma/d\Omega dE')/\sigma_{\text{Mott}}$, in GeV^{-1}, vs. q^2 for $W = 2$, 3 and 3.5 GeV. The lines drawn through the data are meant to guide the eye. Also shown is the cross section for elastic $e-p$ scattering divided by σ_{Mott}, $(d\sigma/d\Omega)/\sigma_{\text{Mott}}$, calculated for $\theta = 10°$, using the dipole form factor. The relatively slow variation with q^2 of the inelastic cross section compared with the elastic cross section is clearly shown.

dence arising from the photon propagator. The q^2 dependence that remains is related primarily to the properties of the target system. Results from 10° are shown in the figure for each value of W. As W increases, the q^2 dependence appears to decrease. The striking difference between the behavior of the deep inelastic and elastic cross sections is also illustrated in this figure, where the elastic cross section, divided by the Mott cross section for $\theta = 10°$, is shown.

When the experiment was planned, there was no clear theoretical picture of what to expect. The observations of Hofstadter[5] in his pioneering studies of elastic electron scattering from the proton showed that the proton had a size of about 10^{-13} cm and a smooth charge distribution. This result, plus the theoretical framework that was most widely accepted at the time, suggested to our group when the experiment was planned that the deep inelastic electron proton cross-sections would fall rapidly with increasing q^2.

(2) Scaling

The second surprising feature in the data, scaling, was found by following a suggestion by Bjorken.[6] To describe the concept of scaling, one has to introduce the general expression for the differential cross section for unpolarized electrons scattering from unpolarized nucleons with only the scattered electrons detected.[7]

$$\frac{d^2\sigma}{d\Omega dE'} = \sigma_{\text{Mott}}\left[W_2 + 2W_1\tan^2\frac{\theta}{2}\right].$$

The functions W_1 and W_2 are called structure functions and depend on the properties of the target system. As there are two polarization states of the virtual photon, transverse and longitudinal, two such functions are required to describe this process. In general, W_1 and W_2 are each expected to be functions of both q^2 and v, where v is the energy loss of the scattered electron. However, on the basis of models that satisfy current algebra, Bjorken conjectured that in the limit of q^2 and v approaching ∞, the two quantities $v W_2$ and W_1 become functions only of the ratio $\omega = 2Mv/q^2$; that is

$$2MW_1\ (v, q^2) \longrightarrow F_1(\omega)$$
$$vW_2\ (v, q^2) \longrightarrow F_2(\omega).$$

The scaling behavior of the structure functions is shown in Fig. 2, where experimental values of vW_2 and $2MW_1$ are plotted as a function of ω for values of q^2 ranging from 2 to 20 GeV^2. The data demonstrated scaling within experimental errors for $q^2 > 2$ GeV^2 and $W > 2.6$ GeV.

The dynamical origin of scaling was not clear at that time, and a number of models were proposed to account for this behavior and the weak q^2 dependence of the inelastic cross section. While most of these models were firmly imbedded in S-matrix and Regge pole formalism, the experimental results caused some speculation regarding the existence of a possible point-like structure in the proton. In his plenary talk at the XIV International Conference on High Energy Physics held in Vienna in 1968, where preliminary results on the weak q^2 dependence and scaling were first presented, Panofsky[2] reported "... theoretical speculations are focused on the possibility that these data might give evidence on the behavior of point-like charged structures in the nucleon." However, this was not the prevailing point of view. Even if one had proposed a constituent model at that time it was not

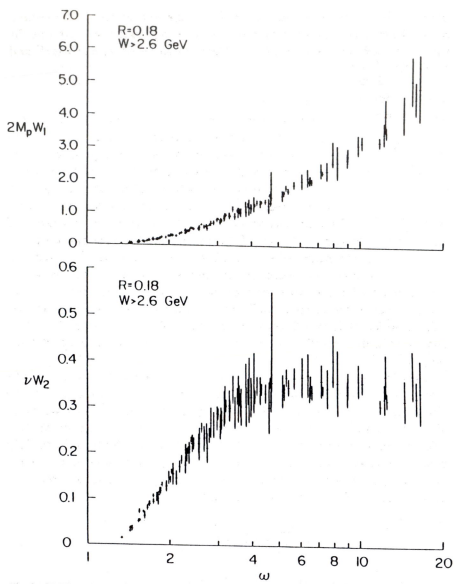

Fig. 2: $2MW_1$ and νW_2 for the proton as functions of ω for $W > 2.6$ GeV, $q^2 > 1(\text{GeV}/c^2)$, and using $R = 0.18$. Data from Ref. [34]. The quantity R is discussed in the section of this paper entitled Non-Constituent Models.

clear that there were reasonable candidates for the constituents. Quarks, which had been proposed independently by Gell-Mann[8] and Zweig[9] as the building blocks of unitary symmetry[10] in 1964, had been sought in numerous accelerator and cosmic ray investigations and in the terrestrial environment without success. Though the quark model provided the best available tool for understanding the properties of the many recently discovered hadronic resonances, it was thought by many to be merely a mathematical representation of some deeper dynamics, but one of heuristic value. Considerably more experimental and theoretical results had to be accumulated before a clear picture emerged. More detailed descriptions of the development of the deep inelastic program and its early results are given in the written versions of the 1990 Physics Nobel Lectures of R. E. Taylor[11] and H. W. Kendall.[12]

NON-CONSTITUENT MODELS

The initial deep inelastic measurements stimulated a flurry of theoretical work, and a number of non-constituent models based on a variety of theoretical approaches were put forward to explain the surprising features of the data. One approach related the inelastic scattering to forward virtual Compton scattering, which was described in terms of Regge exchange[13–17] using the Pomeranchuk trajectory, or a combination of it and non-diffractive trajectories. Such models do not require a weak q^2 dependence, and scaling had to be explicitly inserted. Resonance models were also proposed to explain the data. Among these was a Veneziano-type model[18] in which the density of resonances increases at a sufficiently rapid rate to compensate for the decrease of the contribution of each resonance with increasing q^2. Another type of resonance model[19] built up the structure functions from an infinite series of N and Δ resonances. None of these models was totally consistent with the full range of data accumulated in the deep inelastic program.

One of the first attempts[20] to explain the deep inelastic scattering results employed the Vector Dominance Model, which had been used to describe photon-hadron interactions over a wide range of energies. This model, in which the photon is assumed to couple to a vector meson which then interacts with a hadron, was extended, using ρ meson dominance, to deep inelastic electron scattering. It reproduced the gross features of the data in that νW_2 approached a function of ω for ν much greater than M_ρ, the mass of the ρ meson. The model also predicted that

$$R = \frac{\sigma_S}{\sigma_T} = \left(\frac{\varepsilon q^2}{M_\rho^2}\right)\left(1 - \frac{q^2}{2M\nu}\right) ,$$

where R is the ratio of σ_S and σ_T, the photo-absorption cross-sections of longitudinal and transverse virtual photons, respectively, and ε is the ratio of the vector meson-nucleon total cross sections for vector mesons with polarization vectors respectively parallel and perpendicular to their direction of motion. Since the parameter ε is expected to have a value of about 1 at high energies, this theory predicted very large values of R for values of $q^2 \gg M_\rho^2$. The ratio R can be related to the structure functions in the following way

$$R = \frac{W_2}{W_1}\left(1 + \frac{\nu^2}{q^2}\right) - 1.$$

The measurements of deep inelastic scattering over a range of angles and energies allowed W_1 and W_2 to be separated and R to be determined

experimentally. Early results for R and the predictions of the vector dominance model are shown in Fig. 3. The results showed that R is small and
does not increase with q^2. This eliminated the model as a possible description of deep inelastic scattering.

Various attempts[21] to save the vector meson dominance point of view
were made with the extension of the vector meson spectral function to
higher masses, including approaches which included a structureless continuum of higher mass states. These calculations of the Generalized Vector
Dominance model failed in general to describe the data over the full
kinematic range.

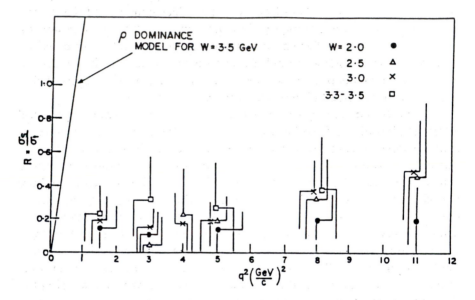

Fig. 3: Measured values of $R = \sigma_S/\sigma_T$ as a function of q^2 for various values of W. The ρ meson
dominance prediction is also shown, calculated for $W = 3.5$ (see Ref. [20]).

CONSTITUENT MODELS

The first suggestion that deep inelastic electron scattering might provide
evidence of elementary constituents was made by Bjorken in his 1967
Varenna lectures.[22] Studying the sum rule predictions derived from current
algebra,[23] he stated, "...We find these relations so perspicuous that, by an
appeal to history, an interpretation in terms of elementary constituents is
suggested." In essence, Bjorken observed that a sum rule for neutrino
scattering derived by Adler[24] from the commutator of two time components
of the weak currents led to an inequality[25] for inelastic electron scattering,

$$\int_{q^2/2M}^{\infty} d\nu \left[W_2^p (\nu, q^2) + W_2^n (\nu, q^2) \right] \geq \frac{1}{2},$$

where W_2^p and W_2^n are structure functions for the proton and neutron,
respectively.

This is equivalent to:

$$\lim_{E \to \infty} \left[\frac{d\sigma_{ep}}{dq^2} + \frac{d\sigma_{en}}{dq^2} \right] \geqslant \frac{2\pi\alpha^2}{q^4} .$$

The above inequality states that as the electron energy goes to infinity the sum of the electron-proton plus electron-neutron total cross sections (elastic plus inelastic) at fixed large q^2 is predicted to be greater than one-half the cross section for electrons scattering from a point-like particle. Bjorken also derived a similar result for backward electron scattering.[26] These results were derived well before our first inelastic results appeared. In hindsight, it is clear that these inequalities implied a point-like structure of the proton and large cross sections at high q^2, but Bjorken's result made little impression on us at the time. Perhaps it was because these results were based on current algebra, which we found highly esoteric, or perhaps it was that we were very much steeped in the physics of the time, which suggested that hadrons were extended objects with diffuse substructures.

The constituent model which opened the way for a simple dynamical interpretation of the deep inelastic results was the parton model of Feynman. He developed this model to describe hadron-hadron interactions,[27] in which the constituents of one hadron interact with those of the other. These constituents, called partons, were identified with the fundamental bare particles of an unspecified underlying field theory of the strong interactions. He applied this model to deep inelastic electron scattering after he had seen the early scaling results that were to be presented a short time later at the 14th International Conference on High Energy Physics, in Vienna, in the late-summer of 1968. Deep inelastic electron scattering was an ideal process for the application of this model. In electron-hadron scattering the electron's interaction and structure were both known, whereas in hadron-hadron scattering neither the structures nor the interactions were understood at the time.

In this application of the model the proton is conjectured to consist of point-like partons from which the electron scatters. The model is implemented in a frame approaching the infinite momentum frame, in which the relativistic time dilation slows down the motions of the constituents nearly to a standstill. The incoming electron thus "sees" and incoherently scatters from partons which are noninteracting with each other during the time the virtual photon is exchanged. In this frame the impulse approximation is assumed to hold, so that the scattering process is sensitive only to the properties and momenta of the partons. The recoil parton has a final state interaction in the nucleon, producing the secondaries emitted in inelastic scattering. A diagram of this model is shown in Fig. 4.

Consider a proton of momentum P, made up of partons, in a frame approaching the infinite momentum frame. The transverse momenta of any parton is negligible and the ith parton has the momentum $P_i = x_i P$, where x_i is a fraction of the proton's momentum. Assuming the electron scatters from a point-like parton of charge Q_i (in units of e), leaving it with the same mass and charge, the contribution to $W_2(v,q^2)$ from this scattering is

$$W_2^{(i)}(v, q^2) = Q_i^2\, \delta(v - q^2/2Mx_i) = \frac{Q_i^2 x_i}{v}\, \delta(x_i - q^2/2Mv).$$

The expression for vW_2 for a distribution of partons is given by

$$vW_2(v, q^2) = \sum_N P(N) \left(\sum_{i=1}^{N} Q_i^2\right) x f_N(x) = F_2(x)$$

where

$$x = \frac{q^2}{2Mv} = \frac{1}{\omega}$$

and where $P(N)$ is the probability of N partons occurring. The sum

$$\left(\sum_{i=1}^{N} (Q_i)^2\right)$$

is the sum of the squares of the charges of the N partons, and $f_N(x)$ is the distribution of the longitudinal momenta of the charged partons.

It was clear that the parton model, with the assumption of point-like constituents, automatically gave scaling behavior. The Bjorken scaling variable ω was seen to be the inverse of the fractional momentum of the struck parton, x, and vW_2 was shown to be the fractional momentum distribution of the partons, weighted by the squares of their charges.

In proposing the parton model, Feynman was not specific as to what the partons were. There were two competing proposals for their identity.

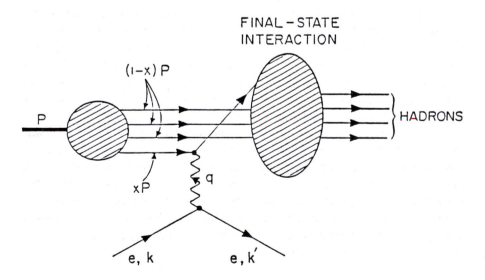

Fig. 4: A representation of inelastic electron nucleon scattering in the parton model. k and k' are the incident and final momenta of the electron. The other quantities are defined in the text.

Applications of the parton model identified partons with bare nucleons and pions,[28-30] and also with quarks.[31-33]. However, parton models incorporating quarks had a glaring inconsistency. Quarks required strong final state interactions to account for the fact that these constituents had not been observed in the laboratory. Before the theory of Quantum Chromodynamics (QCD) was developed, there was a serious problem in making the "free" behavior of the constituents during photon absorption compatible with the required strong final state interaction. One of the ways to get out of this difficulty was to assign quarks very large masses but this was not considered totally satisfactory. This question was avoided in parton models employing bare nucleons and pions because the recoil constituents are allowed to decay into real particles when they are emitted from the nucleon.

Drell, Levy and Yan[28] derived a parton model, in which the partons are bare nucleons and pions, from a canonical field theory of pions and nucleons with the insertion of a cutoff in transverse momenta. The calculations showed that the free point-like constituents which interact with the electromagnetic current in each order of perturbation theory and to leading order in logarithms of $2Mv/q^2$ are bare nucleons making up the proton and not the pions in the pion cloud.

A further development of the approach that identified bare nucleons and pions as partons was a calculation by Lee and Drell[30] that provided a fully relativistic generalization of the parton model that was no longer restricted to an infinite momentum frame. This theory obtained bound state solutions of the Bethe-Salpeter equation for a bare nucleon and bare mesons, and connected the observed scale invariance with the rapid decrease of the elastic electromagnetic form factors.

When the quark model was proposed in 1964 it contained three types of quarks, up (u), down (d), and strange (s), having charges $2/3$, $-1/3$, and $-1/3$, respectively, and each of these a spin $1/2$ particle. In this model the nucleon (and all other baryons) is made up of three quarks, and all mesons consist of a quark and an antiquark. As the proton and neutron both have zero strangeness, they are (u,u,d) and (d,d,u) systems respectively. Bjorken and Paschos[31] studied the parton model for a system of three quarks, commonly called valence quarks, in a background of quark-antiquark pairs, often called the sea, and suggested further tests for the model. A more detailed description of a quark-parton model was later given by Kuti and Weisskopf.[32] Their model of the nucleon contained, in addition to the three valence quarks, a sea of quark-antiquark pairs, and neutral gluons, which are quanta of the field responsible for the binding of the quarks. The momentum distribution of the quarks corresponding to large ω was given in terms of the requirements of Regge behavior. Decisive tests of these models were provided by extensive measurements with hydrogen and deuterium targets that followed the early results.

MEASUREMENTS OF PROTON AND NEUTRON STRUCTURE FUNCTIONS

The first deep inelastic electron scattering results[3] were obtained in the period 1967—1968 from a hydrogen target with the 20 GeV spectrometer set at scattering angles of 6° and 10°. By 1970 the proton data[34] had been extended to scattering angles of 18°, 26° and 34° with the use of the 8 GeV spectrometer. The measurements covered a range of q^2 from 1 GeV^2 to 20 GeV^2, and a range of W^2 up to 25 GeV^2. By 1970 data[35] had been also obtained at scattering angles of 6° and 10° with a deuterium target. Subsequently, a series of matched measurements[36-38] with better statistics and covering an extended range of q^2 and W^2 were done with hydrogen and deuterium targets, utilizing the 20 GeV, the 8 GeV, and the 1.6 GeV spectrometers. These data sets provided, in addition to more detailed information about the proton structure functions, a test of scaling for the neutron. In addition, the measured ratio of the neutron and proton structure functions provided a decisive tool in discriminating among the various models proposed to explain the early proton results.

Neutron cross sections were extracted from measured deuteron cross sections using the impulse approximation along with a procedure to remove the effects of Fermi motion. The method used was that of Atwood and West,[39] with small modifications[40] representing off-mass-shell corrections. In this method the measured proton structure functions, W_1 and W_2, were kinematically smeared over the Fermi momentum distribution of the deuteron and combined to yield the smeared proton cross section σ_{ps}. Subtracting the smeared proton cross section from the measured deuteron cross section yielded the smeared neutron cross section $\sigma_{ns} = \sigma_d - \sigma_{ps}$. With the use of a deconvolution procedure[37] on σ_{ns}, the unsmeared neutron cross section σ_n was obtained. From this and the measured value of the proton cross section σ_p the ratio σ_n/σ_p, which is free of kinematic smearing, was determined. The results were insensitive to the choice of the deuteron wave function used to calculate the momentum distribution of the bound nucleons, as long as the wave functions were consistent with the known properties of the deuteron and the $n-p$ interaction.

The conclusions that were derived from the analysis of these extensive data sets were the following:

(1) The deuterium and neutron structure functions showed the same approximate scaling behavior as the proton. This is shown in Fig. 5 which presents νW_2 for the proton, neutron, and deuteron as a function of x for data ranging in q^2 from 2 GeV^2 to 20 GeV^2.

(2) The values of R_p, R_n and R_d were equal within experimental errors. This is shown in Fig. 6, where the difference of R_d and R_p is plotted.

(3) The ratio of the neutron and proton inelastic cross sections falls continuously as the scaling variable x approaches 1. From a value of about 1 near $x = 0$, the experimental ratio falls to about 0.3, in the neighbor-

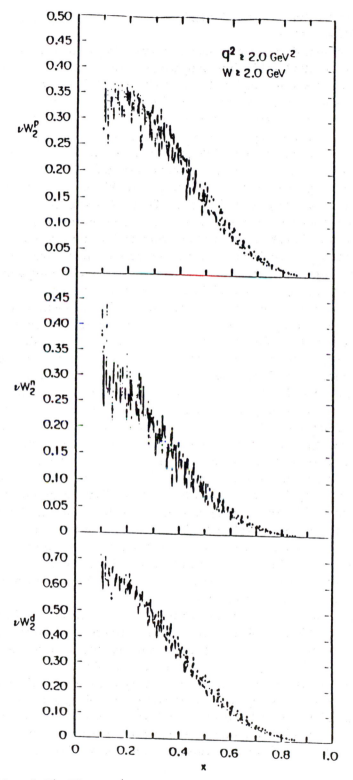

Fig. 5: Values of νW_2^p, νW_2^n and νW_2^d plotted against x. Data from Ref. [36].

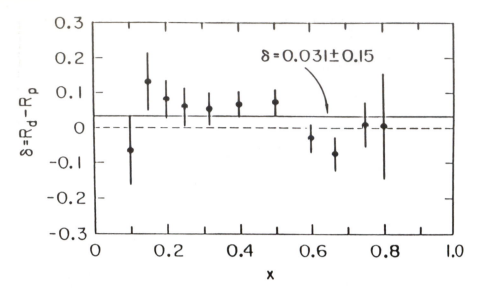

Fig. 6: Average values of the quantity $\delta = R_d - R_p$ for each of the 11 values of x studied. Errors shown are purely random. The systematic error in δ is 0.036. Data from Refs. [36] and [37].

hood of $x = 0.85$. This is shown in Fig. 7 in which σ_n/σ_p is plotted as a function of x. These results put strong constraints on various models of nucleon structure, as discussed later.

SUM RULE RESULTS

A sum rule generally relates an integral of a cross section (or of a quantity derived from it) and the properties of the interaction hypothesized to produce that reaction. Experimental evaluations of such relations thus provide a valuable tool in testing theoretical models. Sum rule evaluations within the framework of the parton model provided an important element in identifying the constituents of the nucleon. The early evaluations of weighted integrals of $\nu W_2(\omega)$ with respect to ω were based on the assumption that the nucleon's momentum is, on the average, equally distributed among the partons. Two important sum rules, which were evaluated for neutrons and protons, were:

$$I_1 = \int_1^\infty \nu W_2(\omega) \frac{d\omega}{\omega^2} = \sum_N P(N) \frac{\left(\sum_{i=1}^N Q_i^2\right)}{N}$$

$$I_2 = \int_1^\infty \nu W_2(\omega) \frac{d\omega}{\omega} = \sum_N P(N) \sum_{i=1}^N Q_i^2,$$

where I_2 is the weighted sum of the squares of the parton charges and I_1[31,41] is the mean square charge per parton. The sum I_2 is equivalent to a sum rule derived by Gottfried[42] who showed that for a proton which consists of three nonrelativistic point-like quarks I_2^p equals 1 at a high q^2. The experimental

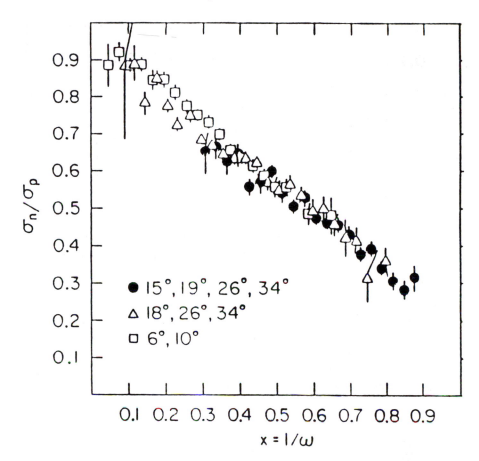

Fig. 7: Values of σ_n/σ_p as a function of x determined from the results presented in Refs. [36] and [37].

value of this integral when integrated over the range of the MIT-SLAC data gave:

$$I_2^P = \int_1^{20} \frac{d\omega}{\omega} \nu W_2^p = 0.78 \pm 0.04$$

where the integral was cut off for $\omega > 20$ because of insufficient information about R_p. Since the experimental values of νW_2 at large ω did not exclude a constant value (see Fig. 2), there was some suspicion that this sum might diverge. This would imply that in the quark model scattering occurs from a infinite sea of quark-antiquark pairs as ν approaches ∞. Table 1 gives a summary of the early comparisons of the experimental values of the sum rules with the predictions of various models. Unlike I_2, the experimental value of I_1 was not very sensitive to the behavior of νW_2 for $\omega > 20$. The experimental value was about one-half the value predicted on the basis of the simple three-quark model of the proton, and it was also too small for a proton having three valence quarks in a sea of quark-antiquark pairs. The Kuti-Weisskopf model[32] which included neutral gluons, in addition to the

	Expected Value[e]		Measurement	$\omega_m{}^f$	$q^2(GeV/c)^2$
	3 Quark	3 Quark + "Sea"			
I_1^p	$\dfrac{1}{3}$	$\dfrac{2}{9} + \dfrac{1}{3\langle N \rangle}$	0.159 ± 0.005	20	1.0
			0.165 ± 0.005	20	1.5
			0.172 ± 0.009^d	20^d	1.5^d
			0.154 ± 0.005	12	2.0
I_1^n	$\dfrac{2}{9}$	$\dfrac{2}{9}$	0.120 ± 0.008	20	1.0
			0.115 ± 0.008	20	1.5
			0.107 ± 0.009	12	2.0
I_2^p	1	$\dfrac{1}{3} + \dfrac{2\langle N \rangle}{9}$	0.739 ± 0.029	20	1.0
			0.761 ± 0.027	20	1.5
			0.780 ± 0.04^d	20^d	1.5^d
			0.607 ± 0.021	12	2.0
I_2^n	$\dfrac{2}{3}$	$\dfrac{2\langle N \rangle}{9}$	0.592 ± 0.051	20	1.0
			0.584 ± 0.050	20	1.5
			0.429 ± 0.036	12	2.0
$I_2^p - I_2^n$		$\dfrac{1}{3}$	0.147 ± 0.059	20	1.0
			0.177 ± 0.057	20	1.5
			0.178 ± 0.042	12	2.0

TABLE 1:   Early Sum Rule Results[a] — Theory[b] and Measurements[c]

[a] From J. I. Friedman and H. W. Kendall, **Ann. Rev. Nucl. Sci. 22**, 203 (1972). Excerpts from this publication are used in the present paper.
[b] Reference [31].
[c] Calculated from preliminary results, later published as Refs. [35,36], except where noted.
[d] Data from Ref. [3].
[e] $\langle N \rangle$ expectation value of number of quarks.
[f] ω_m is upper limit of integral.

valence quarks and the sea of quark-antiquark pairs, predicted a value of I_1^p that was compatible with this experimental result.

The difference $I_2^p - I_2^n$ was of great interest because it is presumed to be sensitive only to the valence quarks in the proton and the neutron. On the assumption that the quark-antiquark sea is an isotopic scalar, the effects of the sea cancel out in the above difference, giving $I_2^p - I_2^n = 1/3$. Unfortunately, it was difficult to extract a meaningful value from the data because of the importance of the behavior of νW_2 at large ω. Extrapolating $\nu W_2^p - \nu W_2^n$

toward $\omega \to \infty$ for $\omega > 12$, with the asymptotic dependence $(1/\omega)^{\frac{1}{2}}$ expected on the basis of Regge theory, we obtained a rough estimate of $I_2^p - I_2^n = 0.22 \pm 0.07$. This was compatible with the expected value, given the error and the uncertainties in extrapolation. The difference $\nu W_2^p(x) - \nu W_2^n(x)$, plotted in Fig. 8 shows a peak, which would be expected in theoretical models[31,32] involving quasi-free constituents.

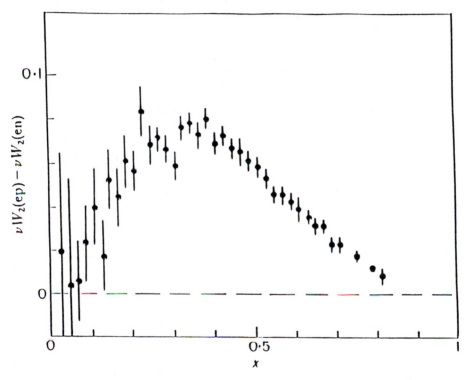

Fig. 8: Values of $\nu W_2^p - \nu W_2^n$ as a function of x.

The Bjorken inequality previously discussed, namely,

$$\int_{q^2/2M}^{\infty} d\nu \left[W_2^p (\nu, q^2) + W_2^n (\nu, q^2) \right] \geq \frac{1}{2}$$

was also evaluated. This inequality was found to be satisfied at $\omega \simeq 5$.

Extensions of the quark-parton model allowed the weighted sum

$$\int \frac{d\omega}{\omega^2} \nu W_2$$

to be theoretically evaluated without making the assumption that the momentum of the nucleon is equally distributed among different types of partons. If $u_p(x)$ and $d_p(x)$ are defined as the momentum distributions of up and down quarks in the proton then $F_2^p(x)$ is given by

$$F_2^p (x) = \nu W_2^p (x) = x \left[(Q_u^2 (u_p(x) + \bar{u}_p(x)) + Q_d^2 (d_p (x) + \bar{d}_p(x))) \right]$$

where $\bar{u}_p(x)$ and $\bar{d}_p(x)$ are the distributions for anti-up and anti-down quarks, and Q_u^2 and Q_d^2 are the squares of the charges of the up and down quarks, respectively. The strange quark sea has been neglected.

Using charge symmetry it can be shown that

$$\frac{1}{2}\int_0^1 \left[F_2^p(x) + F_2^n(x)\right] dx = \left[\frac{Q_u^2 + Q_d^2}{2}\right] \int_0^1 x \left[u_p(x) + \bar{u}_p(x) + d_p(x) + \bar{d}_p(x)\right] dx .$$

The integral on the right-hand side of the equation is the total fractional momentum carried by the quarks and antiquarks, which would equal 1.0 if they carried the nucleon's total momentum. On this assumption the expected sum should equal

$$\frac{Q_u^2 + Q_d^2}{2} = \frac{1}{2}\left[\frac{4}{9} + \frac{1}{9}\right] = \frac{5}{18} = 0.28 .$$

The evaluations of the experimental sum from proton and neutron results over the entire kinematic range studied yielded

$$\frac{1}{2}\int \left[F_2^p(x) + F_2^n(x)\right] dx = 0.14 \pm 0.005 .$$

This again suggested that half of the nucleon's momentum is carried by neutral constituents, gluons, which do not interact with the electron.

IDENTIFICATION OF THE CONSTITUENTS OF THE NUCLEON AS QUARKS

The confirmation of a constituent model of the nucleon and the identification of the constituents as quarks took a number of years and was the result of continuing interplay between experiment and theory. By the time of the XVth International Conference on High Energy Physics held in Kiev in 1970 there was an acceptance in some parts of the high energy community of the view that the proton is composed of point-like constituents. At that time we were reasonably convinced that we were seeing constituent structure in our experimental results, and afterwards our group directed its efforts to trying to identify these constituents and making comparisons with the last remaining competing models.

The electron scattering results which played a crucial role in identifying the constituents of protons and neutrons or which ruled out competing models were the following:

(1)   Measurement of R

At the Fourth International Symposium on Electron and Photon Interactions at High Energies held in Liverpool in 1969, MIT-SLAC results were presented which showed that R was small and was consistent with being independent of q^2. The subsequent measurements,[36,37] which decreased the errors, were consistent with this behavior.

The experimental result that R was small for the proton and neutron at large values of q^2 and v required that the constituents responsible for the scattering have spin $1/2$, as was pointed out by Callan and Gross.[43] These results ruled out pions as constituents but were consistent with the constituents being quarks or bare protons.

(2)   The σ_n/σ_p Ratio

As was discussed in a previous section σ_n/σ_p decreased from 1 at about $x = 0$ to 0.3 in the neighborhood of $x = 0.85$. The ratio σ_n/σ_p is equivalent to W_2^n/W_2^p for $R_p = R_n$, and in the quark model a lower bound of 0.25 is imposed on W_2^n/W_2^p. While the experimental values approached and were consistent with this lower bound, Regge and resonance models had difficulty at large x, as they predicted values for the ratio of about 0.6 and 0.7, respectively, near $x = 1$, and pure diffractive models predicted 1.0. The relativistic parton model in which the partons were associated with bare nucleons and mesons predicted a result for W_2^n/W_2^p which fell to zero at $x = 1$ and was about 0.1 at $x = 0.85$, clearly in disagreement with our results.

A quark model in which up and down quarks have identical momentum distributions would give a value of $W_2^n/W_2^p = 2/3$. Thus, the small value observed experimentally requires a difference in these distributions and quark-quark correlations at low x. To get a ratio of 0.25, the lower limit of the quark model, only a down quark from the neutron and an up quark from the proton can contribute to the scattering at the value of x at which the limit occurs.

(3)   Sum Rules

As previously discussed, several sum rule predictions suggested point-like structure in the nucleon. The experimental evaluations of the sum rule related to the mean square charge of the constituents were consistent with the fractional charge assignments of the quark model provided that half the nucleon's momentum is carried by gluons.

EARLY NEUTRINO RESULTS

Neutrino deep inelastic scattering produced complementary information that provided stringent tests of the above interpretation. Since charged-current neutrino interactions with quarks were expected to be independent of quark charges but were hypothesized to depend on the quark momentum distributions in a manner similar to electrons, the ratio of the electron and neutrino deep inelastic scattering was predicted to depend on the quark charges, with the momentum distributions cancelling out.

That is

$$\frac{\frac{1}{2}\int \left[F_2^{ep}(x) + F_2^{en}(x) \right] dx}{\frac{1}{2}\int \left[F_2^{vp}(x) + F_2^{vn}(x) \right] dx} = \frac{Q_u^2 + Q_d^2}{2}$$

where $1/2 \ (F_2^{\nu p}(x)+F_2^{\nu n}(x))$ is the F_2 structure function obtained from neutrino-nucleon scattering from a target having an equal number of neutrons and protons. The integral of this neutrino structure function over x is equal to the total fraction of the nucleon's momentum carried by the constituents of the nucleon that interact with the neutrino. This directly measures the fractional momentum carried by the quarks and antiquarks because gluons are not expected to interact with neutrinos.

The first neutrino and anti-neutrino total cross-sections were presented in 1972 at the XVI International Conference on High Energy Physics held at Fermilab and the University of Chicago. The measurements were made at the CERN 24 GeV Synchrotron with the use of the large heavy-liquid bubble chamber "Gargamelle." At this meeting Perkins,[44] who reported these results, stated that, "...the preliminary data on the cross-sections provide an astonishing verification for the Gell-Mann/Zweig quark model of hadrons."

These total cross section results, presented in Fig. 9, demonstrate a linear dependence on neutrino energy for both neutrinos and anti-neutrinos that is a consequence of Bjorken scaling of the structure functions in the deep inelastic region. By combining the neutrino and anti-neutrino cross-sections

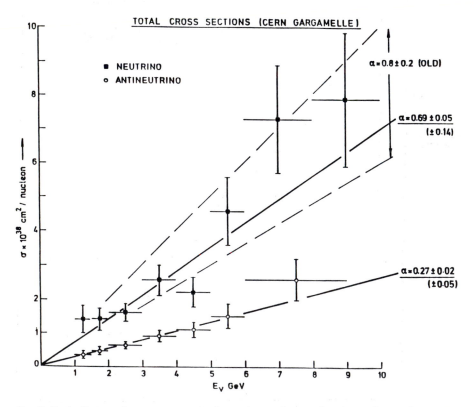

Fig. 9: Early Gargamelle measurements of neutrino nucleon and anti-neutrino nucleon cross sections as a function of energy. These results were presented at the XVI International Conference on High Energy Physics, NAL-Chicago, 1972, Ref. [44].

the Gargamelle group was able to show that

$$\frac{1}{2}\int\left(F_2^{\nu p}(x) + F_2^{\nu n}(x)\right) dx = \int x\left[u_p(x) + \bar{u}_p(x) + d_p(x) + \bar{d}_p(x)\right] dx = 0.49 \pm 0.07$$

which confirmed the interpretation of the electron scattering results that suggested that the quarks and antiquarks carry only about half of the nucleon's momentum. When this result was compared with

$$\frac{1}{2}\int\left[F_2^{ep}(x) + F_2^{en}(x)\right] dx$$

they found that the ratio of neutrino and electron integrals was 3.4 ± 0.7 as compared to the value predicted for the quark model, $18/5 = 3.6$. This was a striking success for the quark model.

Within the next few years additional neutrino results solidified these conclusions. The results presented[45] at the XVII International Conference on High Energy Physics held in London in 1974 demonstrated that the ratio $18/5$ was valid both as a function of x and neutrino energy. Figure 10, taken

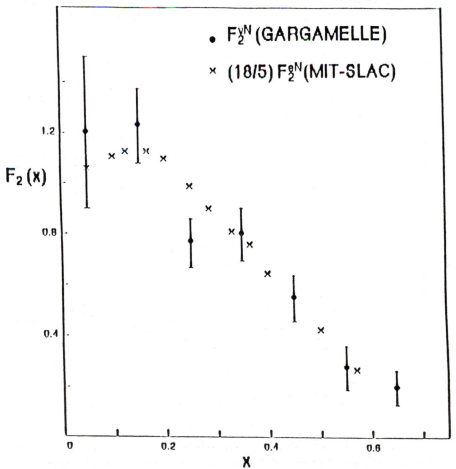

Fig. 10: Early Gargamelle measurements of $F_2^{\nu N}$ compared with $(18/5)F_2^{eN}$ calculated from the MIT-SLAC results.

from Gargamelle data, shows a comparison of $F^{vN}(x)$ and $18/5\ F_2^{eN}$, where F_2^{vN} and F_2^{eN} each represents an average of proton and neutron structure functions, and Fig. 11 shows the ratio of the integrals of the two structure

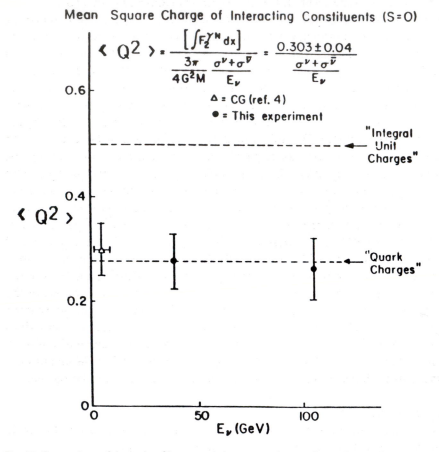

Fig. 11: Comparison of the ratio of integrated electron-nucleon and neutrino-nucleon structure functions to the value 5/18 expected from quark charges. The open triangle data point is from Gargamelle and the filled-in circles are from the CIT-NAL Group. From Ref. [45]. The quantity $\langle Q^2 \rangle$ is the mean square charge of the quarks in a target consisting of an equal number of protons and neutrons.

functions as a function of neutrino energy calculated from Gargamelle and CIT-NAL data. In addition, the Gargamelle group evaluated the Gross-Llewellyn Smith sum rule[46] for the F_3 structure function, which uniquely occurs in the general expressions for the inelastic neutrino and antineutrino nucleon cross sections as a consequence of parity non-conservation in the weak interaction. This sum rule states that

$$\int F_3^{vN}(x)\ dx = (\text{number of quarks}) - (\text{number of antiquarks})$$

which equals 3 for a nucleon in the quark model. Obtaining values of $F_3^{vN}(x)$ from the differences of the neutrino and anti-neutrino cross sections, the

Gargamelle group found the sum to be 3.2 ± 0.6, another significant success for the quark model.

GENERAL ACCEPTANCE OF QUARKS AS CONSTITUENTS

After the London Conference in 1974, with its strong confirmation of the constituent quark model, a general change of view developed with regard to the structure of hadrons. The bootstrap approach and the concept of nuclear democracy were in decline, and by the end of the 1970's, the quark structure of hadrons became the dominant view for developing theory and planning experiments. A crucial element in this change was the general acceptance of QCD,[47, 48] which eliminated the last paradox, namely, why are there no free quarks? The infra-red slavery mechanism of QCD provided a reason to accept quarks as physical constituents without demanding the existence of free quarks. The asymptotic freedom property of QCD also readily provided an explanation of scaling, but logarithmic deviations from scaling were inescapable in this theory. These deviations were later confirmed in higher energy muon and neutrino scattering experiments at FNAL and CERN. There were a number of other important experimental results reported in 1974 and the latter half of the decade which provided further strong confirmations of the quark model. Among these were the discovery of Charmonium[49, 50] and its excited states,[51] investigations of the total cross section for $e^+e^- \rightarrow$ hadrons,[52] and the discoveries of quark jets[53] and gluon jets.[54] The constituent quark model, with quark interactions described by QCD, became the accepted view of the structure of hadrons. This picture which is one of the foundations of the Standard Model has not been contradicted by any experimental evidence in the intervening years.

ACKNOWLEDGMENTS

There were many individuals who made essential contributions to this work. An extensive set of acknowledgments is given in Ref. [55].

REFERENCES

1. D. H. Coward *et al.*, *Phys. Rev. Lett.* **20**, 292 (1968).
2. W. K. H. Panofsky, in *Proceedings of 14th International Conference on High Energy Physics* Vienna (1968) 23. The experimental report, presented by the author, is not published in the Conference Proceedings. It was, however, produced as a SLAC preprint.
3. E. D. Bloom *et al. Phys. Rev. Lett.* **23**, 930 (1969); M. Breidenbach *et al. Phys. Rev. Lett.* **23**, 935 (1969).
4. The Mott cross-section $\sigma_{Mott} = \dfrac{e^4}{4E^2} \dfrac{\cos^2\theta/2}{\sin^4\theta/2}$
5. R. W. McAllister and R. Hofstadter, *Phys. Rev.* **102**, 851 (1956).
6. J. D. Bjorken, *Phys. Rev.* **179**, 1547 (1969); In a private communication, Bjorken told the MIT-SLAC group about scaling in 1968.
7. S. D. Drell and J. D. Walecka, *Ann. Phys.* (NY) **28**, 18 (1964).

8. M. Gell-Mann, *Phys. Lett.* **8,** 214 (1964).
9. G. Zweig, CERN preprint 8182/TH 401 (1964); CERN preprint 8419/TH 412.
10. M. Gell-Mann, Caltech Synchrotron Laboratory Report CTSL-20 (1961); Y. Neeman, *Nucl. Phys.* **26,** 222 (1961).
11. R. E. Taylor, Deep Inelastic Scattering: The Early Years. *Les Prix Nobel 1990.*
12. H. W. Kendall, Deep Inelastic Scattering: Experiments on the Proton and the Observation of Scaling. *Les Prix Nobel 1990.*
13. H. D. Abarbanel, M. L. Goldberger and S. B. Treiman, *Phys. Rev. Lett.* **22,** 500 (1969).
14. H. Harari, *Phys. Rev. Lett.* **22,** 1078 (1969); *Phys. Rev. Lett.* **24,** 286 (1970).
15. T. Akiba, *Lett. Nuovo Cimento* **4,** 1281 (1970).
16. H. Pagels, *Phys. Rev.* **D3,** 1217 (1971).
17. J. W. Moffat and V. G. Snell, *Phys. Rev.* **D30,** 2848 (1971).
18. P. V. Landshoff and J. C. Polkinghorne, DAMPT 70/36 (1970).
19. G. Domokos, S. Kovesi-Domokos and E. Shonberg, *Phys. Rev.* **D3,** 1184 (1971); *Phys. Rev.* **D3,** 1191 (1971).
20. J. J. Sakurai, *Phys. Rev. Lett.* **31B,** 22 (1970); J. Chou and J. J. Sakurai, *Phys. Lett.* **31B,** 22 (1970).
21. For a review of the Vector Dominance and Generalized Vector Dominance Models, see T. H. Bauer, R. E. Spital, D. R. Yennie and F. M. Pipkin, *Rev. Mod. Phys.* **50,** 261 (1978).
22. J. D. Bjorken, *Proceedings of the International School of Physics "Enrico Fermi", Course XLI: Selected Topics in Particle Physics,* J. Steinberger, ed. (Academic Press, New York, 1968).
23. M. Gell-Mann, *Phys. Rev.* **125,** 1062 (1962); For a review of current algebra see: J. D. Bjorken and M. Nauenberg, *Ann. Rev. Nucl. Sci.* **18,** 229 (1968).
24. S. L. Adler, *Phys. Rev.* **143,** 1144 (1966).
25. J. D. Bjorken, *Phys. Rev. Lett.* **16,** 408 (1966).
26. J. D. Bjorken, *Phys. Rev.* **163,** 1767 (1967).
27. R. P. Feynman, *Phys. Rev. Lett.* **23,** 1415 (1969); *Proceedings of the III International Conference on High Energy Collisions,* organized by C. N. Yang *et al.* (Gordon and Breach, New York, 1969).
28. S. Drell, D. J. Levy and T. M. Yan, *Phys. Rev.* **187,** 2159 (1969); *Phys. Rev.* **D1,** 1035, 1617 (1970).
29. N. Cabbibo, G. Parisi, M. Testa and A. Verganelakis, *Lett. Nuovo Cimento* **4,** 569 (1970).
30. T. D. Lee and S. D. Drell, *Phys. Rev.* **D5,** 1738 (1972).
31. J. D. Bjorken and E. A. Paschos, *Phys. Rev.* **185,** 1975 (1969).
32. J. Kuti and V. F. Weisskopf, *Phys. Rev.* **D4,** 3418 (1971).
33. P. V. Landshoff and J. C. Polkinghorne, *Nucl. Phys.* **B28,** 240 (1971).
34. G. Miller *et al.,* *Phys. Rev.* **D5,** 528 (1972).
35. J. S. Poucher *et al.,* *Phys. Rev. Lett.* **32,** 118 (1974).
36. A. Bodek *et al.,* *Phys. Rev. Lett.* **30,** 1087 (1973); *Phys. Lett.* **51B,** 417 (1974); *Phys. Rev.* **D20,** 1471 (1979).
37. E. M. Riordan *et al.,* *Phys. Rev. Lett.* **33,** 561 (1974); *Phys. Lett.* **52B,** 249 (1974).
38. W. B. Atwood *et al.,* *Phys. Lett.* **64B,** 479 (1976).
39. W. B. Atwood and G. B. West, *Phys. Rev.* **D7,** 773 (1973).
40. A Bodek, *Phys. Rev.* **D8,** 2331 (1973).
41. C. G. Callan and D. J. Gross, *Phys. Rev. Lett.* **21,** 311 (1968).
42. K. Gottfried, *Phys. Rev. Lett.* **18,** 1174 (1967).
43. C. G. Callan and D. J. Gross, *Phys. Rev. Lett.* **22,** 156 (1969).
44. D. H. Perkins, in *Proceedings of the XVI International Conference on High Energy Physics,* Chicago and NAL, Vol. 4, 189 (1972).
45. *Proceedings of the XVII International Conference on High Energy Physics,* London, (1974), M. Haguenauer, p. IV-95; F. Sciulli, p.IV-105; D. C. Cundy, p. IV−131.

DEEP INELASTIC SCATTERING

ACKNOWLEDGEMENTS

Jerome I. Friedman

Henry W. Kendall

Richard E. Taylor

The physics experiments (Reference 1) cited in 1990 by the Royal Swedish Academy of Sciences were a study of the deep inelastic scattering of electrons from the nucleon. The program, carried out by personnel from MIT and SLAC, was a group effort and we are grateful to our collaborators, all of whom played essential roles in the program. D. Coward and H. DeStaebler were with the experiments from the beginning and made indispensable contributions throughout their course. Other collaborators, whose efforts made the program possible, were: W. B. Atwood, E. Bloom , A. Bodek, M. Breidenbach, G. Buschhorn, R. Cottrell, R. Ditzler, J. Drees, J. Elias, G. Hartmann, C. Jordan, M. Mestayer, G. Miller, L. Mo, H. Piel, J. Poucher, C. Prescott, M. Riordan, L. Rochester, D. Sherden, M. Sogard, S. Stein, D. Trines and R. Verdier. Valuable help with computing was provided by D. Dubin, R. Early, A. Gromme, and E. Miller.

The inelastic experiments were part of a larger program of electron scattering carried out at the linear accelerator. Many of our colleagues in the other experiments made contributions of direct relevance to the development of the inelastic experiments. B. Barish, K. Brown, P. Kirk, J. Litt, S. Loken, J. Mar, A. Minten, C. Peck, and J. Pine made contributions at the outset of the program, while C. Sinclair provided assistance in the later experiments. We are grateful for help from A. Boyarski, F. Bulos, R. Diebold, E. Garwin, R. S. Larsen, R. Miller, B. Richter, and D. Ritson.

This work could not have been done without the SLAC laboratory director, W. K. H. Panofsky, who established and led an outstanding laboratory that provided us with a superb accelerator and the opportunity to do these experiments. R. Neal and the SLAC Technical Division played a critical role in the building, implementation, and operation of the accelerator. We owe him and his division a deep debt of gratitude. We also thank J. Ballam and the SLAC Research Division, along with F. Pindar and Administrative Services for many helpful contributions to the program. J. I. Friedman and H. W. Kendall wish also to acknowledge aid and support provided by their many MIT colleagues, including W. Buechner (now deceased), P. Demos, M. Deutsch, F. Eppling, H. Feshbach, A. Hill, and V. F. Weisskopf.

We benefitted greatly from the willingness of J. D. Bjorken to help us understand both his own crucial works on inelastic scattering and those of other theorists. Our understanding of a number of physics issues associated with this program was also advanced by discussions with S. Drell, M. Gell-Mann, F. Gilman, K. Gottfried, R. Jaffe, K. Johnson, J. Kuti, F. Low, P. Tsai, V. Weisskopf, and G. West.

The spectrometer hardware was designed and constructed by a large team of engineers and technicians under the direction of E. Taylor. Among others, the group included M. Berndt, L. Brown, M. Brown, J. Cook, W. Davies-White, S. Dyer, R. Eisele, A. Gallagher, N. Heinen, E.K.Johnson, T. Lawrence, J. Mark, R. (Lou) Paul, and R. Pederson.

We gratefully acknowledge the suppport of this work provided by the US Department of Energy and its predecessor agencies.

REFERENCES

1. R. E. Taylor, Deep Inelastic Scattering: The Early Years. H. W. Kendall, Deep Inelastic Scattering: Experiments on the Proton and the Observation of Scaling, and J. I. Friedman, Deep Inelastic Scattering: Comparisons with the Quark Model. *Les Prix Nobel 1990*.